Semiconductor Laser Theory

Semiconductor Laser Theory

Prasanta Kumar Basu
University of Calcutta, Institute of Radio Physics and Electronics
Kolkata, West Bengal, India

Bratati Mukhopadhyay
University of Calcutta, Institute of Radio Physics and Electronics
Kolkata, West Bengal, India

Rikmantra Basu
National Institute of Technology Delhi
Electronics & Communication Engineering Department
Delhi, India

CRC Press
Taylor & Francis Group
Boca Raton London New York

CRC Press is an imprint of the
Taylor & Francis Group, an **informa** business

CRC Press
Taylor & Francis Group
6000 Broken Sound Parkway NW, Suite 300
Boca Raton, FL 33487-2742

© 2016 by Taylor & Francis Group, LLC
CRC Press is an imprint of Taylor & Francis Group, an Informa business

No claim to original U.S. Government works

Printed on acid-free paper
Version Date: 20150316

International Standard Book Number-13: 978-1-4665-6191-5 (Hardback)

Library of Congress Cataloging-in-Publication Data

Basu, P. K. (Prasanta Kumar), author.
  Semiconductor laser theory / Prasanta Kumar Basu, Bratati Mukhopadhyay, and Rikmantra Basu.
      pages cm
  Includes bibliographical references and index.
  ISBN 978-1-4665-6191-5
  1. Semiconductor lasers. I. Mukhopadhyay, Bratati, author. II. Basu, Rikmantra, author. III. Title.

QC689.55.S45B37 2016
621.36'61--dc23   2015000204

Visit the Taylor & Francis Web site at
http://www.taylorandfrancis.com

and the CRC Press Web site at
http://www.crcpress.com
Printed and bound in Great Britain by CPI Group (UK) Ltd, Croydon, CR0 4YY

To
Dilip, Pradip, Chhabi, Baby, Karabi, Kaberi,
Chittaranjan, Arundhati, Goutam, Uttam and Rupam (by Prasanta)

To
Late Jiban Ratan Mukhopadhyay (father)
and
Smt. Nilima Mukhopadhyay (mother)
(by Bratati)

To
Chitrani (by Rikmantra)

Contents

Preface

The concept of stimulated emission was introduced by Einstein in 1917. The real use of this process came in 1954 with the announcement of the ammonia laser. The race to obtain light amplification by the stimulated emission of radiation (LASER) action then started. It took another 7–8 years to realize laser action in semiconductors, when four groups announced the p-n junction laser almost simultaneously; their discoveries were published in September and October 1962 issues of different journals.

Since its first announcement, research and development on the semiconductor laser has never stopped. After some periods of uncertainty, the first room temperature double heterostructure laser working in continuous wave mode was developed. After that, there was no looking back. The double heterostructure laser has now been replaced by the quantum well laser. Intensive research is being conducted to utilize the benefits of quantum wire and quantum dot lasers. Today, quantum dot lasers are considered to be a serious competitor for quantum wire lasers.

The manufacture of semiconductor lasers is a very big industry. The all-pervading nature of laser diodes in day-to-day life is manifested in the use of bar-code readers, printers, optical mouses in computers, laser torches and pointers, optical disc readers, and many other gadgets. Semiconductor lasers also find use in medicine, military, environment, surveillance, and lighting; their most important application is long- and short-haul fiber-optic communications and networks.

All undergraduate and graduate courses in electrical engineering, physics, and materials science include in brief the basic properties, structures, and applications of semiconductor lasers. There are a number of advanced texts solely on semiconductor lasers, which form part of graduate- and PhD-level course work. The available texts are divided into two categories. The first type deals with sophisticated theories of lasers: quantum mechanical treatment of electron–photon interactions, density matrix formalism, and the like. The other type employs the semiclassical theory of radiation and its interaction with electrons, rate equation models, and the like. The second approach is simpler and caters better to the needs of students in electrical engineering. However, the texts in this category, except one or two, deal more on principles; the structure, applications, and the like are given somewhat less emphasis.

In a rapidly changing scenario, there is a need to constantly upgrade the textbooks covering recent developments in the field. Some of the earlier texts, though useful for a clear understanding, do not include recent progress, such as diode lasers using quantum wells, dots, quantum cascade lasers, nitride lasers, and so on.

The present text has been developed by considering the recent advances. It belongs to the second category, that is, it uses a semiclassical approach. The target audience comprises graduate students and recently graduated post-doctorate workers.

The book is divided into two parts. After giving a brief overall introduction including the historical development of semiconductor lasers in Chapter 1, the basic concepts are developed in Chapters 2 through 6. Chapter 2 discusses the basic semiconductor physics that is needed to understand the operation of lasers. It also includes p-n junction theory. Chapter 3 considers alloys, heterostructures, and quantum nanostructures. Chapter 4 develops **k.p** theory. Chapter 5 covers waveguides, resonators, and filters. Chapter 6 discusses optical processes.

The actual lasers are covered from Chapter 7 to Chapter 15. The double heterostructure laser is treated in Chapter 7. Though it is hardly used, the theory forms the basis for a discussion of all modern lasers. Chapter 8 discusses quantum wire lasers. Without discussing quantum wire lasers, Chapter 9 deals with quantum dot lasers. Quantum cascade lasers are introduced in Chapter 10, followed by vertical-cavity surface-emitting lasers in Chapter 11. Chapter 12 treats single-mode and tunable lasers. The next three chapters attempt to introduce the readers to the latest developments. Chapters 13 through 15 cover nitride lasers, Group IV lasers, and transistor lasers, respectively. However, these three chapters are by no means self-sufficient. There is rapid progress in all these areas. The intention is to acquaint the readers with the developments in these important fields.

It is difficult to cover all the important aspects of lasers in a volume like this. A noteworthy omission is the noise properties of lasers. To do justice to this topic, a detailed introduction to the theory, Langevin's equation, and so on would be needed. It was felt that such an inclusion would excessively increase the volume of the text.

The text grew out of a lecture on materials given by the authors to undergraduate and graduate students and the knowledge gathered through the research topics set for PhD students. Existing textbooks with similar titles to this book, invited reviews and tutorials, and special issues on semiconductor lasers published by professional societies served as invaluable sources for the present text. These sources are duly acknowledged in the reading lists and references. Numerous examples have been included in each chapter to illustrate the theory. A Problems section is also included in most of the chapters.

Despite the efforts of the authors, there may be errors and omissions, which may be more serious in the treatment of more recent topics. The authors offer their sincerest regrets.

The authors are indebted to a numbers of people and agencies for both financial and moral support, encouragement, and active help. They are duly acknowledged in the proper place.

P. K. Basu
Calcutta and Kharagpur

Bratati Mukhopadhyay
Calcutta

Rikmantra Basu
Delhi

Acknowledgments

Prasanta Kumar Basu has spent many years at the Institute of Radio Physics and Electronics, University of Calcutta, giving lectures on the topics covered in this book. He would like to express his gratitude to his numerous undergraduate and graduate students, in particular his PhD students, who through their interest in semiconductors urged him to keep abreast of the recent developments in semiconductor lasers. He would also like to thank all his colleagues at the Institute for their affection and cooperation throughout his long career.

The program was undertaken while the author was a University Grants Commission–Basic Scientific Research faculty fellow. Partial financial support was provided by the University Grants Commission, New Delhi, for this purpose, which is duly acknowledged. The last part of this book was written when the author joined the Indian Institute of Technology Kharagpur as a visiting professor. The author is thankful to the authorities of the E&EC Engineering Department for providing the environment to complete the book.

Lastly, the author is grateful to his family members, Chitrani (wife), Kaberi (sister), and Rik (son) for their constant encouragement and support. He fondly remembers the care and affection of his late elder brother, Pradip, during his studies and in the later period of his life.

Bratati Mukhopadhyay would first like to acknowledge her coauthor Professor Prasanta Kumar Basu, who convinced her to accept the task of writing a book. She would like to thank Professor Basu as her PhD supervisor from whom she received rigorous training as well as continuous inspiration in her PhD work; it was at this time she became interested in semiconductor optoelectronics. She is happy to acknowledge the contribution of her teacher and colleague Professor Gopa Sen, who shared her scheduled work in the department during this project and provided constant inspiration and encouragement to complete this work. She is indebted to all her students, PhD scholars, and colleagues for their heartfelt support.

Bratati Mukhopadhyay would like to offer her gratitude to her mother (Smt. Nilima Mukhopadhyay) and her elder sister (Dr. Bulbul Mukhopadhyay), who provided much-needed moral support in completing the work. As always, it is the never-ending patience, unstinting support, and continuous encouragement from her husband, Dr. Saibal Bhattacharyya, that have helped her so much in completing the work. Finally, she would like to thank her daughter, Sucheta, for not following through on her intentions to help her mother at work.

Rikmantra Basu obtained his academic degrees from the University of Calcutta. He started working on this book at the end of his doctoral study in 2012. For the last year, he has been attached as a faculty member to the Electrical and Electronics Engineering Department of the Birla Institute of Technology and Science Pilani. He is thankful to his thesis supervisor (Bratati) and Vedatrayee at the Institute of Radio Physics and Electronics, University of Calcutta, for their constant support. Rikmantra acknowledges the cooperation received from all his colleagues at the Birla Institute of Technology and Science Pilani, especially the head of department, Dr. Anu Gupta, and the authorities of the Birla Institute of Technology and Science Pilani, for providing all kinds of support, and also for giving him the OPERA research award. Partial support from the Council of Scientific & Industrial Research, India, is also acknowledged. He is thankful to Professor Ajay Sharma, Director of NIT Delhi, where he is currently employed, for his encouragement.

Rikmantra would like to thank his friends Soumyadeep Misra (Laboratoire de Physique des Interfaces et des Couches Minces, École Polytechnique, France) and Drs. Subrata and Sanchita Ghosh (Faculties, Mechanical Engineering, Birla Institute of Technology and Science Pilani) for providing useful references and constant support and encouragement during the writing of this book.

Finally, Rikmantra is grateful to his mother (Chitrani), father (Prasanta), and aunt (Kaberi) for their constant encouragement and support during his academic career.

1

Introduction to Semiconductor Lasers

In this chapter, a brief history of the development of semiconductor lasers will be given. The basic concept of absorption and emission processes will be presented by considering the interaction of light with a two-level atomic system, and the working principle of a laser will be explained. The semiconductor laser and its structure, principle of operation, threshold current density, and power output will be introduced. After a discussion of the usual structures and the need for their use, the materials issue will be addressed. Finally, some of the application areas of semiconductor lasers will be presented.

1.1 Brief History

LASER is an acronym for Light Amplification by Stimulated Emission of Radiation. As it implies, stimulated emission is the key phenomenon involved in the operation of the laser. This process was first conceived by Einstein as early as 1917 [1]. The work by Bose laid a solid foundation for the understanding of the nature of photons and the statistics obeyed by the particles [2]. However, even 35 years after this concept was introduced, no practical devices using the phenomenon had been realized. The first device exploiting the stimulated emission of electromagnetic (EM) radiation was a maser (Microwave Amplification by Stimulated Emission of Radiation) by Townes and coworkers [3,4]. It has now been discovered that von Neumann, in an unpublished note, proved that laser action might be achieved in a p-n junction [5]. The proposal for an optical maser was put forward by Townes and Schawlow in 1958 [6]. Soon after this, gas lasers and solid-state (ruby) lasers were announced by Javan et al. [7] and Maiman [8], respectively.

The possibility of achieving laser action was examined by several workers, notably by Aigrain [9,10], Basov et al. [11,12], Boyle and Thomas [13,14], and Pankove [15]. Lax considered transition between quantized Landau levels [16]. Research conducted by workers in various locations established the fact that direct band-gap semiconductors are needed to achieve the desired action [17,18]. The seminal paper by Bernard and Duraffourg pointed out the need for high injection to achieve stimulated emission [19]. Finally, in 1962, four groups almost simultaneously reported laser emission from a forward-biased p-n junction by using different material combinations [20–23].

The early p-n junction lasers, which are homojunction types, could not work continuously above temperatures well below room temperature. Kroemer in 1963 pointed out the advantages of using heterostructures to improve the performance of diode lasers [24]. The group led by Alferov worked on a similar idea and made important developments in the growth of the structures [25,26]. Efforts by Hayashi and Panish significantly contributed to the development of the first continuous-wave (CW) room-temperature diode lasers [27]. All these works led to high-volume production of diode lasers and improvement of their performance in terms of power output, efficiency, and modulation bandwidth. New materials are being studied to extend the wavelength range of operation.

Semiconductor heterojunctions may be considered to be a simple example of how the band structure of a device can be engineered. A very important idea of band structure engineering came from Esaki and Tsu, who introduced the concept of the superlattice [28]. At about the same time, a proposal for the unipolar laser involving transitions between quantized energy levels came from Kazarinov and Suris, though the structure utilizing this concept came much later [29]. Soon after the Esaki and Tsu paper, two papers by Dingle et al. [30] and Chang et al. [31], dealing with energy quantization in heterostructures, appeared in the literature. The term *quantum well* (QW) was then coined. A QW laser was proposed in the patent application by Dingle and Henry [32]. The first optically pumped QW laser was announced by van der Ziel et al. [33], followed by the electrically pumped QW laser by Dupuis et al. [34]. At almost the same time, Iga proposed the vertical-cavity surface-emitting laser (VCSEL) and began extensive work on it [35].

The effect of further quantization in a QW structure is to yield quantum-wire (QWR) and quantum-box/dot (QB/QD) structures, and the advantages of using such structures in lasers were established by Arakawa and Sakaki [36].

Arrays of VCSELs were developed and reported by Jewell et al. [37], and further improvements were introduced by Coldren et al. [38] and Chang-Hasnain et al. [39].

A report on the successful growth and operation of QD lasers appeared in the literature in 1994 [40].

Unipolar laser action involving transitions between subbands in a QW was first reported by Faist et al. in 1994, more than two decades after the proposal by Kazarinov and Suris [41]. These lasers, known as *quantum cascade lasers* (QCLs), were then reported for different wavelengths and using different materials systems by a number of workers. Blue-green lasers using nitride-based compounds and alloys were reported by Nakamura in 1996 [42,43]. These lasers have been greatly improved since their discovery. The concept of tunnel injection in QW lasers was outlined by Bhattacharyya et al. in the same year, and the advantages were demonstrated [44].

The first QCL operating in the so-called terahertz region was announced in 2002, and room-temperature operation with high power was achieved by Evans et al. [45,46].

A new kind of laser termed a *transistor laser* was reported in 2004 by a group led by Feng and Holonyak [47]. It is basically a heterojunction bipolar transistor (HBT), and stimulated emission from the QWs in the base produces light; at the same time, it acts as an amplifying device for electrical signals, like a normal transistor.

Si, being an indirect band gap semiconductor, cannot be employed as an efficient light emitter. However, an Si light emitter and other photonic devices integrated with electronic circuits on the same Si platform can provide a myriad of advantages. Attempts to make this cherished dream a reality were made by many workers for over four decades. Finally, the breakthrough came in the form of the Raman laser, announced by Rong et al. in 2005 [48]. It was pointed out later that application of tensile strain in Ge, another indirect-gap group IV semiconductor, may induce an indirect-to-direct band crossover. An optically pumped Ge laser grown on Si was reported in 2010 by Liu et al., and the electrically pumped version was reported in 2012 [49,50].

The above paragraphs give a brief account of the important milestones achieved in the area of semiconductor lasers. However, this report is by no means complete. Furthermore, it is based on publications in available journals. The account, therefore, may not be free of errors.

1.2 Principle of Lasers

The working principle of lasers is best understood by considering a hypothetical two-level atomic system and its interaction with EM waves. In this section, the interaction mechanisms and the conditions to be satisfied for amplification of EM waves are described.

1.2.1 Absorption and Emission

There are three basic processes through which EM radiation can interact with matter: (1) spontaneous emission, (2) absorption, and (3) stimulated emission. These processes were first identified by Einstein in 1917.

The processes are illustrated in Figure 1.1 using a hypothetical atom having only two energy levels: E_1 (lower level) and E_2 (upper level). The respective population densities in these levels are N_1 and N_2. A collimated monochromatic beam of light with angular frequency $\omega = (E_2 - E_1)/\hbar$ may interact with the atoms in the following three ways.

1.2.1.1 Absorption

Under normal conditions, all materials absorb light. The absorption process is illustrated by Figure 1.1a, in which a photon of energy $\hbar\omega = E_2 - E_1$ is

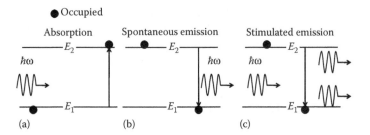

FIGURE 1.1
Three basic processes of interaction of radiation with a two-level atomic system: (a) absorption, (b) spontaneous emission, and (c) stimulated emission. In each of (a), (b), and (c), the left half represents the initial state of occupancy of a level, while the right half represents the state of occupancy after the process.

absorbed, transferring an electron from the lower level to the upper level. The left side of Figure 1.1a represents the initial state of the atom, while the right side depicts the states after the transition. The intensity of the incident light is attenuated due to absorption, but the direction of propagation and polarization of light remains unaltered. The rate of absorption is given by

$$\frac{dN_1(t)}{dt} = -B_{12}N_1\rho(\omega) = -\frac{dN_2(t)}{dt} \tag{1.1}$$

where:
   B_{12} is the Einstein B-coefficient for absorption
   ρ is the energy density of the incident photon flux

1.2.1.2 Spontaneous Emission

In this process, as shown in Figure 1.1b, an atom from the upper level jumps down spontaneously to the lower level, and a photon of energy $\hbar\omega$ is emitted. The direction of propagation and polarization of emitted radiation is arbitrary. The corresponding rate equation is

$$\frac{dN_1(t)}{dt} = -A_{21}N_2(t) = -\frac{dN_2(t)}{dt} \tag{1.2}$$

where A_{21} is called the Einstein A-coefficient and is related to the spontaneous emission lifetime by $\tau = (A_{21})^{-1}$.

1.2.1.3 Stimulated Emission

In this case, as illustrated in Figure 1.1c, the incident photon induces the atom to make a downward transition from the upper to the lower level. The energy is released in the form of a photon, which has the same frequency, phase, polarization, and direction of propagation as the incident photon stimulating the transition. The corresponding rate equation is

$$\frac{d N_2(t)}{dt} = -B_{21}N_2\rho(\omega) = -\frac{d N_1(t)}{dt} \tag{1.3}$$

where B_{21} is the Einstein B-coefficient for stimulated emission.

1.2.2 Absorption and Emission Rates

We may write the rates of spontaneous emission, stimulated emission, and absorption from Equations 1.1 through 1.3 as

$$R_{spon} = A_{21}N_2, \quad R_{stim} = B_{21}N_2\rho(\omega), \quad \text{and} \quad R_{abs} = B_{12}N_1\rho(\omega) \tag{1.4}$$

In thermal equilibrium, atomic densities obey Boltzmann statistics, and accordingly

$$N_2/N_1 = \exp[-(E_2 - E_1)/k_BT] = \exp(-\hbar\omega / k_BT) \tag{1.5}$$

where:
   k_B is the Boltzmann constant
   T is the absolute temperature

Under steady state, the rates for upward and downward transitions should be equal, leading to

$$A_{21}N_2 + B_{21}N_2\rho(\omega) = B_{12}N_1\rho(\omega) \tag{1.6}$$

Using Equation 1.5 in Equation 1.6, the spectral density may be expressed as

$$\rho(\omega) = \frac{A_{21} / B_{21}}{(B_{12} / B_{21}) \exp(\hbar\omega / k_BT) - 1} \tag{1.7}$$

In thermal equilibrium, the radiation spectral density must be identical with that for blackbody radiation, given by the Planck formula

$$\rho_{BB}(\omega) = \frac{2\hbar\omega^3}{\pi c^3} \frac{1}{\exp(\hbar\omega/k_BT) - 1} \tag{1.8}$$

Comparing Equations 1.7 and 1.8 gives the relations

$$A_{21} = (2\hbar\omega^3 / \pi c^3)B_{21} \quad \text{and} \quad B_{12} = B_{21} \tag{1.9}$$

The ratio of stimulated to spontaneous emission rates is therefore

$$R_{\text{stim}} / R_{\text{spon}} = [\exp(\hbar\omega / k_{\text{B}}T) - 1]^{-1} \tag{1.10}$$

Example 1.1: Let us calculate the ratio of the stimulated emission rate to the spontaneous emission rate for wavelength 0.8 μm at room temperature. For this wavelength, $\hbar\omega = 1.55$ eV and $k_{\text{B}}T \approx 25$ meV at room temperature. The ratio is 8.43×10^{-26} from Equation 1.10.

It appears from Equation 1.10 and Example 1.1 that in the visible and infrared region ($\hbar\omega \sim 1$ eV), spontaneous emission always dominates over stimulated emission around room temperature ($k_{\text{B}}T \approx 25$ meV).

It also appears from Equations 1.4 and 1.10 that $R_{\text{stim}} > R_{\text{abs}}$ only when $N_2 > N_1$. The system then operates away from thermal equilibrium. This condition is referred to as population inversion, and it may be created by a process known as pumping. The population inversion is a necessary condition for the operation of a laser.

In the above discussion, the atomic energy levels are assumed to be sharp. The levels are, however, broadened due to interatomic collisions and several other processes. The emitted radiation is therefore no longer monochromatic, but has a linewidth corresponding to a lineshape function, which may be Lorentzian or Gaussian in nature. The total number of stimulated emission per unit volume per second considering a lineshape function $L(\omega)$, as the atomic levels become broadened due to collision and other effects, may be written as

$$W_{21} = N_2 \int B_{21}\rho(\omega)L(\omega)d\omega = N_2 \frac{\pi^2 c^3}{\hbar\tau_{\text{sp}}n_{\text{r}}^3} \int \frac{\rho(\omega)}{\omega^3} L(\omega)d\omega \tag{1.11}$$

1.2.3 Absorption and Amplification of Light in a Medium

Consider now the propagation of a monochromatic beam of light in an atomic medium, in which the atoms have only two energy levels, as assumed so far. Let z denote the direction of propagation of the beam. Consider two planes P_1 and P_2, each having an area $= S$, perpendicular to this direction and located at z and $z + dz$, respectively, as shown in Figure 1.2. The number of stimulated emissions per unit time in the volume element Sdz is $W_{21}Sdz$, and the energy added to the beam due to this process is $W_{21}\hbar\omega Sdz$. Similarly, the energy given out by the beam due to absorption is $W_{12}\hbar\omega Sdz$. The net rate of absorption of energy in the volume element in the frequency interval between ω and $\omega + d\omega$ is the difference between the two.

The intensity of light at planes P_1 and P_2 is, respectively, $I_\omega(z)$ and $I_\omega(z + dz)$. Therefore, the total energy leaving the volume element per unit time is

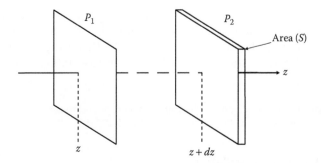

FIGURE 1.2
Propagation of radiation along the *z*-direction.

$$\left[I_\omega(z+dz)-I_\omega(z)\right]S = \frac{\partial I_\omega}{\partial z}Sdz \tag{1.12}$$

Equating this with the net absorption rate within the volume element, one obtains

$$\frac{\partial I_\omega}{\partial z}Sdz = -\left(W_{12}-W_{21}\right)\hbar\omega Sdz \tag{1.13}$$

Using the expression for W_{21} as in Equation 1.11, and considering a small frequency interval $d\omega$, one obtains

$$\frac{\partial I_\omega}{\partial z} = -\frac{\pi^2 c^3}{n_r^2 \tau_{sp}\omega^2}L(\omega)\rho(\omega)\left(N_1-N_2\right) \tag{1.14}$$

The energy density and the intensity are related by $I_\omega = (c/n_r)\rho(\omega)$, where *c* is the velocity of light in free space, and n_r is the refractive index. Equation 1.14 may therefore be expressed as

$$\frac{\partial I_\omega}{\partial z} = -\frac{\pi^2 c^2}{\omega^2 \tau_{sp} n_r^2}\left(N_1-N_2\right)L(\omega)I_\omega = -\alpha_\omega I_\omega \tag{1.15}$$

Integrating Equation 1.15, and assuming that the populations in two states do not depend on the intensity, one arrives at the well-known expression

$$I_\omega(z) = I_\omega(0)\exp\left(-\alpha_\omega z\right) \tag{1.16}$$

where:
$I_\omega(0)$ is the intensity at the surface of the medium
α_ω is the absorption coefficient of the medium at an angular frequency ω

Since $N_1 > N_2$, under usual circumstances, there is always an absorption of EM waves in a medium. However, if a situation is created such that $N_2 > N_1$, that is, a state of population inversion is created, then the EM waves will be amplified. In that case, the stimulated emission rate will exceed the rate of absorption.

> **Example 1.2:** Let $\alpha = 5$ cm^{-1}. The intensity is reduced to 0.6064 $I(0)$ as the beam traverses 1 mm. The reduction is thus 2.172 dB.

1.2.4 Self-Sustained Oscillation

Although the basic phenomenon of interest is light amplification, and the device based on this is named a laser, in most cases the laser supports self-sustained oscillation. It is well known that in order to convert an amplifying medium into an oscillator, one needs to introduce a positive feedback. The basic structure of a laser therefore consists of three different parts: (1) *a gain medium*, in which the condition for population inversion is established by (2) a *pump*, and (3) *a mechanism to provide feedback*. The feedback is provided by using a Fabry–Perot (FP) resonator consisting of a pair of plane-parallel mirrors, as shown in Figure 1.3.

The plane-parallel mirrors form a resonant cavity, and as such, the cavity selectively supports different modes, that is, patterns of EM field or power distribution. Simply put, the EM waves suffer reflections from the end mirrors and form a standing wave pattern. The separation between the mirrors, L, is such that it contains an integral multiple of half wavelengths of the EM field. The reflectivity of one mirror is 100%, while that of the other is purposely made lower (90% in the figure) so that light will come out of this mirror.

The gain medium has a well-defined range of frequency or energy over which positive gain is obtained. Let the medium enclosed by the mirrors have a gain coefficient g and an absorption (loss) coefficient α. Let R_1 and

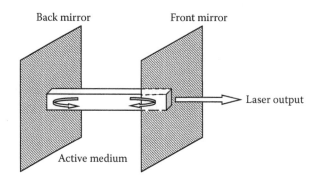

FIGURE 1.3
FP resonator. EM field distribution of three longitudinal modes is shown.

R_2 denote the reflectivities of the two mirrors. If I_0 is the initial intensity, then after a round trip of length $2L$ within the cavity, the intensity becomes

$$I(2L) = I_0 R_1 R_2 \exp\left[(g-\alpha)2L\right] \tag{1.17}$$

When $g > \alpha$, the wave is amplified. Self-sustained oscillation results when $I_0(2L) = I_0$ and the corresponding gain reaches its threshold value g_{th}, which may be expressed as follows:

$$g_{th} = \alpha + \frac{1}{2L} \ln\left(\frac{1}{R_1 R_2}\right) \tag{1.18}$$

1.3 Semiconductor Laser

A semiconductor laser is a special kind of laser in which a pump creates the condition for population inversion and an FP resonator provides the positive feedback needed. We shall describe in this section the structure of lasers, pumping mechanisms, and some basic properties of this kind of laser.

1.3.1 Early p-n Junction Laser

The early semiconductor laser is a simple p-n junction made up of a direct band gap semiconductor such as GaAs. Both p- and n-regions of the diode are heavily doped so that the Fermi levels lie above the conduction band in the n-region and below the valence band in the p-region. The equilibrium band diagram of the structure without any external bias has the Fermi level aligned in both the regions, and a barrier potential appears near the junction. When the junction is forward biased, this barrier potential, as shown in Figure 1.4, is reduced from the equilibrium contact potential difference. Electrons and holes are now injected across the junction in sufficient numbers. As shown in Figure 1.4, the condition of population inversion is established in the narrow region called the active region, where there are sufficient electrons (occupied states) in the upper conduction band (CB) and sufficient holes (unoccupied states) in the lower valence band (VB).

Since the material has a direct gap, there is a high probability of radiative recombination, as will be explained in Chapter 2. It will also be proved in Chapter 7 that if the injected carrier concentrations are high enough, the stimulated recombination rate may exceed the absorption rate, so that the active region is converted into a gain medium. The condition for achieving

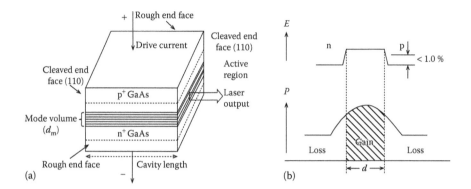

FIGURE 1.4
(a) Active region, mode volume, cleaved facets; (b) refractive index profile and gain and loss regions.

population inversion is expressed by the following relation, known as the Bernard–Duraffourg condition, which will be derived in Chapter 7:

$$E_g \leq \hbar\omega \leq F_n - F_p \tag{1.19}$$

where F_n and F_p are the respective quasi-Fermi levels in CB and VB, as shown in Figure 1.5.

The structure is enclosed in an FP resonator; actually, the cleaved facets of the diode structure act as the mirrors. Laser oscillations occur when the

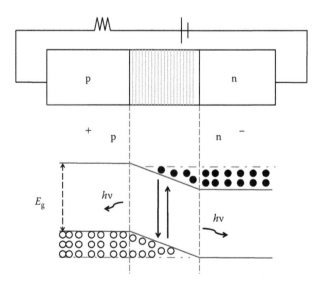

FIGURE 1.5
p-n junction band diagram under a forward bias.

round-trip gain exceeds the total losses in the medium and the mirrors. In semiconductors, the losses are due to free-carrier absorption and scattering at optical inhomogeneities in the material.

> **Example 1.3:** The length of a p-n junction laser is $L=400$ μm, the mirror reflectivities are 32% and 98%, and the material loss is 5 cm^{-1}. The required gain to overcome the total losses may be calculated from Equation 1.18 and is $g_{th}=19.5$ cm^{-1}.

> **Example 1.4:** The bandgap of GaAs is 1.43 eV. The forward bias in a GaAs p-n junction injects sufficient electron–hole pairs so that the quasi-Fermi level F_n is 0.03 eV above the CB edge, and F_h is 0.007 eV below the VB edge. From Equation 1.19, the gain spectrum covers the range from 1.43 eV to $(1.43+0.03+0.007) = 1.467$ eV.

A simplified structure of the p-n junction laser with the orientation of the cleaved facets is shown in Figure 1.4a. The radiation generated within the active volume spreads out in the surrounding lossy GaAs material. The presence of large carriers in the active region leads to a higher refractive index (RI) there, as shown in Figure 1.4b. Some kind of waveguiding of the radiation is achieved by the changed RI, and the light is effectively confined within the mode volume.

The typical light output power versus junction current of a semiconductor laser is shown in Figure 1.6. For low current, the emission is due to spontaneous recombination of electron–hole pairs (EHPs). As the current reaches the threshold, there is a break in the curve, and thereafter the power increases

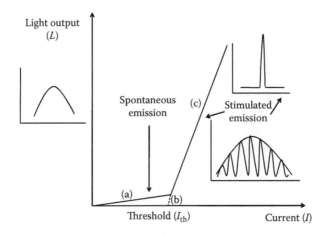

FIGURE 1.6
L-I curve of laser: (a) spontaneous emission; (b) onset of threshold; (c) higher current. The corresponding spectra are also shown.

rapidly with a slight increase in the current. The spectral characteristic of the laser is quite broad below threshold; at the onset of threshold, the structure supports a few of the longitudinal modes, and for larger currents, there is only one mode with a very narrow linewidth.

1.3.2 Threshold Current Density

The theory of lasers has been developed in Section 1.2 by considering discrete energy levels. It is therefore applicable to gas lasers as well as solid-state lasers such as the ruby laser (Cr-doped Al_2O_3), in which the impurities introduced lead to discrete energy levels. Since the transitions in semiconductors occur between bands, the above theory is not applicable to semiconductor lasers. Also, the current density or injected current is more meaningful in semiconductor lasers, rather than the threshold population inversion. Nevertheless, we shall attempt to use the theory to calculate the threshold current, assuming that the electrons and holes lie in two discrete energy levels.

We note that the gain coefficient is the negative absorption coefficient. Therefore, putting $-\alpha_\omega = g$ into Equation 1.15,

$$N_{th} = (N_2 - N_1)_{th} = \frac{\omega^2 g_{th}\tau_{sp}n_r^2}{\pi^2 c^3 L(\omega)} \tag{1.20}$$

The lasing condition is achieved when the lineshape function $L(\omega)$ is a maximum corresponding to $\omega = \omega_0$. Using the approximation $L(\omega) = \Delta\omega^{-1}$ in Equation 1.20, the threshold population becomes

$$N_{th} = \frac{\omega_0^2 g_{th}\tau_{sp}\Delta\omega n_r^2}{\pi^2 c^2} \tag{1.21}$$

It should be noted that the portion of the mode that leaks out of the active volume suffers loss and thereby reduces the gain of the mode propagating through the active layer. The reduction is taken into account by assuming that the effective population inversion in the active volume of $d_m LW$ is reduced by a factor d/d_m, as shown in Figure 1.7. The threshold condition is reached at a higher level, and thus Equation 1.21 is modified to

$$N_{th} = \frac{d_m}{d}\frac{\omega_0^2 g_{th}\tau_{sp}\Delta\omega n_r^2}{\pi^2 c^2} \tag{1.22}$$

We now assume that $N_{th} = (N_2)_{th}$, ignoring the population in the lower level. The number of electrons injected into the active volume of width d is J/e, when the current density is denoted by J, so that the density of electrons injected per unit time is J/ed. The recombination rate is N_2/τ, and

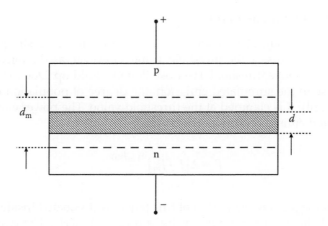

FIGURE 1.7
Active-region width and the width for mode volume.

it should be equal to the injection rate under steady state. One therefore writes

$$\frac{J_{th}}{ed} = \frac{(N_2)_{th}}{\tau_r} \tag{1.23}$$

Using Equation 1.22, the threshold current density is expressed as

$$J_{th} = \frac{ed_m}{\tau_r} \frac{\omega_0^2 g_{th} \tau_{sp} \Delta\omega n_r^2}{\pi^2 c^2} \tag{1.24}$$

The quantity τ_r is the total recombination lifetime of the electron expressed as $\tau_r^{-1} = \tau_{sp}^{-1} + \tau_{nr}^{-1}$, where the subscripts refer to radiative (sp) and nonradiative (nr) recombination processes. The ratio $\tau_r/\tau_{sp} = \eta_{int}$, the internal quantum efficiency. It stands for the fraction of the injected electrons that take part in the radiative recombination process. Using now the expression for g_{th} as given in Equation 1.18, the expression for the threshold current density becomes

$$J_{th} = \frac{\omega_0^2 e d_m \Delta\omega n_r^2}{\pi^2 c^2 \eta_{int}} \left[\alpha + \frac{1}{2L} \ln\left(\frac{1}{R_1 R_2}\right) \right] \tag{1.25}$$

Example 1.5: The threshold current density in a GaAs laser is now calculated by assuming $\lambda_0 = 0.84$ μm, $\Delta\omega_0 = 10^{14}$, $\alpha = 10^3$ m^{-1}, $n_r = 3.6$, $L = 400$ μm, $d_m = 2$ μm, and $\eta_{inj} = 1$.
The reflectivity is $R = [(n_r - 1) - (n_r + 1)]^2 = 0.32$. Using $R_1 = R_2$, the threshold gain $g_{th} = 3850$ m^{-1}. Also, $\omega_0 = 2\pi c/\lambda_0 = 22.44 \times 10^{14}$. Hence, $J_{th} = 9.04 \times 10^6$ A m$^{-2} = 904$ A cm^{-2}.

1.3.3 Power Output and Efficiency

The light power output in a semiconductor laser increases sharply as the current density crosses the threshold value. However, as the injection current increases above threshold, laser oscillations build up. Due to the resulting increase in the stimulated emission, the degree of population inversion decreases until it is clamped at the threshold value. The power emitted due to stimulated emission is then expressed as

$$P = A[J - J_{th}] \frac{\eta_{int} \hbar \omega}{e} \tag{1.26}$$

where A is the junction area. Part of this power is dissipated inside the laser cavity, and the rest comes out via the end crystal facets. The two components are proportional to α and $(1/2L) \ln (1/R_1 R_2)$, respectively, as indicated in Equation 1.18. Denoting the second component by the symbol α_m, the mirror loss, we may write the output power as

$$P = \frac{A[J - J_{th}] \eta_{int} \hbar \omega}{e} \frac{\alpha_m}{\alpha + \alpha_m} \tag{1.27}$$

The ratio of the rate of increase in the number of output photons to the rate of injection of number of carriers is defined as the external differential quantum efficiency, and thus

$$\eta_{ext} = \frac{d(P_0 / \hbar \omega)}{d\left(\frac{A}{e}[J - J_{th}]\right)} = \eta_{int} \frac{\alpha_m}{\alpha + \alpha_m} \tag{1.28}$$

where Equation 1.27 has been used to obtain the last equality. From the experimentally determined value of η_{ext} on L, the value of internal quantum efficiency may be obtained. The value for GaAs is in the range 0.7–1.0, which is quite high.

> **Example 1.6:** The value of $\alpha = 5$ cm^{-1} in Example 1.2 and the calculated value of $\alpha_m = 14.5$ cm^{-1}. Therefore, from Equation 1.28, $\eta_{ext} = 0.74 \, \eta_{int}$.
>
> If a forward-bias voltage V_f is applied to the diode, the power input is $V_f AJ$, and the efficiency of the laser in converting electrical input power into light output power is
>
> $$\eta = \frac{P}{V_f AJ} = \eta_{int} \left[\frac{J - J_{th}}{J} \right] \left(\frac{\hbar \omega}{eV_f} \right) \frac{\alpha_m}{\alpha + \alpha_m} \tag{1.29}$$

Using the approximation that $eV_f = \hbar \omega$ and $\alpha_m \gg \alpha$ and working well above threshold ($J \gg J_{th}$), the conversion efficiency η approaches η_{int}, which is quite high. The conclusion is that semiconductor lasers possess very high power-conversion efficiency: higher than gas or other types of lasers.

1.3.4 Heterojunction Lasers

It may be noted from Equation 1.25 that the optical radiation spreads out of the active region due to diffusion (large d_m), and there is also loss outside the active region. The threshold current density in a simple p-n junction laser is therefore quite high. A significant decrease in the threshold current density and a rise in efficiency may be obtained if a heterojunction is used instead. A heterojunction is a junction between two dissimilar semiconductors having different bandgaps, permittivities, and so on. The properties of heterojunctions, the band alignment, and different types will be presented in Chapter 3. In the following, we discuss qualitatively how such structures are useful for the reduction of threshold current density.

The properties of heterostructure lasers that give rise to lower threshold current density, permitting CW operation at room temperature, may be understood better by considering a double heterostructure (DH), shown in Figure 1.8b. In this structure, a lower-gap GaAs layer is sandwiched between two higher-gap AlGaAs layers, which have a higher RI than that of GaAs. Both N-p-P and P-n-N heterostructures show the same behavior, where the upper-case symbols N or P refer to the higher-gap AlGaAs material.

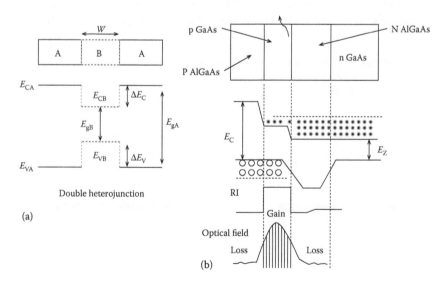

FIGURE 1.8
Double heterojunction: (a) band diagram; (b) illustrating carrier and mode confinement.

As may be noticed from the band diagram, the band gap difference between GaAs and AlGaAs gives rise to steps in the CB and VB, which act as potential barriers to the electrons and holes injected into the middle GaAs layer. The injected carriers are thus prevented from diffusing out of the GaAs layer, and the EHP recombination becomes more efficient there. This phenomenon is known as carrier confinement. The thickness d of the active GaAs layer may be reduced, leading to lower J_{th} (see Equation 1.15). In addition, the step changes of RI at the two heterojunctions lead to waveguiding action, resulting in more optical confinement. As the optical power is more or less confined to the active GaAs layer, and as the part of it leaking into the AlGaAs layer is not absorbed due to wider bandgap, the loss coefficient α in Equation 1.15 is lower than in homojunction lasers. All these factors lead to smaller values for J_{th}.

In actual practice, it is the threshold current that needs to be reduced. A small threshold current eases the design of power supply, especially when the bias current is modulated at a high bit rate. Threshold current is reduced by reducing the emissive area; the current along the junction plane is restricted to a narrow stripe. The schematic diagram of these stripe geometry lasers is shown in Figure 1.9.

There are many different ways to prepare the stripe geometry. Two illustrative examples are shown in Figure 1.9. In Figure 1.9a, the highly resistive regions on the two sides of the structure are obtained by proton bombardment, and current flows only through the low-resistivity region forming the stripe. This is an example of gain-guided structure. In Figure 1.9b, a mesa structure has been formed by etching, and AlGaAs layers are regrown in the etched regions. This structure, a buried heterostructure laser, forms an example of index-guided structure. Here, the active region is surrounded by

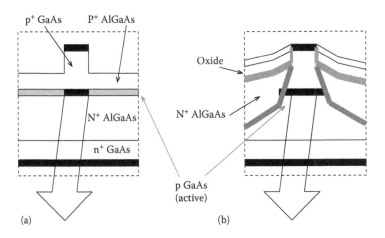

FIGURE 1.9
Stripe geometry laser: (a) gain guided; (b) index guided.

higher-RI AlGaAs both in the direction of layer growth and in the direction normal to the growth. The stripe region thus forms a two-dimensional (2-D) waveguide, and better optical mode confinement is achieved.

Stripe geometry lasers have further advantages. Their emission area matches the cross-sectional area of the core of optical fibers, and the coupling becomes easier. Also, they do not show the kinks in the light output power versus current curves that appear in broad-area lasers.

1.3.5 Distributed Feedback Lasers

The active region of the stripe geometry lasers in which an FP resonator provides feedback may be considered as a dielectric rectangular cavity. In general, a cavity supports different modes of EM radiation. In simple terms, each mode is characterized by three integers (i,j,k), where the integers represent the number of maxima occurring in the power distribution of EM radiation along (x,y,z)-directions. The kth longitudinal mode therefore satisfies the condition

$$k\frac{\lambda_k}{2} = n_r L \qquad (1.30)$$

where:
   L is the length of the cavity (separation between mirrors)
   λ_k is the free-space wavelength of the kth mode
   n_r is the RI of the material within the cavity

Similar equations can be written for transverse mode number (i) along the thickness d of the active layer, and for lateral mode number (j) along the width w of the stripe. Usually d and w are small; therefore only the fundamental transverse and lateral modes are supported by the cavity. However, L being large (~200 μm), there are very many longitudinal modes, and the overall linewidth of emission is usually large. A large linewidth does not favor long-distance communication with very high bit rates. To increase the bit rate–length product for fiber-optic communication, a single longitudinal mode (SLM) laser is needed.

> **Example 1.7:** Consider a DH laser with GaAs as the active material. Let the gain profile cover the range E_g to $E_g+0.1$ eV. The frequency range covered is $\Delta f = 0.1e/h = 2.4 \times 10^{13}$ Hz. From Equation 1.30, the frequency of the kth mode is $f_k = kc/2n_r L$, and the frequency separation between adjacent modes is $\delta f = f_k - f_{k-1} = c/2n_r L = 8.33 \times 10^{10}$ Hz for $L = 500$ μm and $n_r = 3.6$. The total number of longitudinal modes within the gain profile is $\Delta f/\delta f \approx 288$.

The SLM feature is exhibited by a laser having a distributed feedback (DFB) mechanism, or with a distributed Bragg reflector (DBR), in place of the

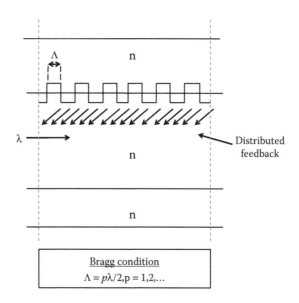

FIGURE 1.10
DFB laser.

FP resonator. The DFB laser is illustrated in Figure 1.10, which shows that a periodic modulation of RI having a period Λ is introduced in the upper cladding layer of the DH structure. The EM wave suffers feedback from the Bragg reflector, as indicated by the arrows. When the reflected waves add up due to constructive interference, only the particular modes with wavelengths λ that satisfy the Bragg condition

$$\Lambda = p\lambda / 2, \quad p = 1, 2, \ldots \tag{1.31}$$

are supported. In most cases, the mode with $p=1$ is dominant. The Bragg reflector thus acts as a filter. Note that the feedback is distributed, contrary to feedback in FP resonators, in which feedback is localized at the position of the two mirrors. Detailed theory of distributed feedback will be developed in Chapter 12.

For all types of semiconductor lasers, the threshold current density increases with temperature. It is difficult to find an expression for temperature dependence of J_{th} that is valid for all materials and temperature ranges, since many factors contribute to the temperature dependence. The following empirical expression for temperature dependence of J_{th} around 0°C is used in the literature:

$$J_{th}(T) = J_{th}(0) \exp(T / T_0) \tag{1.32}$$

where T_0 is called the *characteristic temperature*, and its ideal value should be infinity. For DH lasers, the value is quite low: ~70°C –100°C.

1.4 Materials for Semiconductor Lasers

The first and foremost requirement for a semiconductor for its use in lasers is that the bandgap should be direct. Only for direct-gap semiconductors is there a high probability for transition from a conduction band state to an empty valence band state. On the other hand, in indirect-gap semiconductors such as Si or Ge, electrons and holes cannot recombine without the participation of a phonon or without involving a trap or defect state in the forbidden gap. The corresponding transition probability is extremely low. Therefore, indirect-gap semiconductors are not suitable for use in lasers or even in light-emitting diodes. The calculation of transition probability will be described in Chapter 6.

Most III–V and II–VI compound semiconductors and their alloys have a direct bandgap and are used to fabricate lasers. The lasers invariably employ double heterostructures, and an important requirement is that the layers of two semiconductors should be lattice matched. Lattice-mismatched hetero-structures give rise to misfit dislocations and defects at or near the heterointerfaces that act as nonradiative centers. Too many defects therefore degrade the emission characteristics and, more seriously, hinder laser action. Some degree of lattice mismatch may be allowed in strained-layer growth, and lasers have been produced with lattice-mismatched layers. However, the mismatch should be restricted to within 0.1%. Another requirement for the structures is that the layers should be doped to the desired level to form p-n junctions, and the RI variation from one layer to the other should be appropriate for efficient waveguiding.

Different material systems are used to achieve emission at different wavelengths. The wavelength is related to the bandgap of the active layer by the following relation:

$$\lambda\,(\mu m) = \frac{1.24}{E_g\,(eV)} \tag{1.33}$$

The chosen wavelength may be obtained by using a semiconductor material with the proper value of band gap energy. In many cases, a wide range of wavelength may be achieved by using a ternary or a quaternary alloy material. A common example of a ternary alloy is $Al_xGa_{1-x}As$, which is lattice matched with GaAs for all values of alloy composition x. A useful quaternary alloy is $In_{1-x}Ga_xAs_yP_{1-y}$, which may be lattice matched with InP substrate provided $y\sim2.2x$.

> **Example 1.8:** The bandgap of $Al_xGa_{1-x}As$ in the direct-gap region ($x<0.45$) is given as $E_g(x)=1.424+1.266x+0.266x^2$. Calculate the composition of $Al_xGa_{1-x}As$ to emit $\lambda=0.82\ \mu m$.
> From Equation 1.32, $E_g=1.51$ eV. This gives $x=0.07$. The required composition is $Al_{0.07}Ga_{0.93}As$.

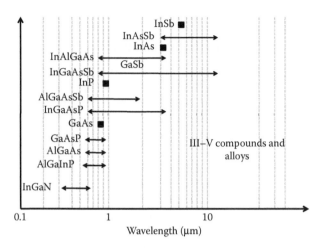

FIGURE 1.11
Emission wavelengths for III–V compounds and their alloys.

> **Example 1.9:** Since AlGaAs is lattice matched to GaAs, assuming that
> the alloy bandgap is direct up to $x=0.45$, we may calculate the range of
> wavelength covered by the alloy. For $x=0$, the gap of 1.242 gives emis-
> sion at 0.871 μm, while for $x=0.45$, the gap is 2.048 eV and the emission is
> at 0.606 μm. The range is from 0.606 to 0.871 μm.

Figure 1.11 shows the III–V compounds and their alloys used to fabricate
lasers; the corresponding wavelength ranges are also indicated. Some of the
alloys do not have a lattice-matching partner. For example, $In_xGa_{1-x}As$ does
not lattice match with $Al_xGa_{1-x}As$. However, it can be combined with AlGaAs
or GaAs to produce strained QW lasers emitting at 0.9–1.0 μm. InGaAsSb
and $InAs_{1-x-y}P_xSb_y$, which have lattice matching with InAs and GaSb, are
useful for lasers emitting in the 1.7–4.4 μm range. Different alloys are used
for lasers emitting at visible wavelengths. Red-emitting lasers in the 0.6 μm
band are commercialized as sources for optical disk memory, such as digital
versatile disks (DVDs).

Candidates for lasers emitting in green, blue, and violet regions are the
II–VI compounds CdS, CdSe, ZnS, ZnSe, and their alloys and the III–V compound semiconductors such as GaN, InN, AlN, and their alloys. $In_xGa_{1-x}N$
QW lasers on sapphire substrate are the currently used sources for the
blue region. IV–VI compounds such as PbSe, PbTe, and PbS and alloys
such as PbSnTe and PbSSe produce injection lasers in the 3–34 μm range.
The emission wavelengths of some II–VI compounds and their alloys are
shown in Figure 1.12. A few IV–VI compounds and their alloys used to
produce lasers are shown in Figure 1.13 along with the corresponding
wavelength ranges.

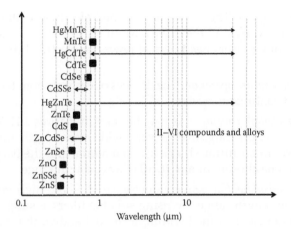

FIGURE 1.12
Emission wavelengths for II–VI compounds and their alloys.

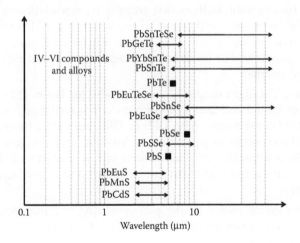

FIGURE 1.13
Emission wavelengths for IV–VI compounds and their alloys.

1.5 Special Features

Semiconductor lasers exhibit many features and advantages over other forms of lasers:

1. *Compactness:* Almost all semiconductor lasers are tiny, with size below 1 mm³. Even if a heat sink and a power supply are included, the whole system is compact and lightweight.

2. *Excitation by bias:* The lasers are pumped electrically; the bias voltage is a few volts, and the drive current may be a few milliamps. In contrast, other lasers need optical power or electrical discharge for pumping.

3. *Room-temperature operation:* The devices operate at room temperature and emit continuous waves.

4. *Wide wavelength coverage:* As may be seen in Figures 1.11–1.13, semiconductor lasers can cover a very wide wavelength range, from ultraviolet to far infrared. The active material and the barrier materials need be chosen for a particular wavelength range.

5. *Wide gain–bandwidth*: Semiconductor lasers show high gain over a wide wavelength range. By using suitable filters, a single semiconductor laser can be tuned within the gain–bandwidth range. This is also true for semiconductor laser amplifiers.

6. *Direct modulation*: One of the important features of semiconductor lasers is that their intensity can be modulated directly by changing the bias current. Both analog and digital modulation is possible. Also, the frequency and phase of the emerging EM wave can be modulated. The modulation bandwidth is quite high, reaching a few tens of gigahertz. For other lasers, a separate modulating device or system is needed.

7. *Coherence*: The light emitted by semiconductor lasers shows a high degree of spatial and temporal coherence. The radiation has the fundamental lateral mode. With DFB or DBR laser structures, a single longitudinal mode of operation is feasible, and the linewidth of emission is a fraction of a nanometer (a few megahertz). Using an external cavity, the linewidth can be reduced to a few hundreds of kilohertz.

8. *Ultrashort optical pulses*: Using gain switching or mode locking, it is possible to generate ultrashort optical pulses of subnanosecond to picosecond width.

9. *Mass producibility*: The use of lithography and planar-processing technique has led to mass production of the devices.

10. *High reliability*: The devices are stable. For many lasers, the wear and tear and fatigue problems have been solved. The devices show a long life of ~10,000 hours or more with no substantial degradation.

11. *Monolithic integration*: By using a common substrate, integration of many lasers is possible. In addition, integration of lasers, photodetectors, modulators, and other electronic devices, including passive waveguides, on the same substrate has been achieved.

TABLE 1.1

Outline of Development of Semiconductor Laser

Year	Work	Reference
1917	Einstein's proposal of stimulated emission	1
1924	Derivation of Planck's radiation formula by S. N. Bose	2
1953	Von Neumann's suggestion for laser action in p-n junction	5
1954	Ammonia maser by Townes et al.	3, 4
1958	Optical maser by Townes and Schawlow	6
1960	He–Ne laser by Javan and ruby laser by Maiman	7, 8
1956–1961	Possibility of laser action in semiconductors: Aigrain, Basov, Boyle, Thomas	9–14
1961	Lax: transition between Landau levels	16
1961	Work by Dumke, Lasher, Nathan on direct gap	17, 18
1961	Paper by Bernard and Duraffourg	19
1962	Announcement of semiconductor p-n junction laser by four groups	20–23
1963	Proposal by Kroemer on heterojunctions	24
1967	Work by Alferov et al.	25, 26
1966–1969	Parallel work by Hayashi and Panish	27
1970	Concept of 1-D superlattice by Esaki and Tsu	28
1971	Unipolar laser by Kazarinov and Suris	29
1973	QW by Dingle et al.	30
1973	QW by Chang et al.	31
1974	QW laser: patent application by Dingle and Henry	32
1976	Optically pumped QW laser demonstrated by Van der Ziel et al.	33
1979	Electrically pumped QW laser by Dupuis et al.	34
1977	VCSEL by Iga	35
1982	Concept of QWR and QD lasers by Arakawa and Sakaki	36
1991	VCSEL by Jewell et al.	37
1993	QD lasers	40
1994	QCL lasers: Faist et al.	41
1990–1996	Blue laser by Akasaki, Amano and Nakamura	42, 43
1996	Tunneling injection laser: Bhattacharyya	44
2002	Terahertz QCL by Köhler	45
2004	QCL: 300 K, high power by Evans	46
2004	Transistor laser by Feng and Holonyak	47
2005	Si Raman laser by Intel	48
2010	Ge on Si laser by Liu et al. (optical pump)	49
2012	Ge on Si laser (electrical pump)	50

There are some disadvantages or drawbacks of using semiconductor lasers:

1. *Temperature instability:* The performance of semiconductor lasers depends on temperature. The emission wavelength, threshold current, and output power change with change in temperature (see e.g. Equation 1.32). This is quite troublesome in optical communication, in which a high level of stability is needed. The changes are compensated by proper control circuits, which add to the complexity of the system.

2. *Noise characteristics:* Since high density of carriers is involved in the lasing process, a fluctuation causes a change in RI of the active region. In addition, due to short resonator length and low reflectivity of mirrors, the perturbations caused by external feedback affect the oscillation characteristics. Semiconductor lasers thus suffer from various noise and instability problems.

3. *Output beam divergence:* The thin waveguide allows the output beam to diverge significantly. An external lens is needed to obtain a collimated beam.

1.6 Applications

1.6.1 Optical Communication

The advent of semiconductor lasers has revolutionized the field of optical communication. Historically, GaAs/AlGaAs lasers were the first to be used as the source for fiber-optic communication. As the next generations used first 1.3 μm and then 1.55 μm as the emission wavelength, $In_{1-x}Ga_xAs_yP_{1-y}$/InP lasers were developed. These lasers have double heterostructures, which in turn gave way to QW and strained QW lasers. The performance of QD lasers is now being investigated for introduction in communication systems.

1.6.2 WDM and DWDM: Tunable Lasers: DFB/DBR

For high-bit-rate long-haul communication, a single-mode laser with narrow linewidth is needed. The DFB and DBR lasers have been developed to meet these requirements. The information-carrying capacity of optical communication systems is increased enormously with the introduction of wavelength division multiplexing (WDM) and dense WDM (DWDM). In these systems, widely tunable lasers are needed. The DFB and DBR lasers and their modifications have been developed for this purpose.

1.6.3 Short-Pulse Lasers: Soliton

For characterization of semiconductors and for the study of ultrafast processes in materials, very-short-pulse lasers are required. These are realized by using mode locking and other techniques.

1.6.4 QCL: Pollution, Atmospheric Monitoring, Terahertz Communication

A less explored region of the electromagnetic spectra is the terahertz region. Communication in this region may offer information transfer with a higher bit rate. Sources emitting in this region may be used for pollution monitoring and for body scanning at airports, as this is safer than x-ray checking. Unfortunately, compact, tiny, low-power sources such as lasers are not fully developed yet. QCLs may provide a useful solution in this regard.

1.6.5 Diode Pump Lasers for EDFA, Raman Amplifiers

Erbium-doped fiber amplifiers (EDFAs) and Raman amplifiers need optical pumping to create population inversion. High-power diode lasers to work as pumps at 980 nm and 1540 nm for EDFAs have already been developed. For Raman amplifiers, the pump wavelength should be higher than the wavelength of the signal to be amplified. Several diode lasers satisfying the condition have been or are being developed.

1.6.6 Semiconductor Optical Amplifiers (SOAs): Amplifiers, Logic Elements, Wavelength Converters

Remarkable developments have also been made in semiconductor laser amplifiers and nonlinear optic-wavelength conversion in semiconductor lasers. Amplifiers, logic gates, and wavelength converters are being investigated for use in future optical networks. The current emphasis is on QW-SOAs, but a considerable amount of research is being done on QD-SOAs also.

1.6.7 Consumer Products: Laser Printer, CD/DVD, Bar-Code Reader

Semiconductor lasers find widespread applications in optical disk memories, in which the prime requirements are low noise and high stability. To record the high density of data, a short-wavelength low-cost laser is needed. $Al_xGa_{1-x}As$ lasers were adopted as a light source for compact-disk (CD) pickup heads, becoming the first laser that penetrated widely into the home. For DVD systems, $(Al_{1-x}Ga_x)_yIn_{1-y}P$ red semiconductor lasers have been adopted. Blue and blue-green lasers are more useful for storing larger amounts of data. These lasers are already in use in DVD systems.

Other applications in optical information processing that are in practical use include laser printers, image scanners, and bar-code readers.

1.6.8 Medical Applications

High-power lasers are in use for various surgical operations, such as removal of cataracts and tumors.

1.6.9 QD and VCSEL for Quantum Computing and Quantum Information Processing

VCSELs have already created a sizable market, as they are used in optical mice for computers. In addition, QD lasers and VCSELs are being widely investigated for use in quantum computing and quantum information processing.

1.6.10 Optoelectronic Integrated Circuits and Photonic Integrated Circuits

In an optoelectronic integrated circuit (OEIC), lasers and electronic circuit elements are monolithically integrated. Photonic integrated circuits (PICs) contain optoelectronic elements such as lasers, modulators, and photodetectors. Many integrated devices are employed in applications for optical communication systems, optical interconnection in computer systems, and optical measurements and sensing.

PROBLEMS

1.1. Calculate A_{21} for $\tau = 0.1$ ps.

1.2. Find the temperature for which $R_{spon} = R_{stim}$. Assume photon energy $= 1$ eV.

1.3. Prove Equation 1.11.

1.4. Discuss the nature of variation of the threshold current in a laser diode with decreasing stripe width. Consider the decrease in the lateral confinement factor when the width is too low. Briefly discuss a few methods to obtain the stripe-geometry configuration.

1.5. Calculate the difference of population in a two-level atomic system to obtain gain $= 1$ m^{-1}. The wavelength is 1.0 μm, $\tau_{sp} = 1$ ps, $n_r = 2$, and $\Delta\omega = 10^{14}$.

1.6. Calculate reflectivity to give $g_{th} = 2$ m^{-1}. $L = 500$ μm, $\alpha = 0.1$ m^{-1}. Assume equal reflectivity for both mirrors.

1.7. Calculate the value of the reflectivity (equal for both mirrors) to give $J_{th} = 500$ A cm^{-2}. All other parameter values are the same as in Example 1.5.

1.8. Calculate the output power of a laser emitting at 1.55 μm. The operating current is 5 mA, the threshold current is 2 mA, the internal and mirror losses are, respectively, 5 cm⁻¹ and 15 cm⁻¹, and the internal quantum efficiency is 70%.

1.9. Assume that the threshold current in a DH laser is 5 mA and all other parameters are the same as in Problem 1.8. Plot the values of conversion efficiency for different operating currents, assuming that the forward voltage $V_f = \hbar\omega$.

1.10. Find the value of grating period needed for a DFB laser to emit at 1.3 μm.

1.11. The dimensions of a stripe-geometry DH laser are 300 μm × 5 μm × 0.2 μm. Find the mode separation for the transverse, lateral, and longitudinal directions. The operating wavelength is 1.3 μm, and the RI is 3.5.

1.12. Calculate the thickness of the active layer that will support the fundamental transverse mode. The emission wavelength is 1.55 μm, and the RI is 3.4.

Reading List

Agrawal, G. P. and N. K. Dutta, *Semiconductor Lasers*, 2nd edition. Van Nostrand Reinhold, New York, 1993.

Basu, P. K., *Theory of Optical Processes in Semiconductors*. Oxford University Press, Oxford, 1997.

Bhattacharyya, P., *Semiconductor Optoelectronic Devices*. Prentice-Hall, Upper Saddle River, NJ, 1997.

Casey, H. C. and M. B. Panish, *Heterostructure Lasers: Fundamentals and Principles*. Academic, New York, 1979.

Chuang, S. L., *Physics of Optoelectronic Devices*. Wiley, New York, 1995.

Gowar, J., *Optical Communication Systems*, Prentice-Hall International Series. Prentice Hall, New York, 1993.

Palais, J. C., *Fiber Optic Communication*. Pearson Prentice Hall, Upper Saddle River, NJ, 2005.

Saleh, B. E. A. and M. C. Teich, *Fundamentals of Photonics*. Wiley, New York, 1991.

Senior, J. M., *Optical Fiber Communication: Principle and Practice*, 2nd edition. Pearson Prentice-Hall, Englewood Cliffs, NJ, 2007.

Singh, J., *Optoelectronics: An Introduction to Materials and Devices*. McGraw-Hill, New York, 1996.

Wilson, J. and J. Hawkes, *Optoelectronics: An Introduction*, 3rd edition. Prentice-Hall PTR, Harlow, 2012.

References

1. Einstein, A., On the quantum theory of radiation, *Physikalische Zeitschrift* 18, 121–128, 1917.
2. Bose, S. N., Planck's Gesetz und Lichtquantenhypothese, *Zs. Physik* 26, 178–181, 1924.
3. J. P. Gordon, H. J. Zeiger, and C. H. Townes, Molecular microwave oscillator and new hyperfine structure in the microwave spectrum of NH_3, *Phys. Rev.* 95, 282, 1954.
4. Gordon, J. P., H. J. Zeiger, and C. H. Townes, The maser: New type of microwave amplifier, frequency standard, and spectrometer, *Phys. Rev.* 99, 1264, 1955.
5. von Neumann, J., Notes on the photon-disequilibrium-amplification scheme (JvN), September 16, 1953, *IEEE J. Quantum Electron.* QE-23, 659–673, 1987.
6. Schawlow, A. L. and C. H. Townes, Infrared and optical masers, *Phys. Rev.* 112, 1940–1949, 1958.
7. Javan, A., W. K. Bennet, Jr., and D. R. Herriot, Population inversion and continuous optical maser oscillation in a gas discharge containing a He–Ne mixture, *Phys. Rev. Lett.* 6, 106–110, 1961.
8. Maiman, T. H., Stimulated optical radiation in ruby, *Nature*, 187, 493–494, 1960.
9. Aigrain, P., Unpublished lecture at the International Conference on Solid State Physics in Electronics and Telecommunications, Brussels, 1958.
10. Aigrain, P., Masers a semi-conducteurs. In *Quantum Electronics, Proceedings of the Third International Congress*, ed. Grivet, P. and Bloembergen, N., Columbia University Press, New York, 1761–1767, 1964.
11. Basov, N. G., B. M. Vul, and Y. M. Popov, Unpublished paper registered with the Committee on Discoveries and Inventions of the Council of Ministers of the U.S.S.R. dated July 7, 1958.
12. Basov, N. G., B. M. Vul, Y. M. Popov, Quantum-mechanical semiconductor generators and amplifiers of electromagnetic oscillations, *Zh. Eskp. Theo. Fiz.* 37, 587–588, 1959 (*Sov. Phys. JETP*, 10, 416, 1959).
13. Boyle, W. S. and D. G. Thomas, *Optical Maser*, U.S. Patent 3059117 filed January 11, 1960 and granted October 16, 1962. Reissued as patent RE. 25,632 on August 18, 1964.
14. Thomas, D. G. and Hopfield, J. J., Fluorescence in CdS and its possible use for an optical maser, *J. Appl. Phys.* 33, 3243–3249, 1962.
15. Pankove, J., 1967 semiconductor laser conference—Introduction, *IEEE J. Quantum Electron.* QE-4, 109–110, 1968.
16. Lax, B., Cyclotron resonance and impurity levels in semiconductors. Presented at the *1st International Conference on Quantum Electronics*, 1959; published in *Quantum Electronics*, ed. C. H. Townes, Columbia University Press, New York, 428–449, 1960.
17. Dumke, W. P., Interband transitions and maser action, *Phys. Rev.* 127, 1559–1563, 1962.
18. Basov, N. G., Inverted populations in semi-conductors. In *Quantum Electronics, Proceedings of the Third International Congress*, ed. Grivet, P. and Bloembergen, N., Columbia University Press, New York, 1769–1785, 1964.
19. Bernard, M. G. A. and G. Duraffourg, Laser conditions in semiconductors, *Phys. Stat. Sol.* 1, 699–703, 1961.

20. Hall, R. N., G. E. Fenner, J. D. Kingsley, T. J. Soltys, and R. O. Carlson, Coherent light emission from GaAs junctions, *Phys. Rev. Lett.* 9, 366–368, 1962.

21. Holonyak, Jr., N. and S. F. Bevacqua, Coherent (visible) light emission from Ga(AsP) junctions, *Appl. Phys. Lett.* 1, 82–83, 1962.

22. Nathan, M. I., W. P. Dumke, G. Burns, F. H. Dill, Jr., and G. Lasher, Stimulated emission of radiation from GaAs p-n junctions, *Appl. Phys. Lett.* 1, 62–64, 1962.

23. Quist, T. M., R. H. Rediker, R. J. Keyes, W. E. Krag, B. Lax, A. L. McWhorter, and H. J. Zeiger, Semiconductor maser of GaAs, *Appl. Phys. Lett.* 1, 91–92, 1962.

24. Kroemer, H., A proposed class of heterojunction injection lasers, *Proc. IEEE*, 51, 1782–1783, 1963.

25. Alferov, Z. I. and R. F. Kazarinov, Semiconductor laser with electric pumping (in Russian), Inventor's Certificate No. 181737, Appl. No. 950 840, March 30, 1963.

26. Alferov, Z. I., V. M. Andreev, E. L. Portnoi, and M. K. Trukan, AlAs–GaAs heterojunction injection lasers with a low room-temperature threshold, *Fiz. Tekh. Poluprov*, 3, 1328–1332, 1969.

27. Hayashi, I., P. B. Panish, P. W. Foy, and S. Sumski, Junction lasers which operate continuously at room temperature, *Appl. Phys. Lett.* 17, 109–111, 1970.

28. Esaki, L. and R. Tsu, Superlattice and negative differential conductivity in semiconductors, *IBM J. Res. Develop.*, 14, 61–65, 1970.

29. Kazarinov, R. and R. A. Suris, Possibility of the amplification of electromagnetic waves in a semiconductor with a superlattice, *Fiz. Tekh. Poluprov.* 5, 797–800, 1971; trans. in *Sov. Phys. Semicond.* 5, 707–709, 1971.

30. Dingle, R., W. Wiegmann, and C. H. Henry, Quantum states of confined carriers in very thin AlGaAs–GaAs–Al GaAs heterostructures, *Phys. Rev. Lett.* 33, 827–830, 1974.

31. Chang, L. L., L. Esaki, and R. Tsu, Resonant tunneling in semiconductor double barriers, *Appl. Phys. Lett.* 24, 593–595, 1974.

32. Dingle, R. and C. H. Henry, Quantum effects in heterostructure lasers, U.S. Patent 3 982 207, September 21, 1976.

33. Van der Ziel, J. P., R. Dingle, R. C. Miller, W. Wiegmann, and W. A. Nordland, Jr., Laser oscillation from quantum states in very thin GaAs–AlGaAs multilayer structures, *Appl. Phys. Lett.* 26, 463–465, 1975.

34. Dupuis, R. D., P. D. Dapkus, R. Chin, N. Holonyak, Jr., and S. W. Kirchoefer, Continuous 300K laser operation of single-quantum-well AlGaAs–GaAs heterostructure diodes grown by metalorganic chemical vapor deposition, *Appl. Phys. Lett.* 34, 265–267, 1979.

35. Iga, K., *Laboratory Notebook of P&I Laboratory 1977 Issue*. Tokyo Institute of Technology, Tokyo, 1977.

36. Arakawa, Y. and H. Sakaki, Multidimensional quantum well laser and temperature dependence of its threshold current, *Appl. Phys. Lett.* 40, 77–78, 1982.

37. Jewell, J. L., J. P. Harbison, A. Scherer, Y. H. Lee, and L. T. Florez, Vertical-cavity surface-emitting lasers: Design, growth, fabrication, characterization, *IEEE J. Quantum Electron.* 27, 1332–1346, 1991.

38. Geels, R. S., S. W. Corzine, J. W. Scott, D. B. Young, and L. A. Coldren, Low threshold planarized vertical-cavity surface-emitting lasers, *IEEE Photon. Technol. Lett.* 2, 234–236, 1990.

39. Chang-Hasnain, C. J., J. P. Harbison, C. E. Zah, M. W. Maeda, L. T. Florez, N. G. Stoffel, and T. P. Lee, Multiplewavelength tunable surface emitting laser arrays, *IEEE J. Quantum Electron.* 27, 1368–1376, 1991.

40. Kirstaedter, N., N. N. Ledentsov, M. Grundmann, D. Bimberg, V. M. Ustinov, S. S. Ruvimov, M. V. Maximov, et al., Low threshold, large T_0 injection laser emission from (InGa)As quantum dots, *Electron. Lett.* 30, 1416–1417, 1994.
41. Faist, J., F. Capasso, D. L. Sivco, C. Sirtori, A. L. Hutchinson, and A. Y. Cho, Quantum cascade laser, *Science*, 264, 553–556, 1994.
42. Amano, H., T. Asahi, and I. Akasaki, Stimulated emission near ultraviolet at room temperature from a GaN film grown on sapphire by MOVPE using an AlN buffer layer, *Jpn. J. Appl. Phys.* 29, L205, 1990.
43. Nakamura, S., M. Senoh, S. Nagahama, N. Iwasa, T. Yamada, T. Matsushita, H. Kiyoku, and Y. Sugimoto, InGaN-based multi-quantumwell-structure laser diodes, *Jpn. J. Appl. Phys. B* 35, L74–76, 1996.
44. Bhattacharya, P., J. Singh, H. Yoon, X. Zhang, A. Gutierrez-Aitken, and Y. Lam, Tunneling injection lasers: A new class of lasers with reduced hot carrier effects, *IEEE J. Quantum Electron.* 32, 1620–1629, 1996.
45. Köhler, R., A. Tredicucci, F. Beltram, H. Beere, E. Linfield, A. Davies, D. Ritchie, R. Iotti, and F. Rossi, Terahertz semiconductor-heterostructure laser, *Nature* 417, 156–159, 2002.
46. Evans, A., J. S. Yu, J. David, L. Doris, K. Mi, S. Slivken, and M. Razeghi, High-temperature, high power, continuous-wave operation of buried heterostructure quantum cascade lasers, *Appl. Phys. Lett.* 84, 314–316, 2004.
47. Walter, G., N. Holonyak, Jr., M. Feng, and R. Chan, Laser operation of a heterojunction bipolar light-emitting transistor, *Appl. Phys. Lett.* 85, 4768–4770, 2004.
48. Rong, H., R. Jones, A. Liu, O. Cohen, D. Hak, A. Fang, and M. Paniccia, A continuous wave Raman laser: An all-silicon Raman laser, *Nature* 433, 725–728, 2005.
49. Liu, J., X. Sun, R. Camacho-Aguilera, L. C. Kimerling, and J. Michel, Ge-on-Si laser operating at room temperature, *Optics Lett.* 35, 679–681, 2010.
50. Camacho-Aguilera, R. F., Y. Cai, N. Patel, J. T. Bessette, M. Romagnoli, L. C. Kimerling, and J. Michell, An electrically pumped germanium laser, *Opt. Express*, 20, 11316–11320, 2012.

2

Basic Theory

2.1 Introduction

The basic structure and the principle of operation of semiconductor lasers have been introduced in Chapter 1. However, the discussions presented there are rudimentary and do not give the readers a complete and quantitative understanding of the physical processes involved, the design rules, and so on. For example, the band structure of semiconductors, the electronic transport and optical processes in semiconductors, the band structure and band alignment of heterojunctions, expressions for the optical gain and absorption, and many other properties of bulk semiconductors, their alloys, and heterojunctions need a detailed discussion. In addition, detailed knowledge is needed of the passive components used, for example, the resonators, distributed feedback structures, distributed Bragg reflectors, optical waveguides, and so on. In the present and following chapters, we will endeavor to discuss these.

The present chapter deals with the basic physics of semiconductors. Full discussions are given in many excellent textbooks listed at the end of this chapter. We will present here the theory and equations essential for understanding the various physical phenomena occurring in a bulk semiconductor. We begin with a qualitative discussion of the band formation in a crystalline solid, introducing the E–k diagrams for direct and indirect band semiconductors, followed by the concept of effective mass. After introducing the density of states function, expressions for the carrier density in intrinsic and extrinsic semiconductors are derived. A brief discussion of excess carriers and the diffusion and different recombination processes then follows, and the continuity equation is developed. Finally, the basic theory for the current–voltage (I-V) characteristics of p-n junction diodes is presented.

It may be mentioned here that for a clear understanding of the various optical processes in semiconductors, knowledge of the E–k diagram is necessary. The use of the E–k diagram, particularly its nature near the band extrema, which is obtained by k·p perturbation theory, will be discussed in Chapter 4.

2.2 Band Structure

The semiconductor materials that we are interested in are single crystals, which consist of a space array of atoms or molecules constituted by the regular repetition of a certain basic structural unit, called a *unit cell*. The periodic arrangement of atoms is called the *lattice*. Most of the semiconductors have a diamond or zinc-blende structure, where each atom is surrounded by four of its nearest neighbors. However, as the electrons in the outer orbit of one atom feel the influence of neighboring atoms, the discrete energy levels for a single free atom do not apply to the same atom in the crystal. When isolated atoms are brought together to form a solid, various interactions occur between neighboring atoms. The forces of attraction and repulsion between the atoms find a balance at the proper interatomic spacing for the crystal.

The formation of bands in a crystalline solid is usually illustrated by considering a one-dimensional chain of atoms to represent the periodic structure. When two atoms are completely isolated from each other, so that there is no interaction of electron wave functions between them, they can have identical electronic structures. As the spacing between the two atoms decreases, their electron wave functions begin to overlap. According to the exclusion principle, no two electrons in a given interacting system can occupy the same quantum states. Thus, there must be a splitting of the discrete energy levels of isolated atoms into new levels belonging to the pair rather than to individual atoms. Since many atoms are brought close together in a solid, the number of split energy levels equals the number of atoms in the chain. If there are N number of atoms, there are N of these discrete levels, but the difference in energy between two adjacent levels is too small. The levels, therefore, essentially form a continuous band of energy.

The simplest models to understand the formation of bands in a one-dimensional chain of atoms are the Kronig–Penney model and the tight binding model. Indeed, the models show that there are several allowed bands separated by a forbidden bandgap. The electrons are free to move in the allowed bands with a momentum $\hbar k$, called the *crystal momentum*. The allowed values for **k** are restricted within a zone called the *reduced Brillouin zone*.

The calculation of the band structure is a very involved problem. It must consider many valence electrons and ions and their interactions. The many electron problem is reduced to a single electron problem by using various approximations. The wave function of an electron is written as

$$\Psi(\mathbf{r}, \mathbf{k}) = e^{i\mathbf{k}\cdot\mathbf{r}} U(\mathbf{r}, \mathbf{k}) \tag{2.1}$$

where:
    Ψ is the wave function
    k   is the wave vector
    r   is the position coordinate of the electron
    U is an envelope function having the periodicity of the lattice

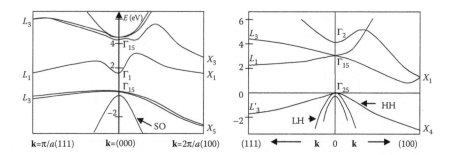

FIGURE 2.1
Band structure of (a) GaAs and (b) Si, calculated by the empirical pseudopotential method.

Many theoretical models are available for the calculation of the band structure of any semiconductor. In most cases, the results obtained by Cohen and Chelikowsky [1] using the empirical pseudopotential method are quoted. Figure 2.1 shows the band diagram of GaAs and Si obtained by this method.

The diagrams show the plot of the energy (E) versus the wave vector (**k**) along a few highly symmetric directions. The energy of the electron increases upward, while that for holes increases downward.

The band structure of GaAs is shown in Figure 2.1a. Both the valence band maximum and the conduction band minimum occur at the same point, that is, **k**=0. GaAs is a member of the direct-bandgap semiconductors. The valence band is degenerate at **k**=0, and comprises two bands, called the heavy-hole (HH) and light-hole (LH) bands. A third valence band appears at an energy of about 0.34 eV below the HH and LH bands. This band is called the split-off (SO) band and is pushed downward due to spin–orbit interactions.

The lowest conduction band minimum occurs at the zone center (Γ valley). The next higher conduction band valleys, the four L valleys, have an energy separation $E_{\Gamma L}$=0.29 eV and the next highest X valleys appear with an energy difference $E_{\Gamma X}$=0.48 eV.

The band diagram of Si, shown in Figure 2.1b, indicates that the maximum of the uppermost filled band, the valence band, occurs at the Γ point where **k**=0. The maxima of the lowest empty (conduction) band occurs at a point $0.85(2\pi/a)$ away from the zone center along the [100] direction. This is an example of an indirect band semiconductor.

2.3 *E*–**k** Diagram and Effective Mass

Figure 2.2 shows the simplified band diagrams for a direct and an indirect-gap semiconductor. The *E*–**k** relationship, near the band edges, can be approximated by a quadratic equation:

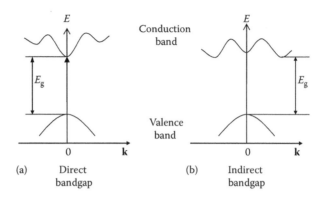

FIGURE 2.2
(a,b) Simple E–k diagram for a direct- and indirect-bandgap semiconductor.

$$E(k) = \frac{\hbar^2 k^2}{2m^*} \tag{2.2}$$

where m^* is termed the *effective mass*. Its value differs from the free electron mass, m_0, due to its interaction with the periodic potential of the lattice. To consider these particles as essentially free, their rest mass must be altered to take into account the effect of the crystal lattice. Equation 2.2 corresponds to a band of a single scalar mass, which does not depend on the direction of the k-vector. The equal energy surfaces are spherical. Equation 2.2 applies to the conduction band and the SO band in GaAs and many other direct-gap semiconductors.

On the other hand, the dispersion relation for some semiconductors can be written as

$$E(k) = \frac{\hbar^2}{2} \left[\frac{(k_x - k_{0x})^2}{m_x} + \frac{(k_y - k_{0y})^2}{m_y} + \frac{(k_z - k_{0z})^2}{m_z} \right] \tag{2.3}$$

This represents a band with ellipsoidal surfaces of equal energy. It can be used for the L and X conduction band valleys in Ge, Si, and GaAs. In this case, the effective mass is dependent on the direction of k, and one may define an effective mass tensor with components:

$$m_{ij} = \frac{\hbar^2}{\left(\partial^2 E / \partial k_i \partial k_j \right)}, \quad i, j = x, y, z \tag{2.4}$$

As an illustration, in Figure 2.3 we show six conduction band valleys (X valleys) in Si having ellipsoidal constant energy surfaces. Considering

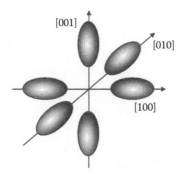

FIGURE 2.3
Constant energy surfaces in X valleys of Si in the k-space.

two ellipsoids whose major axes are along the [100] direction, the energy of the electrons in the conduction band can be expressed as

$$E - E_C = \frac{\hbar^2 (k_x - k_{x0})^2}{2m_l} + \frac{\hbar^2 k_y^2}{2m_t} + \frac{\hbar^2 k_z^2}{2m_t}$$

For these two ellipses, one can define two effective masses, one along its major axis, which is known as the *longitudinal effective masses* (m_l); and the other along either of its two minor axes, called the *transverse effective masses* (m_t).

The electron effective masses of Si, Ge, and GaAs are listed in Table 2.1.

The E–k relation for the valence bands of most semiconductors may be expressed as [2]

$$E(k) = E_v - \frac{\hbar^2}{2m_0} \left[Ak^2 \pm \left\{ B^2 k^4 + C^2 \left(k_x^2 + k_y^2 + k_z^2 \right) \right\}^{1/2} \right] \qquad (2.5)$$

The parameters A, B, and C will be defined in Chapter 4. The plus and minus signs correspond to the HH and the LH. The effective masses are different along different directions and the constant energy surfaces are called *warped surfaces*. The three valence bands are shown in Figure 2.4a and an

TABLE 2.1

Effective Mass, Bandgap, and Temperature Coefficients of Bandgap

	Electron Effective Mass			Hole Effective Mass		Gap at 0 K	$(10^{-4}\,eV$ $K^{-2})$	(K)
	Longitudinal	Transverse	Isotropic					
	m_l	m_t	m_e	m_{hh}	m_{lh}	E_{g0}	α	β
Si	0.92	0.19		0.53	0.16	1.170	4.73	636
Ge	1.64	0.082		0.35	0.043	0.744	4.77	235
GaAs			0.067	0.62	0.074	1.519	5.405	204

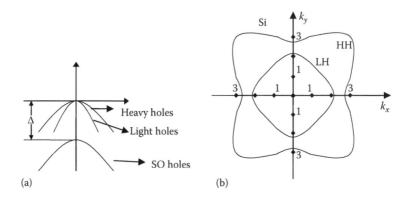

FIGURE 2.4
(a,b) E–**k** relation for valence bands in Si and GaAs.

illustration of the section of warped surface on the k_x–k_y plane in Si is given in Figure 2.4b.

The hole effective masses for Si, Ge, and GaAs in the valence band are also listed in Table 2.1.

The E–**k** relationship given by Equation 2.2 is the dispersion relation for a parabolic band with isotropic effective mass. The E–**k** relation is sometimes taken as nonparabolic [3] and is expressed as

$$E(k)\left[1+\alpha.E(k)\right] = \frac{\hbar^2 k^2}{2m^*} \tag{2.6}$$

The nonparabolicity parameter $\alpha = 0.67$ eV^{-1} for GaAs.

2.4 Density of States

To calculate the electrical properties of a semiconductor and analyze the behavior of semiconductor devices, it is necessary to know the carrier distribution over the available energy states. In this context, a concept of great importance, the *density of states function*, $N(E)dE$, gives the number of available quantum states in the energy interval between E and $E+dE$.

2.4.1 Derivation of Three-Dimensional Density of States

To develop the expression for the density of state function for a bulk semiconductor, we consider a cubic region of a crystal with dimension L along three perpendicular directions. The wave function of the electron will be of the form:

$$\Psi_k(r) = U_k(r)\sin(k_x x)\sin(k_y y)\sin(k_z z) \qquad (2.7)$$

where $U_k(r)$ is a periodic function.

The boundary condition states that $\Psi_k(r)$ becomes zero at the boundaries of the cube, which leads to $k_x L = 2\pi n_x$, $k_y L = 2\pi n_y$, and $k_z L = 2\pi n_z$, where n_x, n_y, and n_z are integers.

Therefore, each allowed value of $\mathbf{k}$ with components k_x, k_y, and k_z occupies a volume $(2\pi/L)^3$ in $\mathbf{k}$-space. Consider two concentric spheres of radii k and $k+dk$ in $\mathbf{k}$-space. The volume in the $\mathbf{k}$-space between these two spheres is $4\pi k^2 dk$. Hence, the total number of states in that volume, taking into account two possible values of spin, is given by

$$dN = 2 \times \frac{4\pi k^2 dk}{(2\pi/L)^3} = \pi k^2 dk \left(\frac{V}{\pi^3}\right) \qquad (2.8)$$

where $V = L^3$ is the volume of the crystal. From the dispersion relation, we can write the following equations for the electrons in the conduction band of a semiconductor:

$$k^2 = \frac{2m_{de}E}{\hbar^2} \quad \text{and} \quad kdk = \frac{2m_{de}dE}{\hbar^2}$$

where m_{de} is the density of state effective mass of electrons. Using the expression for dk derived from Equation 2.8, one obtains

$$dN = \frac{V}{2\pi^2}\left(\frac{2m_{de}}{\hbar^2}\right)^{3/2} E^{1/2} dE \qquad (2.9)$$

Finally, we can write the density of states per unit volume of the crystal, taking into account the number of equivalent minima (g_v) in the conduction band, as

$$N(E)dE = \frac{g_v}{2\pi^2}\left(\frac{2m_{de}}{\hbar^2}\right)^{3/2} E^{1/2} dE \qquad (2.10)$$

A plot of this function is schematically shown in Figure 2.5. A similar equation holds for the density of states for holes in the valence band.

Example 2.1: With the mass values given in Table 2.1, the conductivity effective mass of electrons in Si is 0.26 m_0 and the density of states effective mass is 1.06 m_0

Example 2.2: With the mass values given in Table 2.1, the density of states effective mass for holes in GaAs is 0.64 m_0.

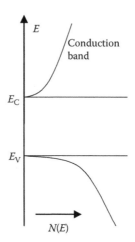

FIGURE 2.5
Density of state functions in the conduction and valence band.

2.5 Carrier Concentration

2.5.1 Electron and Hole Concentration

The number of electrons in an occupied conduction band level can be evaluated by the total number of states, $N(E)$, multiplied by the occupancy, $f(E)$, integrated over the conduction band:

$$n = \int_{E_C}^{\infty} N(E) f(E) dE \tag{2.11}$$

where E_C represents the bottom of the conduction band and $N(E)$ can be written from Equation 2.10 as

$$N(E) = \frac{g_v}{2\pi^2} \left(\frac{2m_{de}}{\hbar^2} \right)^{3/2} (E - E_C)^{1/2} \tag{2.12}$$

The occupancy of an energy level E is expressed by the Fermi–Dirac distribution function, which is given as

$$f(E) = \frac{1}{1 + \exp\left[(E - E_F)/k_B T \right]} \tag{2.13}$$

where E_F is called the *Fermi level*.

The integral of Equation 2.11 can be evaluated to

$$n = N_C F_{1/2}\left(\frac{E_F - E_C}{k_B T}\right) \tag{2.14}$$

where N_C is the effective density of states in the conduction band and is given by

$$N_C = 2g_v \left(\frac{2\pi m_{de} k_B T}{h^2}\right)^{3/2} \tag{2.15}$$

The quantity $F_{1/2}(\eta)$ is the Fermi–Dirac integral. This integral has different values for different positions of the Fermi level with respect to the conduction band edge. For further detailed study of the Fermi–Dirac integral, readers are referred to Sze (see Reading List). For nondegenerate semiconductors, the doping concentration is smaller than the density of state and the Fermi integral takes an exponential form, which results in

$$n = N_C \exp\left[\frac{-(E_C - E_F)}{k_B T}\right] \tag{2.16}$$

Example 2.3: The prefactor N_C in Equation 2.16 is called the *effective density of states*. From Table 2.1, the electron effective mass of GaAs is $m_e = 0.067\, m_0$. The value of N_C becomes 0.42×10^{18} cm^{-3}.

Similarly, for p-type semiconductors, the hole concentration can be derived as

$$p = N_V \exp\left[\frac{-(E_F - E_V)}{k_B T}\right] \tag{2.17}$$

where N_V is the effective density of states in the valence band and is given by

$$N_V = 2\left(\frac{2\pi m_{dh} k_B T}{h^2}\right)^{3/2}, \quad (m_{dh})^{3/2} = (m_{hh})^{3/2} + (m_{lh})^{3/2} \tag{2.18}$$

For degenerate semiconductors, where concentrations n or p are near or beyond the effective density of states, the density of electrons is given by Equation 2.14. Defining a parameter $\eta = (E_F - E_C)/k_B T$, we get

$$F_{1/2}(\eta) = \left(\frac{2}{\sqrt{\pi}}\right)\int_0^\infty \frac{x^{1/2}dx}{\left[1+\exp(x-\eta)\right]} \tag{2.19}$$

The above Fermi integral can be approximated under extreme limits. For $\eta \ll 1$, $F_{1/2}(\eta) = \exp(\eta)$, which is the usual result that is valid for Boltzmann statistics. On the other hand, when $\eta \gg 1$:

$$F_{1/2}(\eta) = 4\eta^{3/2}/\left(3\sqrt{\pi}\right) \tag{2.20}$$

which is valid for extreme degeneracy. For $-10 < \eta < 10$, the Fermi integral can be interpolated as follows:

$$F_{1/2}(\eta) = \exp\left(-0.32881 + 0.74041\eta - 0.045417\eta^2\right.$$

$$\left. -8.797 \times 10^{-4}\eta^3 + 1.5117 \times 10^{-4}\eta^4\right) \tag{2.21}$$

Example 2.4: The position of the Fermi level above the conduction band edge in $In_{0.53}Ga_{0.47}As$ having a carrier density of 10^{18} cm^{-3} as obtained from Equation 2.16 is 0.074 eV.

2.5.2 Joyce–Dixon Formula

In many cases, it is necessary to find the Fermi level, E_F, when the carrier concentration is given. This is trivial for Boltzmann statistics; however, for the degenerate case, the value can only be found by numerical method. Approximate expressions are available in the literature, relating Fermi energy with the carrier density (see Reading List). We quote here a formula by Joyce and Dixon [4] that has been used in a number of cases. The expression is

$$\eta = \ln(n/N_C) + \sum_1^4 A_i(n/N_C)^i \tag{2.22}$$

$A_1 = +3.53553 \times 10^{-1}; A_2 = -4.95009 \times 10^{-3}; A_3 = +1.48386 \times 10^{-4}; A_4 = -4.42563 \times 10^{-6}$.

A similar equation can be written for p-type semiconductors, with the effective density of states, N_V, given by Equation 2.18.

Example 2.5: For $In_{0.53}Ga_{0.47}As$, using $m_e = 0.042\ m_0$, the value $N_C = 0.21 \times 10^{18}$ cm^{-3} is obtained.

Applying the Joyce–Dixon formula, the position of the Fermi level above the conduction band edge in having a carrier density of 10^{18} cm^{-3} is 0.072 eV.

2.6 Intrinsic and Extrinsic Semiconductor

2.6.1 Intrinsic Semiconductor

A perfect semiconductor crystal with no impurities or defects is called an *intrinsic semiconductor*. In such a material, the valence band is filled and the conduction band is empty at 0 K. As the temperature increases, electron–hole pairs (EHPs) are generated due to the breaking of covalent bonds. As a result, the electrons from the valence band are excited to the conduction band. These EHPs are the only carriers in an intrinsic semiconductor. Since the electrons and holes are created in pairs, we may write $n = p = n_i$, where n_i is called the *intrinsic carrier concentration*.

2.6.1.1 Carrier Concentration in Intrinsic Semiconductor

If E_i denotes the position of the Fermi level in an intrinsic semiconductor, then from Equations 2.16 and 2.17, we have

$$E_i = \frac{E_C + E_V}{2} - \frac{k_B T}{2} \ln\left(\frac{N_C}{N_V}\right) \tag{2.23}$$

If $N_C = N_V$, the intrinsic Fermi level would be located exactly at the center of the bandgap. In practical situations, the second term of Equation 2.23 gives a very small value (few millielectronvolts).

Again, from Equations 2.16 and 2.17, we can have

$$n_i p_i = n_i^2 = N_C N_V \exp\left[-\left(\frac{E_C - E_V}{k_B T}\right)\right] = N_C N_V \exp\left[-\left(\frac{E_g}{k_B T}\right)\right] \tag{2.24}$$

Therefore,

$$n_i = (N_C N_V)^{1/2} \exp\left[-\left(\frac{E_g}{2k_B T}\right)\right] \tag{2.25}$$

In Equation 2.25, in addition to the presence of temperature T as a variable, N_C, N_V, and E_g are functions of temperature. Combining Equations 2.15 and 2.18 with Equation 2.25, we have

$$n_i(T) = 2\left(\frac{2\pi k_B T}{h^2}\right)^{3/2} (m_{de} m_{dh})^{3/4} \exp\left[-\left(\frac{E_g(T)}{2k_B T}\right)\right] \tag{2.26}$$

Thus, n_i increases rapidly with temperature, the increase being almost exponential in nature.

The temperature dependence of the bandgap may be described by the following phenomenological equation [5]:

$$E_g(T) = E_{g0} - \alpha T^2/(T+\beta) \qquad (2.27)$$

where:

E_{g0} is the gap at $T=0$ K
α and β are empirical parameters

The bandgap at 0 K and the temperature coefficients of Si, Ge, and GaAs are included in Table 2.1. Using the relevant expressions, the intrinsic carrier concentration in these semiconductors can be calculated for different temperatures. The values are plotted in Figure 2.6 as a function of 1000/*T*.

> **Example 2.6:** Using Equation 2.27 and the values of the parameters given in Table 2.1, the bandgap of GaAs decreases from 1.519 eV at 0 K to 1.425 eV at 300 K, and to 1.07 eV at 1000 K.

2.6.2 Extrinsic Semiconductor

In addition to thermally generated carriers, it is possible to create extra carriers in the semiconductor by introducing impurities into the crystal. This process is known as *doping*. By doping, the crystal can be made to have electrons or holes predominantly as the primary charge carrier. When a crystal is doped such that the equilibrium concentrations of electrons (n_0) or holes

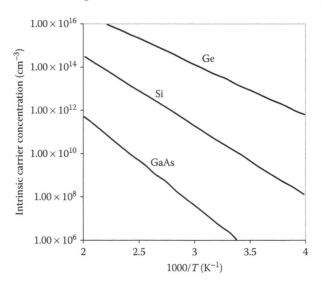

FIGURE 2.6
Intrinsic carrier concentration of Ge, Si, and GaAs as a function of inverse temperature.

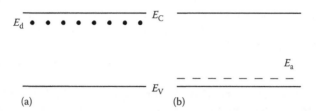

FIGURE 2.7
Formation of (a) donor and (b) acceptor energy levels in an extrinsic semiconductor.

(p_0) are different from the intrinsic carrier concentration (n_i), the semiconductor is known as *extrinsic*.

Doping creates additional energy levels within the bandgap. For example, in Si or Ge, Group V elements (P, As, Sb) introduce energy levels (E_d) very near to the conduction band, as depicted in Figure 2.7a. This energy level is filled with electrons at 0 K, and with very little thermal energy these electrons can be excited into the conduction band. This process is known as *ionization*. Such an impurity is therefore called a *donor* as it donates electrons and E_d is referred to as the donor energy level. The difference in energy between the conduction band and the donor energy level, $(E_C - E_d)$ in Figure 2.7a, is known as *donor ionization energy*. In this type of semiconductor, even at low temperature, there is a large concentration of electrons in the conduction band, that is, $n \gg n_i, p$. Such a semiconductor is called an *n-type semiconductor*.

Similarly, if an element of Group III, for example, B, Al, or In, is used to dope Si or Ge, an energy level, E_a, is created very close to the valence band, as shown in Figure 2.7b. This level is empty at 0 K; however, with little thermal energy, electrons from the valence band can be excited to this level, creating a large number of holes in the valence band. As this level accepts electrons, this type of dopant is known as *acceptors* and the material is called *p-type*. In this case, $p \gg n_i, n$. Here, the ionization energy is $E_a - E_V$.

The donor binding (ionization) energy may be calculated using a simple model. Consider a pentavalent atom (say P) replacing an Si atom. Four of the outermost electrons of a P atom will take part in the covalent bonding; the fifth electron will be loosely bound to the parent nucleus. It revolves around its nucleus whose effective charge is +e. The energies of the bound electron may be calculated using Bohr's theory of H-atom. There are, however, two important differences: Firstly, the fifth electron revolves within the Si crystal and experiences forces due to all other electrons and nuclei. Its mass is the effective mass. Secondly, the Coulomb force between the fifth electron and the nucleus acts in a semiconductor medium of permittivity, ε_s, rather than in free space. Therefore,

$$E_C - E_d = \left(\frac{\varepsilon_0}{\varepsilon_s}\right)^2 \left(\frac{m_0}{m_e}\right) \frac{e^4 m_0}{32\pi^2 \varepsilon_0^2 \hbar^2} = 13.6 \left(\frac{\varepsilon_0}{\varepsilon_s}\right)^2 \left(\frac{m_0}{m_e}\right) \text{ (eV)} \qquad (2.28)$$

This equation gives values of correct order, but the experimental values are higher. As shown in Figure 2.7, the energy levels of donor atoms are discrete and near the conduction band edge. These are the energy levels of shallow donors. The levels for shallow acceptors are slightly above the valence band edge.

The ionization energies of different dopants in Si, Ge, and GaAs are shown in Sze (see Reading List). It appears that only a few impurities behave as shallow donors or acceptors even at room temperature. Some impurity atoms, deep donors or deep acceptors, have energy levels well separated from the band edges (more than a few k_BT/e) even at room temperature.

It is clear that Group V impurities act as donors and Group III impurities act as acceptors in Si or Ge. Following the same arguments, it may be stated that Group II impurities, such as C, Be, Mg, and Zn, act as acceptors when they replace the Group III atoms. Also, certain impurity atoms, for example, Si in GaAs, may behave as donors if they replace Ga atoms, but act as acceptors if they occupy the sites of As atoms. These impurities are called *amphoteric* and show complicated behavior.

> **Example 2.7:** Let us assume $\varepsilon_s = 12$ and $m^* = 0.26\ m_0$, which correspond approximately to Si. The ionization energy as calculated above is nearly 25 meV.

2.6.2.1 Positions of Fermi Level in Extrinsic Semiconductors

From Equations 2.14, 2.16, and 2.24, we obtain

$$np = N_C N_V \exp\left[-\left(\frac{E_C - E_V}{k_B T}\right)\right] = N_C N_V \exp\left[-\left(\frac{E_g}{k_B T}\right)\right] = n_i^2 \qquad (2.29)$$

One may easily show that $n = n_i \exp[(E_F - E_i)/k_B T]$ and a similar equation for p and obtain

$$E_F = E_i + k_B T \ln\left(\frac{n}{n_i}\right) = E_i - k_B T \ln\left(\frac{p}{n_i}\right) \qquad (2.30)$$

For an n-type semiconductor, $n \gg n_i$, and E_F is above E_i. For a p-type semiconductor, $p \gg n_i$, and E_F is below E_i and comes closer to the valence band.

> **Example 2.8:** The bandgap of GaAs is 1.42 eV. Since the intrinsic level is located approximately at the center of the bandgap, we can write the position of the intrinsic level as $E_g/2 = 0.71$ eV above the valence band edge. The position of the Fermi level with $N_D = 10^{17}$ cm^{-3} can be determined from Equation 2.30 as 1.35 eV.

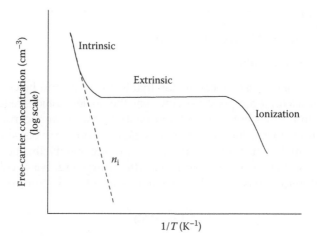

FIGURE 2.8
Variation of free-carrier concentration with temperature in a doped sample.

2.6.2.2 Effect of Temperature on Extrinsic Semiconductor

The variation of the electron density in an n-type semiconductor as a function of inverse temperature is shown in Figure 2.8. At low temperatures, not all donors are ionized, but the number of ionized donors increases exponentially with temperature. This explains the exponential rise of the free electron density, n, in the *ionization region*. With a further rise of temperature, when 100% ionization is achieved, the free electron concentration would be equal to the donor concentration and would remain at this value for a range of temperatures. This is known as the *extrinsic region*, where $n \gg n_i$. With a further increase of temperature, n_i would increase exponentially, and eventually exceed n. In this region, called the *intrinsic region*, n_i increases exponentially with temperature, and the sample loses its extrinsic nature. In the extrinsic region, the device can retain its usability and for a good semiconductor this range should be as wide as possible. Ideally, it should extend well up to and even beyond the maximum temperature at which the device will be operated.

> **Example 2.9:** For device operation, the semiconductor must be either n- or p-type. This defines a maximum temperature of operation of any device at which temperature the extrinsic carrier density equals the intrinsic carrier density. Therefore, $N_D = \sqrt{N_C N_V}\, \exp\left(-E_g / 2k_B T_{max}\right)$ and thus $T_{max} = \left[E_g(T_{max})/2k_B\right]/\ln\left(\sqrt{N_C N_V}/N_D\right)$. Let $N_D = 10^{15}$ cm^{-3}, $N_C = 5 \times 10^{18}$ cm^{-3}, $N_V = 10^{19}$ cm^{-3}, and $E_g = 1$ eV; these are assumed to be constant. This gives $T_{max} = 409$ K $= 136$ C.

2.7 Transport of Charge Carriers

2.7.1 Low-Field Mobility

The charge carriers in a semiconductor, that is, electrons and holes, are in constant motion at a finite temperature. However, the carriers collide frequently with lattice ions, impurities both ionized and neutral, other charge carriers, defects, and so on. If we follow the motion of a single electron, it travels some distance due to its thermal motion, then it suffers a scattering and its direction of motion is changed, and the process repeats itself. After a large number of collisions, the net drift of the electron is zero. The current density J of an electron is expressed as

$$J = -nev_d \tag{2.31}$$

where:

    n is the electron density
    v_d is the drift velocity
    v_d is therefore zero, as expected, since there is no electric field.

The presence of an electric field, F, slightly deviates the electron paths between collisions and produces a net displacement of the electron in a direction opposite to the direction of the electric field. The effective shift of carriers per unit time is the drift velocity v_d.

In the presence of an electric field, the electrons experience a force, $P = -eF$, and have an acceleration expressed as $P = m_e(dv/dt)$. For a small electric field, the acceleration may be written as v_d/τ, where τ is called the *relaxation time*. Thus

$$-eF = m_e v_d / \tau \tag{2.32}$$

or, the drift velocity is now

$$v_d = -\frac{e\tau}{m_e} F = -\mu_n F \tag{2.33}$$

The proportionality constant $\mu_n = e\tau/m_e$ between the drift velocity and the electric field is called the *mobility of electrons* and it signifies the drift velocity per unit electric field. The smaller the effective mass, the larger the mobility.

The quantity τ has been defined as the relaxation time. It is alternatively termed the *mean scattering time* and it represents the mean time between two successive collisions of the electron. As mentioned already, an important collision process in semiconductors is between electrons (or holes) and vibrating lattice atoms; it is usually termed *lattice scattering*. The process is also termed *electron–phonon interaction*, as the vibration of lattice atoms is described by phonons. The reciprocal of the relaxation time denotes the probability of scattering by the corresponding process. Apart from phonon scattering, electrons are scattered by impurities, defects, carriers, and so on. Assuming that the total probability of scattering is the sum of the individual probabilities, one may write the total probability as

$$\frac{1}{\tau} = \frac{1}{\tau_L} + \frac{1}{\tau_{imp}} + \frac{1}{\tau_{ee}} + \cdots \qquad (2.34)$$

where the subscripts L, imp, and ee stand for the scattering time for lattice, impurity, and electron-electron scattering, respectively. Using the relation between the mobility and relaxation time, one may thus write

$$\frac{1}{\mu} = \sum_{i=1}^{N} \frac{1}{\mu_i} \qquad (2.35)$$

where:
μ_i is the mobility limited by the ith scattering mechanism
μ is the overall mobility

The above simple rule is known as Matthiessen's rule. In general, however, the calculation of μ is not so simple.

The lattice scattering is the result of the interaction of electrons with vibrating atoms. Since the vibration increases with temperature, the frequency of this collision increases also, making mobility decrease with increasing temperature. On the other hand, at low temperatures, lattice scattering is less important, but the thermal motion of the carriers is also slower. Since a slower electron is more strongly influenced by an ionized impurity, the impurity scattering becomes more dominant as the temperature is decreased. These arguments qualitatively explain why $\mu_L \propto T^{-x}$, $0 < x < 3$, and $\mu_{imp} \propto T^{3/2}$. The approximate nature of the variation of the mobility of electrons in the presence of lattice and impurity scattering processes is shown in Figure 2.9.

The ionized impurity scattering may be important even at room temperature when the doping density is quite high. The variation is illustrated in Figure 2.10.

> **Example 2.10:** Assuming that in a GaAs sample, there are two scattering mechanisms of relaxation time, 0.1 and 0.2 ps, the values of the mobility due to these two scattering mechanisms are 2624.2 and 5248.4 cm²/V–s, respectively. Using Mathiessen's rule, the net mobility is 1749.5 cm²/V–s.

2.7.2 High-Field Effect

Equations 2.31 through 2.33 indicate that the current density is proportional to the applied electric field, $J = \sigma F$, where the conductivity is $\sigma = ne\mu$, which

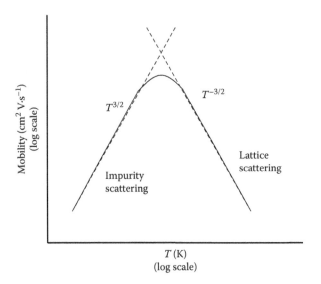

FIGURE 2.9
Approximate variation of the mobility with temperature.

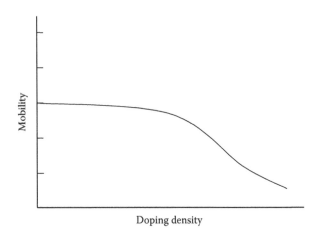

FIGURE 2.10
Approximate variation of the mobility with doping density at a fixed temperature.

ensures that the Ohm's law is valid. The constant of proportionality, the mobility, is independent of the applied field F. However, when the applied field is quite large, ~1 kV cm^{-1}, the mobility becomes a function of the field. In this situation, the drift velocity of the carriers becomes comparable with the thermal velocity. This means that the temperature, T_e, of the electrons, which is a measure of the mean kinetic energy of the electrons, is higher than the temperature of the lattice $T_L (T_e > T_L)$. The effect is termed the *hot carrier effect*.

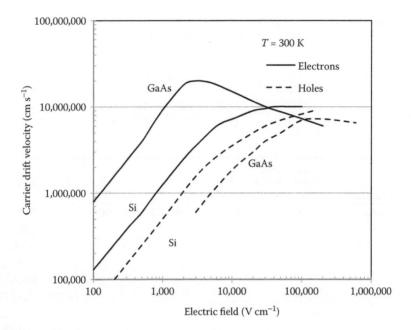

FIGURE 2.11
Variation of the drift velocity in Si and GaAs with an electric field.

The qualitative variations of the drift velocity of electrons and holes in Si and GaAs are shown in Figure 2.11. For both electrons and holes, the drift velocity first increases linearly with the field, satisfying Ohm's law, but then it increases sublinearly and finally attains a saturation value. The variation in n-type GaAs is somewhat different. At low fields there is a linear increase of the drift velocity; however, after a critical field, v_d decreases and then increases almost linearly with a lower slope and finally assumes a saturation value.

Example 2.11: Illustration of the hot electron effect. The following empirical expressions are used to give the values of electron and hole drift velocities at different temperatures:

$$v_d = v_1 \frac{F}{F_c} \left[\frac{1}{1 + (F / F_c)^\beta} \right]^{1/\beta}$$

The values for Si at 300 K are as follows:

	Electron	Hole
v_1 (cm s^{-1})	1.07×10^7	8.34×10^6
F_c (V cm^{-1})	6.91×10^3	1.45×10^4
B	1.11	2.637

Example 2.12: The empirical expression for mobility variation in Groups III–V [6]. The low-field mobility is expressed as

$$\mu = \mu_{min} + \cfrac{\mu_{max}\left(300\ \text{K}\right)+\left(300\ \text{K}/T\right)^{\theta_1} - \mu_{min}}{1 + \cfrac{N}{\left(N_{ref}\left(300\ \text{K}\right)\left(T/300\ \text{K}\right)^{\theta_2}\right)^{\lambda}}}$$

Material	μ_{max} (cm²/ V–s)	μ_{min} (cm²/ V–s)	N_{ref} (cm⁻³)	λ	θ_1	θ_2	T (K)	N (cm⁻³)	μ (cm²/ V–s)
GaAs	9,400	500	6×10^{16}	0.394	2.1	3.0	300	2×10^{17}	3913.85
								1×10^{18}	2708.56
								1×10^{19}	1546.23
							70	2×10^{17}	20468.21
								1×10^{18}	11614.55
								1×10^{19}	5140.77
In$_{0.53}$Ga$_{0.47}$As	14,000	300	1.3×10^{17}	0.48	1.59	3.68	300	2×10^{17}	6444.22
								1×10^{18}	4040.49
								1×10^{19}	1815.32
							70	2×10^{17}	8573.70
								1×10^{18}	4245.56
								1×10^{19}	1631.37

2.8 Excess Carriers

Excess carriers, making the carrier concentration exceed the thermal equilibrium concentration, can be created by optical excitation or electron bombardment, or they can be injected across a forward-biased p-n junction. In this section, we will investigate the excess carrier creation in semiconductors by optical excitation. This phenomenon is called *optical absorption*. We shall study next the mechanism of annihilation of EHPs, known as *recombination*. Finally, we shall discuss the *diffusion* process of excess carriers.

2.8.1 Optical Absorption

In the optical absorption experiment, photons of a selected wavelength are directed at the sample and their relative transmission is measured. The transmitted intensity (I) of a beam of photons of wavelength λ through a sample of thickness l is given by

$$I = I_0 e^{-\alpha l} \qquad\qquad (2.36)$$

where:
  I_0 is the incident intensity
  α is the absorption coefficient, which is a function of the wavelength of
      the photons and varies with the material

If a beam of photons with energy $\hbar\omega > E_g$ falls on a semiconductor, there will be a significant amount of absorption, since the valence band contains many electrons and the conduction band contain many unoccupied states into which the electrons can move. Electrons excited to the conduction band and holes created in the valence band may initially have energy much higher than the band-edge energies. By the process of scattering, electrons and holes lose their energy until their equilibrium energies equal the band-edge energies E_C and E_V, respectively. The EHPs are called the *excess carriers*, and as they are out of balance, they would eventually recombine. However, while these excess carriers remain in their respective bands, they contribute to the current conduction process and the conductivity of the material increases. This type of conductivity is referred to *photoconductivity*.

On the other hand, if the energy of the photon is low, the resulting absorption is negligible. In this case, the entire radiation passes through the sample without getting absorbed as the photons are too weak to create EHPs by breaking the covalent bonds of the crystal.

2.8.2 Recombination

The electrons in the conduction band of a semiconductor may make transitions to the valence band, that is, recombine with holes in the valence band. Energy lost by an electron in making the transition is given up as a photon. Commonly, there are four types of recombination processes observed in a semiconductor. We will briefly discuss these processes in the following sections.

2.8.2.1 Direct Band-to-Band Recombination

This type of recombination occurs when an electron from the conduction band falls to an empty state in the valence band, thus annihilating the EHP. This process is schematically shown in Figure 2.12. In a direct-bandgap semiconductor (e.g., GaAs), the chance of this recombination process is high as the conduction band minima and the valence band maxima occur for the same value of k. The direct recombination process occurs spontaneously, that is, the probability that an electron and a hole will recombine is constant in time.

In the conduction band, the net rate of change of the electron concentration $n(t)$ at any time t is given by

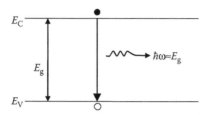

FIGURE 2.12
Direct band-to-band recombination processes annihilating an EHP.

$$\frac{d}{dt}n(t) = \alpha_r n_i^2 - \alpha_r n(t)p(t) \tag{2.37}$$

where α_r is the recombination coefficient. In Equation 2.37, the first term is the thermal generation rate and the second term gives the recombination rate. Let us assume that the initial excess electron and hole concentrations are equal, that is, $\Delta n = \Delta p$, and are created by a short flash of light at $t = 0$. Then, as the electrons and holes recombine in pairs, the instantaneous excess carrier concentrations $\delta n(t)$ and $\delta p(t)$ are also equal. Thus, the total electron and hole concentrations can be expressed in terms of the equilibrium values n_0 and p_0 as

$$n(t) = n_0 + \delta n(t) \tag{2.38a}$$

and

$$p(t) = p_0 + \delta p(t) \tag{2.38b}$$

Using Equation 2.38a and b, Equation 2.37 can be expressed as

$$\frac{d}{dt}\delta n(t) = -\alpha_r \left[\left(n_0 + p_0 \right) \delta n(t) + \delta n^2(t) \right] \tag{2.39}$$

For low-level injection, the excess carrier concentration at any instant of time is much smaller than the equilibrium majority carrier concentration. For example, if we consider a sufficiently extrinsic p-type semiconductor, we can make two assumptions: $\delta n, \delta p \ll p_0$ and $n_0 \ll p_0$, and Equation 2.39 becomes

$$\frac{d}{dt}\delta n(t) = -\alpha_r p_0 \delta n(t) \tag{2.40}$$

The solution to this equation is an exponential decay from the original excess carrier concentration Δn:

$$\delta n(t) = \Delta n \exp\left(-\frac{t}{\tau_n}\right) \tag{2.41}$$

Excess minority carriers, that is, electrons in a p-type semiconductor, recombine with a decay constant $\tau_n = (\alpha_r p_0)^{-1}$; τ_n is called the *recombination lifetime* and is often called the *minority carrier lifetime*. Similarly, the holes in an n-type semiconductor decay with the lifetime of $\tau_p = (\alpha_r n_0)^{-1}$. It should be mentioned here that a more general expression for carrier lifetime for not sufficiently extrinsic samples can be given by $\tau = [\alpha_r (n_0 + p_0)]^{-1}$, where the assumptions that $n_0 \gg p_0$ or $p_0 \gg n_0$ cannot be made.

> **Example 2.13:** Assume that the coefficient α_r is $10^{-16}\,\mathrm{m^3\,s^{-1}}$ in a direct-gap p-type material with $p_0 = 10^{23}\,\mathrm{m^{-3}}$. The radiative lifetime will be 100 ns.

2.8.2.2 Indirect Recombination through Trap States

In indirect-bandgap materials, the probability of direct band-to-band recombination is extremely low, as both momentum and energy change is involved. For recombination to happen in these materials, defect states are to be created by impurity atoms, which are either deliberately introduced or present spontaneously due to the semiconductor fabrication technology. The defect states form an allowed energy level within the bandgap. The defect states are classified into one of two categories: The first one is the *recombination centers* where the defect states lie near the middle of the bandgap and assist in the recombination process of the carriers. The second category is the *trap* where the defect states lie farther away from either of the band edges. These traps capture and release carriers from and to the bands without assisting in recombination. For traps, the carriers make the transition from the bands to the defect centers, but then get reexcited back to their respective bands instead of recombining with the opposite type of carriers. It must be stated that the position of the defect center within the bandgap alone does not exclusively predict whether the defect center is a trap or a recombination center, but it is definitely one of the main criterion. There are other parameters that determine this property.

In indirect-bandgap semiconductors, recombination takes place with the assistance of recombination centers in two steps: the state first captures a carrier from one of the bands and then captures the opposite type of carrier from the other band, thus annihilating the pair, with the recombination occurring at the defect centers. So, there are four basic processes associated with the recombination process: electron capture, electron release, hole capture, and hole release. These processes are explained schematically in Figure 2.13. An

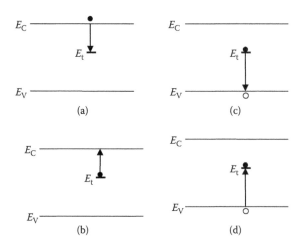

FIGURE 2.13
Four basic processes of the indirect recombination process with the help of trapping level (E_t).
(a) Electron capture, (b) electron release, (c) hole capture, and (d) hole release.

extended theory of this type of recombination was proposed by Shockley, Hall, and Read [7–9]; this process is known as the *SHR recombination process*.

2.8.2.3 Auger Recombination

In the recombination process shown in Figure 2.14, an electron and a hole recombine directly without involving trap levels, and the released energy is transferred to another carrier. Figure 2.14a and b illustrate the process for heavily doped n-type materials; the energy released by recombination (a) is taken up by an electron, thereby moving to a higher energy state in the conduction band (see Figure 2.14b). Here, two electrons and one hole are involved and the process is known as an *e-e-h process*. The process for heavily doped p-type materials is shown in Figure 2.14c and d and can be easily understood; it is an h-h-e process. No electromagnetic radiation

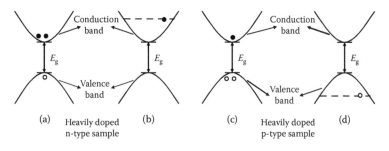

FIGURE 2.14
The e-e-h process (a) before recombination and (b) after recombination; and the h-h-e process (c) before recombination and (d) after recombination.

is emitted by this recombination process; instead, the released energy is absorbed by a third carrier, which moves higher up in its respective bands. Since two electrons for n-type materials and two holes in p-type materials are involved in this process, it is highly unlikely to occur except in heavily doped materials.

2.8.2.4 Surface Recombination

The recombination process also occurs at the semiconductor surface. The discontinuity of the lattice structure at the semiconductor surface introduces a large number of energy states in the forbidden energy gap, called *surface states*. The presence of these states greatly enhances the recombination rate at the surface. In addition, there may be adsorbed ions, molecules, or damage in the surface layer, which increase the recombination rate. Thus, the diffusion flux of minority carriers at the surface is determined by the surface recombination process. For example, if the holes are assumed to be the minority carriers, the surface recombination rate at the surface per unit area can be described by the following equation:

$$U_s = S(p_n - p_{n0})$$ (2.42)

where p_{n0} is the equilibrium minority carrier concentration present in the sample; and S has the same dimension as the velocity and is called the *surface recombination velocity*, even though it has no relation with the actual velocity. It should be mentioned here that this process does not need to have carriers of opposite types for recombination to take place.

2.8.2.5 Expression for Recombination Lifetime

The discussions presented in this section point out that the recombination of EHPs can occur in different ways. There may be both radiative and non-radiative processes, and radiative recombination may occur spontaneously or by stimulated processes. All these phenomena can be captured by the following expression for the recombination rate:

$$R(n) = An + Bn^2 + Cn^3 + R_{st}N_{ph}$$ (2.43)

In Equation 2.43, the linear term An is due to the nonradiative process, and the quadratic term Bn^2 is due to spontaneous recombination wherein an electron in the conduction band recombines with a hole in the valence band spontaneously. The cubic term Cn^3 is due to Auger recombination. The last term is due to stimulated recombination and is proportional to the photon density N_{ph}. Neglecting the stimulated recombination for the present, the expression for recombination lifetime becomes

$$\frac{1}{\tau} = A + Bn + Cn^2 \tag{2.44}$$

Example 2.14: Assume that the values of the coefficients A, B, and C are 3×10^7 and $0.12 \times 10^{-10}\,\mathrm{cm^3\,s^{-1}}$ and $9.6 \times 10^{-29}\,\mathrm{cm^6\,s^{-1}}$, respectively. The value of τ becomes 31 ns with $n = 1 \times 10^{17}\,\mathrm{cm^{-3}}$.

2.8.3 Diffusion Process

In the preceding section, the excess carriers are assumed to be uniform in space. When the excess carriers are generated nonuniformly in a semiconductor, the electron and hole concentrations vary with the position in the sample. Any such spatial variation in the carrier concentration results in a net motion of carriers from the high carrier concentration region to the low carrier concentration region. This type of motion is called *diffusion* and has many important impacts on the charge transport phenomena of semiconductors. Considering electrons as the carrier, the flow or flux of carriers can be expressed as

$$\left.\frac{dn}{dt}\right|_x = -D_n \frac{dn}{dx} \tag{2.45}$$

D_n is the proportionality constant and is known as the *diffusion coefficient* or *diffusivity*.

This flux of carriers constitutes a diffusion current given by

$$J_n = eD_n \frac{dn}{dx} \tag{2.46}$$

Similarly for holes, the diffusion current is

$$J_p = -eD_p \frac{dp}{dx} \tag{2.47}$$

In the sample (n-type), the zero net current necessitates that the drift current exactly balances the diffusion current, that is,

$$en\mu_n \mathcal{E} = -eD_n \frac{dn}{dx} \tag{2.48}$$

In this case, the electric field ($\mathcal{E}$) is created due to nonuniform doping and we have $E = (1/e)(dE_C/dx)$ with the Fermi level (E_F) constant for equilibrium. Using Equation 2.16 for n, we obtain

$$\frac{dn}{dx} = -\frac{e}{k_B T} \mathcal{E} N_C \exp\left(-\frac{E_C - E_F}{k_B T}\right) = -\frac{en}{k_B T} \mathcal{E} \tag{2.49}$$

Substituting this into Equation 2.48, we get

$$\frac{D_n}{\mu_n} = \frac{k_B T}{e} \tag{2.50}$$

and for a p-type sample, $D_p/\mu_p = k_B T/e$.

This important equation is known as Einstein's relation. It allows the calculation of either D or mobility μ from a measurement of the other. It should be mentioned here that this relation is valid only for nondegenerate semiconductors.

> **Example 2.15:** The electron mobility of GaAs is 8500 cm² V·s⁻¹. So the diffusion coefficient of an electron at 300 K will be 221 cm² s⁻¹.

2.8.4 Concept of Quasi-Fermi Level

When excess carriers are generated in a sample, the concept of the equilibrium Fermi level is no longer valid as the law of mass action relation, $np = n_i^2$, does not hold. However, to retain the same form of expressions for the electron and hole concentrations, we define separate quasi-Fermi levels, F_e and F_h, for electrons and holes, respectively, for nonequilibrium cases. As in Equations 2.16, 2.17, and 2.25, we can write the expressions for the electron and hole concentrations for the nonequilibrium case in steady state as

$$n = n_i \exp\left(\frac{F_e - E_i}{k_B T}\right) = N_C \exp\left(-\frac{E_C - F_e}{k_B T}\right) \tag{2.51}$$

$$p = n_i \exp\left(\frac{E_i - F_h}{k_B T}\right) = N_V \exp\left(-\frac{F_h - E_V}{k_B T}\right) \tag{2.52}$$

The product of these two equations yields

$$np = n_i^2 \exp\left(\frac{F_e - F_h}{k_B T}\right) \tag{2.53}$$

From Equation 2.53, we may come to three conclusions: when $F_e > F_h$, $np > n_i^2$ excess carrier injection takes place; when $F_e < F_h$, $np < n_i^2$ carrier extraction occurs; and when $E_{Fn} = E_{Fp}$ and $np = n_i^2$, equilibrium with a single Fermi level is reached.

Example 2.16: Consider a p-type semiconductor at $T=300$ K with a carrier concentration of $p_0=10^{15}$, $n_i=1.5\times10^{10}$, and $n_0=2.25\times10^5$ cm^{-3}. In nonequilibrium, assume that the excess carrier concentrations are $\delta n=\delta p=5\times10^{12}$ cm^{-3}. So, the position of the quasi-Fermi level for an electron will be 0.151 eV above the intrinsic level and that for a hole will be 0.289 eV below the intrinsic level.

Example 2.17: Assume a GaAs sample with $n=p=1\times10^{18}$ cm^{-3}. The quasi-Fermi level for electrons will be 0.041 eV above the conduction band and 0.055 eV below the valence band using the Joyce–Dixon relationship at $T=300$ K.

2.9 Diffusion and Recombination: The Continuity Equation

In the previous discussion of the diffusion of excess carriers, we have neglected the effect of recombination. As recombination causes the variation in the carrier distribution, its effect must be included in the conduction process. In this section, we consider the overall effect of drift, diffusion, and generation as well as the recombination of carriers occurring in a semiconductor. The combined effect is expressed by the *continuity equation*.

Let us consider a bar of semiconductor with a cross-sectional area A as shown in Figure 2.15. The volume of an infinitesimal element of thickness dx located at x is Adx. Thus, the rate of increase in the number of electrons in that infinitesimal element can be expressed as

$$Adx\frac{\partial n}{\partial t}=-\frac{A}{e}J_n(x)+\frac{A}{e}J_n(x+dx)+(G_n-R_n)Adx \qquad (2.54)$$

In Equation 2.54, the first term gives the rate of flow of electrons into the infinitesimal element at x; the second term expresses the negative rate of flow of electrons out of the element at $x+dx$; and the third term is the rate at which electrons are generated in the element and negative of the rate at which they

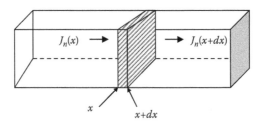

FIGURE 2.15
A semiconductor bar with cross-sectional area A.

recombine in it, with G_n and R_n being the generation and recombination rates of electrons, respectively.

As dx is extremely small, we can write

$$J_n(x+dx) = J_n(x) + \frac{\partial J_n}{\partial x} dx \tag{2.55}$$

Substituting Equation 2.55 into Equation 2.54, we get

$$\frac{\partial n}{\partial t} = \frac{1}{e} \frac{\partial J_n}{\partial x} + (G_n - R_n) \tag{2.56}$$

Similarly for the flow of holes, we have

$$\frac{\partial p}{\partial t} = -\frac{1}{e} \frac{\partial J_p}{\partial x} + (G_p - R_p) \tag{2.57}$$

Equations 2.56 and 2.57 are referred to as the *continuity equations* for electrons and holes, respectively. As the total electron or hole current densities are the summation of the drift and the diffusion current, we can write the continuity equations for minority carriers (n_p for electrons in p-type semiconductors and p_n for holes in n-type semiconductors) as

$$\frac{\partial n_p}{\partial t} = n_p \mu_n \frac{\partial \mathcal{E}}{\partial x} + \mu_n \mathcal{E} \frac{\partial n_p}{\partial x} + D_n \frac{\partial^2 n_p}{\partial x^2} + G_n - \frac{n_p - n_{p0}}{\tau_n} \tag{2.58}$$

$$\frac{\partial p_n}{\partial t} = -p_n \mu_p \frac{\partial \mathcal{E}}{\partial x} - \mu_p \mathcal{E} \frac{\partial p_n}{\partial x} + D_p \frac{\partial^2 p_n}{\partial x^2} + G_p - \frac{p_n - p_{n0}}{\tau_p} \tag{2.59}$$

There are three unknown quantities to be evaluated (n, p, and $\mathcal{E}$) in Equations 2.58 and 2.59. In principle, these two equations along with Poisson's equation with appropriate boundary conditions are solved simultaneously to obtain a unique solution. However, as the equations are nonlinear, analytical solutions are not possible but some simplifying assumptions can be made. If we consider that there is no electric field ($\mathcal{E}=0$) and no other generation of carriers in addition to the thermal generation ($G_n=G_p=0$), we can express the continuity equations as

$$\frac{\partial n_p}{\partial t} = D_n \frac{\partial^2 n_p}{\partial x^2} - \frac{n_p - n_{p0}}{\tau_n} \tag{2.60}$$

$$\frac{\partial p_n}{\partial t} = \frac{\partial^2 p_n}{\partial x^2} - \frac{p_n - p_{n0}}{\tau_p} \tag{2.61}$$

Equations 2.60 and 2.61 can be modified in terms of excess electron and hole concentrations as

$$\frac{\partial \delta n}{\partial t} = D_n \frac{\partial^2 \delta n}{\partial x^2} - \frac{\delta n}{\tau_n} \tag{2.62}$$

$$\frac{\partial \delta p}{\partial t} = D_p \frac{\partial^2 \delta p}{\partial x^2} - \frac{\delta p}{\tau_p} \tag{2.63}$$

Under the steady-state distribution of excess carriers, the diffusion equations become

$$\frac{\partial^2 \delta n}{\partial x^2} = \frac{\delta n}{D_n \tau_n} = \frac{\delta n}{L_n^2} \tag{2.64}$$

$$\frac{\partial^2 \delta p}{\partial x^2} = \frac{\delta p}{D_p \tau_p} = \frac{\delta p}{L_p^2} \tag{2.65}$$

where $L_n = \sqrt{D_n \tau_n}$ is called the *electron diffusion length* and similarly $L_p = \sqrt{D_p \tau_p}$ is the *hole diffusion length*.

Solving the steady-state continuity equation for electrons as given in Equation 2.64, we get the excess electron concentration as

$$\delta n(x) = \Delta n \exp(-x/L_n) \tag{2.66}$$

where $\Delta n = \delta n|_{x=0}$. This equation implies that the injected excess electron concentration dies out exponentially along x due to recombination, and the diffusion length L_n represents the distance at which the excess electron distribution is reduced to $1/e$ of its value Δn at the point of injection. The steady-state distribution of excess electrons causes diffusion, and therefore an electron current, in the direction of increasing concentration. The electron diffusion current can be expressed as

$$J_n(x) = -\frac{eD_n}{L_n} \delta n(x) \tag{2.67}$$

In a similar manner, the hole diffusion current may be written as

$$J_p(x) = -\frac{eD_p}{L_p} \delta p(x) \tag{2.68}$$

2.10 Basic p-n Junction Theory

Up to this point in the chapter, we have considered the properties of semi-conductor material. We now wish to consider the situation where a p-type and an n-type semiconductor are brought into contact to form a junction. Most semiconductor devices have at least one junction between the p-type and n-type semiconductor regions. Depending on the doping profile, device geometry, and biasing condition, a p-n junction can perform various terminal functions. Therefore, considerable attention has to be given to the basic p-n junction theory, which serves as the foundation of the physics of semiconductor devices. In this section, we derive the expressions for built-in potential, depletion layer width, and depletion layer capacitances for p-n junctions with different doping profiles.

2.10.1 Abrupt Junction

First, we consider the abrupt junction in which the doping concentration is uniform in each region and there is an abrupt change in doping at the junction. Figure 2.16a shows an abrupt junction where the acceptor impurities, N_a,

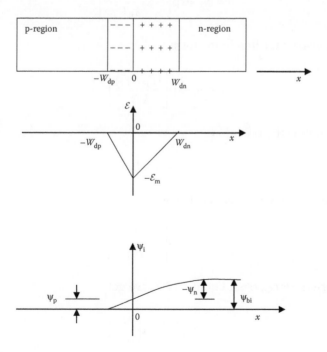

FIGURE 2.16
Potentials and fields in an abrupt junction.

change abruptly to donor impurities, N_d. In particular, if $N_a \gg N_d$ (or $N_d \gg N_a$), the junction becomes a one-sided abrupt p$^+$n (or n$^+$p) junction.

When no voltage is applied across the junction, then the junction is in thermal equilibrium and the Fermi level is constant throughout the entire system. Electrons in the conduction band of the n-region see a potential barrier in trying to move into the conduction band of the p-region. This potential barrier is referred to as the built-in potential (V_{bi}). It maintains equilibrium between the majority carrier electrons in the n-region and the minority carrier electrons in the p-region, and also the majority carrier holes in the p-region and the minority carrier holes in the n-region.

The intrinsic Fermi level is equidistant from the conduction band edge through the junction, thus the built-in potential barrier can be determined as the difference between the intrinsic Fermi levels in the p- and n-regions. From Figure 2.16b, we have

$$\Psi_{bi} = \Psi_{Bn} + \Psi_{Bp} \tag{2.69}$$

In the n-region, the electron concentration in the conduction band is given by

$$n = N_C \exp\left[-\frac{(E_C - E_F)}{k_B T}\right] \tag{2.70}$$

which can also be written in the form

$$n = n_i \exp\left[\frac{(E_F - E_i)}{k_B T}\right] \tag{2.71}$$

We may define the potential

$$e\Psi_{Bn} = E_i - E_F \tag{2.72}$$

Equation 2.71 may then be written as

$$n = n_i \exp\left[-\frac{(e\Psi_{Bn})}{k_B T}\right] \tag{2.73}$$

Considering $n = N_d$, from Equation 2.73 we get

$$\Psi_{Bn} = \frac{-k_B T}{e} \ln\left(\frac{N_d}{n_i}\right) \tag{2.74}$$

Similarly, we can define the potential Ψ_{Bp} in the p-region as

$$\Psi_{Bp} = \frac{+k_B T}{e} \ln\left(\frac{N_a}{n_i}\right) \tag{2.75}$$

Finally, the built-in potential barrier for the abrupt junction is found to be

$$\Psi_{bi} = \frac{k_B T}{e} \ln\left(\frac{N_a N_d}{n_i^2}\right) \tag{2.76}$$

Since at equilibrium, $n_{n0} p_{n0} = n_{p0} p_{p0} = n_i^2$,

$$\Psi_{bi} = \frac{k_B T}{e} \ln\left(\frac{p_{p0}}{p_{n0}}\right) = \frac{k_B T}{e} \ln\left(\frac{n_{n0}}{n_{p0}}\right) \tag{2.77}$$

This gives the relationship between the carrier densities on either side of the junction.

> **Example 2.18:** An abrupt p-n junction in Si is doped with $N_D = 10^{15}\ cm^{-3}$ in the n-region and $N_A = 2 \times 10^{18}\ cm^{-3}$ in the p-region. Using Equation 2.76, the built-in potential will be 0.775 V at $T = 300$ K.

Next, we will calculate the field and the potential distribution inside the depletion region considering the depletion approximation. Since in thermal equilibrium the electric field in the neutral regions of the semiconductor must be zero, the total negative charge per unit area in the p-side must be equal to the total positive charge per unit area in the n side:

$$N_a W_{dp} = N_d W_{dn} \tag{2.78}$$

From Poisson's equation, we have

$$-\frac{d^2 \Psi_i}{dx^2} = \frac{d\mathcal{E}}{dx} = \frac{\rho(x)}{\varepsilon_s} = \frac{e}{\varepsilon_s}\left[N_d^+(x) - n(x) - N_a^-(x) + p(x)\right] \tag{2.79}$$

Inside the depletion region, $n(x) \approx p(x) \approx 0$ and assuming complete ionization:

$$\frac{d^2 \Psi_i}{dx^2} \approx \frac{eN_a}{\varepsilon_s} \quad \text{for} \quad -W_{dp} \leq x \leq 0 \tag{2.80a}$$

$$-\frac{d^2 \Psi_i}{dx^2} \approx \frac{eN_d}{\varepsilon_s} \quad \text{for} \quad 0 \leq x \leq W_{dn} \tag{2.80b}$$

The electric field can be evaluated as shown in Figure 2.16c:

$$\mathcal{E}(x) = -\frac{eN_a(x + W_{dp})}{\varepsilon_s} \quad \text{for} \quad -W_{dp} \leq x \leq 0 \tag{2.81a}$$

$$\mathcal{E}(x) = -\frac{eN_d(W_{dn} - x)}{\varepsilon_s} \quad \text{for} \quad 0 \leq x \leq W_{dn} \tag{2.81b}$$

where the maximum electric field existing at $x=0$ is given by

$$|\mathcal{E}_m| = \frac{eN_d W_{dn}}{\varepsilon_s} = \frac{eN_a W_{dp}}{\varepsilon_s} \tag{2.82}$$

The potential distribution can be obtained by once again integrating Equation 2.81a and b:

$$\Psi_i(x) = \frac{eN_a}{2\varepsilon_s}(x + W_{dp})^2 \quad \text{for} \quad -W_{dp} \leq x \leq 0 \tag{2.83a}$$

$$\Psi_i(x) = \Psi_i(0) + \frac{eN_d}{\varepsilon_s}\left(W_{dn} - \frac{x}{2}\right)x \quad \text{for} \quad 0 \leq x \leq W_{dn} \tag{2.83b}$$

With these, the potentials across different regions can be found as

$$\Psi_p = \frac{eN_a W_{dp}^2}{2\varepsilon_s} \tag{2.84a}$$

and

$$|\Psi_n| = \frac{eN_d W_{dn}^2}{2\varepsilon_s} \tag{2.84b}$$

Therefore, we get

$$\Psi_{bi} = \Psi_p + |\Psi_n| = \Psi_i(W_{dn}) = \frac{|\mathcal{E}_m|}{2}(W_{dp} + W_{dn}) \tag{2.85}$$

where $\mathcal{E}_m$ can be expressed as

$$|E_m| = \left(\frac{2eN_a \Psi_p}{\varepsilon_s}\right)^{1/2} = \left(\frac{2eN_d |\Psi_n|}{\varepsilon_s}\right)^{1/2} \tag{2.86}$$

The depletion widths are calculated to be

$$W_{dp} = \left(\frac{2\varepsilon_s \Psi_{bi}}{e} \frac{N_d}{N_a (N_a + N_d)} \right)^{1/2} \tag{2.87a}$$

$$W_{dn} = \left(\frac{2\varepsilon_s \Psi_{bi}}{e} \frac{N_a}{N_d (N_a + N_d)} \right)^{1/2} \tag{2.87b}$$

$$W_d = W_{dp} + W_{dn} = \left(\frac{2\varepsilon_s}{e} \left(\frac{N_a + N_d}{N_a N_d} \right) \Psi_{bi} \right)^{1/2} \tag{2.88}$$

For a one-sided abrupt junction, Equation 2.88 reduces to

$$W_d = \left(\frac{2\varepsilon_s \Psi_{bi}}{eN} \right) \tag{2.89}$$

where N is N_a or N_d depending on whether $N_a \gg N_d$ or vice versa and

$$\Psi_i(x) = |\mathcal{E}_m| \left(x - \frac{x^2}{2W_d} \right) \tag{2.90}$$

Furthermore, when a voltage V is applied to the junction, the total electrostatic potential variation across the junction is given by $(\Psi_{bi} - V)$ where V is positive for forward bias and negative for reverse bias. The depletion layer width will be obtained by replacing Ψ_{bi} in Equation 2.88 with $(\Psi_{bi} - V)$.

> **Example 2.19:** To calculate the maximum electric field and the width of the depletion region at zero bias for an abrupt GaAs p-n junction with $N_D = 10^{15}$ and $N_A = 10^{18}$ cm^{-3} at room temperature, we first determine the built-in potential as 1.22 V. The depletion layer width and the maximum electric field come out as 13.35 μm and 1.83 V cm^{-1}, respectively. For this calculation, we take the intrinsic carrier concentration of GaAs as 1.79×10^6 cm^{-3} and the permittivity as 13.2.

2.11 I-V and Capacitance–Voltage Characteristics of p-n Junction

2.11.1 Static Current–Voltage Characteristics of a p-n Junction

At zero bias condition, there is no net transfer of carriers across the junction. When an external voltage is applied to the p-n junction, the electron

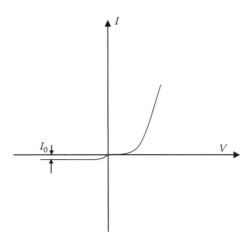

FIGURE 2.17
Current–voltage characteristic curve of a p-n junction.

and hole concentrations deviate from their equilibrium values. The diode current is closely related to these variations. Practically all of the applied voltage is dropped across the space charge region as this region is devoid of mobile carriers. The band structure of a biased p-n junction is shown in Figure 2.17. With the application of forward bias, the potential barrier gets lowered and electrons (holes) from the n-region (p-region) are able to diffuse to the p-region (n-region). Thus, the diffusion current in a forward bias junction is quite large under forward-biased condition. An opposite situation takes place under reverse-biased condition when the current flow should occur due to the electrons (holes) from the p-region (n-region) crossing the junction. As a result, we get the reverse current, which is extremely low. To derive an analytical relation for the I-V characteristics, some assumptions are made.

1. The current flow across the junction is one-dimensional.
2. The drift current due to minority carriers is neglected as the electric field in the neutral p- and n-regions is negligible.
3. There is no generation and recombination of carriers in the space charge region and the current in this region is constant throughout.
4. The low-level injection condition is maintained.
5. The drift and diffusion currents balance each other in the depletion region even when a bias is applied.

In the forward bias condition, the minority carriers (holes) are injected from the p-region to increase the hole concentration at the surface adjacent to the space charge region. Neglecting the minority carrier drift current

(Assumption 2), the steady-state continuity equation in the n-region can be written as

$$\frac{d^2 p_n}{dx^2} - \frac{(p_n - p_{n0})}{L_p^2} = 0 \qquad (2.91)$$

The boundary conditions are as follows:

1. At $x = W_{dn}$, $p_n = p_{ne} = p_{n0} \exp(qV/k_B T)$, where p_{ne} is the hole concentration at the edge of the depletion region.
2. At $x = \infty$, $p_n = p_{n0}$.

Solving Equation 2.91, we obtain

$$p_n - p_{n0} = p_{n0} \left[\exp\left(\frac{eV}{k_B T}\right) - 1 \right] \exp\left[-\frac{x - W_{dn}}{L_p} \right] \qquad (2.92)$$

Similarly, the electron concentration in the p-region can be represented by

$$n_p - n_{p0} = n_{p0} \left[\exp\left(\frac{eV}{k_B T}\right) - 1 \right] \exp\left[\frac{x + W_{dp}}{L_n} \right] \qquad (2.93)$$

Now, the hole diffusion current density at the edge of the depletion layer is given by

$$J_p(W_{dn}) = -eD_p \left. \frac{dp_n}{dx} \right|_{x=W_{dn}} = \frac{eD_p p_{n0}}{L_p} \left[\exp\left(\frac{eV}{k_B T}\right) - 1 \right] \qquad (2.94)$$

Similarly, the electron diffusion current at the edge of the depletion region is given by

$$J_n(-W_{dp}) = -eD_n \left. \frac{dn_p}{dx} \right|_{x=-W_{dp}} = \frac{eD_n n_{p0}}{L_n} \left[\exp\left(\frac{eV}{k_B T}\right) - 1 \right] \qquad (2.95)$$

According to Assumption 3, there is no generation or recombination of carriers in the depletion region. So, the two current components as given in Equations 2.94 and 2.95 remain constant throughout the depletion layer. The total current flowing through the p-n junction will be as follows:

$$J = J_p(W_{dn}) + J_n(-W_{dp}) = e \left(\frac{D_p p_{n0}}{L_p} + \frac{D_n n_{p0}}{L_n} \right) \left[\exp\left(\frac{eV}{k_B T}\right) - 1 \right] \qquad (2.96)$$

or

$$J = J_0 \left[\exp\left(\frac{eV}{k_B T} \right) - 1 \right] \tag{2.97}$$

where J_0 is the *reverse saturation current density* of the diode and is given by

$$J_0 = e \left(\frac{D_p p_{n0}}{L_p} + \frac{D_n n_{p0}}{L_n} \right) \tag{2.98}$$

Figure 2.17 shows the I-V characteristics of the diode. In the forward bias condition, the current increases exponentially with the applied voltage, while when the diode is reverse biased, the current ($=J \times$ area of the diode) is limited to $I = -I_0$.

> **Example 2.20:** Consider an Si diode with area 10^{-4} cm^2. If $N_A = 2 \times 10^{16}$ and $N_D = 3 \times 10^{15}$ cm^{-3}, the values of $n_{p0} = 1.125 \times 10^4$ and $p_{n0} = 7.5 \times 10^5$ cm^{-3}. Assume that $D_n = 32.5$ and $D_p = 13$ cm^2 s^{-1}, and $L_n = 57$ and $L_p = 36$ μm. The reverse saturation current will be 44.4 fA.

2.11.2 Capacitance–Voltage Characteristics of a p-n Junction

The charge storage effects in a p-n junction lead to capacitive effects. There are basically two types of capacitance associated with a p-n junction; one is diffusion capacitance and the other is depletion capacitance. The minority carriers injected across a forward-biased diode is a very strong function of the applied bias. Thus, the injected minority carrier charges in the two neutral regions along with their associated voltage dependence would contribute to a capacitive effect. This capacitance is known as *diffusion capacitance*, which becomes dominant under the forward bias condition of the junction. The simplest way to account for this capacitance is through a *charge control approach*, which is described in any textbook. On the other hand, the space charge layer can be considered as a dipole layer of positive donor and negative acceptor ions. The width of the layer changes with the change in the bias voltage as the majority carriers flow in or out of the layer. This can be visualized as the charging and discharging of a capacitor. This type of capacitance is known as *depletion capacitance* and becomes effective under the reverse bias condition. The dependence of the depletion capacitance on the reverse bias voltage gives an important capacitance–voltage (C-V) characteristic, which can be used to measure the value of the built-in potential and the doping concentration of the junction.

The depletion capacitance (also known as the *junction capacitance*) can be expressed as $C_d = |dQ_d/dV|$, where Q_d represents the depletion charges on each side of the junction and is given by

$$Q_d = eAW_{dn}N_d = eAW_{dp}N_a \tag{2.99}$$

With the help of Equations 2.11.23 through 2.11.25, we can write the following equations:

$$W_{dn} = \frac{N_a W_d}{(N_a + N_d)} \tag{2.100a}$$

and

$$W_{dp} = \frac{N_d W_d}{(N_a + N_d)} \tag{2.100b}$$

where W_d is the depletion layer width under zero bias condition and its expression under a given voltage V can be reproduced as

$$W_d = \left(\frac{2\varepsilon_s}{e} \left(\frac{N_a + N_d}{N_a N_d} \right) (\Psi_{bi} - V) \right)^{1/2} \tag{2.101}$$

V is positive for forward bias and negative for reverse bias.

Therefore, combining Equations 2.99–2.101, the charge Q_d on each side of the dipole can be given by

$$Q_d = A \left[2e\varepsilon_s (\Psi_{bi} - V) \frac{N_a N_d}{N_a + N_d} \right]^{1/2} \tag{2.102}$$

The depletion capacitance is thus given by

$$C_d = \frac{C_{d0}}{\left(1 - \dfrac{V}{\Psi_{bi}} \right)^{1/2}} \tag{2.103}$$

C_{d0} is defined as the *zero bias depletion capacitance* of the junction with $V = 0$ and is given by

$$C_{d0} = A \left(\frac{e\varepsilon_s}{2\Psi_{bi}} \frac{N_a N_d}{N_a + N_d} \right)^{1/2} \tag{2.104}$$

Figure 2.18 shows the variation of $1/C_d^2$ with different bias voltage V. It gives a straight line with the slope equal to

$$\left(-\frac{2(N_a + N_d)}{A^2 e\varepsilon_s N_a N_d} \right)$$

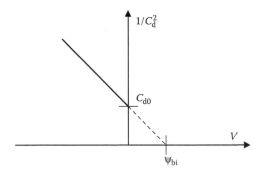

FIGURE 2.18
C-V characteristic curve for a reverse biased p-n junction.

For a one-sided abrupt p^+n junction, the slope becomes equal to

$$\left(-\frac{2}{A^2 e \varepsilon_s N_d}\right)$$

Thus, for a given value of A, the doping concentration of donor ions can easily be calculated. For $\Psi_{bi} = V$, $1/C_d^2$ will be equal to zero. Therefore, extrapolating the $1/C_d^2$ versus V curve for positive values of V, we get the value of the built-in potential Ψ_{bi} from the intercept of the curve with the v-axis.

> **Example 2.21:** The depletion capacitance for the junction in Example 2.20 with an applied reverse bias of 3 V will be $1.29 \times 10^{-2}\,\mu F$, where the total area of the diode has been considered as $100 \times 100\,\mu m$.

PROBLEMS

2.1. Obtain the expression for the electron density when the Fermi function is boxlike.

2.2. Obtain the expression for the density of states for electrons in Si. Show that Equation 2.10 is valid if m_{de} is replaced by $6^{2/3}(m_l m_t^2)^{1/3}$.

2.3. Calculate the total number of states in silicon between E_C and $E_C + k_B T$ at 300 K.

2.4. Assume that a semiconductor becomes degenerate when the Fermi level touches the band edge. Assume also that Equation 2.16 is valid. Calculate the donor density that is needed to make the electron system in GaAs degenerate at $T = 300$ K. Take $m_e = 0.067m_0$.

2.5. Calculate the maximum temperature of operation of an Si device doped with 10^{16} donor atoms cm^{-3}. Use the expression for the bandgap given by Equation 2.27.

2.6. Using the expression for the temperature-dependent bandgap, calculate the gap at 300 and 1200 K. Take $E_g(0) = 1.17$ eV, $\alpha = 4.73 \times 10^{-4}$ eV K^{-1}, and $\beta = 636$ K.

2.7. Obtain the general expression for electron density given as

$$n = N_C F_{1/2}(\eta)$$

where:

$$F_{1/2}(\eta) = (2 / \sqrt{\pi}) \int\limits_0^\infty \frac{x^{1/2} dx}{[1 + \exp(x - \eta)]}$$

$$\eta = (E_F - E_C) / k_B T.$$

2.8. Show that when $\eta \ll -1$, $F_{1/2}(\eta) = \exp(\eta)$, the electron density as expressed by Equation 2.15 is valid for nondegenerate semiconductors.

2.9. When $\eta \gg 1$, $F_{1/2}(\eta) \approx [4\eta^{3/2} / 3\sqrt{\pi}]$. Using this, express Fermi energy in terms of electron density.

2.10. Obtain the expressions for the electron and hole densities when both the donors and acceptors are present.

2.11. Calculate the wavelength of emission for transition from 2s to 1s level of P impurity in Si. The donor binding energy is 25 meV and the Bohr model is valid.

2.12. Calculate the Bohr radius for donor ions in Ge using $m_e = 0.12m_0$ and $\varepsilon_r = 16$. Calculate the approximate number of Ge atoms within the volume defined by the Bohr radius ($a_{Ge} = 5.66$ A). Does this number justify the use of effective mass in Equation 2.28?

2.13. Calculate the resistivity of intrinsic Ge at 300 K.

2.14. The mobilities of two scattering processes vary with temperature T as AT and B/T. Prove that the combined mobility will show a maximum when $T^2 = B/A$.

2.15. The electrons in a semiconductor are scattered by impurities ($\mu_{imp} = AT^{3/2}$) and by phonons ($\mu_{ph} = BT^{-2.5}$). Using Mathiessen's rule, show that the inverse mobility attains a minimum at a certain temperature. Obtain the expression for the temperature giving minimum $1/\mu$.

2.16. The current density for electrons in Si along the x-direction is $J_x = ne\mu_{xx}F_x$, where F is the field along the $x = (100)$ direction and $\mu_{xx} = e\langle\tau\rangle/m_{xx}$. Consider all six valleys and show that $J_x = ne^2\langle\tau\rangle F_x/m_c$, where m_c is the conductivity effective mass.

2.17. Prove that the impurity scattering limited mobility follows the $T^{3/2}$ law.

2.18. Using kinetic theory for the electron gas, prove that $D/\mu = k_B T/e$.

2.19. Prove that τ_n is the mean time spent by an excess electron before being lost by recombination.

2.20. The field-dependent mobility is expressed as $\mu(F) = \mu_0/[1 + (F/F_c)]$. How would you obtain the expression for the saturation drift velocity?

2.21. In the empirical expression for the drift velocity at high fields given in Example 2.11, find the values of low-field mobility and saturation drift velocity for electrons in Si.

2.22. Under a high electric field, the electrons become "hot" with an effective temperature $T_e > T_L$, T_L being the lattice temperature. The energy balance equation $e\mu F^2 = (3/2)k_B(T_e - T_L)/\tau_e$ is used to determine the electron temperature. Prove that when the electric field is along the $\langle 100 \rangle$ direction, the two conduction band valleys having longitudinal mass along the field direction will be colder than the remaining four valleys.

2.23. Calculate the value of the injected electron density that will raise the quasi-Fermi level 0.01 eV above the conduction band edge of GaAs.

2.24. The Fermi level is constant in a p-n junction under equilibrium. Using this, show that the built-in potential in the junction is expressed by Equation 2.77.

2.25. Obtain the expressions for the potential, field, and junction capacitance of a linearly graded p-n junction.

2.26. The depletion layer of a p-n junction acts as a parallel plate capacitor in which the −ve (+ve) charges are assumed to be placed at x_{p0} (x_{n0}): the edges of the depletion layer on the p(n) side. Let $N_a = 10^{18}$ and $N_d = 10^{17}$ cm^{-3} and $x_{p0} = 0.1$ μm. Calculate the depletion layer capacitance per unit area [$\varepsilon_0 = 8.84 \times 10^{-12}$ F m^{-1}].

2.27. Give a sketch of the variation of electron and hole densities in a forward-biased p-n junction, infinitely long in both sides.

2.28. From the diode equation, prove that the dynamic resistance of a forward-biased p-n junction diode is $dV/dI = 26\ \Omega$ for the forward current of 1 mA.

2.29. Prove that in a p$^+$-n junction, the forward-bias current is predominantly due to holes injected into the n-region.

2.30. Prove that the stored charge in the n-region due to injected holes is $Q_p = I\tau_p$. Obtain the expression for the small signal diffusion capacitance dQ_p/dV.

Reading List

Balkanski, M. and R. F. Wallis, *Semiconductor Physics and Applications*. Oxford University Press, Oxford, 2000.

Nag, B. R., *Electron Transport in Compound Semiconductors*. Springer, Berlin, 1980.

Neamen, D. A., *Semiconductor Physics and Devices: Basic Principles*. McGraw-Hill, New York, 1992.

Ridley, B. K., *Quantum Processes in Semiconductors*, 5th edn. Clarendon Press, Oxford, 2000.

Shur, M., *Physics of Semiconductor Devices*. Prentice-Hall, Upper Saddle River, NJ, 2003.

Singh, J., *Electronic and Optoelectronic Properties of Semiconductor Structures*. Cambridge University Press, Cambridge, 2003.

Smith, R. A., *Semiconductors*. Cambridge University Press, Cambridge, 1964.

Streetmann, B. G. and S. Banerjee, *Solid State Electronic Devices*, 6th edn. Prentice-Hall International, Upper Saddle River, NJ, 2006.

Sze, S. M., *Physics of Semiconductor Devices*. Wiley, New York, 1969.

Yu, P. and M. Cardona, *Fundamentals of Semiconductors*, 4th edn. Springer, Berlin, 2010.

References

1. Cohen, M. L. and J. R. Chelikowsky, Electronic structure and optical properties of semiconductors. In *Springer Series in Solid-State Science 75*, ed. Cardona, M. Springer, Berlin, 1989.

2. Luttinger, J. M. and W. Kohn, Motion of electrons and holes in perturbed periodic fields, *Phys. Rev.* 97, 869–883, 1955.

3. Kane, E. O., Energy band structure in p-type germanium and silicon, *J. Phys. Chem. Solids* 1, 82–99, 1956.

4. Joyce, W. B. and R. W. Dixon, Analytic approximations for the Fermi energy of an ideal Fermi gas, *Appl. Phys. Lett.* 31, 354–356, 1977.

5. Varshni, Y. P., Band-to-band radiative recombination in Groups IV, VI, and III–V semiconductors (I), *Phys. Stat. Sol.* 19, 459–514, 1967.

6. Sotoodeh, M., A. H. Khalid, and A. A. Rezazadeh, Empirical low-field mobility model for III–V compounds applicable in device simulation codes, *J. Appl. Phys.* 87, 2890–2900, 2000.

7. Shockley, W. and W. T. Read, Statistics of recombination of electrons and holes, *Phys. Rev.* 87, 835–842, 1952.
8. Hall, R. N., Electron-hole recombination in germanium, *Phys. Rev.* 152, 387, 1952.
9. Sah, C.-T. and W. Shockley, Electron-hole recombination statistics in semiconductors through flaws with many charge conditions, *Phys. Rev.* 109, 1103–1115, 1958.

3

Heterojunctions and Quantum Structures

3.1 Introduction

As stated in Chapter 1, most semiconductor lasers used today contain a double heterostructure (DH). A heterojunction is made up of two dissimilar semiconductors of different bandgaps, permittivities, and so on. Usually a compound semiconductor and one of its alloys are used to make a heterojunction, the most common example being GaAs and its alloy $Ga_{1-x}Al_xAs$. An important feature of the GaAs/GaAlAs system is that the two materials are lattice matched. Heterojunctions may also be formed by using two materials having a slight degree of lattice mismatch. This gives rise to built-in strain, and the structures are called strained heterostructures. Many heterojunctions under suitable conditions, that is, doping variation in different layers, extremely low thickness of a layer, and so on, give rise to quantum confinement of electrons and holes, and accordingly quantum well (QW), quantum wire (QWR), and quantum dot (QD) structures are realized.

In all the structures mentioned above, there is an alteration of band structure. The common methods of modification of band structures are

1. To use alloys of a binary compound semiconductor
2. To use quantum nanostructures
3. To use lattice-mismatched heterostructures

In this chapter, the basic idea of band structure modification will be presented by considering the above three methods. The band structure of alloys, band alignment in heterojunctions, condition for quantum confinement, band structures in quantum nanostructures, and finally the effect of strain on bandgap will be discussed.

3.2 Alloys

As already mentioned, real progress in semiconductor lasers, especially operation at room temperature, took off with the development of heterostructures using good-quality alloys. The technique of alloying has been known since the Bronze Age. The motivation for the development of semiconductor alloys came from the need to develop a material with a chosen bandgap and a chosen lattice constant.

In most semiconductor alloys, the two or more constituents have the same crystal structure, so that the same structure is maintained in the final product. The ordering of the atoms in an alloy A_xB_{1-x} is also of importance. In a clustered alloy, all the A atoms may be localized in one region, while all the B atoms may be localized in another region. In another situation, A and B atoms may be distributed randomly. In a superlattice, the A and B atoms may form a well-ordered periodic structure. In the present context, we will discuss only random alloys.

Alloys may be classified as binary, ternary, or quaternary, depending on the number of different atoms involved in making the materials. A binary alloy involving two atoms A and B may have the composition A_xB_{1-x}, and a practical example is $Si_{1-x}Ge_x$. Three atoms A, B, and C make a ternary alloy in the form $A_{1-x}B_xC$; the most quoted example is $Ga_{1-x}Al_xAs$, an alloy of GaAs and AlAs. An alloy represented generally by $A_{1-x}B_xC_yD_{1-y}$ is called a quaternary alloy. The alloy $In_{1-x}Ga_xAs_yP_{1-y}$ belongs to this class.

The important parameters of an alloy, such as lattice constant, bandgap, and so on, are dependent on the parameters of the constituent materials. The lattice constant of the alloy $A_{1-x}B_xC$ is given by Vegard's law as

$$a_{ABC} = (1-x)a_{AC} + xa_{BC} \tag{3.1}$$

where the subscripts correspond to the constituent binaries. This linear relationship is violated in other material parameters. The parameter P_{ABC} of a ternary alloy $A_{1-x}B_xC$ may be expressed in terms of the corresponding parameters of binary compounds AC and BC as

$$P_{ABC} = (1-x)P_{AC} + xP_{BC} + x(1-x)P_{AB} = a + bx + cx^2 \tag{3.2}$$

where:
$$a = P_{AC}$$
$$b = P_{BC} - P_{AC} + P_{AB}$$
$$c = -P_{AB}$$

The parameter c, called the bowing parameter, arises due to lattice disorder created by intermixing atoms A and B on the lattice site normally occupied by one kind of atom in a binary compound. The expressions for bandgap for a number of ternary alloys are given in Table 3.1 [1].

TABLE 3.1

Bandgap Energy of Different Alloys

Alloy	Energy Gap (eV)
$In_{1-x}Al_xP$	$1.351 + 2.23x$
$Ga_{1-x}Al_xAs$	$1.424 + 1.247x$
$In_{1-x}Al_xAs$	$0.360 + 2.012x + 0.698x^2$
$Ga_{1-x}Al_xSb$	$0.726 + 1.129x + 0.386x^2$
$In_{1-x}Al_xSb$	$0.172 + 1.621x + 0.43x^2$
$In_{1-x}Ga_xP$	$1.351 + 0.643x + 0.786x^2$
$In_{1-x}Ga_xAs$	$0.36 + 1.064x$
$In_{1-x}Ga_xSb$	$0.172 + 0.139x + 0.415x^2$
$GaAs_{1-x}P_x$	$1.424 + 1.150x + 0.176x^2$
$GaAs_xSb_{1-x}$	$0.726 - 0.502x + 1.2x^2$
$InAs_{1-x}P_x$	$0.360 + 0.891x + 0.101x^2$
$InAs_xSb_{1-x}$	$0.18 - 0.41x + 0.58x^2$
$In_{1-x}Ga_xAs_yP_{1-y}$	$1.35 - 0.72y + 0.12y^2 [x = 0.4526y/(1-0.031y)]$
$Hg_{1-x}Cd_xTe$	$-0.3 + 1.9x$
$Cd_xZn_{1-x}Se$	$2.730 - 1.388x + 0.35x^2$
$Cd_xZn_{1-x}Te$	$2.250 - 0.869x + 0.128x^2$
$Pb_{1-x}Sn_xTe$	$0.19 - 0.543x$ (12 K; $x < 0.35$)
$Si_{1-x}Ge_x$	$E_g^X(x) = 1.155 - 0.43x + 0.206x^2, \quad 0 < x < 0.85$
	$E_g^L(x) = 2.010 - 1.270x, \quad x > 0.85$
$Si_xGe_ySn_{1-x-y}$	$4.185x + 0.7985y - 0.413(1-x-y) - 0.21xy -$
	$13.2x(1-x-y) - 1.94y(1-x-y)$ [2]

Example 3.1: The bandgap of $In_{0.7}Ga_{0.3}P$ ($x = 0.3$) is 1.615 eV, as obtained from the expression given in Table 3.1. If the bowing parameter is neglected, the value is 1.544 eV.

It appears from Table 3.1 that various combinations of binary compounds form ternary alloys. We shall now discuss a few specific alloy systems.

3.2.1 Ternary Alloys: III–V

3.2.1.1 GaAs–AlAs Alloys

The AlGaAs system is the most widely studied. The lattice mismatch between the two extremes AlAs ($a = 0.56612$ nm) and GaAs ($a = 0.565325$ nm) is only 0.14%, and is still lower between GaAs and its alloy AlGaAs. The GaAs–AlGaAs heterostructures have been used in DH lasers, and, in addition, in modulation-doped field-effect transistors (MODFETs), and to form QW, QWR, QD, and superlattice structures. The variation of bandgap at Γ, X, and L points may be expressed as

$$E_{g\Gamma}(x) = 1.424 + 1.247x, \qquad\qquad (0 < x < 0.45)$$

$$= 1.424 + 1.247x + 1.147(x - 0.45)^2 \quad (0.45 < x < 1)$$

(3.3a)

$$E_{gX} = 1.900 + 0.125x + 0.143x^2 \qquad\qquad (3.3b)$$

$$E_{gL} = 1.708 + 0.642x \qquad\qquad (3.3c)$$

The alloy becomes an indirect-gap material for $x \geq 0.45$.

3.2.1.2 InAs–GaAs Alloy

InAs and GaAs have bandgaps of 0.39 eV and 1.42 eV, respectively. The two materials have a lattice mismatch of 7%. The alloy $In_{1-x}Ga_xAs$ is grown on an InP substrate, and the lattice-matching condition is

$$(1 - x)a_{InAs} + xa_{GaAs} = a_{InP}$$

> **Example 3.2:** Taking a(InAs)=0.60584 nm, a(GaAs)=0.56533 nm, and a(InP)=0.58688 nm, x=0.47 for lattice matching.

The alloy $In_{0.53}Ga_{0.47}As$ (E_g=0.86 eV) lattice matched to InP has been used for fabrication of high-speed transistors, lasers, and photodetectors for fiber-optic communication. It is also lattice matched to $In_{0.52}Al_{0.48}As$. Lattice-mismatched structures of $In_{0.53+x}Ga_{0.47-x}As$ on InP or $In_yGa_{1-y}As$ on GaAs (or GaAlAs) are used to produce strained optoelectronic devices or MODFETs. The pair InAs and GaAs, with a high degree of lattice mismatch, is used to produce QD structures.

3.2.1.3 III–V Nitrides and Alloys

GaN and its alloys with InN and AlN are of immense current interest. The heterojunctions and quantum nanostructures made of GaN and its alloys are now routinely used for production of lasers and light-emitting diodes (LEDs) in the blue, blue-green, and green regions of the visible spectrum. The lasers are used in making digital versatile discs (DVDs); suitable mixing of blue and green light with red light leads to emission of white light.

3.2.2 Ternary II–VI Alloys

Semiconducting II–VI compounds offer some unique possibilities that are not usually met by III–V compounds. As may be seen from Figure 1.12, a wide range of bandgap and hence emission wavelength may be covered by some of the alloys. Hg compounds with S, Se, or Te have a bandgap in the infrared region, while Cd or Zn compounds with S, Se, or Te have energy gaps covering the visible region of the electromagnetic (EM) spectrum.

3.2.2.1 HgTe–CdTe Alloys

HgTe is a zero-gap semiconductor; the bandgap of CdTe is about 1.5 eV. The alloy maintains direct-gap nature throughout the composition range. The small gap of HgCdTe (MCT) allows application in infrared systems such as night vision, in thermal imaging for industrial and medical applications, and in imaging in adverse environments, such as in the presence of dense fog, at long wavelength. It may be grown on CdTe, ZnTe, Si, or GaAs substrates.

3.2.2.2 ZnSe–CdSe Alloys

III–V compounds and their alloys do not give high values of direct band-gap, and emission from these materials is mainly in the infrared range. II–VI compounds have larger gaps, reaching about 4.0 eV, so that emission from the blue region may be achieved.

Various combinations of II–VI binary compounds are shown in Table 3.1. ZnCdSe layers are used for room-temperature LEDs and lasers.

3.2.2.3 Dilute Magnetic Alloys

Introduction of a very small quantity of compounds, such as MnTe, containing paramagnetic atoms, such as Mn, in ZnTe or CdTe produces CdMnTe or ZnMnTe alloys. These are dilute magnetic alloys, which, either in bulk form or in the form of quantum nanostructures, give rise to interesting magnetic properties and find use in isolators, Faraday rotators, and so on.

3.2.3 Group IV Alloys

There is ~4% lattice mismatch between Si ($a = 0.5431$ nm) and Ge ($a = 0.566$ nm). Direct growth of Ge on Si is therefore problematic. Si, Ge, and their alloy $Si_{1-x}Ge_x$ are all indirect-gap materials and as such cannot be used for light emission. Recently, there has been much interest in the topic of Group IV photonics and, in particular, in the development of lasers on a Si substrate. For this purpose, new alloys such as GeSn, SiGeSn, SiGeC, and so on have been developed.

3.2.4 Quaternary Alloys

Quaternary alloys are made by four binary compound semiconductors. The introduction of the fourth atom allows more flexibility in tailoring the physical properties of the material. For example, the alloy may be lattice matched to a substrate but the composition may be changed to obtain the desired bandgap. We have seen earlier that $In_{1-x}Ga_xAs$ with $x = 0.47$ is lattice matched to InP. The bandgap is, however, fixed. In order to attain more flexibility in the value of the bandgap, the quaternary alloy $In_{1-x}Ga_xAs_yP_{1-y}$ has been used. Note that the condition for lattice matching to InP is given in Table 3.1 (alternatively $y \approx 2.2x$). This alloy is used for LEDs and lasers for fiber-optic communication at 1.3 and 1.55 μm.

Atoms A^1, A^2, ... belonging to Group III and B^1, B^2, ... belonging to Group V may combine in three different ways to form quaternary alloys:

1. Two A atoms and two B atoms to form $A^1_{1-x}A^2_xB^1_yB^2_{1-y}$ (e.g., $In_{1-x}Ga_xAs_yP_{1-y}$)
2. Three A atoms and one B atom to form $A^1_xA^2_yA^3_{1-x-y}B$ (e.g., $(Al_yGa_{1-y})_xIn_{1-x}P$, used in visible lasers)
3. One A atom and three B atoms in the form $AB^1_xB^2_yB^3_{1-x-y}$ (e.g., InAsSbP)

The material $In_{1-x}Ga_xAs_yP_{1-y}$ is an important quaternary material for devices used in optical fiber communication, and its lattice constant $a(x,y)$ obeys Vegard's law as follows:

$$a(x, y) = xya(GaAs) + x(1-y)a(GaP) + (1-x)ya(InAs) + (1-x)(1-y)a(InP)$$

The lattice constants in angstroms are $a(GaP) = 5.45117$, $a(GaAs) = 5.65325$, $a(InP) = 5.86875$, and $a(InAs) = 6.0584$. Using $a(x,y) = a(InP)$, one obtains

$$x = \frac{0.4526y}{1 - 0.031y} \tag{3.4}$$

For the lattice-matched quaternary alloy grown on InP, the bandgap is expressed as

$$E_g = 1.35 - 0.72y + 0.12y^2 \tag{3.5}$$

Example 3.3: The bandgap corresponding to 1.55 μm wavelength is 0.8 eV. Using this in Equation 3.5, and solving the quadratic equation, one obtains $y = 0.9$. The value of x from Equation 3.4 is 0.42. The composition is $In_{0.58}Ga_{0.42}As_{0.9}P_{0.1}$.

3.3 Heterojunctions

It has been mentioned already that a heterojunction is formed when two semiconductors of different bandgaps, permittivities, electron affinities, and so on are joined together. Ideally, there should be an abrupt change in the bandgap and other characteristics of the constituent semiconductors at the heterointerface. With modern growth techniques, it is possible to ensure this abruptness. In the discussions to follow, we shall assume abrupt heterojunctions and consider lattice-matched pairs.

The most important question to answer for the heterojunctions is how the band edges line up at the heterointerface. Many theoretical models are available in the literature. However, no single model can explain the results obtained for all the material systems. In this subsection, we shall mention the first theory, the electron affinity rule, to illustrate the band discontinuity in a single heterojunction, and then mention the empirical rules.

3.3.1 Electron Affinity Rule

Electron affinity refers to the energy necessary to take an electron from the conduction band edge to the vacuum level. The work function is the energy needed to take an electron from the Fermi level to the vacuum level.

The bandgap (E_g) and electron affinity (χ) of two semiconductors 1 and 2, isolated from each other, are shown in Figure 3.1a. The electron affinity rule states that, when the two semiconductors make an abrupt heterojunction, there is an abrupt discontinuity (band offset) in the conduction band edge given by

$$\Delta E_c = E_{c1} - E_{c2} = \chi_1 - \chi_2 \tag{3.6}$$

It is easy to express the valence band discontinuity as

$$\Delta E_v = (E_{g2} - E_{g1}) - \Delta E_c \tag{3.7}$$

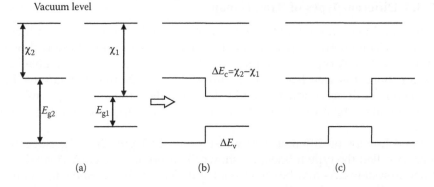

(a)   (b)   (c)

FIGURE 3.1
Illustration of electron affinity rule: (a) bandgaps and electron affinities of two isolated semiconductors; (b) band lineup under electron affinity rule; and (c) band lineup for a double heterojunction.

The band lineup for a single heterojunction using this electron affinity rule is shown in Figure 3.1b. In Figure 3.1c, the corresponding band diagram for a DH is shown.

Unfortunately, this simple rule fails to give the correct values for real semiconductors. The discrepancy may arise due to the fact that, when two semiconductors form the junction, there is a charge transfer across the interface due to the dissimilar nature of the chemical bonds. This short-range dipole at the interface may alter the band offset.

> **Example 3.4:** The value of electron affinities is 4.07 eV for GaAs and 4.07–1.06x for Al$_x$Ga$_{1-x}$As for 0<x<0.45. We consider x=0.3 and ΔE_c = 1.06×0.3=0.318 eV. The measured value is about 0.24 eV.

3.3.2 Empirical Rule

In the absence of an accepted theory for band lineup in heterojunctions, experimental results are used to determine band offsets ΔE_c and ΔE_v. It is now accepted that the ratio $\Delta E_c/\Delta E_v$=60/40 or 65/35 for the GaAlAs/GaAs system. On the other hand, this ratio is approximately 40/60 for InGaAsP/InP pairs.

> **Example 3.5:** For the Ga$_{0.7}$Al$_{0.3}$As/GaAs system, the bandgap difference is 1.798–1.424=0.374 eV. Taking the 65/35 ratio, ΔE_c=0.243 eV and ΔE_v=0.131 eV.

A simple theory, called the model solid theory [3], is useful in predicting the band offsets between various semiconductor pairs.

3.3.3 Different Types of Band Lineup

The band lineup for a double heterojunction with lower-gap semiconductor A sandwiched between two layers of higher-gap semiconductor B is shown in Figure 3.2a. A typical example is an AlGaAs–GaAs–AlGaAs double heterojunction. It is seen from both Figures 3.1c and 3.2a that a square well is formed in both the conduction and valence bands of the lower-gap semiconductor B in the BAB structure. This type of band lineup is commonly termed *type I*.

Band lineups of different nature also occur. In Figure 3.2b the staggered lineup, called the type II lineup, is shown. This occurs in GaInAs/GaAsSb or Si/SiGe systems. As may be seen from Figure 3.2b, both the band edges of one semiconductor are shifted in the same direction relative to those of the other. It is also evident that in this type II alignment electrons and holes accumulate in different materials, whereas in type I structure both the carriers are trapped in the lower-gap semiconductor. The extreme case is shown in Figure 3.2c, and is

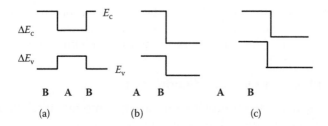

FIGURE 3.2
Different types of band alignment: (a) type I, (b) type II, and (c) broken-gap type II band alignment.

known as type II misaligned. This is found in the InAs/GaSb pair. In this case, the conduction band edge of one semiconductor, for example InAs, is below the valence band edge of the other semiconductor, GaSb.

> **Example 3.6:** Two examples of type II alignment are given in Figure 3.3 [4]. Here $E_1=0.453$ eV and $E_2=0.630$ eV are the values for the spatially indirect gap in Figure 3.3a and Figure 3.3b, respectively. The band-gap for GaAsSb is $E_{g2}=0.813$ eV. The conduction band offsets are $\Delta E_{c2}=E_{g2}-E_1=0.36$ eV and $\Delta E_{c3}=E_{g2}-E_2=0.18$ eV.

3.4 Quantum Structures

Most of the present electronic and optoelectronic devices are based on nano-structures in which at least one dimension of the layer of interest is only a few nanometers thick. The behavior of particles in the nanostructure differs grossly from their behavior in bulk materials, and the motion of particles is

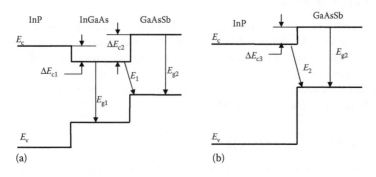

FIGURE 3.3
Band alignment in (a) InP–In$_{0.52}$Ga$_{0.48}$As–GaAs$_{0.5}$Sb$_{0.5}$ and (b) InP–GaAs$_{0.5}$Sb$_{0.5}$ heterostructures.

quantum mechanically confined along the direction concerned. The confinement gives rise to electron or hole gas of lower dimensions. We consider in this section the conditions necessary for quantum confinement.

3.4.1 Conditions for Quantum Confinement

The electrons in a bulk semiconductor have free motion in all three dimensions, as indicated by the E–k relationship expressed by

$$E(k) = \frac{\hbar^2 k^2}{2m_e} \quad k \equiv k_x, k_y, k_z \tag{3.8}$$

Here, m_e is the associated effective mass of an electron. If the electron is confined in a one-dimensional (1-D) potential well having width comparable to its de Broglie wavelength, the free motion of the electron is inhibited in that direction. The electron waves are then standing waves in nature, and the energy becomes quantized. The electron is, however, free to move along the other two dimensions, and, accordingly, a two-dimensional (2-D) electron gas results. The electron wavelength $\lambda = 2\pi/k$ is related to its energy E by

$$\lambda = \frac{2\pi\hbar}{\sqrt{2m_e E}} \tag{3.9}$$

Example 3.7: Let us assume that the energy of the electron is 25 meV and the effective mass of the electron is $m_e = m_l = 0.91\ m_0$: the longitudinal mass in the Si conduction band. Then, $\lambda \approx 8$ nm. The value for GaAs will be ≈ 30 nm using $m_e = 0.07\ m_0$.

3.5 Quantum Wells

In this section and in Section 3.6, the following sections, the structures that support the low-dimensional systems will be discussed. To start with, we consider a QW structure and associated electronic properties. The expressions for quantized energy will be derived by using a simple model, followed by the derivation of the density of states function. Further refinement of the theory will then be presented.

3.5.1 Simplified Energy Levels

The one-dimensional potential well described above may be realized in a double heterojunction. Consider the simplest case of a double heterojunction made by GaAs and its alloy $Ga_{1-x}Al_xAs$, as shown in Figure 3.4.

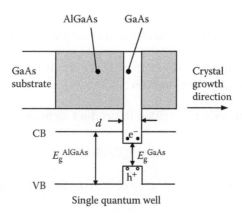

FIGURE 3.4
The quantum well structure formed by sandwiching a GaAs layer between two AlGaAs layers.

In the simplest approximation, the difference in the bandgap of the two semiconductors, $E_g^{AlGaAs} - E_g^{GaAs}$, is consumed by discontinuities (steps) in the conduction and valence bands in the two heterointerfaces. The heterointerfaces are assumed to be abrupt, which condition is more or less satisfied in present-day structures grown by molecular beam epitaxy (MBE) or metal organic chemical vapor deposition (MOCVD). It appears from the figure that rectangular potential wells with abrupt walls at the two heterointerfaces exist in the valence and the conduction bands of the lower-gap GaAs sandwiched between two AlGaAs layers. If the thickness d of GaAs along the growth direction is comparable to the de Broglie wavelength, quantum confinement of electrons and holes occurs within the well. The structure shown in Figure 3.4 is called a square QW structure. In this example of QW structure, the GaAs layer is called the well material and the two AlGaAs layers surrounding the well are called the barrier layers.

The calculation of the energy levels in a square potential well is presented in all elementary text books on quantum mechanics (please see Reading List). The energy levels in the conduction and valence bands shown in Figure 3.4 are calculated by using the effective mass theory. To simplify matters, we assume that the lattice constant and the effective mass of electrons in the two materials are equal. Since the motion of electrons is confined along the growth direction (the z-direction), electrons are free along the $(x-y)$ plane. The effective mass Schrödinger equation in the presence of a potential along the z-direction, $V(z)$, may be written as

$$\left[-\frac{\hbar^2}{2m_{||}} \frac{\partial^2}{\partial r^2} - \frac{\hbar^2}{2m_z} \frac{\partial^2}{\partial z^2} + V(z) \right] \psi(\mathbf{r},\mathbf{k}) = E\psi(\mathbf{r},\mathbf{k}) \tag{3.10}$$

where:

m_{II} and m_z are, respectively, the electron effective mass along the layer and along the z-direction

$V(z)$ denotes the energy at the bottom of the conduction band

E is the total energy of the electron

The total wave function is written in product form as

$$\psi(\mathbf{r}, \mathbf{k}) = \chi(\mathbf{r})\phi_n(z) \tag{3.11}$$

Using this in Equation 3.10, we may obtain the following equation for $\phi_n(z)$:

$$\left[-\frac{\hbar^2}{2m_z}\frac{d^2}{dz^2} + V(z)\right]\phi_n(z) = E_n\phi_n(z) \tag{3.12}$$

where n stands for different quantized energy levels.

The solution of Equation 3.12 is greatly simplified under the infinite well approximation, having $V=0$ within the well and $V=\infty$ outside the well. Specifically, for a well of width d,

$$V(z) = 0, \quad |z| < d/2, \tag{3.13}$$
$$= \infty, \quad |z| > d/2$$

Since the electron cannot penetrate into a semi-infinite region of infinite potential energy, the envelope function must satisfy the condition

$$\phi_n\left(\frac{d}{2}\right) = \phi_n\left(\frac{-d}{2}\right) = 0 \tag{3.14}$$

Equations 3.15a and 3.15b are two independent envelope functions satisfying the effective mass equations

$$\phi_n(z) = (2/d)^{1/2}\sin(k_n z) \tag{3.15a}$$

$$\phi_n(z) = (2/d)^{1/2}\cos(k_n z) \tag{3.15b}$$

where

$$k_n = \left(\frac{2m_z E_n}{\hbar^2}\right)^{1/2} \tag{3.16}$$

From the boundary condition, Equation 3.14, k_n must satisfy

$$k_n = \frac{n\pi}{d} \tag{3.17}$$

where n is a positive integer, with even values applying to Equation 3.15a and odd values to Equation 3.15b. The energy eigenvalues are given by

$$E_n = \frac{\hbar^2}{2m_z}\left(\frac{\pi n}{d}\right)^2 ; \quad n = 1,2,3,\ldots \tag{3.18}$$

Using $\hbar = 1.054 \times 10^{-34}$ J s^{-1}, $m_0 = 9.11 \times 10^{-31}$ kg, and 1 eV $= 1.6 \times 10^{-19}$ J, and expressing d in nanometers, we may obtain for the eigenvalue

$$E_n = \left(\frac{m_0}{m_z}\right)\left(\frac{n}{d}\right)^2 \times 0.376\ \text{eV}; \quad n = 1,2,3,\ldots \tag{3.18a}$$

A diagram showing the first few energy levels, or subbands, as they are called, and envelope functions is given in Figure 3.5. The total energy including the translational energy parallel to the interface is now expressed as

$$E(\mathbf{k}, n) = E_n + \frac{\hbar^2 k^2}{2m_{II}} \tag{3.19}$$

where $\mathbf{k}$ is the in-plane (2-D) wave vector. The energy dispersion relation is shown in Figure 3.6. The dispersion relation for holes is expressed identically to Equation 3.19 by using the appropriate hole effective mass.

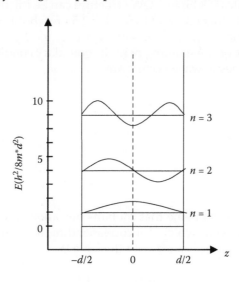

FIGURE 3.5
Energy levels and envelope functions for the infinite square well.

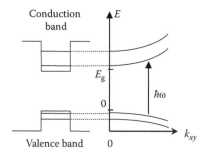

FIGURE 3.6
E–k diagram of two-dimensional electron and hole gas. Two subbands for each of the conduction and valence bands are shown.

> **Example 3.8:** Consider two semiconductors GaAs and $Ga_{0.7}Al_{0.3}As$. The bandgaps of the two are, respectively, 1.43 eV and 1.8 eV. If it is assumed that the bandgap difference is partitioned as $\Delta E_c : \Delta E_v \equiv 60{:}40$, where ΔE_c and ΔE_v are the steps in conduction and valence bands at the heterointerface, then $\Delta E_c = 222$ meV and $\Delta E_v = 148$ meV.

> **Example 3.9:** Let the width of GaAs QW $d = 10$ nm. Then the energy of the first subband in the conduction band is $E_1 = 56.1$ meV and of the second subband $E_2 = 225$ meV, by using $m_z = m_e = 0.067 m_0$.

3.5.2 Density of States in Two Dimensions

The density of states (DOS) in a QW structure can be calculated following a similar approach to that followed in Section 2.5.1 for the bulk (3-D) case. Here we assume that the lengths of the sample are L_x and L_y, respectively, along the *x*- and *y*-directions. Assuming periodic boundary conditions along *x* and *y*, the allowed values of wave vectors are

$$k_x = \frac{2\pi m_x}{L_x}, \quad k_y = \frac{2\pi m_y}{L_y} \tag{3.20}$$

where:
m_x and m_y are integers
$k = (k_x^2 + k_y^2)^{1/2}$

The area in the (k_x, k_y) plane that contains one allowed value of *k* is $(2\pi)^2 / L_x L_y$. The area between two circles corresponding to *E* and *E+dE* is $2\pi k \, dk$. The number of states associated with this area is

$$dN = \frac{2\pi k \, dk}{(2\pi)^2} L_x L_y = \frac{k \, dk}{2\pi} L_x L_y \tag{3.21}$$

The change in energy is as obtained from Equation 3.19

$$dE = \frac{\hbar^2 k dk}{m_{\mathrm{II}}} \tag{3.22}$$

The DOS is therefore

$$\frac{dN}{dE} = \frac{m_{\mathrm{II}}}{2\pi\hbar^2} L_x L_y \tag{3.23}$$

Introducing a factor 2 for spin and normalizing to unit area, one obtains the 2-D density of states function as

$$\rho_{2D} = \frac{m_{\mathrm{II}}}{\pi\hbar^2} \tag{3.24}$$

which is independent of energy. In a QW, each subband contributes this amount to the total DOS. The profile is like a staircase, as shown in Figure 3.7.

Example 3.10: We calculate the DOS for GaAs. Let $m_e = 0.067 m_0$. The DOS is 1.75×10^{36} m^{-2} J^{-1}. If we consider an energy interval $\Delta E = 10$ meV, then the total number of states will be $\rho_{2D} \times \Delta E = 2.8 \times 10^{15}$ m^{-2}.

Example 3.11: The expression for the density of a 2-D electron gas may be obtained easily from the following relation (see Equation 2.15 for bulk semiconductors):

$$n_{2D} = \int_{E_1}^{\infty} \rho_{2D}(E) f(E) dE$$

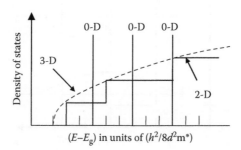

FIGURE 3.7

The staircase-like DOS function for 2-D electrons. The dashed curve represents 3-D DOS function. The DOS for quantum dot (0-D) structures is also included.

where $f(E)$ is the Fermi function. The integration is easily performed to yield

$$n_{2D} = \frac{m_e}{\pi\hbar^2}\int\limits_{E_1}^{\infty}\frac{dE}{1+\exp\left[(E-E_F)/k_BT\right]} = \frac{m_e k_B T}{\pi\hbar^2}\ln\left\{1+\exp\left(E_F - E_1\right)/k_BT\right\} \quad (3.25)$$

Example 3.12: Let $T=0$ K. Since the exponential term $\gg 1$, the ln term in Example 3.11 is $(E_F - E_1)/k_B T$. Let $(E_F - E_1) = 10$ meV. Then $n_{2D} = \rho_{2D}(E_F - E_1)$. Using the value in Example 3.10, $n_{2D} = 2.8 \times 10^{15}$ m^{-2}.

3.5.3 Finite Quantum Well

In a real QW structure, the height of the potential well at the heterointerfaces is never infinitely high. The finite QW is characterized by a potential V_0, shown in Figure 3.8, at the heterointerface.

The potential $V(z)$ may now be defined as

$$\begin{aligned} V(z) &= 0, && |z| < d/2 \\ &= V_0, && |z| > d/2 \end{aligned} \quad (3.26)$$

The envelope functions in the barriers are no longer zero; rather, they penetrate into the two barrier layers due to the finiteness of the barrier height V_0. The solutions of the effective mass equation take the form

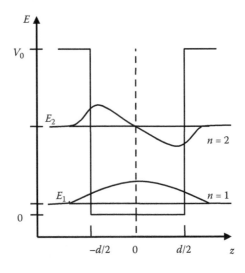

FIGURE 3.8
Potential profile for a finite quantum well structure. The envelope functions penetrate into the barrier materials.

$$\phi_n(z) = C_1 \cos k_w z, \quad |z| < d/2 \tag{3.27a}$$

$$= C_2 \exp\left[-k_b(z - d/2)\right], \quad z > d/2 \tag{3.27b}$$

$$= C_2 \exp\left[k_b(z + d/2)\right], \quad z < -d/2 \tag{3.27c}$$

The solutions given above are even functions. Cs are constants, and the subscripts w and b refer, respectively, to the well and the barrier. The odd solutions can also be written similarly by replacing cos by sin in Equation 3.27a. The exponential functions in the other two are unchanged. The energy eigenvalues for both the cases may be expressed as

$$E_n = \frac{\hbar^2 k_w^2}{2m_w} \quad \text{or} \quad E_n = V_0 - \frac{\hbar^2 k_b^2}{2m_b} \tag{3.28}$$

The values of k_w and k_b are determined by the boundary conditions at $\pm d/2$ that ϕ_n and $(1/m)d\phi_n/dz$ are continuous. For even solutions, one obtains

$$C_1 \cos(k_w d/2) = C_2 \tag{3.29a}$$

$$(k_w/m_w)C_1 \sin(k_w d/2) = k_b C_2/m_b \tag{3.29b}$$

The nontrivial solution of Equation 3.29 is

$$(k_w/m_w)\tan(k_w d/2) = (k_b/m_b) \tag{3.30}$$

From Equation 3.28, one obtains the following by eliminating E_n:

$$\frac{\hbar^2 k_w^2}{2m_w} + \frac{\hbar^2 k_b^2}{2m_b} = V_0 \tag{3.31}$$

Equations 3.30 and 3.31 may be solved graphically or numerically for k_w and k_b, and their values are then substituted in Equation 3.28 to give the energy eigenvalues.

The above simple model for the calculation of energy eigenvalues and envelope functions must be modified in real situations. Here, we point out some of the modifications that should be introduced for the refinement of the theory.

First, the above simple picture of calculation applies to bands having isotropic effective mass. The simple picture is to be modified for semiconductors having anisotropic effective mass (e.g., Si, Ge, and their alloys), where the subband energy depends on the mass along the z-direction, that is, the direction of quantization. The in-plane kinetic energy, however, is governed by the mass in the (x, y) plane. The subband energies therefore depend on the direction of growth of the well. Moreover, the calculation for hole subbands is further complicated due to anisotropic effective mass and the warped nature of the constant energy surfaces. Even by considering isotropic effective mass, one may note from Equation 3.18 that heavy-hole (HH) and light-hole (LH) bands will have different subband energies. Thus, the degeneracy of LH and HH bands at $\mathbf{k}=0$ in the bulk material is lifted in a QW. The in-plane dispersion relation is more complicated. The HH subbands show lighter mass along the plane, and the opposite is the case for LH subbands. The band mixing should be considered even for the simple case of an infinite potential well. In a real situation, the band structure is calculated by using $\mathbf{k}.\mathbf{p}$ perturbation theory, outlined in Chapter 4. It should be noted also that the potential in the above calculation has been taken to be constant even in the finite well problem. In an actual situation, one or both of the semiconductors forming the heterojunctions may be doped, and there is transfer of carriers from one to the other. As a result, space charges develop at the heterojunctions. The shape of the potential becomes fairly complicated then, and the energy eigenvalues and envelope functions of different subbands are totally dependent on the shape of $V(z)$. The carriers distribute themselves among different subbands according to the carrier statistics, and this distribution also governs the shape of the potential $V(z)$. The determination of the energies and eigenfunctions therefore reduces to the solution of the coupled Schrödinger and Poisson equation self-consistently. When the electron density is large, the effect of electron–electron interaction is to be included in determining the self-consistent potentials and eigenvalues. The reader is referred to different suggested texts, papers, and monographs discussing these problems (please see Reading List).

> **Example 3.13:** An approximate value for subband energies may be calculated for a GaAs/Al$_{0.3}$Ga$_{0.7}$As system. As noted earlier, the value of $\Delta E_c = 230$ meV. The penetration depth of the electron wave function into the barrier is $\delta = [\hbar^2/2m_B(\Delta E_c - E_n)]^{1/2} \approx (\hbar^2/2m_B\Delta E_c)^{1/2}$, for the lowest subband, where m_B is the electron effective mass in the barrier. Taking $m_B = 0.093\, m_0$, $\delta = 1.335$ nm. The lowest subband energy in GaAs is written as $E_1 = 5.61(d+2\delta)^{-2}$ eV $= 35$ meV for $d = 10$ nm. The ideal value is 56.1 meV.

The following approximate expression for the subband energies has been obtained by Makino [5] by solving the transcendental Equation 3.30 and a similar equation replacing tan by cot:

$$E_{xn} = \frac{\left[\dfrac{(n+1)\pi}{2} \dfrac{a_x}{d_x + \Delta d_x}\right]^2}{\left[1 + \left\{\dfrac{(n+1)\pi}{2}\right\}^2 b_x \left(\dfrac{\Delta d_x}{d_x + \Delta d_x}\right)^3\right]}, \quad n = 0,1,2,\ldots \qquad (3.32)$$

where:

$$\Delta d_x = a_x / \sqrt{b_x \Delta E_x}$$

$$a_x = 2\hbar / \sqrt{2m_{xw}}$$

$$b_x = m_{xw} / m_{xb}$$

$$x = \text{e or } h$$

Here w and b refer to well and barrier, and ΔE is the band offset. It may easily be established that when ΔE becomes infinitely large, the expression for E_{xn} given by Equation 3.18 is recovered.

> **Example 3.14:** We use Equation 3.32 to calculate the ground subband energy for electrons in a GaAs QW of width $d = 10$ nm, sandwiched between $Al_{0.3}Ga_{0.7}As$ layers. We have $m_w = 0.067m_0$, $m_b = 0.093m_0$, and $\Delta E_c = 230$ meV. We obtain $a_e = 0.603 \times 10^{-18}$, $b_e = 0.72$, and $\Delta d_e = 3.71$ nm. Using the calculated values in Equation 3.32, $E_0 = 28.8$ meV. This value is lower than 35 meV, calculated in Example 3.13. The exact value is 28 meV.

3.5.4 Different Band Alignment

In all our preceding discussions, we have concentrated on the GaAs–AlGaAs QW. This is a member of a type I family, where the electron or the hole gas is formed in the same layer, that is, the middle–lower bandgap material. There are different types of heterointerfaces, and they are divided into four different kinds: type I, type II misaligned, type II staggered, and type III, depending on the band offsets and bandgaps of the two constituent materials. The band diagrams of these different types of heterojunctions are shown in Figure 3.2. The same diagram applies to describing the different types of band alignment in QW structures. We give here a few more examples.

In a type I junction, the potential steps ΔE_C and ΔE_V appear at both the conduction and valence bands in going from a smaller bandgap material A to the larger gap material B. Also, $\Delta E_C + \Delta E_V = E_{gB} - E_{gA} = E_g$. Interfaces formed by III–V compounds either with different group III elements, such as GaAs–AlAs, or with different group V elements, such as GaAs–GaP, belong to type I. The common examples are InAlAs–InGaAs and InGaAsP–InP systems.

Interfaces formed by III–V compounds with both different group III elements and different group V elements belong to type II. If the GaSb–InAs system is considered, a type II misaligned interface is formed. Here the conduction band edge E_{CA} of InAs lies below the valence band edge E_{VB} of GaSb. Electrons are thus attracted to the InAs side and holes to the GaSb side. The result is that a 2-D electron gas (2DEG) is formed in InAs and a 2-D hole gas (2DHG) in GaSb. In this case, $\Delta E_g = |\Delta E_C - \Delta E_V|$.

The type II staggered interface can be obtained by mixing group III elements or group V elements or both to the lower-gap material. Common examples are InGaAs–GaSbAs and AlInAs–InP systems. The type III alignment is observed between zero gap HgTe and CdTe.

3.5.5 Multiple Quantum Wells and Superlattices

When two semiconductors A and B are grown alternately, with A serving as the well and B as the barrier, the structure BAB is a single QW. A repetition of the layer sequence BAB, such as BABAB...AB, leads to a multiple quantum well (MQW) or a superlattice (SL) structure. The main difference between an MQW and an SL structure arises due to the difference in well-to-well coupling. The individual QWs in an MQW are uncoupled and therefore the energy levels remain discrete. On the other hand, in an SL, the widths of the barrier layers are too small, so that electronic wave function in a well tunnels into the adjacent wells. For two coupled wells, the coupling leads to two closely spaced energy levels centered on the original degenerate energy level. The presence of N coupled wells gives rise to N closely spaced energy levels. When the energy separation between the levels is small, the levels are almost continuously distributed and an energy band called the miniband is formed.

The periodic potential in a superlattice is shown in Figure 3.9. The simplest method of obtaining the eigenvalues and envelope functions for this case is to employ the well-known Kronig–Penney model [6] or the tight binding model treated in almost all textbooks on solid state physics. The models consider rectangular potentials as shown above, with different atoms spaced at a distance of a_0, the interatomic distance. Both the models prove the existence of energy bands separated by gaps occurring at wave vectors $k_z = \pm \pi / a_0$.

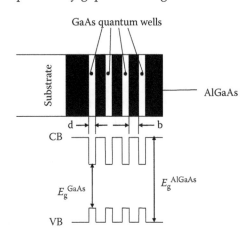

FIGURE 3.9
Schematic diagram of a semiconductor multiple quantum well (MQW) or a superlattice.

In the present situation the periodicity is $a=d+b$, where d and b are, respectively, the well and barrier widths. As the lattice spacing $a \gg a_0$, the structure is rightly called the superlattice. The E–k_z diagram obtained from the analysis is shown in Figure 3.10. The energy bands now exist in the minizone ranging from $-\pi/a$ to $+\pi/a$. Note that $a \gg a_0$, and therefore the bands are termed *minibands*. The band diagram shows gaps, called minigaps, at the edges of the minizone.

Example 3.15: The barrier width in a superlattice should be small so that the wave functions from adjacent wells penetrating into the common barrier will have some overlap. We consider a GaAs–Ga$_{0.7}$Al$_{0.3}$As superlattice, in which the barrier $\Delta E_c = 230$ meV. The penetration length of the electron wave function into the barrier is approximately $\delta = (\hbar^2/2m_B\Delta E_c)^{1/2}$. Taking $m_B = 0.093\ m_0$, where m_B is the effective mass of an electron in the GaAlAs barrier, one obtains $\delta = 1.335$ nm. The wave function in the barrier decreases exponentially, and we assume that it is zero at a distance of 4δ. Under this approximation, a barrier width of $8\delta = 10.68$ nm will not lead to any coupling between adjacent wells. A barrier width less than this will couple the wave functions in two wells and lead to a superlattice structure. A typical value for well (barrier) width is 5 nm (5 nm).

Example 3.16: Let us assume that the QW width $d = 3$ nm and the barrier width $b = 5$ nm. The period of the superlattice is $a = d + b = 8$ nm, and it is quite large compared with the lattice constant of common semiconductors, which is approximately 0.5 nm. The minizone of the superlattice is therefore confined in the range $\pm\pi/a = \pm 3.9 \times 10^8$ m$^{-1} \ll \pm\pi/a_0$ $(= 6.3 \times 10^9$ m$^{-1})$.

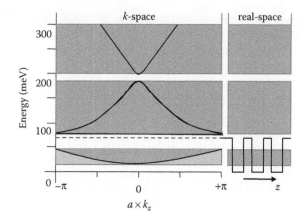

FIGURE 3.10

Miniband (shaded region) and minigap (clear region) in a superlattice, shown in both real and k-space.

3.6 Quantum Wires and Quantum Dots

In Section 3.5, we have seen that confinement of electrons or holes in a narrow 1-D potential well, formed along the z-direction, say, leads to 2-D electron and hole gas. The structure is called a QW. In this section, we shall consider the effect of introducing further confinement to the motion of electrons, and briefly study the relevant subband structures and a few electronic properties. The simple theory with infinite potential barriers will be presented.

3.6.1 Subbands and DOS in Quantum Wires

Assume first that the confinement is extended into yet another dimension, that is, the confinement occurs now along the y- and the z-directions. Extending the earlier arguments, the electron motion is now confined in two directions, y and z; the only direction the electrons are free to move along is the x-direction. We now encounter a one-dimensional electron gas (1DEG). Because of the nature of the potential, the structure supporting 1-D electrons (or holes) is called the quantum wire (QWR) structure. A schematic diagram of the structure is shown in Figure 3.11.

Following the arguments given in Section 3.5.1, the envelope function for electrons in a 2-D confining potential $V(y,z)$ along y and z should satisfy the effective mass equation

$$\left[-\frac{\hbar^2}{2m_e} \left(\frac{d^2}{dy^2} + \frac{d^2}{dz^2} \right) + V(y, z) \right] \phi(y, z) = E_{mn} \phi(y, z) \tag{3.33}$$

The effective mass has been assumed to be isotropic. In the simplest situation of infinite barriers along both the dimensions, $\phi(y, z) = \phi(y)\phi(z)$, and, as noted earlier, both of them may be expressed by sinusoidal functions. The subband energies may be expressed as

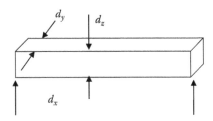

FIGURE 3.11
Schematic structure of a quantum wire.

$$E_{mn} = \frac{\hbar^2}{2m_e}\left[\left(\frac{m\pi}{d_y}\right)^2 + \left(\frac{n\pi}{d_z}\right)^2\right] \tag{3.34}$$

Here ds are the widths of the well in respective directions, and m and n are integers including zero, but both cannot be zero. The dispersion relation is now

$$E(k_x) = E_{mn} + \frac{\hbar^2 k_x^2}{2m_e} \tag{3.35}$$

indicating free motion along the x-direction: the axis of the QWR.

The expression for the DOS function for a 1-D electron may be worked out along the same lines, and it takes the form

$$\rho_{1D} = \frac{2d_x}{\pi\hbar}\left[\frac{m_e}{2(E - E_{mn})}\right]^{1/2} \tag{3.36}$$

where d_x is the length of the sample. It follows therefore that the DOS for 1-D shows singularities at the subband edges.

3.6.2 Quantum Dots

Suppose now that barriers are created along all the three directions of the sample. To visualize this, let us consider a cubic box of GaAs, having a width of a few nanometers in all directions, that is surrounded by AlGaAs, as shown in Figure 3.12.

There exists now a 3-D potential well for both electrons and holes. The motion of the carriers will now be confined in all three dimensions, leading to the zero-dimensional (0-D) carrier systems.

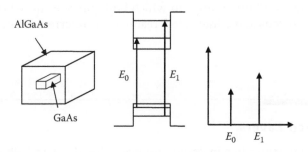

FIGURE 3.12
Schematic diagram of a quantum box and the DOS function.

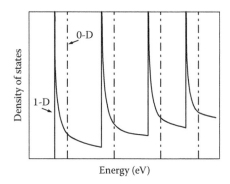

Energy (eV)

FIGURE 3.13
Density of states function of a quantum wire (1-D) and a quantum dot (0-D) structure.

Such a structure is referred to as a quantum box, or a QD, or a superatom if the shape is spherical.

Considering the barrier height to be infinite, the energy levels can easily be expressed as

$$E_{n_x,n_y,n_z} = \frac{\hbar^2 \pi^2}{2m_e} \left(\frac{n_x^2}{d_x^2} + \frac{n_y^2}{d_y^2} + \frac{n_z^2}{d_z^2} \right) \tag{3.37}$$

where, as usual, ds are the dimensions in the respective directions. For a spherical dot, the subband energies may be expressed in terms of three parameters describing levels for hydrogen atoms. The energy levels are now discrete. The DOS in the ideal case consists of delta functions centered on each discrete level. The nature of DOS for both QWR and QD is shown in Figure 3.13.

For a spherical dot, the situation is different. In this case, there exists a critical radius that is different for holes and electrons. This difference gives rise to three different situations: (i) a range of radius R in which no particles are quantized, (ii) a range in which an electron possesses a bound state while a hole does not, and (iii) a range in which both particles have bound states. For detailed discussions, interested readers are referred to the suggested readings.

3.7 Strained Layers

Introduction of strain in a material changes its bandgap and alters the effective mass of carriers, which are used to obtain better electronic and photonic devices. The strained-layer epitaxy is related to the growth of an epitaxial layer with a slight degree of lattice mismatch from and on a thick substrate. The lattice mismatch produces a strain on the grown layer and alters its physical properties. However, under controlled growth, the crystallinity and long-range order are maintained throughout the structure, including the grown overlayer. As the crystal morphology is maintained throughout, the growth is known as pseudomorphic growth.

3.7.1 Pseudomorphic Growth

When an epitaxial layer of lattice constant different from that of the substrate is grown, misfit dislocations or voids appear in the interface. After the growth of a few monolayers containing the defects, the epitaxial layer grows freely with its native lattice constant. However, when the mismatch between the lattice constants of the epilayer and the substrate is sufficiently small, the epitaxial layer may grow on the substrate with a lattice constant equal to that of the substrate. The epilayer is thus under strain, and the misfit dislocations are completely absent. The condition to be maintained for such pseudomorphic growth is that the epilayer thickness must be below a critical layer thickness. If the layer thickness exceeds this critical value, misfit dislocations will appear. The reason for such behavior lies in the competition between strain energy and energy of formation of misfit dislocation. Below critical thickness, strain energy is lower than that for misfit dislocation, and therefore pseudomorphic growth is preferred. The strain energy increases linearly with increasing thickness, but the formation energy for misfit dislocation increases superlinearly with increasing layer thickness and then saturates. At critical thickness, the two energies are equal. With further increase in thickness, the strain energy becomes larger, and therefore growth with dislocations takes place.

The growth of a lattice-mismatched epilayer on a substrate is illustrated in Figure 3.14. In the top part of Figure 3.14a, the epitaxial layer having a lattice constant larger than that of the substrate is shown isolated. When the heterojunction is formed, as shown in the bottom part of Figure 3.14a, the film is compressively strained. An example is the growth of $In_{0.3}Ga_{0.7}As$ on GaAs. In the top part of Figure 3.14b, the epilayer and the substrate are shown isolated; in this case, however, the lattice constant of the epilayer is smaller. When the epilayer forms on the substrate, as in the bottom part of Figure 3.14b, it experiences tensile strain. This situation arises for the growth of GaAsP on GaAs.

3.7.2 Expression for Critical Thickness

Several authors have given expressions for the critical layer thickness. The following is the expression given by Voisin [7]:

$$h_c = \frac{1-v/4}{4\pi(1+v)} b\varepsilon^{-1} \left[\ln(h_c/b) + \theta \right] \tag{3.38}$$

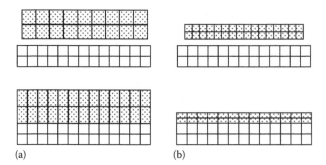

FIGURE 3.14
Schematic diagram depicting growth of lattice-mismatched epilayer on a substrate. Top: epilayer and substrate shown isolated; bottom: growth of strained layer with in-plane lattice matching. (a) Compressively strained and (b) tensile strained growth. The effect of perpendicular strain is also illustrated.

where:

    ν is the Poisson ratio
    b is the dislocation Burgers vector ($\approx$4 Å)
    ε is the in-plane strain (to be defined later)
    θ is a constant (~1)

The calculated value of h_c is 9 nm for ε = 1%.

When a film of lattice constant a_f is grown on a substrate of lattice constant a_s, the strain between the two materials is defined as

$$\varepsilon = \frac{a_s - a_f}{a_s} \tag{3.39}$$

A more simplified form of the critical layer thickness is given in terms of this strain, as follows:

$$h_c \approx \frac{a_s}{2|\varepsilon|} \tag{3.40}$$

> **Example 3.17:** We estimate the critical layer thickness for growth of $In_{0.2}Ga_{0.8}As$ layer on GaAs substrate. The lattice constant of the alloy is given by Vegard's law as $a(InGaAs) = 0.2a(InAs) + 0.8a(GaAs) = 0.2(6.058) + 0.8(5.653) = 5.734$ Å. The strain is ε = (5.653−5.743)/5.653 = −0.016. The critical layer thickness is 176.7 Å = 17.7 nm.

3.7.3 Strain-Symmetric Structures and Virtual Substrates

Pseudomorphically grown strained layers are used for tailoring the optoelectronic properties of the semiconductors. However, epitaxial layers exceeding critical layer thickness are also grown to provide virtual substrates for different layers in a multilayered structure. In this case, the dislocations are confined near the interface between the substrate and the thick epilayer. This kind of virtual substrate is needed to grow strain-symmetric superlattice structures, in which one layer is tensile strained and the next layer is compressively strained. The buffer or virtual substrate has a lattice constant intermediate between the lattice constants of the constituent materials A and B making the superlattice.

Example 3.18: We consider growth of alternate layers of $In_{0.4}Ga_{0.6}As$ (A) and $In_{0.65}Ga_{0.35}As$ (B) on InP substrate. The lattice constants are $a(InP) = 5.8686$ Å, $a(InAs) = 6.058$ Å, and $a(GaAs) = 5.653$ Å. Using Vegard's law, $a(A) = 5.815$ Å and $\varepsilon(A) = +0.0092$, $a(B) = 5.9163$ Å, and $\varepsilon(B) = -0.0081$. For strain compensation, the thickness should be $d_A/d_B = \varepsilon(B)/\varepsilon(A) = 1.136$. Thus, if $d_A = 5$ nm, then $d_B = 5.68$ nm.

3.7.4 Strain, Stress, and Elastic Constants

In most situations, the stress, that is, the force per unit area on a face of a crystal, is described by a stress tensor, having elements σ_{ij}, $i, j = x, y, z$. The non-diagonal elements, σ_{ij}, $i \neq j$, represent shear components and cause the crystal to rotate. The presence of an equal and opposite component, the shear stress, deforms a cubic lattice into a nonrectangular shape. As a result, the crystal axes become nonorthogonal. In typical semiconductors, this type of deformation is rare, and we may safely assume that $\sigma_{ij} = 0$, for $i \neq j$. The normal components of stress expand or contract the crystal along the crystal axes, but the deformed lattice remains rectangular. The six faces of the cubic crystal are now acted on by three normal forces, σ_{11}, σ_{22}, and σ_{33}, which we denote from now on by σ_1, σ_2, and σ_3. If the forces are directed outwards, they are reckoned as positive.

Due to the application of stress, the crystal is deformed or strained. We may now introduce a strain tensor, which is diagonal, having three elements: ε_1, ε_2, and ε_3. If $\varepsilon_i > 0$, the crystal is elongated along the ith axis, and if $\varepsilon_i < 0$, the crystal is compressed along the ith axis.

Example 3.19: Consider growth of $In_{0.2}Ga_{0.8}As$ on GaAs. Using previous parameters, $a(InGaAs) = 5.734$ Å and $a(GaAs) = 5.653$ Å. Therefore $\varepsilon_i < 0$ and the epitaxial layer is compressed.

In an isotropic medium, stress and strain are related by Hooke's law as $\sigma = C\varepsilon$, where C is the Young's (or rigidity) modulus. However, in a crystal, we have to consider the tensor nature of both stress and strain. There are six independent components of stress, which are related to the strain components by Hooke's law as

$$[\sigma] = [C] \cdot [\varepsilon] \tag{3.41}$$

The components C_{ij} of the tensor C are known as the elastic stiffness coefficients or elastic moduli. In general, there are 36 such elements, but for cubic crystals the numbers reduce drastically, so that one may write

$$\begin{bmatrix} \sigma_1 \\ \sigma_2 \\ \sigma_3 \end{bmatrix} = \begin{bmatrix} C_{11} & C_{12} & C_{13} \\ C_{12} & C_{11} & C_{12} \\ C_{12} & C_{12} & C_{11} \end{bmatrix} \begin{bmatrix} \varepsilon_1 \\ \varepsilon_2 \\ \varepsilon_3 \end{bmatrix} \tag{3.42}$$

The above elastic moduli tensor applies to a crystal with cubic symmetry, having all off-diagonal components equal as well as having all diagonal elements equal. There are only two elastic moduli to consider: C_{11} and C_{12}. For crystals with less symmetry, there are more C_{ij} components. If shear is to be considered, another component, C_{44}, is to be included in the elastic moduli tensor even for cubic crystals. In common semiconductors, $C_{11} > C_{12}$, both the moduli being positive and usually described in units of 10^{11} dynes cm^{-2}.

Consider now the growth of a film, the natural lattice constant of which is denoted by a_f, on a substrate of lattice constant a_s. Assume also that the directions 1,2,3 are coincident with the x-, y-, and z-directions, respectively, and the growth is along the z-direction. As the film adjusts its lattice constant with that of the substrate, it is easy to visualize that the film is under biaxial stress, with stresses along all the four x and y faces, but no stress is applied to the z faces. When $a_f < a_s$, the film or epilayer will be under biaxial tensile strain, and the stress components are directed outward from the four x and y faces ($\sigma_1 = \sigma_2 > 0$). Similarly, for $a_f > a_s$, the film will experience biaxial compressive strain, and the stress components will be directed inwardly through x and y faces ($\sigma_1 = \sigma_2 < 0$). By symmetry, the stress and strain components along the x- and y-directions must be equal, and this sets $\sigma_1 = \sigma_2$ and $\varepsilon_1 = \varepsilon_2$. With $\sigma_3 = 0$, one obtains from the first and third equations in Equation 3.42

$$\sigma_1 = C_{11}\varepsilon_1 + C_{12}\varepsilon_1 + C_{12}\varepsilon_3 \tag{3.43a}$$

$$0 = C_{12}\varepsilon_1 + C_{12}\varepsilon_1 + C_{11}\varepsilon_3 \tag{3.43b}$$

The strain along the z-direction is immediately obtained as

$$\varepsilon_3 = -(2C_{12} / C_{11})\varepsilon_1 \tag{3.44}$$

which implies that the deformation along z will be opposite to that along the x- and y-directions, since both C_{11} and C_{12} are positive quantities. Using Equation 3.43a, we may write

$$\sigma_1 = C_{11}\varepsilon_1 \left[1 + C_{12}/C_{11} - 2(C_{12}/C_{11})^2 \right] \qquad (3.45)$$

Note that if $C_{12} = 0$, then $\sigma_1 = C_{11}\varepsilon_1$. This implies that the strain perpendicular to the stress plane, that is, along the z-direction, is zero, and one gets back the scalar relation.

Now consider that the crystal is under uniform stress applied equally to all the six faces of the cubic crystal. This amounts to a uniform pressure change, dP, directed inward. We can then set $\sigma_1 = \sigma_2 = \sigma_3 = -dP$. Adding up all the three equations resulting from Equation 3.42, we find

$$-3dP = (C_{11} + 2C_{12})(\varepsilon_1 + \varepsilon_2 + \varepsilon_3) \qquad (3.46)$$

The volume of the crystal is now $V + dV = a^3(1 + \varepsilon_1)(1 + \varepsilon_2)(1 + \varepsilon_3)$, where the unstrained volume is $V = a^3$ and a is the lattice constant. The fractional change in the volume of the crystal, neglecting second-order terms such as $\varepsilon_1\varepsilon_2$ and so on, is given by

$$dV/V \approx \varepsilon_1 + \varepsilon_2 + \varepsilon_3 \qquad (3.47)$$

3.7.5 Band Structure Modification by Strain

The presence of biaxial strain lifts the degeneracy of the HH and LH bands. The expressions showing the shift of the HH and LH bands are given below. The method of calculating the shifts will be outlined in Chapter 4. The shifts are given by

$$\Delta E_{hh}(\varepsilon) = \left[2a\left(\frac{C_{11} - C_{12}}{C_{11}} \right) + b\left(\frac{C_{11} + 2C_{12}}{C_{11}} \right) \right]\varepsilon \qquad (3.48a)$$

$$\Delta E_{lh}(\varepsilon) = \left[2a\left(\frac{C_{11} - C_{12}}{C_{11}} \right) - b\left(\frac{C_{11} + 2C_{12}}{C_{11}} \right) \right]\varepsilon \qquad (3.49b)$$

where a and b are deformation potentials. The shifts given in Equations 3.48a and 3.48b are measured with respect to the conduction band, which is assumed to have a fixed energy.

It follows easily from Equations 3.48a and 3.48b that, depending on the sign of strain, we can have a situation in which either the HH band is on the top of the valence band or the LH band is at the top. The former arises for a compressively strained layer, while the latter is found for a tensile-strained layer. This situation is shown in Figure 3.15.

An important consequence of the splitting between the HH and LH bands is a dramatic decrease of the DOS mass as a function of stress. This

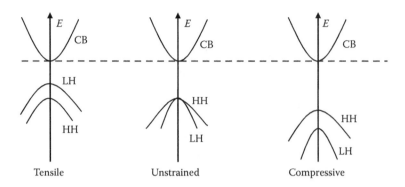

FIGURE 3.15
Effect of compressive and tensile strains on the heavy-hole (HH) and light-hole (LH) band edges.

has an impact on the reduction of the threshold current density in the laser.

> **Example 3.20:** We calculate the HH–LH splitting for $In_{0.2}Ga_{0.8}As$ layer grown on a (001) GaAs substrate. The deformation potential is $b=-2.0$ eV. The moduli are $C_{11}=11.5\times10^{11}$ dynes cm^{-2} and $C_{12}=5.5\times10^{11}$ dynes cm^{-2}. The splitting produced is $\Delta E_{hh}(\varepsilon)-\Delta E_{lh}(\varepsilon) = 2b[(C_{11} + 2C_{12})/C_{11}]\varepsilon = -5.91\varepsilon$. The strain is $a(GaAs) - a(InGaAs)/a(GaAs)=5.653 - 5.734/5.653=-0.014$. The splitting is $\Delta E_{hh}(\varepsilon)-\Delta E_{lh}(\varepsilon) = 0.085$ eV.

PROBLEMS

3.1. Find the wavelength range over which GaAlAs alloy can be used as a laser.

3.2. For a DH laser, you need type I alignment. Find the maximum value of x for GaAlAs that ensures that $E_g\Gamma$ in GaAlAs will be at least 0.2 eV below E_{gX} or E_{gL}.

3.3. Show that $In_{0.52}Ga_{0.48}As$ is lattice matched to $In_{0.53}Ga_{0.47}As$.

3.4. Find the composition of $In_{1-x}Ga_xAs_yP_{1-y}$ lattice matched to InP for emission at 1.3 and 1.55 μm.

3.5. Show that the conduction band offset in the GaAs/AlGaAs heterojunction increases with x, attains a maximum, and then decreases. Use the 65:35 ratio for calculating the band offsets.

3.6. Obtain the expression for the mean free path of a 2DEG at 0 K in terms of the sheet carrier density and the mobility, and state the condition for quantization. Calculate the value of the mean free path if $n_s=10^{11}$ cm^{-2} and $\mu=10^6$ m^2 (V·s)$^{-1}$.

3.7. The quantized energy levels in a QW of infinite barrier are broadened due to scattering. Assume that the scattering times in both the first and second subbands are 1 ps. Calculate the well width for which the two broadened energy levels will touch each other. Assume that the lineshape for each level is rectangular.

3.8. Show that the wave function in the barrier of a rectangular QW of width L may be expressed as

$$\phi(z) = \left[\frac{1}{\alpha} + \frac{L/2}{\cos^2(kL/2)} + \frac{m_w \alpha}{m_b k^2} \right]^{-1/2} \exp\left[\alpha\left(\frac{L}{2} \pm z\right) \right]$$

where:
$k^2 = 2m_w E/\hbar^2$
$\alpha 2 = 2m_b(\Delta E - E_1)/\hbar^2$
w and b refer to well and barrier, respectively

3.9. Calculate the values of the Fermi level and the areal electron densities in different subbands in a GaAs QW at 0 K and 300 K, where the sheet carrier density is 10^{12} cm^{-2}.

3.10. Prove that the DOS at the subband energies E_n equals that in the bulk, that is, $\rho_{3D}(E_n)L = \rho_{2D}(L)$.

3.11. Using the expressions for the subband energies for the infinite barriers, prove that $\rho_{2D}(E_n)L \rightarrow \rho_{3D}$, as the well width increases to a very large value.

3.12. The screening constant in a 2-D system is expressed as $q_S = (e^2/2\varepsilon)$ (dN_S/dE_F). Obtain an expression for the screening constant at very low temperature.

3.13. The lowest subband energy for a QW of width L is approximately given by [8]

$$E_1 = \frac{\Delta E_c}{2} - \frac{\pi^2 \hbar^2}{4m_e L^2}\left[\left(1 + \frac{32\Delta E_c^2 L^4 m_e^2}{\pi^6 \hbar^4}\right)^{1/2} - 1\right]$$

Calculate E_1 versus the well width in a GaAs–Ga0.7Al0.3As QW and compare the results with exact values. Assume $\Delta E_c/\Delta E_v = 60/40$.

3.14. Give plots showing the nature of variation of the two sides in Equation 3.24, and explain how you would obtain the values of subband energies from the graphs.

3.15. In Example 3.13, the subband energy E_1 is neglected in calculating δ, and the subband energy is calculated in this approximation. Use the value of 56.1 meV for a 10 nm wide QW of infinite barrier to calculate E_1, and then use this value to recalculate δ and E_1. Repeat this iteration. Compare your final value with the value calculated by using Equation 3.32.

3.16. Work out the steps to obtain Equation 3.36: the expression for the DOS function of 1DEG.

3.17. The DOS in 0-D is $\rho_{0D} = \Sigma_{m,n,l}\, \delta(E - E_{mnl})$. Using m, n, l as continuous variables, and the expressions for subband energies for infinite barriers, show that the expressions for the DOS functions for 1-D, 2-D, and 3-D can successively be obtained.

3.18. Show that the density of electrons in a 1-D wire under nondegenerate condition may be expressed as $n_{1D} = (2m_e k_B T / \pi\hbar^2)^{1/2} \exp[(E_{F-}E_i)/k_B T]$.

3.19. Show that at low temperature, when the Fermi function is box-like, the 1-D carrier density is proportional to $(E_F - E_i)^{1/2}$. Obtain the complete expression for n_{1D}.

3.20. Solve the effective mass Schrödinger equation for a quantum box by using separation of variables. Hence show that the energy states are given by Equation 3.30. The barrier potentials are infinite.

3.21. The Luttinger parameters for GaAs (InAs) are $\gamma_1 = 6.8(20.4)$, $\gamma_2\ 1.9(8.3)$. Calculate the lowest subband energies for HH and LH in an In0.53Ga0.47As layer, assuming that the barrier height is infinite.

3.22. Consider the problem of confinement in a spherical dot of radius R. The potential is a square well of height ΔE_c for $r > R$ and 0 for $r < R$. Consider the radial part of wave function and zero angular momentum (please see Reading List). Show that there exists a critical radius $R_c^2 > \pi^2\hbar^2/8m_e\Delta E_c$ for which an electron possesses a bound state.

3.23. Calculate the width of the QW of $In_{0.2}Ga_{0.8}As$ grown on GaAs that will make the HH and LH degenerate. Assume infinite barrier height, parameter values given in Example 3.20, and the following hole masses for GaAs(InAs): $m_{hh} = 0.50(0.40)\ m_0$, $m_{lh} = 0.087(0.026)\ m_0$.

3.24. The strain-balanced condition is expressed by some workers as

$$\sum_{i=1}^{n} \frac{A_i L_i \varepsilon^i}{a_i} = 0, \quad \text{with } A_i = C_{11}^{(i)} + C_{12}^i - \frac{2C_{12}^{(i)2}}{C_{11}^i}$$

where:
  L_i is the thickness of the ith layer
  ε_i is the strain
  a_i is the lattice constant of the ith layer

The $In_zGa_{1-z}As$ well thickness is 100 Å and GaAs barrier thickness is 90 Å. Find the composition z for the well. Use the linear interpolation technique with the values of the lattice constants and elastic constants given in Appendix II.

3.25. Calculate the HH–LH splitting in $GaAs_{0.7}P_{0.3}$ grown on GaAs by using the following parameters for GaAs(GaP): $a = 5.6533(5.4505$ Å$)$, $C_{11} = 11.5(14.05) \times 10^{11}$ dynes cm^{-2}, $C_{12} = 5.5(6.2) \times 10^{11}$ dynes cm^{-2}, $b = -1.7(-1.8)$ eV.

Reading List

Ando, T., A. B. Fowler, and F. Stern, Electronic properties of two-dimensional systems, *Rev. Mod. Phys.* 54, 437–672, 1982.

Balkanski, M. and R. F. Wallis, *Semiconductor Physics and Applications*. Oxford University Press, Oxford, 2000.

Bastard, G., *Wave Mechanics Applied to Semiconductor Heterostructures*. Les Editions de Physique, Les Ulis, 1988.

Basu, P. K., *Theory of Optical Processes in Semiconductors: Bulk and Microstructures*. Clarendon Press, Oxford, 2003.

Chuang, S. L., *Physics of Optoelectronic Devices*. Wiley, New York, 1995.

Coldren, L. and S. W. Corzine, *Diode Lasers and Photonic Integrated Circuits*. Wiley, New York, 1995.

Datta, S., *Quantum Phenomena*. Addison-Wesley, Reading, MA, 1989.

Davies, J. H., *The Physics of Low-Dimensional Semiconductors: An Introduction*. Cambridge University Press, Cambridge, 1998.

Harrison, P., *Quantum Wells, Wires and Dots: Theoretical and Computational Physics*. Wiley, Chichester, 2000.

Manasreh, O., *Semiconductor Heterojunctions and Nanostructures*. McGraw Hill, New York, 2005.

Mitin, V., M. A. Strocio, and V. A. Kochelap, *Quantum Heterostructures*. Wiley, New York, 1999.

Ridley, B. K., *Quantum Processes in Semiconductors*, 5th edition. Clarendon Press, Oxford, 2000.

Roblin, P. and H. Rohdin, *High-Speed Heterostructure Devices*. Cambridge University Press, Cambridge, 2002.

Shur, M. S., *Physics of Semiconductor Devices*. Prentice-Hall, Englewood Cliffs, 1990.

Singh, J., *Electronic and Optoelectronic Properties of Semiconductor Structures*. Cambridge University Press, Cambridge, New York, 2003.

Weisbuch, C. and B. Vinter, *Quantum Semiconductor Structures*. Academic, San Diego, 1991.

References

1. Vurgaftman, I. and J. R. Meyer, Band parameters for III–V compound semiconductors and their alloys, *J. Appl. Phys.* 89, 5815–5875, 2001.
2. D'Costa, V. R., Y.-Y. Fang, J. Tolle, J. Kouvetakis, and J. Menéndez, Tunable optical gap at a fixed lattice constant in Group-IV semiconductor alloys, *Phys. Rev. Lett.* 102, 107403(1–4), 2009.
3. Van de Walle, C. G., Band lineups and deformation potentials in the model-solid theory, *Phys. Rev. B* 39, 1871–1883, 1989.
4. Hu, J., G. Xu, J. A. H. Stotz, S. P. Watkins, A. E. Curzon, and M. L. W. Thewalt, Type II photoluminescence and conduction band offsets of GaAsSb/InGaAs and GaAsSb/InP heterostructures grown by metalorganic vapor phase epitaxy, *Appl. Phys. Lett.* 73, 2799–2801, 1998.
5. Makino, T., Analytical formulas for the optical gain of quantum wells, *IEEE J. Quantum Electron.* 32, 493–501, 1996.
6. Mukherji, D. and B. R. Nag, Band structure of semiconductor superlattices, *Phys. Rev. B* 12, 4338–4342, 1975.
7. Voisin, P., Heterostructures of lattice mismatched semiconductors: Fundamental aspects and device perspectives. In *Quantum Wells and Superlattices in Optoelectronic Devices and Integrated Optics*, SPIE Proceedings, Vol. 861, p. 88, SPIE, Bellingham, WA, 1988.
8. Singh, J., Analytical closed form expressions for the effective band edges in shallow quantum wells, *Appl. Phys. Lett.* 64, 2694, 1994.

4

Band Structures

4.1 Introduction

The formation of bands in a periodic crystalline solid was discussed qualitatively in Chapter 2. It was also mentioned that the calculation of the
band structure in semiconductors involves the solution of the one-electron
Schrödinger equation in the presence of a periodic potential. Most textbooks on solid-state physics and semiconductors discuss the Kronig–Penney
model, the nearly free-electron model, or the tight binding model to illustrate the formation of bands in crystals. The band structure is described by
the E–$\mathbf{k}$ diagram, and a brief introduction to the diagram has been given in
Chapter 2.

The quantitative understanding of the electronic and optical processes and
analysis of experimental data depend on the accurate E–$\mathbf{k}$ relationship of the
semiconductor under study. Though powerful techniques, such as the pseudopotential method, exist to predict the band structure, there are approximate methods that can satisfactorily explain most of the important electronic
and optoelectronic behavior of semiconductors. The most widely used
method, the $\mathbf{k.p}$ perturbation method, will be described in this chapter in
relation to the band structure in bulk semiconductors first. The modification
for quantum wells (QWs) and strained layers will be described thereafter.

4.2 Band Theory: Bloch Functions

A solid consists of many atoms and electrons. The total energy of the system
is therefore the sum of the kinetic energies of all the nuclei and electrons and
all the potential energies, which arise due to nuclear forces, interaction of
electrons in the field of nuclei due to electron–electron interactions, and the
magnetic energy associated with the spin and the orbit. The total Hamiltonian
of the system may be constructed accordingly. It follows naturally, therefore,
that the solution of the resultant Schrödinger equation is a formidable task.

The problem is bypassed by introducing several approximations. Since the motion of nuclei is slow, the electrons instantaneously adjust their motion to that of the ions. The total wave function is then written in terms of a wave function for ions, $\phi(\mathbf{R})$, and that for all electrons, $\psi(\mathbf{r},\mathbf{R})$, instantaneously dependent on all ionic positions $\mathbf{R}$. An approximation, known as an *adiabatic approximation*, is introduced to decouple the Schrödinger equation into a purely ionic and a purely electronic equation. As a result, the following two expressions are obtained:

$$H_L \phi(\mathbf{R}) = E_L \phi(\mathbf{R}) \tag{4.1}$$

and

$$H_e \psi(\mathbf{r},\mathbf{R}) = E_e \psi(\mathbf{r},\mathbf{R}) \tag{4.2}$$

In Equations 4.1 and 4.2, the subscripts L and e denote, respectively, the ionic (lattice) and electronic quantities, and $\mathbf{r}$ denotes the electronic coordinates [1,2].

The electron potential energy is due to electron–electron and electron–ion interactions. If a suitable average is found for the potential for electron–electron interaction, a constant repulsive contribution can be added to the electron energy, and then each electron becomes independent. The one-electron Schrödinger equation then takes the form

$$H_{ei} \psi_i(\mathbf{r}_i,\mathbf{R}) = E_{ei} \psi_i(\mathbf{r}_i,\mathbf{R}) \tag{4.3}$$

where

$$H_{ei} = \frac{p_i^2}{2m_0} + \sum_i V(\mathbf{r}_i,\mathbf{R}_i) \tag{4.4}$$

and p_i is the momentum of the ith electron. The Hamiltonian still depends on the fluctuating position of the ion. In the next approximation, the ions are assumed to lie in their equilibrium position, and the effect of ionic vibration is taken as a perturbation. Thus, the problem reduces to solving the following equation:

$$\frac{p^2}{2m_0} + \sum_i V(\mathbf{r} - \mathbf{R}_{l0})\psi(\mathbf{r}) = E\psi(\mathbf{r}) \tag{4.5}$$

The ionic potential V is periodic, and the eigenfunctions are Bloch functions expressed as

$$\psi_{nk}(\mathbf{r}) = U_{nk}(\mathbf{r})\exp(j\mathbf{k}.\mathbf{r}) \tag{4.6}$$

where the cell periodic part $U(\mathbf{r})$ obeys the relation

$$U_{nk}(\mathbf{r}+\mathbf{R}) = U_{nk}(\mathbf{r}) \tag{4.7}$$

In the above equations, $\mathbf{R}$ is a vector of a Bravais lattice, n denotes the band index, and k is a wave vector of the electron in the first Brillouin zone. From Equations 4.6 and 4.7,

$$\psi_{nk}(\mathbf{r}+\mathbf{R}) = \psi_{nk}(\mathbf{r})\exp(j\mathbf{k}.\mathbf{R})$$

The Bloch functions are eigenfunctions of the one-electron Schrödinger equation, and therefore they are orthogonal to one another. Thus,

$$\int \psi_{n'k'}\psi_{nk}d^3r = \delta_{n'n}\delta_{k'k} \tag{4.8}$$

The wave functions are also normalized over the volume V of the crystal, and therefore

$$\psi_{nk} = V^{-1/2}U_{nk}(\mathbf{r})\exp(j\mathbf{k}.\mathbf{r}) \tag{4.9}$$

4.3 The k.p Perturbation Theory Neglecting Spin

A complete solution of the Schrödinger equation is needed to obtain complete knowledge of the band structure of a semiconductor and an accurate description of the *E–k* dispersion relation. This is a rather difficult task, since the form of the periodic potential $V(\mathbf{r})$ must be accurately specified. Fortunately, in most semiconductors, electrons and holes lie within a fraction of an electronvolt from the respective band edges, and a knowledge of the behavior of electrons and holes is sufficient to give correct electronic and optoelectronic properties. Thus, if the wave functions and energies of the carriers are known at the band extrema, perturbation methods may be applied to find out the wave functions and energies at other points in the Brillouin zone, leading to a knowledge of the *E–k* relationship. The method, known as **k.p** perturbation theory, is most widely used in the study of transport and optical processes in common semiconductors [3–9].

The **k.p** perturbation theory is based on the fact that the cell periodic parts U_k of the electrons, for any value of k but for different bands, form a complete set. Let us consider the wave functions for electrons having a value $\mathbf{k}$ near the minima in the nth band. For simplicity, we assume that the minima are located at $\mathbf{k}=0$, though the theory is applicable also when the minima are located at $\mathbf{k}=\mathbf{k}_0$. The wave function is given by

$$\psi = U_{nk}(\mathbf{r})\exp(j\mathbf{k}.\mathbf{r}) = \left[\sum_m c_m U_{m0}(\mathbf{r})\right]\exp(j\mathbf{k}.\mathbf{r}) \tag{4.10}$$

since U_{m0} forms a complete orthonormal set. Using this form of ψ in the Schrödinger equation, one obtains

$$\left[-\frac{\hbar^2}{2m_0}\nabla^2 + \frac{\hbar^2}{m_0}\mathbf{k}.\mathbf{p} + \frac{\hbar^2 k^2}{2m_0} + V(\mathbf{r})\right]U_{nk}(\mathbf{r}) = E_n(\mathbf{k})U_{nk}(\mathbf{r}) \tag{4.11}$$

However, U_{m0} is the wave function for $\mathbf{k}=0$ in the nth band satisfying the equation

$$\left[-\frac{\hbar^2}{2m_0}\nabla^2 + V(\mathbf{r})\right]U_{m0}(\mathbf{r}) = E_m(0)U_{m0}(\mathbf{r}) \tag{4.12}$$

We replace $U_{nk}(\mathbf{r})$ in Equation 4.11 by its expanded form given in Equation 4.10 and use Equation 4.12 to obtain

$$\sum_m c_m\left[E_m(0) + \frac{\hbar^2}{2m_0}k^2 + \frac{\hbar^2}{m_0}\mathbf{k}.\mathbf{p}\right]U_{m0}(\mathbf{r}) = \sum_m c_m E_n(\mathbf{k})U_{m0}(\mathbf{r}) \tag{4.13}$$

By multiplying both sides of Equation 4.13 by $U_{l0}^*(\mathbf{r})$ and integrating over a volume of a unit cell (V_c), the following set of linear homogeneous equations is obtained:

$$c_l\left[E_n(\mathbf{k}) - E_l(0) - \frac{\hbar^2}{2m_0}k^2\right] - \sum_m c_m\frac{\hbar}{m_0}(\mathbf{k}.\mathbf{p}_{lm}) = 0 \tag{4.14}$$

where:

$$\mathbf{p}_{lm} = \int_{V_c} U_{l0}^*(\mathbf{r})\mathbf{p}U_{m0}(\mathbf{r})d^3r \tag{4.15}$$

By giving l successive integer values, one obtains the full set of equations.

In the general case, the set of equations has a nontrivial solution if the determinant formed by the coefficients c_l is zero. This condition gives the energy eigenvalues $E_n(\mathbf{k})$ in terms of the quantities $E_m(0)$ and $\mathbf{p}_{lm}$. The relative values of the expansion coefficients c_m are then obtained by using the values of $E_n(\mathbf{k})$. The absolute values of c_m are obtained by imposing a normalization condition on ψ. The accuracy of the calculation is greater if the number of such coefficients is increased. For practical reasons, however, we need to limit our consideration to a few bands. The bands of greatest interest in common semiconductors are conduction (C), heavy-hole (HH), light-hole (LH), and split-off (SO) bands. Each of these four bands has two spin components, so that we need to consider eight bands altogether. Depending on the problem at hand and the degree of accuracy required, we may use some or all of these eight bands. In Sections 4.3.1 and 4.3.2 we present the results for two cases: first, when the number of bands is restricted to only one, and second, considering all four bands.

4.3.1 Single Electron Band

Let us assume that U_{nk} is determined mostly by U_{n0}, and the contributions from other bands are negligible. In other words, we assume that $c_m (m \neq n) \ll c_n$. Then,

$$\psi = \left[c_n U_{n0}(\mathbf{r}) + \sum_m c_m U_{m0}(\mathbf{r}) \right] \exp(j\mathbf{k}.\mathbf{r}). \tag{4.16}$$

Since ψ is normalized, $\sum_m |c_m|^2 = 1$; but we have assumed $c_m \ll c_n$. It follows therefore that $c \approx 1$.

To solve Equation 3.16, stationary perturbation theory is applied. First, by neglecting $c_m (m \neq n)$ in comparison to c_n in the nth equation, the following equation results:

$$E_n(\mathbf{k}) = E_n(0) + \frac{\hbar^2 k^2}{2m_0} + \frac{\hbar}{m_0} \mathbf{k}.\mathbf{p}_{nn} \tag{4.17}$$

Next we put $c_n = 1$, and obtain from Equation 4.14

$$c_m \approx \frac{\hbar}{m_0} \frac{\mathbf{k}.\mathbf{p}_{mn}}{E_n(0) - E_m(0)} c_n \tag{4.18}$$

neglecting the **k.p** term in the denominator. If this expression for c_m is now used in Equation 4.14, we obtain in the second-order approximation

$$E_n(\mathbf{k}) = E_n(0) + \frac{\hbar^2 k^2}{2m_0} + \frac{\hbar}{m_0} \mathbf{k}.\mathbf{p}_{nn} + \sum_{m \neq n} \left(\frac{\hbar}{m_0} \right)^2 \frac{|\mathbf{k}.\mathbf{p}_{nn}|^2}{E_n(0) - E_m(0)} \tag{4.19}$$

Since the extrema occur at $\mathbf{k} = 0$, $\mathbf{p}_{nn} = 0$. Therefore, by choosing a proper coordinate system, one may write

$$E_n(\mathbf{k}) = E_n(0) + \frac{\hbar^2 k^2}{2m_i}$$

It follows from Equation 4.19, therefore, that

$$\frac{1}{m_i} = \frac{1}{m_0} + \frac{2}{m_0{}^2} \sum_m \frac{\left|\mathbf{i}.\mathbf{p}_{nm}\right|^2}{E_n(0) - E_m(0)} \tag{4.20}$$

where $\mathbf{i}$ is a unit vector along the ith coordinate axis. The above equation predicts a parabolic E–$\mathbf{k}$ relation.

The analysis presented in this subsection may be improved by combining the single band considered here with other bands close to it, and treating the effects of the additional bands as small perturbations.

4.3.2 Four Bands

We now consider four bands: the C, HH, LH, and SO bands. Both the C-band minima and the triply degenerate valence-band maxima occur at $\mathbf{k}=0$. In the present analysis, the spin–orbit interaction is neglected. We denote the cell periodic part of the conduction band wave function by U_c and the corresponding parts related to the three valence bands by U_{v1}, U_{v2}, and U_{v3}. Also, E_c is used to denote the energy for conduction band minima and E_v the energy for the valence-band maxima. For any $\mathbf{k}$, the wave function may be written as

$$\psi = (a_k U_c + b_k U_{v1} + c_k U_{v3} + d_k U_{v2}) \exp(j\mathbf{k}.\mathbf{r}) \tag{4.21}$$

in accordance with Equation 4.10. Using the symbol $E' = E - \hbar^2 k^2/2m_0$, the linear homogeneous equations are

$$a_k(E' - E_c) - (\hbar/m_0)\mathbf{k}.(b_k \mathbf{p}_{cv1} + d_k \mathbf{p}_{cv2} + c_k \mathbf{p}_{cv3}) = 0$$

$$-a_k(\hbar/m_0)\mathbf{k}.\mathbf{p}_{cv1} + b_k(E' - E_v) - (\hbar/m_0)\mathbf{k}.(d_k \mathbf{p}_{v1v2} + c_k \mathbf{p}_{v1v3}) = 0$$

$$(-a_k(\hbar/m_0)\mathbf{k}.\mathbf{p}_{cv2} - b_k(\hbar/m_0)\mathbf{k}.\mathbf{p}_{v1v2} - d_k(E' - E_v) - (\hbar/m_0)\mathbf{k}.c_k \mathbf{p}_{v1v3} = 0 \tag{4.22}$$

$$(-a_k(\hbar/m_0)\mathbf{k}.\mathbf{p}_{cv3} - b_k(\hbar/m_0)\mathbf{k}.\mathbf{p}_{v1v3} - d_k(\hbar/m_0)\mathbf{k}.c_k \mathbf{p}_{v2v3}) + c_k(E' - E_v) = 0$$

The quantities $\mathbf{p}_{cv1}$, $\mathbf{p}_{v1v2}$, and so on are defined in Equation 4.15. The matrix elements may be evaluated once the Us are known. Since U_c is an atomic s-like function, and the U_vs are p-like functions, the $\mathbf{k}.\mathbf{p}_{cv1}$ term may be expressed as

$$h_1 = \frac{\hbar^2}{m_0 j} \int U_t^* \left(k_x \frac{\partial}{\partial x} + k_y \frac{\partial}{\partial y} + k_z \frac{\partial}{\partial z} \right) U_m d^3r \quad t, m = s, x, y, z \tag{4.23}$$

Since $(\partial/\partial x)U_s$ is an odd function of x, the matrix element $\int U_j^*(\partial/\partial x)U_s dx$ is nonzero only when $j=x$. The same is true for $\int U_s^*(\partial/\partial x)U_j dx$. The function $(\partial/\partial x)U_y$ is odd in both x and y. Thus,

$$\iint U_j^* \frac{\partial}{\partial x} U_y dxdy = 0, \quad j=s,x,y,z$$

The only nonvanishing matrix elements are the ones defined in Equation 4.24:

$$P = -\frac{\hbar^2}{m_0}\int U_j \frac{\partial}{\partial j}U_s d^3r = -\frac{\hbar^2}{m_0}\int U_s \frac{\partial}{\partial j}U_j d^3r, \quad j=x,y,z \qquad (4.24)$$

Assuming now that **k** is parallel to the z-direction, the four homogeneous equations (Equation 4.22) may be rewritten as

$$a_k(E'-E_c)-c_k Pk = 0$$

$$b_k(E'-E_v) = 0$$

$$-a_k Pkc_k(E'-E_v) = 0 \qquad (4.25)$$

$$d_k(E'-E_v) = 0$$

The energy eigenvalues are thus given by

$$E' = E_c, E_V \quad \text{and} \quad (E'-E_c)(E'-E_v)-P^2k^2 = 0 \qquad (4.26)$$

These equations give the dispersion relations when the conduction bands and valence bands are strongly coupled. Denoting the energy gap by $E_g = E_c - E_v$, we may write Equation 4.26 as

$$(E'-E_c)(E'-E_c+E_g)-P^2k^2 = 0$$

When E' tends to E_c, we may neglect $E'-E_c$ in comparison to E_g and write

$$(E'-E_c)E_g = P^2k^2 \qquad (4.27)$$

If the band-edge effective mass is denoted by m_{e0}, we obtain

$$P^2 = \left(\frac{E-E_c-\hbar^2k^2}{2m_0}\right)\frac{E_g}{k^2} = \hbar^2\left(\frac{1}{m_{e0}}-\frac{1}{m_0}\right)\frac{E_g}{2} \qquad (4.28)$$

4.4 Spin–Orbit Interaction

The electron is a fermion with spin ½ in units of $\hbar$. In classical mechanics, a point particle rotating about an axis has an angular momentum $\mathbf{L} = \mathbf{r} \times \mathbf{p}$. In the quantum picture, the angular momentum of a point particle is quantized, and the intrinsic value of the momentum is called the spin. There is strong interaction between the spin and orbital motion of the electrons. This spin–orbit coupling may be calculated for isolated atoms; however, it is difficult to do so in crystals.

4.4.1 Spin–Orbit Interaction Term

To calculate the interaction, a general form of spin–orbit interaction is assumed, the fitting parameter in which is adjusted to fit experimentally observed effects. The total Hamiltonian in the presence of spin–orbit interaction is written as $H = H_0 + H_{so}$, where H_0 is the Hamiltonian without interaction and H_{so} is the spin–orbit interaction written as

$$H_{so} = \lambda \mathbf{L} \cdot \mathbf{S} \tag{4.29}$$

Here, $\mathbf{L}$ represents the operator for orbital angular momentum, $\mathbf{S}$ is the operator for spin angular momentum, and λ is treated as a constant. The total angular momentum $\mathbf{J}$ may be expressed as

$$J^2 = (\mathbf{L} + \mathbf{S})^2 = L^2 + S^2 + 2\mathbf{L} \cdot \mathbf{S} \tag{4.30a}$$

Thus,

$$\langle \mathbf{L} \cdot \mathbf{S} \rangle = (1/2)\langle J^2 - L^2 - S^2 \rangle = \frac{\hbar^2}{2}\left[j(j+1) - l(l+1) - s(s+1) \right] \tag{4.30b}$$

where j, l, and s are the quantum numbers for the operators $\mathbf{J}$, $\mathbf{L}$, and $\mathbf{S}$, respectively. However, to calculate the spin–orbit interaction energy, one needs the pure angular momentum states to which Equation 4.30 is applicable. We introduce symbols α and β to denote, respectively, the spin-up and spin-down states. One should note that states such as $|X\alpha\rangle$ are mixed states. To illustrate this statement, we express $|X\rangle$ in terms of pure angular momentum states, that is,

$$|X\rangle = \frac{1}{\sqrt{2}}(-\phi_{1,1} + \phi_{1,-1})$$

$$|Y\rangle = \frac{j}{\sqrt{2}}(\phi_{1,1} + \phi_{1,-1}) \qquad (4.31)$$

$$|Z\rangle = \phi_{1,0}$$

The $\phi_{i,j}$s are pure angular momentum states, and the expressions for the lower eigenstates are

$$\phi_{1,\pm 1} = Y_{1,\pm 1}(\theta,\phi) = \mp\sqrt{\frac{3}{8\pi}}\,\sin\theta\,\exp(\pm j\phi)$$

$$\phi_{1,0} = Y_{1,0}(\theta,\phi) = \sqrt{\frac{3}{4\pi}}\,\cos\theta$$

The $\phi_{i,j}$s are eigenfunctions of L^2 and L_z. The respective quantum numbers are $l=I$ and $l_z=j$. For example, $L^2\phi_{1,-1} = \hbar^2\,(1)(1 + 1)\phi_{1,1-1} = 2\hbar^2\phi_{1,-1}$ and $L_z\phi_{1,-1} = -1\hbar\phi_{1,-1}$.

Equation 4.31 is modified if spin is included; for example, the spin-up state $p_x = |X\alpha\rangle$ is expressed as

$$|X\alpha\rangle = \frac{1}{\sqrt{2}}(-\phi_{1,1} + \phi_{1,-1})\alpha$$

This formulation is still in terms of mixed states. To decompose the mixed states into states of pure angular momentum, the spin and orbital angular momentum must be added to obtain the total angular momentum states. The standard Clebsch–Gordan (CG) technique is employed for this addition. The following six equations are obtained as a result:

$$\phi_{3/2,3/2} = \phi_{1,1}\alpha = (-1/\sqrt{2})|(X+jY)\alpha\rangle$$

$$\phi_{3/2,1/2} = \frac{1}{\sqrt{3}}\phi_{1,1}\beta + \frac{2}{\sqrt{6}}\phi_{1,0}\alpha = \frac{-1}{\sqrt{6}}\Big[|(X+jY)\beta\rangle - |2Z\alpha\rangle\Big]$$

$$\phi_{3/2,-1/2} = \frac{1}{6}\phi_{1,0}\beta + \frac{1}{\sqrt{3}}\phi_{1,-1}\alpha = \frac{1}{\sqrt{6}}\Big[|(X-jY)\alpha\rangle + |2Z\beta\rangle\Big] \qquad (4.32)$$

$$\phi_{3/2,-3/2} = \phi_{1,-1}\beta = (1/\sqrt{2})|(X-jY)\beta\rangle$$

$$\phi_{1/2,1/2} = \frac{-1}{\sqrt{3}}\phi_{1,0}\alpha + \frac{2}{\sqrt{6}}\phi_{1,1}\beta = \frac{-1}{\sqrt{3}}\Big[|(X+jY)\beta\rangle + |Z\alpha\rangle\Big]$$

$$\phi_{1/2,-1/2} = \frac{-2}{\sqrt{6}}\phi_{1,-1}\alpha + \frac{1}{\sqrt{3}}\phi_{1,0}\beta = \frac{-1}{\sqrt{3}}\left[|(X - jY)\alpha\rangle - |Z\beta\rangle\right]$$

These six equations are inverted to find states such as $\phi_{1,0}$, and from the resultant equations one obtains states such as $|X\alpha\rangle$, and so on:

$$|X\alpha\rangle = \frac{1}{\sqrt{2}}\left[-\phi_{3/2,3/2} + \frac{1}{\sqrt{3}}\phi_{3/2,-1/2} - \sqrt{\frac{2}{3}}\phi_{1/2,-1/2}\right]$$

$$|X\beta\rangle = \frac{1}{\sqrt{2}}\left[-\frac{1}{\sqrt{3}}\phi_{3/2,1/2} - \frac{2}{\sqrt{3}}\phi_{1/2,1/2} - \phi_{3/2,-3/2}\right]$$

$$|Y\alpha\rangle = \frac{j}{\sqrt{2}}\left[\phi_{3/2,3/2} + \frac{1}{\sqrt{3}}\phi_{3/2,-1/2} - \sqrt{\frac{2}{3}}\phi_{1/2,-1/2}\right] \qquad (4.33)$$

$$|Y\beta\rangle = \frac{j}{\sqrt{2}}\left[\frac{1}{\sqrt{3}}\phi_{3/2,1/2} + \frac{2}{\sqrt{3}}\phi_{1/2,1/2} + \phi_{3/2,-3/2}\right]$$

$$|Z\alpha\rangle = \sqrt{\frac{2}{3}}\phi_{3/2,1/2} - \frac{1}{\sqrt{3}}\phi_{1/2,1/2}$$

$$|Z\beta\rangle = \sqrt{\frac{2}{3}}\phi_{3/2,-1/2} + \frac{1}{\sqrt{3}}\phi_{1/2,-1/2}$$

The phases used in the above expressions for ϕ_{j,m_j} in terms of $|X\alpha\rangle,\ldots|Z\beta\rangle$ are obtained in the standard derivation of CG coefficients. The overall phase of a state is arbitrary and has no effect on the physical predictions. The convention used by Luttinger and Kohn is in widespread use and will be employed here [10]. The states are expressed in terms of CG states as

$$\phi_{3/2,3/2}(\text{LK}) = -\phi_{3/2,3/2}(\text{CG}) = \frac{1}{\sqrt{2}}|(X + jY)\alpha\rangle$$

$$\phi_{3/2,1/2}(\text{LK}) = -j\phi_{3/2,1/2}(\text{CG})$$

$$\phi_{3/2,-1/2}(\text{LK}) = \phi_{3/2,-1/2}(\text{CG}) \qquad (4.34)$$

$$\phi_{3/2,-3/2}(\text{LK}) = j\phi_{3/2,-3/2}(\text{CG})$$

$$\phi_{1/2,1/2}(\text{LK}) = -\phi_{1/2,1/2}(\text{CG})$$

$$\phi_{1/2,-1/2}(\text{LK}) = j\phi_{1/2,-1/2}(\text{CG})$$

The spin–orbit Hamiltonian may be calculated now with the above states. The interaction is

$$H_{so} = \frac{\lambda\hbar^2}{2}\left[j(j+1) - l(l+1) - s(s+1)\right] \tag{4.35}$$

For p-type electron orbitals $l=1$ and $s=1/2$, j is given by the first subscript of ϕ in Equation 4.33. Many terms become zero, as the pure states are orthogonal. We conclude that only the following terms are nonzero:

$$\langle X\alpha|H_{so}|Y\alpha\rangle = \langle Y\alpha|H_{so}|Z\beta\rangle = \langle Y\beta|H_{so}|Z\alpha\rangle = -j\frac{\Delta}{3}$$

$$\langle X\alpha|H_{so}|Z\beta\rangle = \frac{\Delta}{3}; \quad \langle X\beta|H_{so}|Z\alpha\rangle = -\frac{\Delta}{3}; \quad \langle X\beta|H_{so}|Y\beta\rangle = j\frac{\Delta}{3} \tag{4.36}$$

where Δ is a parameter known as *spin-orbit splitting*, given by $\Delta = \Delta_{so} = 3\lambda\hbar^2/2$.

4.4.2 Conduction Band Energy

The calculation of energy levels in the conduction band by including the spin–orbit interaction is easier, and therefore we consider it first. As mentioned previously, we are interested in four basis vectors, the $|S\rangle$ state for the conduction band, and $|X\rangle,|Y\rangle,|Z\rangle$ states for the valence bands. There are four coefficients, $a_k...d_k$ as in Equation 4.21, needed to describe a state. With the inclusion of spin, the number of basis vectors to be considered becomes eight. The secular equation containing the coefficients and basis vectors involves an 8×8 matrix. It turns out that if the basis vectors are arranged in the following manner:

$$|S\alpha\rangle.|(X+jY)\beta\rangle,|Z\alpha\rangle,|(X-jY)\beta\rangle,|S\beta\rangle,|(X-jY)\alpha\rangle,|Z\rangle, \text{ and } |-(X+jY)\alpha\rangle$$

the matrix may be written in the form

$$\begin{bmatrix} H & 0 \\ 0 & H \end{bmatrix}$$

where H is a 4×4 matrix. Using the matrix elements between different states and the earlier elements obtained without the spin–orbit interactions, we may express the different matrix elements in terms of P in the following form:

$$
\begin{array}{c|cccc}
 & |S\alpha\rangle & |(X+jY)\beta\rangle & |Z\alpha\rangle & |(X-jY)\beta\rangle \\
\hline
|S\alpha\rangle & E_s - E' & 0 & -jkP & 0 \\
|(X+jY)\beta\rangle & 0 & E_p - E' - \Delta/3 & \sqrt{2}\Delta/3 & 0 \\
|Z\alpha\rangle & jkP & \sqrt{2}\Delta/3 & E_p - E' & 0 \\
|(Xxx-jY)\beta\rangle & 0 & 0 & 0 & E_p - E' + \Delta/3
\end{array} \qquad (4.37)
$$

To simplify the calculation, we choose the k-vector in Equation 4.37 along the z-direction. Furthermore, to account for the shift of band energies due to spin–orbit interaction, notations E_p and E_s are used. The difference in signs in $E_p - E' \pm \Delta/3$ is due to the fact that $L_z S_z |(X+jY)\beta\rangle = -|(X+jY)\beta\rangle$ while $L_z S_z |(X-jY)\beta\rangle = |(X-jY)\beta\rangle$. Expanding the determinant given by Equation 4.37, one obtains

$$E' = E_p + \Delta/3 \qquad (4.38a)$$

$$\left(E' - E_p + \frac{2\Delta}{3}\right)\left(E' - E_p - \frac{\Delta}{3}\right)(E' - E_s) - k^2 P^2\left(E' - E_p + \frac{\Delta}{3}\right) = 0 \qquad (4.38b)$$

For small values of k^2, the cubic equation can easily be solved by treating the term $k^2 P^2$ as a small perturbation. This yields

$$E_1' = E_s + \frac{k^2 P^2 (E_s - E_p + \Delta/3)}{(E_s - E_p + 2\Delta/3)(E_s - E_p - \Delta/3)} \qquad (4.39)$$

Let $E_v = E_p + \Delta/3 = 0$, $E_c = E_s = E_{g0}$, and $E_c - E_v = E_{g0}$, the direct gap. We then rewrite Equation 4.38 as

$$E'(E' - E_{g0})(E' + \Delta) - k^2 P^2 (E' + 2\Delta/3) = 0 \qquad (4.38c)$$

Taking $E_c = E_{g0}$ in the first approximation,

$$E_c(\mathbf{k}) = E_{g0} + \frac{\hbar^2 k^2}{2m_0} + \frac{k^2 P^2}{3}\left[\frac{2}{E_{g0}} + \frac{1}{E_{g0} + \Delta}\right] \qquad (4.40a)$$

We also obtain by putting $E_v = E_p + \Delta/3 = 0$ in Equation 4.38c

$$E_{v1}(\mathbf{k}) = \frac{\hbar^2 k^2}{2m_0} \qquad (4.40b)$$

Taking $E = 0$,

$$E_{v2} = \frac{\hbar^2 k^2}{2m_0} - \frac{2k^2 P^2}{3E_{g0}} \tag{4.40c}$$

Finally, taking $E' = -\Delta$ as a first approximation,

$$E_{v3} = -\Delta + \frac{\hbar^2 k^2}{2m_0} - \frac{k^2 P^2}{3(E_{g0} + \Delta)} \tag{4.40d}$$

In Equation 4.40, E_c is the energy of the conduction band electrons, while E_{v1}, E_{v2}, and E_{v3}, are, respectively, the energies of the three valence bands. From Equation 4.40a, we may define the band-edge effective mass for conduction band electrons by writing

$$E_c = E_{g0} + \hbar^2 k^2 / 2m_{e0}$$

It then follows that

$$\frac{1}{m_{e0}} = \frac{1}{m_0} + \frac{2P^2}{3\hbar^2} \left(\frac{2}{E_{g0}} + \frac{1}{E_{g0} + \Delta} \right) \tag{4.41}$$

The momentum matrix element, P, which is central to all calculation of transition probabilities from the valence band to the conduction band, may be expressed in terms of m_{e0} as

$$P^2 = \frac{\hbar^2}{2m_{e0}} \frac{E_{g0}(E_{g0} + \Delta)}{E_{g0} + 2\Delta/3} \frac{m_0 - m_{e0}}{m_0} \tag{4.42}$$

4.4.3 Valence-Band Energies

The earlier treatment of the dispersion relation cannot explain the properties of electrons in the valence band. Referring to Equation 4.40, one notices that the energy of the electrons increases with k, which, however, is opposite to what is observed experimentally. To treat the valence-band properties correctly, degenerate perturbation theory is needed [11].

Let the state of the electron in the lth band at $k = 0$ be degenerate, having f-fold degeneracy. It follows from the theory of perturbation of degenerate states that the second-order corrections $\Delta E^{(2)}$ due to the $(\hbar/m_0)(\mathbf{k} \cdot \mathbf{p})$ perturbation are the roots of the secular equation

$$\left[\left(\frac{\hbar^2}{m_0}\right)^2 {\sum_{n,s}}' \frac{\langle l,r'|\mathbf{k}\cdot\mathbf{p}|n,s\rangle\langle n,s|\mathbf{k}\cdot\mathbf{p}|l,r\rangle}{E_l(0)-E_n(0)} - \Delta E^{(2)}\delta_{rr'}\right] = 0 \qquad (4.43)$$

where the primed summation sign indicates that the summation is over all $n \neq l$ and over s. $|l,r\rangle$ and $|l,r'\rangle$ are the unperturbed f-fold degenerate wave functions $(r,r'=1,2,\dots f)$ satisfying Equation 4.43 for the energy eigenvalue $E_l(0)$. The $|n,s\rangle$s are the wave functions for energy level $E_n(0)$. The order of the determinant of the secular equation is equal to the degree of degeneracy of the level $E_l(0)$. In the present situation, the degenerate eigenstates at $k=0$ are the three $l=1$, $m_l=0, \pm 1$ states. To second order in perturbation, the energy in the nth band is

$$E_n(k) = E_n(0) + \frac{\hbar^2 k^2}{2m_0} + \Delta E_n^{(2)}(k) \qquad (4.44)$$

where $\Delta E_n^{(2)}$ is given by Equation 4.43. Hence, we have the set of three equations

$$\sum_{i=1}^{3}\left[{\sum_m}' \frac{\langle j|H_I|m\rangle\langle m|H_I|l\rangle}{E_l(0)-E_m(0)} + \left\{E_l + \frac{\hbar^2 k^2}{2m_0} - E_n(k)\right\}\delta_{j,l}\right]\langle n|k\rangle = 0 \qquad (4.45)$$

Nontrivial solutions of this set of N coupled homogeneous equations occur only if

$$\det\left[|H| - E_n(k)\mathbf{I}\right] = 0$$

where $\mathbf{I}$ is the identity matrix and H is a 3×3 matrix whose elements are

$$H_{jl} = \left[E_l + \frac{\hbar^2 k^2}{2m_0}\right]\delta_{jl} + {\sum_m}' \frac{\langle j|H_I|m\rangle\langle m|H_I|l\rangle}{E_l(0)-E_m(0)}$$

The calculation of the matrix element is first made by ignoring spin–orbit interaction for the present. We take the basis sets as $|X\rangle$, $|Y\rangle$, and $|Z\rangle$. Then,

$$H_{11} = \langle X|H|X\rangle = E_1(0) + \frac{\hbar^2 k^2}{2m_0} + {\sum_m}' \frac{|\langle X|H_1|m\rangle|^2}{E_l(0)-E_m(0)} \qquad (4.46)$$

Since $|X\rangle$ is proportional to $xf(\mathbf{r})$, we may verify that

$$\frac{m_0^2}{\hbar^2}\left|\langle X|H_l|m\rangle\right|^2 = \left|\langle X|p_x|m\rangle\right|^2 k_x^2 + \left|\langle X|p_y|m\rangle\right|^2 k_y^2 + \left|\langle X|p_z|m\rangle\right|^2 k_z^2$$

Thus, we may write

$$H_{11} = E_1 + \sum_{j=x,y,z}\left[\frac{\hbar^2}{2m_0} + \frac{\hbar^2}{m_0^2}\sum'\frac{\left|\langle X|p_j|m\rangle\right|^2}{E_1 - E_m}\right]k_j^2$$

Due to symmetry at $k=0$,

$$\left|\langle X|p_y|m\rangle\right|^2 = \left|\langle X|p_z|m\rangle\right|^2$$

Therefore,

$$H_{11} = E_1 + Ak_x^2 + B\left(k_y^2 + k_z^2\right) \tag{4.47a}$$

where

$$A = \frac{\hbar^2}{2m_0} + \frac{\hbar^2}{m_0^2}\sum_j'\frac{\left|\langle X|p_x|j\rangle\right|^2}{E_1 - E_j} \tag{4.47b}$$

$$B = \frac{\hbar^2}{2m_0} + \frac{\hbar^2}{m_0^2}\sum_j'\frac{\left|\langle X|p_y|j\rangle\right|^2}{E_1 - E_j} \tag{4.47c}$$

The remaining matrix elements can be evaluated in a similar way to give the following Hamiltonian matrix:

$$H = \begin{bmatrix} E_1 + Ak_x^2 + B\left(k_y^2 + k_z^2\right) & Ck_xk_y & Ck_xk_z \\ Ck_xk_y & E_1 + Ak_y^2 + B\left(k_x^2 + k_z^2\right) & Ck_yk_z \\ Ck_xk_z & Ck_yk_z & E_1 + Ak_z^2 + B\left(k_x^2 + k_y^2\right) \end{bmatrix}$$

$$(4.48)$$

where

$$C = \frac{\hbar^2}{m_0^2}\sum_j'\frac{\langle X|p_x|j\rangle\langle j|p_y|Y\rangle + \langle X|p_y|j\rangle\langle j|p_x|Y\rangle}{E_1 - E_j} \tag{4.49}$$

Let us now consider the effect of spin. As noted in Equation 4.30, for $j = \frac{1}{2}$, $\langle \mathbf{L.S} \rangle = -\hbar^2$ and for $j = 3/2$, $\langle \mathbf{L.S} \rangle = +\hbar^2/2$. Thus, the states are split by an amount Δ proportional to $(3/2)\hbar^2$, the doubly degenerate state with $j = 3/2$ moving up by $\Delta/3$ and the single $j = \frac{1}{2}$ state moving down by $2\Delta/3$. Since in the designation $|ls; jm_j\rangle$, $l = 1$ and $s = \frac{1}{2}$, we shall use only the symbol $|jm_j\rangle$ to denote the states. As noted already, the Hamiltonian for the spin–orbit coupling is diagonalized if the states are chosen according to Equation 4.32.

We shall treat the $j = 3/2$ states and $j = \frac{1}{2}$ states separately, since the splitting energy is large. The Hamiltonian matrix $|H|$ now becomes a 4×4 matrix for $j = 3/2$ states and a 2×2 matrix for $j = \frac{1}{2}$ states. We may evaluate the matrix elements for the 4×4 matrix using Equation 4.47. Thus,

$$
\begin{aligned}
H_{11} &= \langle 3/2, 3/2 | H | 3/2, 3/2 \rangle = (1/2)\langle (X + jY)\alpha | H | (X + jY)\alpha \rangle \\
&= (1/2)\left[\langle X\alpha | H | X\alpha \rangle + \langle Y\alpha | H | Y\alpha \rangle + j\langle X\alpha | H | Y\alpha \rangle - j\langle Y\alpha | H | X\alpha \rangle \right] \\
&= E_1 + \frac{A}{2}\left(k_x^2 + k_y^2 \right) + \frac{B}{2}\left(k_x^2 + k_y^2 + 2k_z^2 \right) \\
&= H_{44}
\end{aligned}
\tag{4.50}
$$

where the symbols 1,2,3,4 are used in the order in which the states are written in Equation 4.48. Similarly,

$$
H_{12} = \frac{1}{2\sqrt{3}}\langle (X + jY)\alpha | H | [(X + jY)\beta - 2Z\alpha] \rangle = \frac{1}{\sqrt{3}}(H_{xz} - jH_{yz})
$$

Instead of computing matrix elements $\langle n | p | m \rangle$ from first principles, one replaces them with experimentally determined parameters called Luttinger parameters, defined as

$$
\gamma_1 = -2m_0(A + 2B)/3\hbar^2, \quad \gamma_2 = -m_0(A - B)/3\hbar^2, \quad \gamma_3 = -m_0 C/3\hbar^2 \tag{4.51}
$$

In terms of Luttinger parameters,

$$
H_{11} = E_1 - \frac{\hbar^2 k_z^2}{2m_0}(\gamma_1 - 2\gamma_2) - \frac{\hbar^2(k_x^2 + k_y^2)}{2m_0}(\gamma_1 + \gamma_2) \tag{4.52}
$$

Since, in measurements, the parameters conform to holes, and since the hole energy is positive, we write $H_{11} = -H_{hh}$, and the zero-energy reference is $E_1 = 0$. Repeating the above calculation for the other matrix elements, we obtain for the Luttinger Hamiltonian

$$H = \begin{bmatrix} H_{hh} & -c & -b & 0 \\ -c^* & H_{lh} & 0 & b \\ -b^* & 0 & H_{lh} & -c \\ 0 & b^* & -c^* & H_{hh} \end{bmatrix} \tag{4.53}$$

where

$$H_{lh} = \frac{\hbar^2 k_z^2}{2m_0}(\gamma_1 + 2\gamma_2) - \frac{\hbar^2\left(k_x^2 + k_y^2\right)}{2m_0}(\gamma_1 - \gamma_2)$$

$$c = \frac{\sqrt{3}\hbar^2}{2m_0}\left[\gamma_2\left(k_x^2 - k_y^2\right) - 2j\gamma_3 k_x k_y\right] \tag{4.54}$$

$$b = \frac{\sqrt{3}\hbar^2}{m_0}\gamma_3 k_z(k_x - jk_y)$$

In the vicinity of $k=0$, one may use the axial approximation, where γ_2 and γ_3 are replaced by an effective Luttinger parameter

$$\bar{\gamma} = (1/2)(\gamma_2 + \gamma_3) \tag{4.55}$$

The function c is then expressed as

$$c \cong \frac{\sqrt{3}\hbar^2\bar{\gamma}}{2m_0}(k_x - jk_y)^2 \tag{4.56}$$

The dispersion relation for valence-band holes may be written as

$$\begin{aligned} E_v &= -\frac{\hbar^2}{2m_0}\left[Ak^2 \pm \left\{B^2 k^4 + C^2(k_x^2 k_y^2 + k_x^2 k_z^2 + k_y^2 k_z^2)\right\}^{1/2}\right] \\ &= \frac{\hbar^2}{2m_0}\left[-\gamma_1 k^2 \pm \left\{4\gamma_2^2 k^4 + 12(\gamma_3^2 - \gamma_2^2)(k_x^2 k_y^2 + k_x^2 k_z^2 + k_y^2 k_z^2)\right\}^{1/2}\right] \end{aligned} \tag{4.57}$$

Introducing the spherical polar coordinate system with polar axis along the z-direction, we obtain

$$E_v = \frac{\hbar^2 k^2}{2m_0}\left[A \pm (B^2 + C^2/5)^{1/2}\right]$$

TABLE 4.1

Band Structure Parameters for Ge, Si, and GaAs

	m_e/m_0	γ_1	γ_2	γ_3
Ge	1.58/0.082	13.25	4.20	5.56
Si	0.916/0.191	4.26	0.34	1.45
GaAs	0.067	6.8	2.1	2.9
GaSb		13.1	4.5	6.0
InP	0.077	5.0	1.6	1.7
InAs	0.027	19.7	8,4	9.3
InSb	0.013	33.5	14.5	15.7
ZnSe	0.16	4.30	1.14	1.84
CdSe	0.13	4.95	1.36	1.84

This enables us to define effective masses for heavy and light holes as

$$m_{hh} = \frac{m_0}{A - (B^2 + C^2/5)^{1/2}} \tag{4.58a}$$

$$m_{lh} = \frac{m_0}{A + (B^2 + C^2/5)^{1/2}} \tag{4.58b}$$

The values of Luttinger parameters for common semiconductors are given in Table 4.1.

> **Example 4.1:** Using the values of γ in the table in Appendix II, the band-edge effective masses for Ge are $m_{hh} = 0.33\, m_0$ and $m_{lh} = 0.04\, m_0$; the values for Si are $m_{hh} = 0.56\, m_0$ and $m_{lh} = 0.16\, m_0$. For GaAs, the values are $m_{hh} = 0.059\, m_0$ and $m_{lh} = 0.08\, m_0$.

4.4.4 Momentum Matrix Elements

It follows from Equation 4.41 that the conduction band effective mass is expressed in terms of P and is related to p_{cv}, the momentum matrix element. The momentum matrix element also appears in the calculation of optical absorption coefficient or recombination rate in semiconductors. The conduction band edge state for a direct-gap semiconductor has been found to have s-type symmetry and is denoted by $|S\alpha\rangle$ and $|-S\beta\rangle$. The valence-band states are written in terms of angular momentum spin representation in Equation 4.32.

From symmetry, we find that only the matrix elements of the form

$$\langle X|p_x|S\rangle = \langle Y|p_y|S\rangle = \langle Z|p_z|S\rangle$$

are nonzero. The nonvanishing matrix elements are

$$\langle \pm 3/2|p_x|\pm S\rangle = (1/\sqrt{2})\langle X|p_x|S\rangle, \quad \langle \pm 1/2|p_x|\mp S\rangle = (1/\sqrt{6})\langle X|p_x|S\rangle$$
$$\langle \pm 3/2|p_x|\pm S\rangle = (2/\sqrt{6})\langle X|p_x|S\rangle \tag{4.59}$$

One may define a quantity

$$E_p = \frac{2}{m_0}|\langle X|p_x|S\rangle|^2 = \frac{2}{m_0}p_{cv}^2 \tag{4.60}$$

The values of E_p for different semiconductors are: GaAs (25.7), InP (20.4), InAs (22.2), and CdTe (20.7), where the numbers in the parentheses are in electronvolts. These values are remarkably close.

4.5 Strain-Induced Band Structure

In the following, we shall outline the methods of obtaining the band structure in the presence of strain. Before doing so, it is necessary to define a few quantities, such as stress and strain, and their relationships in a crystal.

4.5.1 Strain, Stress, and Elastic Constants

There are six independent components of stress: X_x, X_y and Y_x, and so on; the capital letters indicate the direction normal to the plane on which the stress acts. The strains are related by Hooke's law as

$$[e] = [s]\cdot[X] \tag{4.61}$$

The inverse relationship is

$$[X] = [c]\cdot[e] \tag{4.62}$$

In general, there are 36 elastic constants s_{ij} and c_{ij}. However, for a cubic crystal, the numbers reduce to only three, and we may write the relationship between stress and strain in the following way:

$$
\begin{bmatrix} X_x \\ Y_y \\ Z_z \\ Y_x \\ Z_x \\ X_y \end{bmatrix} = \begin{bmatrix} c_{11} & c_{12} & c_{12} & 0 & 0 & 0 \\ c_{12} & c_{11} & c_{12} & 0 & 0 & 0 \\ c_{12} & c_{12} & c_{11} & 0 & 0 & 0 \\ 0 & 0 & 0 & c_{44} & 0 & 0 \\ 0 & 0 & 0 & 0 & c_{44} & 0 \\ 0 & 0 & 0 & 0 & 0 & c_{44} \end{bmatrix} \cdot \begin{bmatrix} e_{xx} \\ e_{yy} \\ e_{zz} \\ e_{yz} \\ e_{zx} \\ e_{xy} \end{bmatrix} \tag{4.63}
$$

We now consider the strain tensor in a lattice-mismatched epitaxy. The epilayer is biaxially strained in the plane of the substrate by an amount ε_p and uniaxially strained in the perpendicular direction by an amount ε_z. For a thick substrate, the in-plane strain is determined from the relation

$$
\varepsilon_p = \frac{a_s - a_f}{a_f} = \varepsilon \tag{4.64}
$$

where a_s and a_f are, respectively, the lattice constants of the substrate and the film. As there is no strain in the perpendicular direction, the perpendicular strain ε_n is simply related to the in-plane strain and the Poisson ratio as

$$
\varepsilon_n = -\varepsilon_p / \sigma \tag{4.65}
$$

Thus, for a strained cubic crystal grown along the (001) direction, one may write

$$
\sigma = \frac{c_{11}}{2c_{12}}; \varepsilon_{xx} = \varepsilon_{yy} = \varepsilon_p; \varepsilon_{zz} = -\frac{2c_{12}}{c_{11}} \varepsilon_p; \varepsilon_{xy} = \varepsilon_{yz} = \varepsilon_{zx} = 0 \tag{4.66}
$$

On the other hand, if the growth is along the (111) direction,

$$
\sigma = \frac{c_{11} + 2c_{12} + 4c_{44}}{2c_{11} + 4c_{12} - 4c_{44}} \tag{4.67}
$$

$$
\varepsilon_{xx} = \varepsilon_{yy} = \varepsilon_{zz} = \left[\frac{2}{3} - \frac{1}{3} \left(\frac{2c_{11} + 4c_{12} - 4c_{44}}{c_{11} + 2c_{12} + 4c_{44}} \right) \right] \varepsilon_p \tag{4.68}
$$

$$
\varepsilon_{xy} = \varepsilon_{yz} = \varepsilon_{zx} = \left[-\frac{1}{3} - \frac{1}{3} \left(\frac{2c_{11} + 4c_{12} - 4c_{44}}{c_{11} + 2c_{12} + 4c_{44}} \right) \right] \varepsilon_p \tag{4.69}
$$

From Equations 4.68 and 4.69, one may conclude that the strain tensor is diagonal for the (001) growth but is nondiagonal for the (111) growth.

4.5.2 Strain Hamiltonian

The perturbation to the Hamiltonian caused by strain may be expressed as

$$H_{\varepsilon=}^{\alpha\beta} = \sum_{i,j} D_{ij}^{\alpha\beta} \varepsilon_{ij} \tag{4.70}$$

where:

D_{ij} is the deformation potential operator
$D_{ij}^{\alpha\beta}$ is the matrix element of D_{ij}

The deformation potential constants are usually obtained by fitting to the experimental data. The number of deformation potential constants is drastically reduced for highly symmetric structures, just like the number of elastic constants. Some of the important results are as follows:

4.5.2.1 Conduction Band at k=0

The effect of strain is to produce a shift in energy given by

$$\delta E^{000} = D_{xx} \left(\varepsilon_{xx} + \varepsilon_{yy} + \varepsilon_{zz} \right) \tag{4.71}$$

where $D_{xx} = E_d^{(000)}$ is the dilation deformation potential constant for the conduction band (000) valley, and $\delta E_c = 2E_d^{(000)} \left(c_{11} - c_{12} / c_{11} \right) \varepsilon$.

4.5.2.2 Conduction Band States along the (100) Direction in k-Space

There are two deformation potential constants: E_d due to dilation and E_u due to uniaxial pressure. The change in energy is

$$\delta E^{100} = E_d^{(100)} \left(\varepsilon_{xx} + \varepsilon_{yy} + \varepsilon_{zz} \right) + E_u^{(100)} \varepsilon_{xx} \tag{4.72}$$

By symmetry, $\delta E^{100} = \delta E^{010} = \delta E^{001}$

4.5.2.3 Conduction Band States along the (111) Direction in k-Space

The change in energy is given by

$$\delta E^{111} = D_{xx} \left(\varepsilon_{xx} + \varepsilon_{yy} + \varepsilon_{zz} \right) + 2D_{xy} \left(\varepsilon_{xy} + \varepsilon_{yz} + \varepsilon_{zx} \right) \tag{4.73}$$

$$D_{xx} = E_d^{(111)} + (1/3)E_u^{111} \tag{4.74}$$

4.5.2.4 Degenerate Valence Bands

The valence-band states are defined by $|X\rangle$, $|Y\rangle$, and $|Z\rangle$ basis states. Consider a matrix element H_ε^{xx} given by

$$H_\varepsilon^{xx} = \langle X|H_\varepsilon|X\rangle = \sum_{i,j} D_{ij}^{xx}\varepsilon_{ij} \tag{4.75}$$

It may be proved that the deformation potentials are

$$D_{xx}^{xx} = l, \; D_{yy}^{xx} = m, \; D_{xy}^{xy} = n \tag{4.76}$$

The following new set of deformation potential constants is now introduced:

$$a = \frac{l+2m}{3}; b = \frac{l-m}{3}; d = \frac{n}{\sqrt{3}} \tag{4.77}$$

In terms of these, we may write

$$m = a - b; \; a + 2b; \; n = \sqrt{3}d \tag{4.78}$$

Instead of using the $|X\rangle$, $|Y\rangle$, and $|Z\rangle$ basis, we now use the angular momentum basis for evaluating the various matrix elements. We may thus write

$$\left\langle \frac{3}{2},\frac{3}{2}\Big|H_\varepsilon\Big|\frac{3}{2},\frac{3}{2}\right\rangle = \frac{1}{2}\left[\langle X - iY|H_\varepsilon|X + iY\rangle\right]$$

$$= \frac{1}{2}\left[\langle X|H_\varepsilon|X\rangle + i\langle X|H_\varepsilon|Y\rangle - i\langle Y|H_\varepsilon|X\rangle + \langle Y|H_\varepsilon|Y\rangle\right]$$

$$= \frac{1}{2}\left[l\varepsilon_{xx} + m\left(\varepsilon_{yy} + \varepsilon_{zz}\right) + l\varepsilon_{yy} + m\left(\varepsilon_{zz} + \varepsilon_{xx}\right)\right]$$

$$= \frac{1}{2}\left[\varepsilon_{xx}(l+m) + \varepsilon_{yy}(l+m) + 2m\varepsilon_{zz}\right]$$

$$= \frac{1}{2}\left[\varepsilon_{xx}(2a+b) + \varepsilon_{yy}(2a+b) + \varepsilon_{zz}(2a-2b)\right] \tag{4.79}$$

Restricted to the hh and lh states, the strain Hamiltonian can be written as

$$H_\varepsilon = \begin{bmatrix} \dfrac{3}{2},\dfrac{3}{2} & \dfrac{3}{2},-\dfrac{1}{2} & \dfrac{3}{2},\dfrac{1}{2} & \dfrac{3}{2},-\dfrac{3}{2} \\[2mm] H^\varepsilon_{hh} & H^\varepsilon_{12} & H^\varepsilon_{13} & 0 \\[1mm] H^{\varepsilon*}_{12} & H^\varepsilon_{lh} & 0 & H^\varepsilon_{13} \\[1mm] H^{\varepsilon*}_{13} & 0 & H^\varepsilon_{lh} & -H^\varepsilon_{12} \\[1mm] 0 & H^{\varepsilon*}_{13} & -H^{\varepsilon*}_{12} & H^\varepsilon_{hh} \end{bmatrix} \tag{4.80}$$

The matrix elements are given by

$$H^\varepsilon_{hh} = a\left(\varepsilon_{xx} + \varepsilon_{yy} + \varepsilon_{zz}\right) - b\left[\varepsilon_{zz} - \frac{1}{2}\left(\varepsilon_{xx} + \varepsilon_{yy}\right)\right] \tag{4.81a}$$

$$H^\varepsilon_{lh} = a\left(\varepsilon_{xx} + \varepsilon_{yy} + \varepsilon_{zz}\right) + b\left[\varepsilon_{zz} - \frac{1}{2}\left(\varepsilon_{xx} + \varepsilon_{yy}\right)\right] \tag{4.81b}$$

$$H^\varepsilon_{12} = -d(\varepsilon_{xz} - i\varepsilon_{yz}) \tag{4.81c}$$

$$H^\varepsilon_{13} = \frac{\sqrt{3}}{2}b\left(\varepsilon_{yy} - \varepsilon_{xx}\right) + id\varepsilon_{xy} \tag{4.81d}$$

For growth along the (001) direction,

$$\varepsilon_{xx} = \varepsilon_{yy} = \varepsilon \quad \text{and} \quad \varepsilon_{zz} = -\left(2c_{12}/c_{11}\right)\varepsilon \tag{4.82}$$

The matrix element may therefore be written as

$$\left\langle \frac{3}{2},\frac{3}{2} \middle| H_\varepsilon \middle| \frac{3}{2},\frac{3}{2} \right\rangle = H^\varepsilon_{hh} = 2a\left(\frac{c_{11} - c_{12}}{c_{11}}\right)\varepsilon + b\left(\frac{c_{11} + 2c_{12}}{c_{11}}\right)\varepsilon \tag{4.83a}$$

and

$$\left\langle \frac{3}{2},\pm\frac{1}{2} \middle| H_\varepsilon \middle| \frac{3}{2},\pm\frac{1}{2} \right\rangle = H^\varepsilon_{lh} = 2a\left(\frac{c_{11} - c_{12}}{c_{11}}\right)\varepsilon - b\left(\frac{c_{11} + 2c_{12}}{c_{11}}\right)\varepsilon \tag{4.83b}$$

All other matrix elements reduce to zero. The complete Hamiltonian is now written as

$$H = H + H_\varepsilon = \begin{bmatrix} H_{hh} - \frac{1}{2}\delta & b & c & 0 \\ b* & H_{lh} + \frac{1}{2}\delta & 0 & c \\ c* & 0 & H_{lh} + \frac{1}{2}\delta & -b \\ 0 & c* & -b* & H_{hh} - \frac{1}{2}\delta \end{bmatrix} \quad (4.84)$$

where δ is the strain-induced separation between the hh and the lh states in the bulk material, which may be calculated from Equations 4.80 and 4.81. The relation between deformation potentials and the values measured from hydrostatic pressure is

$$E_d^{(000)} = E_{hyd}^{(000)} + a; \ E_d^{(100)} = E_{hyd}^{(100)} - \frac{1}{3}E_u^{(100)} + a; \ E_d^{(111)} = E_{hyd}^{(111)} - \frac{1}{3}E_u^{111} + a$$

If the material is uniformly compressed along all the directions (hydrostatic compression), the change in the bandgap is given by

$$\delta E_g(\mathbf{k}) = E_{hyd}^{(\mathbf{k})}\left(\varepsilon_{xx} + \varepsilon_{yy} + \varepsilon_{zz}\right) \quad (4.85a)$$

where

$$E_{hyd}^{(\mathbf{k})} = \frac{1}{3}\left(c_{11} + c_{12}\right)\frac{dE_g}{dP} \quad (4.85b)$$

The elastic constants and the hydrostatic pressure coefficient of the bandgap, dE_g/dp, are measured by various experiments, from which the deformation potential constants may be evaluated.

Example 4.2: The following values of parameters are given for InAs(GaAs): $a = -1.3$ (–2.0) eV, $b = -1.7$ (–1.7) eV, $c_{11} = 8.33$ (11.88), $c_{12} = 4.58$ (5.38), both in 10^{11} dyne cm^{-2}. Using linear interpolation, one obtains the following values for In$_{0.2}$Ga$_{0.8}$As: $a = -1.86$, $b = -1.7$, $c_{11} = 11.17$, and $c_{12} = 5.21$. The separation between hh and lh states is -6.57ε eV. The lattice constant of InGaAs film is 5.734, and that of GaAs substrate is 5.653 A. Therefore, $\varepsilon = 0.0142$, and the separation $\delta = 93$ meV.

4.6 Quantum Wells

The subband structures for electrons and holes have been calculated in Chapter 3 by using simple theory. However, for refined calculation, the complete knowledge of the E–k dispersion relation is needed in order to explain the experimental results. Here, we shall give the outline of the theory for valence band states in a QW.

The degenerate nature of the valence bands prompts us to employ the multiband effective mass approximation. The Hamiltonian is written as

$$\sum_{v'} [H_{vv'}(\mathbf{k}) + V(z)\delta_{vv'}] \phi_m^{v'} = E_m \phi_m^v \tag{4.86}$$

The complete wave function for the valence-band hole in the mth subband in the vth valence band $\varphi_{mk}(r)$ is written in terms of the envelope function ϕ_m^v as

$$\varphi_{mk}(r) = \sum_{v=1}^{4} \phi_m^v(\mathbf{k}, z) \exp(j\mathbf{k}.\mathbf{r}) U^v(\mathbf{r}) \tag{4.87}$$

We have assumed as before that the conduction band is decoupled from the valence bands, and only the two top valence bands are considered. We therefore consider four eigenstates:

$$|1\rangle = |3/2, 3/2\rangle, |2\rangle = |3/2, -1/2\rangle, |3\rangle = |3/2, 1/2\rangle, |4\rangle = |3/2, -3/2\rangle.$$

The Luttinger Hamiltonian for the hole states is given by Equation 4.53, and the coefficients are given in Equation 4.54.

For a rectangular QW, $H_{vv'}(\mathbf{k})$ given in the above matrix form should be replaced by $H_{vv'}(\mathbf{k}, -j\partial/\partial z)$, where $\mathbf{k}$ is now the in-plane wave vector. The simple solution for $\mathbf{k}=0$ has been worked out in Chapter 3 (see Equation 3.57).

The in-plane effective masses for holes in different subbands have been calculated by different authors using various degrees of approximation. We assume that the bandgap as well as the separation between the heavy-hole and split-off bands is large, so the Hamiltonian matrix is treated as a 4×4 matrix as before. The eigenvalues are obtained by solving the secular determinant of the 4×4 matrix, and the expression is given by Equation 4.57.

The upper and lower signs correspond, respectively, to heavy- and light-hole bands. The character of the bands becomes increasingly mixed for higher values of $\mathbf{k}$. The effective mass in the xy plane depends, in general, on the direction, and the magnitude of the anisotropy is determined by the difference in γ_2 and γ_3. In many cases, the difference is small, and it is justifiable to take a spherical average. It can be shown that the average $\langle k_x^2 k_y^2 + k_y^2 k_z^2 + k_z^2 k_x^2 \rangle \cong k^4/5$. Putting this in Equation 4.57, we obtain

$$E = \frac{\hbar^2}{2m_0}(-\gamma_1 \pm 2\bar{\gamma})k^2 \tag{4.88}$$

where

$$\bar{\gamma}^2 = (2\gamma_2^2 + 3\gamma_3^2)/5 \tag{4.89}$$

Each of the four eigenfunctions is of the form

$$\varphi = \left[A|3/2,3/2\rangle + B|3/2,-1/2\rangle + C|3/2,1/2\rangle + D|3/2,-3/2\rangle \right] \exp(j\mathbf{k}.\mathbf{r})$$
(4.90)

The four eigenfunctions may be expressed as column matrices.

In the practical situation when the barrier height is finite, a parameter $\bar{\gamma} = (\gamma_2 + \gamma_3)/2$ is introduced. The 4×4 Hamiltonian matrix is transformed to a new matrix $\bar{H}$ by using a unitary matrix U, such that [12]

$$\bar{H} = UHU^+ = \begin{bmatrix} H^U & 0 \\ 0 & H^L \end{bmatrix} \tag{4.91}$$

where

$$H^U = \begin{bmatrix} H_{hh} & R \\ R* & H_{hh} \end{bmatrix} \quad \text{and} \quad H^L = \begin{bmatrix} H_{lh} & R \\ R* & H_{lh} \end{bmatrix} \tag{4.92}$$

$$R = |c| - j|b|$$

The upper and lower blocks are then decoupled. Writing now the upper and lower block envelope functions as

$$\varphi_{mk}{}^U(\mathbf{r}) = \sum_{v=1,2} g_m^{(v)}(\mathbf{k},z) \exp(j\mathbf{k}.\mathbf{r})|v\rangle \tag{4.93a}$$

and

$$\varphi_{mk}{}^L(\mathbf{r}) = \sum_{v=2,4} g_m^{(v)}(\mathbf{k},z) \exp(j\mathbf{k}.\mathbf{r})|v\rangle \tag{4.93b}$$

where $\{|v\rangle\}$ denotes the transformed basis set, and the envelope functions satisfy

$$\sum_{v'=1,2}\left[H^{U}_{vv'}\left(\mathbf{k}.-j\frac{\partial}{\partial z}\right)+V(z)\delta_{vv'}\right]g^{v'}_{m}(\mathbf{k},z)=E^{U}_{m}(\mathbf{k})g^{v}_{m}(\mathbf{k},z) \qquad (4.94a)$$

$$\sum_{v'=1,2}\left[H^{L}_{vv'}\left(\mathbf{k}.-j\frac{\partial}{\partial z}\right)+V(z)\delta_{vv'}\right]g_{m}^{(v'+2)}(\mathbf{k},z)=E^{L}_{m}(\mathbf{k})g_{m}^{(v+2)}(\mathbf{k},z) \qquad (4.94b)$$

Problems

4.1. Expand the determinant given in Equation 4.37 and obtain Equation 4.38b. Work out the remaining steps to arrive at Equation 4.39.

4.2. Calculate the energy separation between lh and hh states in $Ga_xAs_{1-x}P$ grown on GaAs for $x=0.1$, 0.3, and 0.5. The lattice parameters are $a(GaAs)=5.653$, $a(GaAsP)=5.8686-0.417x$. For the alloy $c_{11}=10.22+3.9x$, $c_{12}=5.76+0.49x$ dynes cm^{-2}, $a=-(6.31+2.53x)$ eV, and $b=-(2.0-0.2x)$ eV. The well width is 5 nm, $m_{hh}=0.46m_0$, and $m_{lh}=0.11m_0$. Assume infinite barrier height.

4.3. Calculate the bandgap of $In_{1-x}Ga_xAs$ as a function of x grown on GaAs. Assume that the thickness is below the critical layer value. Include also the effect of quantum confinement.

References

1. Madelung, O., *Introduction to Solid State Theory*. Springer, Berlin, 1978.
2. Ridley, B. K., *Quantum Processes in Semiconductors*. Oxford University Press, Oxford, 1982.
3. Basu, P. K., *Theory of Optical Processes in Semiconductors: Bulk and Microstructures*. Clarendon Press, Oxford, 2003.
4. Singh, J., *Electronic and Optoelectronic Properties of Semiconductor Structures*. Cambridge University Press, Cambridge, New York, 2003.
5. Nag, B. R., *Electron Transport in Compound Semiconductors*. Springer, Berlin, 1980.
6. Wang, S., *Fundamentals of Semiconductor Theory and Device Physics*. Prentice Hall, Englewood Cliffs, NJ, 1989.
7. Anselm, A. I., *Introduction to Semiconductor Theory*, trans. M. M. Samokkvalov. Mir, Moscow, 1981.
8. Kane, E. O., Band structure of indium antimonide, *J. Phys. Chem. Solids* 1, 249–261, 1957.

9. Chuang, S. L., *Physics of Optoelectronic Devices*. Wiley, New York, 1995.

10. Luttinger, J. M. and W. Kohn, Motion of electrons and holes in perturbed periodic fields, *Phys. Rev.* 97, 869–883, 1955.

11. Landau, L.D. and E. M. Lifshitz, *Quantum Mechanics: Non-Relativistic Theory*. Pergamon, Oxford, 1977.

12. Broido, D. A. and L. J. Sham, Effective masses of holes at GaAs–AlGaAs heterojunctions, *Phys. Rev. B*, 31, 888–892, 1985.

5

Waveguides and Resonators

5.1 Introduction

The active layer in a semiconductor laser is a lower bandgap material sandwiched between two layers of a higher-bandgap semiconductor. The relative permittivities of the core and cladding layers are, however, opposite, that is, the refractive index (RI) is higher for the core layer. This difference in RIs leads to the optical confinement of the generated light. The electromagnetic (EM) wave is essentially guided and the core layer acts as a waveguide for the light. The waveguiding may also extend in two directions, as in a buried heterostructure laser.

In addition to waveguiding, a certain amount of feedback is needed to ensure self-sustained oscillation. Usually, this feedback is provided by a Fabry–Perot (FP) resonator, which also supports different modes of EM waves. Feedback may also be provided in a distributed fashion by some gratings, as in a distributed feedback (DFB) or a distributed Bragg reflector (DBR) laser. In vertical-cavity surface-emitting lasers (VCSELs), DBRs act as resonators.

In the present chapter, we shall discuss the principles and the basic properties of the aforementioned passive structures used in lasers. The principle of waveguiding will be discussed first, using two different approaches based on the geometrical ray optic theory and the usual EM theory. Both two-dimensional (2-D) and three-dimensional (3-D) waveguides will be considered. The theory of resonators will be developed next by considering the FP resonator. Then, the reflectivity of Bragg mirrors using dielectric multilayers will be considered. The principle of operation of DFB and DBR structures may be understood using the coupled-mode theory. This theory will be developed in Chapter 11 dealing with single-mode lasers. Finally, a brief description of the ring resonator will be presented.

5.2 Ray Optic Theory

The ray optic theory is useful in understanding the light-guiding property of a dielectric. Consider that a light ray, E_i, strikes the interface of two dielectric media of RIs n_1 and n_2 at an angle θ_1, as shown in Figure 5.1a. The angle of incidence, θ_1, and the angle of refraction, θ_2, are related by Snell's law as

$$n_1 \sin \theta_1 = n_2 \sin \theta_2 \tag{5.1}$$

When $n_1 > n_2$ and $\theta_1 > \theta_c$, the critical angle $\theta_2 = \pi/2$, giving $\theta_c = \sin^{-1}(n_2/n_1)$. For $\theta_1 > \theta_c$, there is total internal reflection (TIR).

Consider now a three-layer structure, as shown in Figure 5.1b, in which medium 1 having RI n_1, called the *core layer*, is sandwiched between two media, each having RI n_2, called the *cladding layers*. A light ray strikes the interface between the core and the upper cladding layer at an angle $\theta > \theta_c$, gets totally reflected, strikes next the core–lower cladding layer interface at an angle θ, again gets totally reflected, and the process continues. In this way, the ray propagates in the core layer, being confined within it. This is light guidance by the core layer.

This simplified approach might tempt anyone to conclude that the waveguide will guide light whenever the incidence angle exceeds the critical angle. This is not, however, true as will be shown next.

5.3 Reflection Coefficients

The complex amplitudes of the electric fields associated with the incident and reflected rays, denoted by E_i and E_r, respectively, in Figure 5.1, are related to each other by

$$E_r = rE_i \tag{5.2}$$

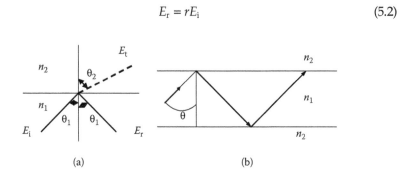

(a)   (b)

FIGURE 5.1
(a) Reflection and refraction in two dielectric media; (b) ray path showing total internal reflection and guidance of mode in a symmetric slab waveguide.

where r is the complex reflection coefficient that depends on both the angle of incidence and the polarization of the light ray. The polarization of a light ray is related to the direction of the electric field vector associated with the EM wave. The transverse electric (TE) condition refers to the situation when the electric fields of the waves are perpendicular to the plane of incidence, that is, the plane containing the wave normal and normal to the interface. Similarly, the transverse magnetic (TM) case occurs when the magnetic field of the wave is normal to the plane of incidence. The reflection coefficients r for the two cases, r_{TE} and r_{TM}, are expressed by the following two Fresnel formulas:

$$r_{TE} = \frac{n_1 \cos \theta_1 - n_2 \cos \theta_2}{n_1 \cos \theta_1 + n_2 \cos \theta_2} \tag{5.3a}$$

$$r_{TM} = \frac{n_2 \cos \theta_1 - n_1 \cos \theta_2}{n_2 \cos \theta_1 + n_1 \cos \theta_2} \tag{5.3b}$$

These two equations may be rewritten using Snell's law (Equation 5.1) as

$$r_{TE} = \frac{n_1 \cos \theta_1 - \sqrt{n_2^2 - n_1^2 \sin^2 \theta_1}}{n_1 \cos \theta_1 + \sqrt{n_2^2 - n_1^2 \sin^2 \theta_1}} \tag{5.4a}$$

$$r_{TM} = \frac{n_2^2 \cos \theta_1 - n_1 \sqrt{n_2^2 - n_1^2 \sin^2 \theta_1}}{n_2^2 \cos \theta_1 + n_1 \sqrt{n_2^2 - n_1^2 \sin^2 \theta_1}} \tag{5.4b}$$

For any angle of incidence less than the critical angle, only partial reflection occurs and the reflection coefficient is real. However, when the critical angle is exceeded, the terms inside the square root in Equation 5.4a and b become negative. As a result, the reflection coefficient becomes complex with magnitude unity and a phase shift is imposed on the reflected wave. The phase shifts related to the TE and TM waves may be expressed as

$$\varphi_{TE} = 2 \tan^{-1} \left[\sqrt{\sin^2 \theta_1 - (n_2/n_1)^2} / \cos \theta_1 \right] \tag{5.5a}$$

$$\varphi_{TM} = 2 \tan^{-1} \left[\sqrt{(n_1/n_2)^2 \sin^2 \theta_1 - 1} / (n_2/n_1) \cos \theta_1 \right] \tag{5.5b}$$

The reflection coefficients r given in Equation 5.3a and b are the field reflection coefficients, that is, the ratios between the electric fields and the incident and reflected waves. Using the definition of Poynting's vector, the power reflection coefficient is simply given by

$$R = \frac{S_r}{S_i} = \frac{E_r^2}{E_i^2} = r^2 \tag{5.6}$$

Example 5.1: The power reflection coefficient between GaAs and air may be calculated for normal incidence $\theta_1 = 0°$. From Equations 5.4 and 5.6, $R = (n_1 - n_2)^2/(n_1 + n_2)^2$. Using $n_1 = 3.6$ for GaAs and $n_2 = 1$ for air, $R = 0.32$.

5.4 Modes of a Planar Waveguide

The symmetric planar waveguide shown in Figure 5.1 is the simplest optical waveguide. A more general structure is the asymmetric waveguide shown in Figure 5.2a in which the RIs of the upper and lower cladding layers, n_3 and n_2, respectively, are different. As shown, the EM wave propagates along the z-direction, but is confined along the y-direction in the guide layer of thickness a. The zigzag path denotes the direction of the wave normal as the waves propagate through the waveguide with wave vector $k (= k_0 n_1)$, where k_0 is the free-space wave vector ($k_0 = 2\pi/\lambda_0$, λ_0 is the free-space wavelength). The components of k along the y- and z-directions are

$$k_z = n_1 k_0 \sin \theta_1$$
$$k_y = n_1 k_0 \cos \theta_1 \tag{5.7}$$

The relation between the propagation constants and the wave normal is shown in Figure 5.2b. The confinement of light makes the EM wave form

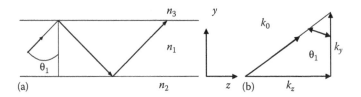

FIGURE 5.2
(a) Ray path and propagation in planar waveguide; (b) propagation constants in y- and z-directions.

a standing wave pattern along the y-direction. The total phase shift that is introduced in making a round trip along the y-direction covering a distance $2a$ may be written as

$$\varphi_t = 2k_0 n_1 a \cos \theta_1 - \varphi_u - \varphi_l \tag{5.8}$$

The first term on the left-hand side of Equation 5.8 is the phase change in traversing a distance $2a$ across the y-direction, and φ_u and φ_l denote, respectively, the phase shifts due to reflection at the core–upper cladding and core–lower cladding interfaces, which may be obtained from Equation 5.5. To sustain the wave along the y-direction, this total phase shift must be an integral multiple of 2π, so that

$$2k_0 n_1 a \cos \theta_1 - \varphi_u - \varphi_l = 2m\pi \tag{5.9}$$

where m is an integer. Since φ_u and φ_l are functions of θ_1, Equation 5.9 is an implicit equation of θ_1, and when solved, it gives the value of θ_1 for a given integral value of m.

The discrete values of θ_1, corresponding to integral values of m, signify that light cannot propagate with any arbitrary value of the angle of incidence. For each allowed solution of Equation 5.9, there will be a definite *mode of propagation* and the mode number is given by the value of the integer m. For each solution of Equation 5.9, there will be corresponding propagation constants along the y- and z-directions, for each polarization. Each mode is characterized by the polarization and the mode number. For example, the first TE mode, the fundamental mode, is denoted as TE_0. Higher-order modes are correspondingly described by using the appropriate value for m.

5.4.1 Symmetric Planar Waveguide

Let us apply Equation 5.9 to the symmetric planar waveguide structure shown in Figure 5.1a, for which the phase shifts due to reflections at the two interfaces are equal, $\varphi_u = \varphi_l$. Using Equation 5.5a for TE polarization, Equation 5.9 becomes

$$2k_0 n_1 a \cos \theta_1 - 4 \tan^{-1} \left[\sqrt{\sin^2 \theta_1 - (n_2 / n_1)^2} / \cos \theta_1 \right] = 2m\pi \tag{5.10}$$

This equation may be rearranged to give

$$\tan \left[(k_0 n_1 a \cos \theta_1 - m\pi)/2 \right] = \left[\sqrt{\sin^2 \theta_1 - (n_2/n_1)^2} / \cos \theta_1 \right] \tag{5.11}$$

The only variable in this equation, θ_1, may be obtained by solving Equation 5.11. The corresponding equation for the TM mode may be written in a similar fashion by using Equation 5.9.

5.4.1.1 Number of Modes Supported

To have an estimate of the total number of modes supported by the symmetric guide, we note that the minimum value of the angle of incidence θ_1 is the critical angle θ_c, corresponding to the highest possible order mode. Putting $\theta_1 = \theta_c$ in Equation 5.11 makes the right-hand side zero and thus

$$(k_0 n_1 a \cos \theta_c - m_{max} \pi)/2 = 0 \tag{5.12}$$

The maximum allowed mode number, m_{max}, is therefore

$$m_{max} = \frac{k_0 n_1 a \cos \theta_c}{\pi} \tag{5.13}$$

In an actual calculation, one should consider the nearest integer $[m_{max}]_{int}$ that is less than the computed value of m_{max}. The total number of modes will actually be $[m_{max}]_{int} + 1$, since the lowest-order mode has a mode number $m = 0$.

> **Example 5.2**: Let $\lambda_0 = 1.0\ \mu m$, $n_1 = 1.5$, $a = 5\ \mu m$. The value of m_{max} turns out to be 7.5. The maximum number of modes to be supported is $[7 + 1] = 8$.

5.4.2 Asymmetric Waveguide

In the asymmetric planar waveguide structure, the cladding layers have different RIs as shown in Figure 5.2, and the phase changes due to reflections at the upper and lower interfaces are unequal. The eigenvalue equation for the TE mode now reads

$$[k_0 n_1 a \cos \theta_1 - m\pi] = \tan^{-1}\left[\sqrt{\sin^2 \theta_1 - (n_2/n_1)^2} \,/ \cos \theta_1\right]$$

$$+ \tan^{-1}\left[\sqrt{\sin^2 \theta_1 - (n_3/n_1)^2} \,/ \cos \theta_1\right] \tag{5.14}$$

For a given mode number m, this equation may be solved numerically or graphically to find the propagation angle θ_1. Note, however, that the critical angles for the two interfaces are different in this case. For TIRs at both boundaries, the angle θ_1 must exceed the larger of these two. It is possible that when the guide thickness, a, is small, the solution to Equation 5.14 ceases to exist for the fundamental mode with number $m = 0$ (see Problem 5.5).

5.4.3 Single-Mode Condition

Depending on the fractional change in RIs and the guide thickness, sometimes it is possible that the guide supports only a single mode. The condition for such a monomode operation will now be derived for a symmetric waveguide using Equation 5.9.

Consider the second mode with mode number $m = 1$. In the limiting condition, $\theta_1 = \theta_c$, the critical angle. Since the incident angle for the second mode is less than that for the fundamental mode, then for all angles greater than this critical angle, the waveguide will be monomode. We may write Equation 5.11 for this situation as

$$\tan[(k_0 n_1 a \cos \theta_c - \pi)/2] = 0 \tag{5.15}$$

This yields

$$\cos \theta_c = \frac{\pi}{k_0 n_1 a} = \frac{\lambda_0}{2 n_1 a} \tag{5.16}$$

Hence, for monomode conditions:

$$\theta_c \leq \cos^{-1}\left(\frac{\lambda_0}{2 n_1 a}\right) \tag{5.17}$$

5.4.4 Effective Index of a Mode

The propagation constants of the ray along the y- and z-directions have been defined in Equation 5.7. The constant k_z is often denoted by the variable β.

We may now define a parameter N called the *effective index of the mode*, in the following way:

$$N = n_1 \sin \theta_1 \tag{5.18}$$

Then, Equation 5.7 becomes

$$k_z = \beta = N k_0 \tag{5.19}$$

We may now state that the ray propagates straight along the z-direction with RI N, *without* following the zigzag path shown in Figures 5.1 and 5.2.

The lower bound on β is determined by the critical angles at the interfaces. As noted already for light guidance in the asymmetric guide, the propagation angle must be greater than the larger of the two critical angles. Since, in most cases, the upper cladding layer is air, the larger critical angle corresponds to the lower cladding. This means that $\theta_1 \geq \theta_{\text{lower}}$. The lower bound on β is given by

$$\beta \geq n_1 k_0 \sin \theta_1 = k_0 n_2 \tag{5.20}$$

The upper bound on β is determined by the maximum value of θ, that is, $90°$, and then $\beta = k = n_1 k_0$. The final result is

$$k_0 n_1 \geq \beta \geq k_0 n_2 \tag{5.21}$$

Using Equation 5.19, it may be written as

$$n_1 \geq N \geq n_2 \tag{5.22}$$

5.5 Wave Theory of Light Guides

We now wish to develop the theory of optical waveguides and of the modes using the EM wave equation.

5.5.1 Wave Equation in a Dielectric

Like all other EM phenomena, the propagation of EM waves along a dielectric waveguide is governed by Maxwell's equations, given by

$$\nabla \times \mathbf{E} + \frac{\partial \mathbf{B}}{\partial t} = 0 \tag{5.23a}$$

$$\nabla \times \mathbf{H} - \frac{\partial \mathbf{D}}{\partial t} = \mathbf{J} \tag{5.23b}$$

$$\nabla \cdot \mathbf{D} = \rho \tag{5.23c}$$

$$\nabla \cdot \mathbf{B} = 0 \tag{5.23d}$$

where $\mathbf{E}$ and $\mathbf{H}$ are the electric and magnetic fields, $\mathbf{D} = \varepsilon \mathbf{E}$, $\mathbf{B} = \mu \mathbf{H}$, and $\mathbf{J}$ and ρ are the current and charge densities, respectively.

Most of the guiding materials are nonmagnetic; therefore, $\mu = \mu_0$. Further, the current density and the free charge densities are assumed to be absent.

The electric polarization is related to the electric field through a susceptibility χ, which is taken to be constant in most cases.

By taking the curl of Equation 5.23a and using Equation 5.23b through d, the wave equation is obtained in the following form:

$$\nabla \times \nabla \times \mathbf{E} = -\mu_0 \varepsilon \varepsilon_0 \frac{\partial^2}{\partial t^2} \mathbf{E} \qquad (5.24)$$

where $\varepsilon = 1 + \chi$ is the relative permittivity of the medium. Now assuming that $\mathbf{E}$ varies sinusoidally as $\mathbf{E} = \mathbf{E} \exp(j\omega t)$, we may write

$$\nabla \times \nabla \times \mathbf{E} = -\varepsilon(\omega^2/c^2)\mathbf{E} \qquad (5.25)$$

where $c^2 = 1/\sqrt{\varepsilon_0\mu_0}$ is the velocity of light in free space. In general, ε is complex and its real and imaginary parts are related to the RI n and the absorption coefficient α (see Chapter 6). We shall now consider that ε is real and equals n^2 because of the low optical loss in the guides and write $\nabla \times \nabla \times \mathbf{E} = \nabla(\nabla \cdot \mathbf{E}) - \nabla^2 \mathbf{E} = -\nabla^2 \mathbf{E}$, since $\nabla \cdot \mathbf{D} = 0$. We finally get

$$\nabla^2 \mathbf{E} + n^2 k_0^2 \mathbf{E} = 0 \qquad (5.26)$$

The symbol k_0 is defined in terms of the free-space wavelength λ_0 by the relation

$$k_0 = \omega/c = 2\pi/\lambda_0 \qquad (5.27)$$

5.5.2 The Ideal Slab Waveguide

The planar slab waveguide structure forms the starting point for all theoretical developments of wave propagation in waveguides. We shall first discuss the origin of the different modes in this structure.

5.5.2.1 Modes in Slab Waveguide

The simple three-layer planar waveguide structure is shown in Figure 5.3. The structure consists of a dielectric layer, denoted as Layer 1 having RI n_1, sandwiched between the upper cladding Layer 3 having RI n_3 and the lower cladding Layer 2 having RI n_2. All the layers are assumed to extend to infinity along the x- and z-directions, and Layers 2 and 3 are semi-infinite also along the y-direction. If the radiation propagates along the z-direction, then a mode should be the solution to the following wave

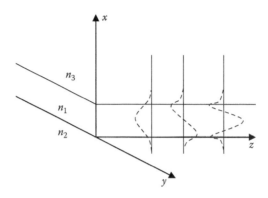

FIGURE 5.3
Three-layer slab waveguide. The electric field distribution of the first three modes is shown.

equation, obtained by modifying Equation 5.28 to account for the spatial variation of the RI:

$$\nabla^2 \mathbf{E(r)} + n^2(\mathbf{r})k_0^2 \mathbf{E(r)} = 0 \tag{5.28}$$

If we now consider the propagation of a uniform plane wave along the z-direction, that is, $\mathbf{E(r)} = \mathbf{E}(x,y)\exp(-j\beta z)$, β being the propagation constant, Equation 5.28 then becomes

$$\partial^2 \mathbf{E}(x,y)/\partial x^2 + \partial^2 \mathbf{E}(x,y)/\partial y^2 + [k_0^2 n^2(r) - \beta^2]\mathbf{E}(x,y) = 0 \tag{5.29}$$

Since the guide extends to infinity along x, we may write Equation 5.29 as

$$\partial^2 \mathbf{E}(x,y)/\partial x^2 + \partial^2 \mathbf{E}(x,y)/\partial y^2 + [k_0^2 n_i^2(r) - \beta^2]\mathbf{E}(x,y) = 0, \quad i = 1,2,3 \tag{5.30}$$

The solutions of Equation 5.30 are either sinusoidal or exponential depending on whether $[k_0^2 n_i^2 - \beta^2]$ is greater than or less than 0. The $E(x,y)$ and its derivative $\partial E/\partial y$ must be continuous at the interfaces between the dielectric layers. The possible modes are shown in Figure 5.3.

Some insight into the mode shape may be obtained from Equation 5.30 for constant frequency ω and $n_1 > n_2 > n_3$. For $\beta > kn_1$, $(1/E)\partial^2 E/\partial y^2$ is positive in all three regions, leading to exponential solutions. The resulting field distribution denoted as (a) in Figure 5.4 is the only solution compatible with the continuity of the field and its derivative. The monotonically increasing wave amplitudes do not represent real waves. For values of β satisfying the condition, $kn_3 < \beta < kn_2$, $(1/E)\partial^2 E/\partial y^2$ is positive in Regions 2 and 3 and negative in Region 1, leading to a sinusoidal solution in Region 1 and decaying exponentials in Regions 2 and 3. We now have mode confinement and guiding in Region 1. The exponential

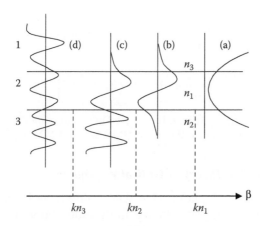

FIGURE 5.4
Possible modes in a planar waveguide.

solutions in Regions 2 and 3 do not give propagating modes, but rather they represent evanescent waves. It follows therefore that for waveguiding, the condition $kn_2 < \beta < kn_1$ must be satisfied. Mode (b) indicated in Figure 5.4 belongs to the family of well-confined guided modes that is generally referred to as TE modes. The mode shown in Figure 5.4 is the TE_1 mode.

If $kn_2 < \beta < kn_3$, mode (c) as shown in Figure 5.4 will result. The mode is confined in the upper cladding layer, for example, air, but is sinusoidal in the lower substrate layer. This is called the *substrate radiation mode*. However, as the mode loses energy from the guiding region to the substrate, it dampens out over a short distance. The mode is not useful in signal transmission, but finds application in the coupler structures.

If $\beta < kn_3$, the solution for $E(x,y)$ is oscillatory in all the three regions as indicated by (d) in Figure 5.4. In this case, energy propagates freely out of the guiding layer and the modes are called *radiation modes*.

An interesting parameter for guided mode propagation is the cutoff parameter. In the following section, we shall see that β can have any value when it is less than kn_2, but only discrete values of β are allowed when $kn_1 < \beta < kn_2$. These discrete values of β correspond to various modes, TE_j, $j = 0, 1, 2...$ (also TM_k, $k = 0, 1, 2...$). The number of modes that can be supported depends on the thickness, a, of the guiding layer, the frequency ω, and the RIs. For given a and n_is there is a cutoff frequency ω_c below which no waveguiding can occur. This frequency leads to a long wavelength cutoff, λ_c. By solving the wave equation, it can be shown that for a given mode number, m, to be supported by an asymmetric guide shown in Figure 5.3, the following condition applies

$$\Delta n = n_2 - n_1 \geq (2m+1)^2 \lambda_0^2 / (32n_2 a^2), \quad m = 0,1,2,...$$ (5.31)

where λ_0 is the wavelength in free space.

Example 5.3: There is no cutoff for the fundamental mode in a symmetric guide. However, an asymmetric guide can support a mode of order m with a wavelength λ_0 when

$$a \geq \left[\left((m+1/2)\lambda_0 / 2\sqrt{n_1^2 - n_2^2} \right) \right].$$

Use $\lambda_0 = 1550$ nm, $n_1 = 3.55$ and $n_2 = 3.17$. Then, single-mode operation is maintained in the range $0.1564\lambda_0 \leq a \leq 0.4692\lambda_0$, or $0.242 \leq a \leq 0.727$ μm.

5.5.2.2 Outline of the Theory of Optical Waveguide

The guided modes in a planar slab waveguide have been considered in Section 5.4 using the ray optic theory. In this subsection, we shall consider guided modes in the step-index 2-D waveguide by employing the EM wave equation developed in Equation 5.26. The basic structure is shown in Figure 5.4. It is called a *2-D waveguide* (or a slab optical waveguide) because light is confined only along the y-direction. We shall assume that the waveguide is a step-index guide in which the RI changes abruptly at the interfaces of the two dielectrics. Layer 3 is the cladding layer, which may be air, and Layer 2 is the other cladding (or substrate) layer. The RIs of the layers are indicated by n_1, n_2, and n_3. However, in some cases, we shall use the symbols n_c, n_g, and n_s, respectively, in place of n_3, n_1, and n_2, where the subscripts refer, respectively, to cladding, guide, and substrate. The cladding Layers 1 and 3 extend to infinity along the $+y$- and $-y$-directions, respectively.

We consider plane waves propagating along the z-direction with propagation constant β. The EM fields vary as

$$E = E(x,y) \exp j(\omega t - \beta z) \tag{5.32a}$$

$$H = H(x,y) \exp j(\omega t - \beta z) \tag{5.32b}$$

where ω is the angular frequency, $\omega = 2\pi\lambda/c$, and $c = 1/\sqrt{\varepsilon_0\mu_0}$ is the velocity of light in free space. In the step-index guide, the fields are independent of x. Therefore, since $\partial/\partial t = j\omega$, $\partial/\partial z = -j\beta$, and $\partial/\partial x = 0$, Equation 5.30 yields two different modes with mutually orthogonal polarization states. One is the TE mode having only E_x as the electric field component. Using Equation 5.23a, the magnetic field components are H_y and H_z. The other mode is the TM mode, which has the components E_y, H_x, and E_z. The wave equations for the TE and TM modes are

$$\text{TE:} \quad \frac{d^2 E_x}{dy^2} + (k_0^2 n^2 - \beta^2)E_x = 0 \tag{5.33a}$$

$$H_y = -\frac{\beta}{\omega\mu_0} E_x, \quad H_z = -\frac{1}{j\omega\mu_0}\frac{dE_x}{dy} \tag{5.33b}$$

$$\text{TE: } \frac{d^2 H_x}{dy^2} + (k_0^2 n^2 - \beta^2)H_x = 0 \tag{5.34a}$$

$$E_z = \frac{\beta}{\omega\varepsilon_0 n^2} H_x, \quad E_y = \frac{1}{j\omega\varepsilon_0 n^2}\frac{dH_x}{dy} \tag{5.34b}$$

Solving for the fields with the boundary conditions at the interface $y = -a$ and $y = 0$, one obtains eigenvalue equations that determine the propagation characteristics of the TE and TM modes.

5.5.2.2.1 Dispersion of the Guided Modes

Here, we consider TE modes only. The corresponding expressions for TM modes may be obtained in a similar fashion. From Equation 5.30, the field solutions may be written as follows

$$
\begin{aligned}
E_x &= E_c \exp(-\gamma_c y); & y &> 0, & \text{in the upper cladding} \\
E_x &= E_g \cos(k_y y + \varphi_1); & -a &< y < 0, & \text{in the guiding film} \\
E_x &= E_s \exp\{\gamma_s(y+a)\}; & y &< -a, & \text{in the substrate}
\end{aligned} \tag{5.35}
$$

The propagation constants in the y-direction are expressed in terms of the following symbols:

$$\gamma_c = k_0\sqrt{N^2 - n_c^2}, \quad k_y = k_0\sqrt{n_g^2 - N^2}, \quad \gamma_s = k_0\sqrt{N^2 - n_s^2} \tag{5.36}$$

where $N = \beta/k_0$ is the effective index. Since the tangential field components E_x and H_z are continuous at the interface $y = 0$, one obtains

$$
\begin{aligned}
E_c &= E_g \cos\phi_1 \\
\tan\phi_1 &= \gamma_c/k_y
\end{aligned} \tag{5.37}
$$

Similarly

$$
\begin{aligned}
E_s &= E_g \cos(k_y a - \phi_1) \\
\tan(k_y a - \phi_1) &= \gamma_s/k_y
\end{aligned} \tag{5.38}
$$

at $y = -a$. Eliminating arbitrary coefficients in Equations 5.37 and 5.38 yields the following eigenvalue equation:

$$k_y a = (m+1)\pi - \tan^{-1}(k_y/\gamma_s) - \tan^{-1}(k_y/\gamma_c) \tag{5.39}$$

where $m = 0,1,2\ldots$ denotes the *mode number*. The value of k_x may be obtained from the values of the RIs and the thickness of the guide. When this value is substituted into Equation 5.36, the value of the effective index N may be obtained. Since the mode number is a positive integer, N must have discrete values lying in the range $n_s < N < n_g$. This means that zigzag rays with certain incident angles can propagate as guiding modes along the guide as shown in Figure 5.5. It is to be noted that the eigenvalue in Equation 5.39 is identical to that in Equation 5.12 if the RIs for the upper cladding and substrate are assumed to be equal.

The transcendental Equation 5.39 may be solved numerically to evaluate the dispersion characteristics of the guided modes. For this purpose, one introduces the following two normalized parameters: *normalized frequency V* and the *normalized guide index b* defined as

$$V = k_0 a \sqrt{n_g^2 - n_s^2} \tag{5.40}$$

and

$$b = \frac{\left(N^2 - n_s^2\right)}{\left(n_g^2 - n_s^2\right)} \tag{5.41}$$

The measure of the asymmetry of the guide is defined by

$$\alpha = \frac{\left(n_s^2 - n_c^2\right)}{n_f^2 - n_s^2} \tag{5.42}$$

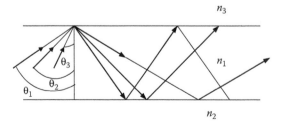

FIGURE 5.5
Different modes incident at different angles at the core–cladding interface.

Thus, when $n_c = n_s$ as in a symmetric waveguide, the asymmetry parameter $\alpha = 0$. In general, waveguides are asymmetric. Using Equations 5.40 and 5.42, Equation 5.39 may be rewritten as

$$V\sqrt{1-b} = (m+1)\pi - \tan^{-1}\sqrt{(1-b)/b} - \tan^{-1}\sqrt{(1-b)/(b+\alpha)} \qquad (5.43)$$

Solving this equation numerically, a plot may be obtained, which is the normalized dispersion curve with m and α as parameters. When material RIs and the guide thickness are given, the effective index of the guided mode may be obtained from the graph.

When the incident angle equals the critical angle related to Layers 1 and 2 (i.e., the guide–substrate interface), the light is no longer confined in the guide layer, but begins to leak into the substrate at the interface $x = -a$. This situation is called the cutoff of the guided mode. In this case, the effective index $N = n_s$, $b = 0$. From Equation 5.43, the value of V_m at the cutoff condition is given by

$$V_m = V_0 + m\pi, \quad V_0 = \tan^{-1}\sqrt{\alpha} \qquad (5.44)$$

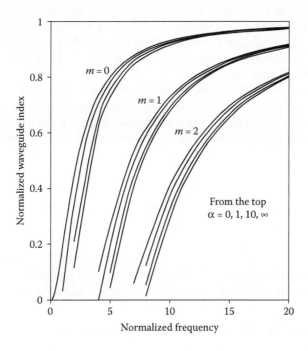

FIGURE 5.6
Normalized plots of V vs. b.

V_0 is the cutoff value for the fundamental mode. When the normalized frequency V lies in between V_m and V_{m+1}, $V_m < V < V_{m+1}$, the TE_0, TE_1,... and TE_m modes are supported, the number of guided modes being $m+1$. For symmetric waveguides, $n_c = n_s$ and $V_0 = 0$. Thus, the fundamental mode is not cut off in a symmetric waveguide.

For a symmetric guide, use of Equations 5.33 and 5.36 indicates that for the waveguiding of a given mode to occur, one should have

$$\Delta n = n_g - n_c > \frac{m^2 \lambda_0^2}{4(n_c + n_g)a^2} \qquad (5.45)$$

A family of curves showing b parameters against V parameters for mode numbers 0, 1, and 2 and for different values of a is given in Figure 5.6. These plots were originally presented by Kogelnik and Ramaswamy [1].

5.5.2.3 Mode Profiles and Confinement Factor

Equation 5.35 allows us to plot the electric field distribution $E_x(y)$ or the intensity distribution $|E_x(y)|^2$. For the fundamental mode, the electric field is maximum at the center of the guide, as shown in Figure 5.3. It then decreases following a cosine distribution. At the two interfaces, the field does not decay to zero but penetrates into the upper and lower claddings following the exponential nature. The field distribution for the higher-order mode, for example, $m=2$, may be qualitatively depicted and the corresponding intensity profile is also shown in Figure 5.3. The number of minima in this profile equals the mode number.

As seen from the field profiles, the power propagating along the guide is not totally confined inside the core of the guide, but rather a part of the power leaks into the two cladding layers. A useful parameter describing the light-guiding property of the waveguide is the *mode confinement factor* defined as follows:

$$\Gamma = \frac{\displaystyle\int_0^a E_x^2(y)dy}{\displaystyle\int_{-\infty}^\infty E_x^2(y)dy} \qquad (5.46)$$

This factor quantifies what fraction of the total power belonging to the particular mode is confined within the guide layer. It is a function of polarization, of the RI difference, of the thickness of the guide, and of the mode number.

5.5.2.4 Approximate Expression for Confinement Factor

We consider the symmetric slab waveguide. The electric fields are given by Equation 5.36, but the decay constants remain the same for both layers.

The interfaces are located at $\pm d/2$, so that $a = 2d$. The electric fields are written as

$$E_x = C_1 \cos(k_y d/2)$$

$$E_x = C_1 \cos(k_y d/2) \exp\left[-\gamma(y-d/2)\right]; \quad y \geq d/2 \qquad (5.47)$$

$$E_x = C_1 \cos(k_y d/2) \exp\left[\gamma(y+d/2)\right]; \quad y \leq -d/2$$

The normalization constant is given by (see Chuang, 1995, Reading List)

$$C_1 = \frac{\left(4\omega/k_z d\right)^{1/2}}{\left[1 + \left(\sin k_x d/k_y d\right) + \left(2/\gamma d\right)\cos^2(k_y d/2)\right]^{1/2}} \qquad (5.47a)$$

The eigen equation is

$$\frac{\gamma d}{2} = \frac{k_y d}{2} \tan\left(\frac{k_y d}{2}\right) \qquad (5.48)$$

The optical confinement factor is

$$\Gamma = \left[1 + \frac{2}{\gamma d}\frac{\cos^2(k_y d/2)}{1 + \sin(k_y d)/k_y d}\right]^{-1} \qquad (5.49)$$

5.6 3-D Optical Waveguides

We now extend the theory for practical waveguides, which are basically 3-D in structure.

5.6.1 Practical Waveguiding Geometries

The symmetric and asymmetric slab waveguide structures discussed in Sections 5.4 and 5.5 form a good starting point for a discussion of light-guiding properties in semiconductor waveguides. In the broad area double heterostructure (DH) laser structure, the active GaAs layer of thickness, approximately a fraction of a micrometer, is sandwiched between two AlGaAs layers. The GaAs guiding layer extends over a few tens of micrometers in the lateral direction and a few hundreds of micrometers along the longitudinal direction. The RI profile makes step changes at the GaAs–AlGaAs interfaces.

The waveguides employed in a stripe geometry laser, or in general, gain-guided, and index-guided lasers as shown in Figure 1.8, are no longer the simple planar waveguides discussed in Sections 5.4 and 5.5. These wave-guides are called *3-D waveguides*. The RI variation now occurs along both the transverse and lateral directions and the structures support mode confinement along these two directions.

In the following, we shall develop and study the properties related to the 3-D waveguides. The practical structures are quite complicated and the actual RI variation depends on the compositions, doping density, and so on of the core and cladding materials. Numerical methods are employed to find the electric field variation, mode confinement factor, and other quantities of interest. The following approximate analyses are also useful in understanding the operations of lasers [4].

5.6.2 Approximate Analyses of Guided Modes

Pure TE and TM modes are not supported in optical 3-D waveguides surrounded by different dielectrics. Instead, two types of hybrid modes exist that are essentially TEM modes polarized along the x- and y-directions. The mode having the main electric field, E_x, is called the E_{pq}^x mode or the TM-like mode, that is, the mode resembles the TM mode in slab waveguide. The subscripts p and q denote the number of modes of the electric field E_x along the x- and y-directions, respectively.

The other class of mode, the E_{pq}^y mode, has the main electric field E_y and is a TE-like mode. In general, the mode pattern in 3-D waveguides can only be obtained numerically. However, approximate analytical methods are available that yield closed-form analytical solutions. The approximate methods Marcatili's method [3] and the effective index method [4,5] are useful when the guided mode is far from cutoff and the aspect ratio W/T, where W is the width and T is the thickness of the guide, is larger than unity. In the following, we give a brief outline of these two methods.

5.6.2.1 Marcatili's Method

Figure 5.7 shows a cross-sectional view of the 3-D guide with a step-index profile surrounded by different dielectric materials. Under the well-guided mode condition, most of the optical power is confined in Region I, while a small amount of power leaks into Regions II, III, IV, and V, where the EM field decays exponentially. Since even less power enters into the four shaded areas, in calculating the field patterns in Region I, field matching at the four

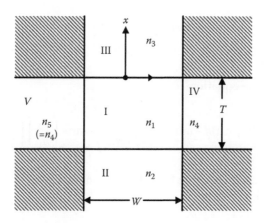

FIGURE 5.7
Cross-sectional view of a 3-D rectangular waveguide.

corners of the shaded regions is not considered. The E_{pq}^x mode with the main field components E_x and H_y is now considered. For this hybrid mode, $H_x=0$ can be put in Maxwell's equations. The wave equations with the field functions $H_y(x,y)$ are then solved analytically by the separation of variables. Using the boundary conditions that H_y is continuous at $x=0$ and $x=-T$ and that E_x is continuous on the interfaces $y=+W/2$ and $-W/2$, the field distribution $H_y(x,y)$ may be written as follows:

Region I: $\quad H_1 \cos(k_x x + \varphi_1) \cos(k_y y + \varphi_2)$

Region II: $\quad H_1 \cos(k_x T - \varphi_1) \cos(k_y y + \varphi_2) \exp\{\gamma_{x2}(x+T)\}$

Region III: $\quad H_1 \cos \varphi_1 \cos(k_y y + \varphi_2) \exp(-\gamma_{x3} x)$ $\qquad$ (5.50)

Region IV: $\quad H_1 \cos(k_y W/2 + \varphi_2) \cos(k_y x + \varphi_1) \exp\left\{-\gamma_{y4}(y - W/2)\right\}$

Region V: $\quad H_1 \cos(k_y W/2 + \varphi_2) \cos(k_x x + \varphi_1) \exp\left\{\gamma_{y4}(y + W/2)\right\}$

The waveguide is assumed to be symmetric along the y-direction, that is, $n_4=n_5$. The phase $\varphi_2=0$ refers to symmetric modes whose fields vary as $\cos(k_y y)$. The condition $\varphi_2=\pi/2$ refers to the existence of antisymmetric modes having a field variation in the form $\sin(k_y y)$. The propagation constants included in Equation 5.50 are expressed as

$$\beta^2 = k_0^2 n_1^2 - k_x^2 - k_y^2$$

$$\gamma_{xi}^2 = (\beta^2 + k_y^2) - k_0^2 n_i^2 \qquad (5.51)$$

where $i=2, 3, 4$. The continuity of E_z on $x=0$ and $-T$ yields the phase shift

$$\varphi_1 = \frac{\pi}{2} - \tan^{-1}\left(\frac{n_3}{n_1}\right)^2 \left(\frac{k_x}{\gamma_{x3}}\right) \qquad (5.52)$$

and the eigenvalue equation

$$k_x T = (p+1)\pi - \tan^{-1}\left(\frac{n_3}{n_1}\right)^2\left(\frac{k_x}{\gamma_{x3}}\right) - \tan^{-1}\left(\frac{n_2}{n_1}\right)^2\left(\frac{k_x}{\gamma_{x2}}\right) \qquad (5.53)$$

where $p=0,1,2,\ldots$. Considering that H_z is continuous at $y = W/2$ and $-W/2$, the following equation obtains

$$k_y W = (q+1)\,\pi - 2\,\tan^{-1}\left(\frac{k_y}{\gamma_{y4}}\right) \qquad (5.54)$$

where $q=0, 2, 4\ldots$ or $1, 3, 5\ldots$ corresponding to symmetric or asymmetric modes, respectively. The solutions of the transcendental Equations 5.51 and 5.54 lead to the value of the propagation constant β of the hybrid modes. The field distribution $H_y(x, y)$ of the fundamental E_{00}^x mode and the first-order E_{01}^x mode are shown as solid and dotted lines, respectively, in Figure 5.8. The E_{pq}^y modes may be analyzed in a similar manner.

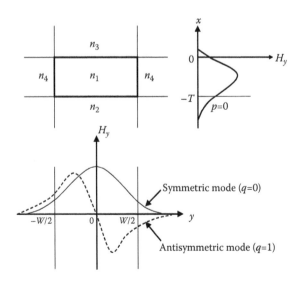

FIGURE 5.8
Mode pattern in a 3-D waveguide.

5.6.2.2 Effective Index Method

In order to introduce the effective index method, let us choose the embedded-strip 3-D waveguide shown in Figure 5.9. This guide consists of a rectangular core of high RI n_f, embedded in a substrate of lower RI n_s. The surrounding cladding region above has an RI n_c.

Guidance in this structure may be viewed as TIRs at the core–cladding and the core–substrate interfaces. In almost all practical situations, the core RI is slightly larger than the substrate RI; therefore, the reflected waves propagate nearly parallel to the z-direction for TIRs to occur. The propagation wave vector for these waves is essentially directed along z. Thus, from Maxwell's equation the electric and magnetic field vectors for each mode are perpendicular to **k** and therefore they are nearly transverse to the z-direction. Assuming two polarizations, the possible solutions are E_y, H_x and E_x, H_y.

As seen from Figure 5.9, in the region $|y| \leq W$, the structure is an asymmetric waveguide, having an RI n_c and thickness T. One cannot, however, directly use the results in Section 5.4 to obtain the propagation constant k_z from the normalized b–V plot, because there is also a field variation along the y-direction.

For now, the field variation along the y-direction is ignored. This means that W is assumed to be large. The structure we need to analyze is shown in Figure 5.9b. Assuming that the electric field is polarized along the y-direction, then the solutions correspond to modes that are TE with respect to the xz-plane. Alternatively, if the electric field is along x, then the magnetic field, **H**, is along y and the solutions are TM with respect to this plane. The normalized dispersion curves discussed in Section 5.4 may now be utilized to find the propagation constant for a given frequency. The normalized frequency parameter may be expressed as follows:

$$v = k_0 T (n_f^2 - n_s^2) \tag{5.55}$$

The asymmetry parameter is α_{TE} or α_{TM}. The symbols n_f and n_s denote, respectively, the RIs for the film (guide) and the substrate. Usually $(n_f - n_s)/n_f$ is small and as discussed in Section 5.5 the same normalization curves may be used for both the TE and TM modes. For each allowed mode p, the corresponding parameter b_p may be defined as

$$b_p = \left[(n_{eff})_p^2 - n_s^2 \right] / \left(n_f^2 - n_s^2 \right) \tag{5.56}$$

The index, $p = 0, 1,...$ indicates the number of half cycles of field variation along x in the core region. An effective RI $(n_{eff})_p$ may now be assigned to each value of b_p. The propagation constant associated with each solution may now be expressed as

$$(k_z)_p = k_0(n_{eff})_p \tag{5.57}$$

This propagation constant is valid for the limit that W is infinite.

It is now easy to find out the corresponding transverse wave vectors in the three regions. The expressions are

$$(\gamma_{cx})_p = \sqrt{k_z^2 - k_c^2} = k_0\sqrt{(n_{eff})_p^2 - n_c^2} \tag{5.58a}$$

$$(\beta_{fx})_p = \sqrt{\beta_f^2 - k_z^2} = k_0\sqrt{n_f^2 - (n_{eff})_p^2} \tag{5.58b}$$

$$(\gamma_{sx})_p = \sqrt{k_z^2 - k_s^2} = k_0\sqrt{(n_{eff})_p^2 - n_s^2} \tag{5.58c}$$

All of the information about guidance in the xz-plane is contained in the effective index $(n_{eff})_p$. Thus, in the region $|y| \leq W/2$, the entire asymmetric slab waveguide could be replaced by an equivalent uniform material of RI $(n_{eff})_p$ or N_I, as shown in Figure 5.9c. The presence of subscript p indicates that different modes need different values of the effective index.

Using the effective index determined by the effective index method, the guidance along the yz-plane may now be analyzed. As shown in Figure 5.9c, we need to consider a symmetric waveguide having a core RI $(n_{eff})_p$ or N_I of width W surrounded by a material of RI n_s. We are now in a position to calculate a new parameter v' analogous to v:

$$v' = k_0W\left[(n_{eff})_p^2 - n_s^2\right]^{1/2} = vb_p^{1/2}W / T \tag{5.59}$$

As we consider a symmetric guide, the asymmetry parameter, α', is equal to zero. Again using the weak guiding approximation, we can use the normalized dispersion curve for the TE modes to calculate a parameter b' from v':

$$(b')_{pq} = \left[(n'_{eff})_p^2 - n_s^2\right] / \left[(n_{eff})_p^2 - n_3^2\right] \tag{5.60}$$

The presence of an additional subscript q in Equation 5.60 signifies that for each value of index p, there are q solutions, $q = 0, 1, ...,$ describing the number of half-cycle variations along y within the core. Using Equations 5.55 and 5.59, the effective index $(n'_{eff})_{pq}$ can be expressed in terms of b_p and b_{pq} as

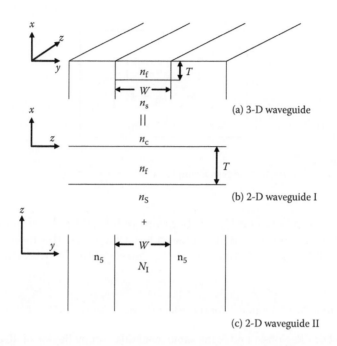

FIGURE 5.9
(a–c) Illustration of the effective index method.

$$(n'_{\text{eff}})_{pq} = \left[n_s^2 + b_p b'_{pq}(n_f^2 - n_s^2) \right]^{1/2} \tag{5.61}$$

Usually the difference in RI, Δn, between RIs of the core and the substrate is small, so that $n_f^2 - n_s^2 \cong 2\Delta n n_s$, and therefore $(n'_{\text{eff}})_{pq} \cong n_s + \Delta n b_p b'_{pq}$. The associated propagation constant for each mode $(k_z)_{pq}$ is given by

$$(k_z)_{pq} = k_0 (n'_{\text{eff}})_{pq} \tag{5.62}$$

The corresponding transverse wave vectors along the y-direction for regions $|y| < W/2$ and $|y| > W/2$ take the following respective forms:

$$(\gamma_y)_{pq} = \sqrt{(k_z)_{pq}^2 - \gamma_s^2} = k_0 \sqrt{(n'_{\text{eff}})_{pq}^2 - n_s^2} \cong k_0 (2\Delta n \, n_s b_p b'_{pq})^{1/2} \tag{5.63}$$

$$(k_y)_{pq} = \sqrt{k_0^2 (n_{\text{eff}})_p^2 - (k_z)_{pq}^2} = k_0 \sqrt{(n_{\text{eff}})_p^2 - (n'_{\text{eff}})_{pq}^2} \cong k_0 \left[2\Delta n \, n_s b_p (1 - b'_{pq}) \right]^{1/2} \tag{5.64}$$

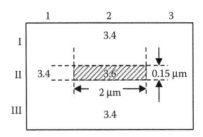

FIGURE 5.10
Calculated values of effective index in Example 5.4.

A comparison between $b'_{pq} - V$ diagrams obtained by the effective index method and computer calculations shows that when the modes are far from cutoff the approximate solutions agree with the exact numerical solution.

Example 5.4: Calculation of the effective index for the 3-D guide using GaAs-AlGaAs.

In both Regions 1 and 3, the same materials occupy Regions I, II, and III. The RI is $\bar{n}_1 = \bar{n}_3 = 3.4$.

We now find the effective index in Region 2. The normalized frequency is

$$V_2 = \frac{2\pi}{\lambda} a \left[n_{II}^2 - n_I^2 \right]^{1/2} = \frac{2\pi}{870} 150 \left[3.6^2 - 3.4^2 \right]^{1/2} = 1.284 \text{ and } \alpha_2 = 0$$

From the plot in Figure 5.6, $b_2 = 0.25$.

$$\bar{n}_2 = \left[b_2(n_{II}^2 - n_{III}^2) + n_{III}^2 \right]^{1/2} = \left[0.25(3.6^2 - 3.4^2) + 3.4^2 \right]^{1/2} = 3.46$$

We now consider the lateral slab waveguide. The V parameter is

$$V = \frac{2\pi}{870} 2000 \left[3.46^2 - 3.4^2 \right]^{1/2} = 9.266$$

Again from the plot in Figure 5.6, we obtain $b = 0.92$ for $V = 9.266$ and $\alpha = 0$. The net effective index is now

$$\bar{n} = \left[b(\bar{n}_2^2 - \bar{n}_3^2) + \bar{n}_3^2 \right]^{1/2} = \left[0.92(3.46^2 - 3.4^2) + 3.4^2 \right]^{1/2} = 3.455$$

5.7 Resonators

The simplest resonator structure used in semiconductor lasers as well as other types of lasers is an FP resonator. Basically, it is made up of two plane-parallel mirrors and it provides proper feedback to the amplifying EM waves in a gain medium, so that self-sustained oscillations occur corresponding to the different modes of the resonator. The feedback is localized as it occurs at the positions of the mirrors. On the other hand, in DFB and DBR structures, the feedback is distributed. In this section, we will discuss the resonant properties of the FP resonator. The DFB and DBR structures will be analyzed by using the coupled-mode theory in Chapter 11. In this chapter, however, the reflectivity pattern of a DBR mirror will be studied using a discrete approach.

5.7.1 Fabry–Perot Resonators

Consider the schematic of an FP interferometer formed by two plane-parallel highly reflecting mirrors, as shown in Figure 5.11. The two mirrors, M_1 and M_2, have (field) reflection and transmission coefficients, denoted by (r_1,t_1) and (r_2,t_2), respectively. Let the electric field incident at the left mirror be E_i. The field enters into the cavity and propagates to the second mirror, where it is partially transmitted and partially reflected. The fields of the various reflected and transmitted beams at the two mirrors are shown by parallel lines. Summing over the electric field of the transmitted rays coming out of mirror M_2, the total transmitted field, E_t, may be expressed as

$$E_t = E_i t_1 t_2 e^{-j\theta} \left[1 + r_1 r_2 e^{-j2\theta} + (r_1 r_2)^2 e^{-j4\theta} + \cdots \right] \tag{5.65}$$

where:

$\theta = kL$
$k = \omega n_g/c$
L is the separation between the two mirrors

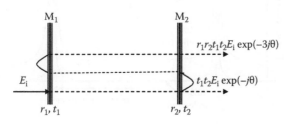

FIGURE 5.11
Schematic diagram of a Fabry–Perot resonator.

The transmittivity of the FP resonator, that is, the ratio of the output to input optical powers of the resonator, may be expressed in terms of the reflectivity of the mirrors as

$$T = \frac{|E_t|^2}{|E_i|^2} = \frac{I_{out}}{I_{in}} = \frac{(1-R_1)(1-R_2)}{(1-\sqrt{R_1 R_2})^2 + 4\sqrt{R_1 R_2}\,\sin^2\theta}$$

$$\theta = (2\pi f n_g/c)L, R_1 = r_1^2 = 1-t_1^2, R_2 = r_2^2 = 1-t_2^2 \tag{5.66}$$

The R's are power reflection coefficients. The transmittivity of the etalon, plotted in Figure 5.12, shows different peaks, the sharpness of which depends on the reflectivities of the mirrors. The peaks occur at the successive longitudinal mode frequencies given by

$$f = f_m = mc/2n_g L \tag{5.67}$$

where:

    n_g is the group index
    m is an integer

The frequency spacing between two successive transmission peaks, known as the *free spectral range* (FSR), is given by

$$\Delta f = \frac{c}{2n_g L} \tag{5.68}$$

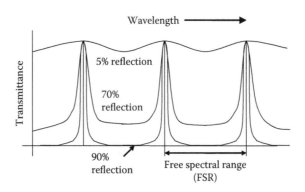

FIGURE 5.12
Transmittance of a Fabry–Perot resonator.

If the etalon is used to filter multichannel signals, the combined bandwidth should satisfy the condition:

$$\Delta f_{sig} = NS_{ch}B \tag{5.69}$$

where:

N is the number of channels
S_{ch} is the normalized channel spacing $\Delta f_{ch}/B$
B is the bit rate

The filter bandwidth, Δf_{FP}, the width of the transmission peak, should be large enough to pass the entire frequency components of the selected channel. Usually, $\Delta f_{FP} = B$. It sets the limit for the total number of channels as

$$N < \frac{\Delta f}{S_{ch}\Delta f_{FP}} = \frac{F}{S_{ch}}, \quad F = \frac{\Delta f}{\Delta f_{FP}} \tag{5.70}$$

where F is known as the finesse of the FP filter, which is related to the reflectivities of the mirrors (assumed equal for both mirrors) as

$$F = \frac{\pi\sqrt{R}}{1-R} \tag{5.71}$$

A particular channel is selected by changing the filter length electronically. Usually the stress applied to a piezoelectric transducer changes the mirror separation by a small amount and the channels are selected one by one. There are examples where the channel is selected by changing the group index of the material forming the cavity by, for example, using the TO effect induced by heating.

The tuned frequency depends on the RI of the material filling the cavity, as seen from Equation 5.68. A slight change in the RI will therefore shift the transmittivity peak.

5.7.2 Dielectric Mirrors

A dielectric mirror is formed by a stack of dielectric layers of alternating RI, as illustrated in Figure 5.13. Here $n_1 < n_2$. The thickness of each layer is a quarter-wavelength. Mirror action is achieved by the constructive interference of the rays reflected from different interfaces, as illustrated in Figure 5.13. With a sufficient number of dielectric layers, the reflectance can reach unity for a particular wavelength.

The reflection coefficient r_{12} for light in Layer 1 being reflected at the 1–2 boundary is $r_{12} = (n_2 - n_1)/(n_2 + n_1)$ and is a positive number indicating no

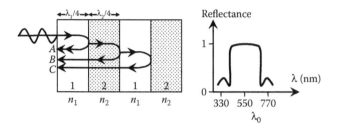

FIGURE 5.13
Schematic illustration of the principle of a dielectric mirror with many low and high RI layers and its reflectance.

phase change. The reflection coefficient for Layer 2 being reflected at the 2–1 boundary is $r_{21} = (n_1 - n_2)/(n_2 + n_1) = -r_{12}$, where the negative sign indicates a phase change of π. Consider two waves A and B in Figure 5.13 that are reflected from two consecutive interfaces. There is already a phase difference of π due to reflections at the two boundaries. Further, wave B travels an additional distance $\lambda_2/2$, and therefore suffers additional phase change of π, while meeting wave A. The two waves A and B therefore interfere constructively. It may be proved similarly that waves B and C also interfere constructively, and so on. After several layers, the transmitted intensity will be very small and the reflected light intensity will be nearly unity. Dielectric mirrors are used in VCSELs and also in realizing an FP cavity.

5.7.3 A Simplified Analysis

The reflectivity of lossless dielectric stacks with combinations of quarter- or half-wavelength thicknesses or both can be obtained by an elegant analytical technique developed by Corzine et al. [2]. Consider a schematic of an FP resonator as shown in Figure 5.14, in which there are two interfaces 0 and 1.

The field reflection coefficients for the two mirrors 0 and 1 are, respectively, r_0 and r_1. The phase shift suffered by the wave in traversing twice the distance between the two interfaces is θ. The overall field reflectivity of the combined interfaces, denoted by Γ_1, is expressed by the well-known formula:

$$\Gamma_1 = \frac{r_1 + r_0 e^{-j\theta}}{1 + r_1 r_0 e^{-j\theta}} \tag{5.72}$$

The trick to obtain the analytical expression for the reflectivity of a multilayer stack lies in the similarity between Equation 5.72 and the following trigonometrical identity:

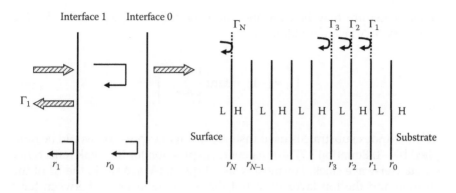

FIGURE 5.14
Multilayer dielectric stack.

$$\tanh(A+B) = \frac{\tanh A + \tanh B}{1 + \tanh A \tanh B} \tag{5.73}$$

Let us first consider the quarter-wave dielectric stack shown in Figure 5.14, in which the RI of each layer alternates sequentially between low (L) and high (H) values, and the optical thickness of each layer is equal to $\lambda/4$ of the incoming light. This ensures that $\theta = \pi$ for all layers. Let us make the following substitution:

$$|r_i| = \tanh s_i \tag{5.74}$$

The variable s_i is the transformed reflectivity of the ith interface. The overall reflectivity, Γ_N, for N layers is calculated by first determining Γ_1. Assuming that the substrate has a higher RI than the first layer, $r_0 = -|r_0|$ and $r_1 = |r_1|$. Then, by setting $\theta = \pi$, and using Equations 5.73 and 5.74, Equation 5.72 becomes

$$\Gamma_1 = \frac{|r_1| + |r_0|}{1 + |r_1||r_0|} = \frac{\tanh s_1 + \tanh s_0}{1 + \tanh s_1 \tanh s_0} = \tanh(s_1 + s_0) \tag{5.75}$$

The reflectivity Γ_2 is now calculated by replacing r_1 and r_0 by r_2 and Γ_1, noting that $r_2 = -|r_2|$, and putting $\theta = \pi$. We obtain

$$\Gamma_2 = -\frac{|r_2| + \Gamma_1}{1 + |r_2|\Gamma_1} = -\tanh(s_2 + s_1 + s_0) \tag{5.76}$$

This procedure may be continued and the following expression for the reflectivity of N layers is obtained:

$$\Gamma_N = (-1)^{N+1} \tanh\left(\sum_{i=0}^{N} s_i\right) \tag{5.77}$$

If a low-index substrate is used instead, the final expression would be multiplied by −1. Equation 5.77 is the general expression for the peak reflectivity of a quarter-wave stack. For most cases of quarter-wave stack, the RI of the substrate and the top layer vary, but the alternate layers in between have low RI n_L and high RI n_H. For this case, all s_i within the stack are equal and Equation 5.77 reduces to

$$|\Gamma_N| = \tanh\left[s_{top} + (N-1)s_{ml} + s_{sub}\right] \tag{5.78}$$

where s_{top} and s_{sub} denote, respectively, the transformed reflectivities of the top layer and the substrate and s_{ml} is the same for a multilayer stack.

We now need to write the reflectivity in terms of the RIs of two sides of a given interface. Let us choose the ith interface which has layers of low RI n_{Li} and high RI n_{Hi}. Then for normal incidence:

$$|r_i| = \frac{n_{Hi} - n_{Li}}{n_{Hi} + n_{Li}} = \frac{1 - n_{Li}/n_{Hi}}{1 + n_{Li}/n_{Hi}} \tag{5.79}$$

From Equation 5.74, we may also write

$$|r_i| = \tanh s_i = \frac{1 - e^{-2s_i}}{1 + e^{-2s_i}} \tag{5.80}$$

Comparing Equations 5.79 and 5.80, we obtain

$$s_i = -\frac{1}{2} \ln\left[\frac{n_{Li}}{n_{Hi}}\right] \tag{5.81}$$

Using Equations 5.81 and 5.74, and expanding the tanh function, we obtain for the quarter-wave stack:

$$|\Gamma_N| = \frac{1-b}{1+b} \tag{5.82a}$$

where

$$b = \prod_{i=0}^{N} \left[\frac{n_{Li}}{n_{Hi}} \right] \qquad (5.82b)$$

We note that the addition rule for the RIs in the transformed domain is replaced by the product rule in the usual domain. For the special case, when the RIs of the top layer and the substrate are different from the RIs of two materials making the stack, we may write

$$b = \left[\frac{n_{top}}{n_H} \right] \left[\frac{n_L}{n_H} \right]^{N-1} \left[\frac{n_L}{n_{sub}} \right] = \left[\frac{n_{top}}{n_{sub}} \right] \left[\frac{n_L}{n_H} \right]^{N} \qquad (5.83)$$

5.7.4 Ring Resonators

Another interesting device that is dependent on coupling between waveguides is a ring resonator [15], a schematic of which is shown in Figure 5.15. The resonator is in the form of a circular ring and it is excited by the straight guide, as a fraction of the input power is coupled to the ring. The light will circulate through the ring and suffer a phase shift $\Delta\varphi = \beta L$, where β is the propagation constant in the ring and L is the optical path length. The device will act as a resonator if this phase shift is an integral multiple of 2π. The resonance condition is given by

$$\Delta\varphi = \beta L = 2m\pi \qquad (5.84)$$

where m is an integer. Using $L = 2\pi R$, where R is the radius of the ring, and expressing β in terms of the effective index N, the resonance condition may be written as

$$\lambda = \frac{2\pi N R}{m} \qquad (5.85)$$

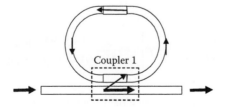

FIGURE 5.15
Structure of a ring resonator.

The analysis of the ring resonator may be performed by using the coupled-mode theory. The input and output electric fields of the straight guide are E_i and E_0, respectively. The electric field coupled into the ring is B_i and that after covering length L is B_0. The steady-state input and output fields are written as

$$E_0 = (1-\gamma)^{1/2}\left[E_i \cos(\kappa L_c) - jB_0 \sin(\kappa L_c)\right] \tag{5.86a}$$

$$B_i = (1-\gamma)^{1/2}\left[-jE_i \sin(\kappa L_c) + B_0 \cos(\kappa L_c)\right] \tag{5.86b}$$

where:
    κ is the mode coupling coefficient
    γ is the intensity insertion loss coefficient
    L_c is the coupling length

It is assumed that the propagation constants are the same, β, in straight and ring guides. If the intensity attenuation coefficient of the ring is denoted by α, B_0 is expressed as

$$B_0 = B_i \exp[(-\alpha L/2) - j\beta L] \tag{5.87}$$

The field transmittance of the ring resonator is then expressed using Equations 5.86 and 5.87 as

$$\frac{E_0}{E_i} = (1-\gamma)^{1/2}\left[\frac{\cos(\kappa L_c) - (1-\gamma)^{1/2}\exp[(-\alpha L/2) - j\beta L]}{1 - (1-\gamma)^{1/2}\cos(\kappa L_c)\exp[(-\alpha L/2) - j\beta L]}\right] \tag{5.88}$$

Introducing new parameters defined as

$$x = (1-\gamma)^{1/2}\exp(-\alpha L/2), \quad y = \cos(\kappa L_c), \quad \text{and} \quad \varphi = \beta L$$

The intensity transmittance of the optical ring resonator may be expressed as

$$T(\varphi) = \left|\frac{E_0}{E_i}\right|^2 = (1-\gamma)\left[1 - \frac{(1-x^2)(1-y^2)}{(1-xy)^2 + 4xy\sin^2(\varphi/2)}\right] \tag{5.89}$$

The transmittance characteristics of the ring resonator show similar maxima and minima as in an FP resonator (see Figure 5.12). The minimum transmission occurs when

$$\varphi = \beta L = 2m\pi \tag{5.90}$$

as noted in Equation 5.84. The maximum and minimum transmittances are given by

$$T_{max} = (1-\gamma)\frac{(x+y)^2}{(1+xy)^2} \tag{5.91a}$$

$$T_{min} = (1-\gamma)\frac{(x-y)^2}{(1-xy)^2} \tag{5.91b}$$

It is noted from these equations that $x \cong y \cong 1$ should be satisfied in order to maximize T_{max} and to minimize T_{min}. The full width at half maximum (FWHM) $\delta\varphi$ and finesse F are given by

$$\delta\varphi = \frac{2(1-xy)}{\sqrt{xy}} \tag{5.92a}$$

$$F = \frac{2\pi}{\delta\varphi} = \frac{\pi\sqrt{xy}}{(1-xy)} \tag{5.92b}$$

The value of T_{min} is zero when $x = y$, or when the condition

$$\cos(\kappa L_c) = (1-\gamma)^{1/2}\exp(-\alpha L/2) \tag{5.93}$$

is satisfied.

The FSR is determined by the spacing of two resonance peaks, for which φ differs by 2π. Writing the wavenumbers as k and $k+\Delta k$, respectively for $\varphi = 2m\pi$ and $\varphi = (2m+1)\pi$ and assuming $\Delta k \ll k$, we obtain from Equation 5.90:

$$\frac{d\beta}{dk}\Delta k = \frac{2\pi}{L} \tag{5.94}$$

Since $\beta = kn$, where n is the effective index, and the frequency shift $\Delta f = (c/2\pi)\Delta k$, we may write

$$\Delta f = \frac{c}{NL} \tag{5.95}$$

In Equation 5.95, $N = n + k(dn/dk)$ is the group index. Using the relation $\delta\varphi = \delta(\beta L) = (d\beta/dk)\delta k \cdot L = 2\pi/F$, the FWHM in terms of frequency may be expressed as

$$\delta f = \frac{c}{FNL} \qquad (5.96)$$

Ring resonators are useful as a wavelength selective filter, the principle of operation of which is illustrated by Figure 5.16. The multiple wavelengths input into Terminal 1 will be partially coupled into the ring through Coupler 1. The optical wave in the ring will be partially coupled into the straight waveguide through Coupler 2 and outputs from Terminal 2, if the wavelength, for example, λ_1, satisfies the resonant condition. The coupling of the wave with wavelength λ_1 will be enhanced and all others will be suppressed. As a result, only λ_1 will be dropped from Terminal 2, while the rest of the wavelengths will come out of Terminal 4.

Since FSR is inversely proportional to the size of the ring resonator, the ring must be small to achieve a high FSR.

The finesse is another key specification of the ring resonator and is dependent on both the internal loss and the coupling (i.e., the external loss) of the resonator. The higher the total losses, the lower the finesse of the resonator; thus, it is advantageous to reduce both the internal and external losses to obtain higher finesse. However, the external loss due to coupling is necessary and cannot be too small for the resonator to operate as an optical filter. If the external loss is smaller than the internal loss, all the coupled power will be lost inside the cavity and no power will be coupled out. Because of these constraints, the ring resonator must use a strongly guided waveguide to minimize the bending loss of a curved waveguide with a very small radius.

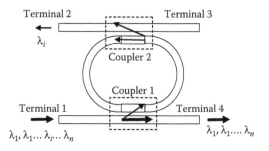

FIGURE 5.16
Schematic diagram of a ring resonator used as a wavelength selective filter.

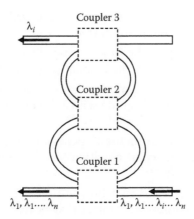

FIGURE 5.17
Schematic diagram of a multiple ring resonator.

5.7.5 Multiple Ring Resonators

To increase the finesse, two or more rings could be used as shown in Figure 5.17. In this case, the combined FSR increases since $FSR = N.FSR_1 = M.FSR_2$ and the total FWHM, in terms of wavelength, $\delta\lambda = \delta\lambda_1\delta\lambda_2/(\delta\lambda_1+\delta\lambda_2)$, decreases, so that there is a net increase of F. The two resonators have to be carefully designed to make sure that both N and M are integers. This way, the Nth peak of Resonator 1 located at the same wavelength as the Mth peak of Resonator 2 results in a sharper peak (high finesse). All the other peaks are blocked by each other.

PROBLEMS

5.1. Light rays from air fall on the core of the guide, which are then confined in the core by TIR. Prove that only the rays within a cone of half-angle α in air will suffer TIR. Calculate the value of α in terms of the RIs of air, core, and cladding.

5.2. Find the value of the numerical aperture and the maximum acceptance angle in degrees for a core layer of RI=1.5 and a cladding layer of RI=1.48.

5.3. From Equation 5.5, derive the expressions for the phase shift for the totally reflected rays.

5.4. Using the definition of Poynting's vector, derive Equation 5.6.

5.5. Set up the equation for the TM mode similar to Equations 5.10 and 5.11. Prove that the eigenvalue equations for both the modes support the fundamental mode with $m=0$.

5.6. Assume the following parameter values for an asymmetric guide: $n_1=1.5$, $n_2=1.49$, $n_3=1.40$, $\lambda_0=1.3$ μm, and the thickness of the guide $=0.3$ μm. Solve Equation 5.14 to establish that the fundamental mode cannot be supported.

5.7. Prove that to support the fundamental TE mode, the thickness of the guide should satisfy the inequality $a < (\lambda_0 / 2n_1\sqrt{2\Delta})$.

5.8. Estimate the number of modes that can be supported in a symmetric planar dielectric guide with a thickness of 100 μm. The RIs are $n_1 = 1.49$ and $n_2 = 1.47$ and $\lambda_0 = 1.55$ μm.

5.9. Assume that the nature of the TE field is similar as in Equation 3.21. Show that the eigenvalue equation for a symmetric planar waveguide is $\gamma = k\tan(ka/2)$.

5.10. Using the electric field variation assumed in Problem 5.11, show that the mode confinement factor may be expressed as

$$\Gamma = \frac{1 + 2\gamma a / D^2}{1 + (2 / \gamma D)}$$

where $D = k_0 (n_1^2 - n_2^2)^{1/2} a$.

5.11. Calculate the mode confinement factor for $In_{0.53}Ga_{0.47}As/InP$ waveguides by solving the wave equation. Also, calculate the approximate value by using the simplified formula [5] and compare the two values. The dielectric constants are: $\varepsilon(InP) = 12.56$, $\varepsilon(InAs) = 15.15$, and $\varepsilon(GaAs) = 12.9$. Assume linear interpolation.

5.12. Prove that the ratio of the transmitted to the incident intensities in an FP resonator filled with a lossy material with loss coefficient α, may be expressed as

$$\frac{I_t}{I_i} = \frac{(1 - R)^2 e^{-\alpha L}}{(1 - R e^{-\alpha L})^2 + 4R e^{-\alpha L} \sin^2 \theta}$$

where θ has been defined in Equation 5.66 and the reflectivities of the two mirrors are assumed equal.

5.13. Express the loss in terms of the ratio of maximum to minimum transmitted intensities.

5.14. Show that the expression for the transmittivity of an FP resonator, Equation 5.66, reduces to the expression given by

$$T = \frac{T_{max}}{1 + \left(\dfrac{4nL}{\pi\Delta\lambda}\right)^2 \sin^2\left(\dfrac{2\pi nL}{\lambda}\right)}$$

5.15. A dielectric mirror using alternate layers of Si and SiO_2 is used to form a resonant cavity for a light-emitting device at 850 nm. Calculate the thickness of each layer; ($n(SiO_2) = 1.5$ and $n(Si) = 3.4$).

5.16. Prove that if the transmission of an FP resonator is represented by a Lorentzian, then the response is $g(f) = [(f - f_0)^2 + (f_0/2Q)^2]^{-1}$, where f is the frequency, f_0 is the resonant frequency, and Q is the quality factor of the resonator.

5.17. The cavity lifetime is defined as $dP/dt = -P/\tau_c$, where $P(t)$ is the power at time t. Prove that power decays as $P(t) = P(0)\exp(-t/\tau_c) = P(0)\exp(\omega t/Q)$. From this, show that the linewidth is $\Delta f = (2\pi\tau_c)^{-1}$ and $\tau_c = \lambda^2/(2\pi c\Delta\lambda)$, where $\Delta\lambda$ is the FWHM.

5.18. Obtain an expression for the photon lifetime in an FP resonator enclosing a medium with loss=α.

5.19. Draw different rays, as shown in Figure 5.11, showing the field intensities coming out of Mirror 2. The FP resonator, however, encloses a medium of gain = g and loss coefficient = α. By summing the field intensities, obtain the condition to prove that the self-sustained laser oscillation condition is $g = \alpha + (1/2L)\ln(1/R_1R_2)$.

5.20. Calculate the reflectivity of a stack consisting of alternate layers of AlAs ($n_L = 3.2$) and GaAs ($n_H = 3.6$) by using air as the top layer and GaAs as the substrate. The value of $N = 10$ is assumed.

Reading List

Agrawal, G. P., *Fiber Optic Communication Systems*, 3rd edn. Wiley, Hoboken, NJ, 2002.

Buckman, A. B., *Guided-Wave Photonics*. Saunders College Publication, Fort Worth, TX, 1992.

Chuang, S. L., *Physics of Optoelectronic Devices*. Wiley, New York, 1995.

Hunsperger, R. G., *Integrated Optics: Theory and Technology*, 3rd edn. Springer, Berlin, 1991.

Lee, D. L., *Electromagnetic Principles of Integrated Optics*. Wiley, New York, 1986.

Nishihara, H., M. Haruna, and T. Suhara, *Optical Integrated Circuits*. McGraw-Hill, New York, 1989.

Okamoto, K., *Fundamentals of Optical Waveguides*. Academic Press, San Diego, CA, 2000.

Ramaswami, R. and K. N. Sivarajan, *Optical Networks: A Practical Perspective*, 2nd edn. Morgan Kaufmann, Burlington, MA, 2002.

Verdeyen, J. T., *Laser Electronics*, 3rd edn. Prentice Hall, Upper Saddle River, NJ, 1995.

Yariv, A., *Quantum Electronics*. Wiley, New York, 1993.

References

1. Kogelnik, H. and V. Ramaswamy, Scaling rules for thin film optical waveguides, *Appl. Opt.* 13, 1857–1862, 1974.
2. Corzine, S. W., R. H. Yan, and L. A. Coldren, A tanh substitution technique for the analysis of abrupt and graded interface multilayer dielectric stack, *IEEE J. Quantum Electron.* 27, 2086–2090, 1991.
3. Marcatili, E. A. J., Dielectric rectangular waveguide and directional coupler for integrated optics, *Bell Syst. Tech. J.* 48, 2071–2102, 1969.
4. Buus, J., The effective index method and its application in semiconductor lasers, *IEEE J. Quantum Electron.* QE-18, 1083–1089, 1982.
5. Botez, D., InGaAsP/InP double-heterostructure lasers: Simple expressions for wave confinement, beamwidth, and threshold current over wide ranges in wavelength (1.1–1.65 µm), *IEEE J Quantum Electron.* QE-17, 178, 1981.

6

Optical Processes

6.1 Introduction

This chapter outlines the theory of optical processes in bulk semiconductors. First of all, the relationship between the absorption coefficient and the refractive index is developed from Maxwell's equations. This is followed by a classification of different absorption processes in semiconductors. The theory of absorption is then developed by using a semiclassical formalism, first for the fundamental band-to-band transition in a direct-bandgap semiconductor and next for the intervalence band absorption (IVBA). A simple classical Drude model is employed to treat the free-carrier absorption (FCA). The opposite process of absorption, that is, recombination and luminescence due to direct interband transition, is then introduced. The theory of Auger recombination is developed for the most important transition, while other recombination processes, trap-assisted recombination and surface recombination, are discussed briefly. A brief description of carrier-induced changes in the bandgap, and the refractive index, then follows. Finally, the theory of excitonic absorption in direct-gap materials is developed.

6.2 Optical Constants

In this section, the relationship between different optical constants in a conductor or a dielectric and their interrelationship will be presented. The starting equations are the following four Maxwell's equations related to the behavior of electromagnetic waves:

$$\nabla \times \mathbf{F} + \frac{\partial \mathbf{B}}{\partial t} = 0 \tag{6.1a}$$

$$\nabla \times \mathbf{H} - \frac{\partial \mathbf{D}}{\partial t} = \mathbf{J} \tag{6.1b}$$

$$\nabla \cdot \mathbf{D} = \rho \tag{6.1c}$$

$$\nabla \cdot \mathbf{B} = 0 \tag{6.1d}$$

where:
   F and **H** are the electric and magnetic fields
   $\mathbf{D} = \varepsilon \mathbf{F}$, $\mathbf{B} = \mu \mathbf{H}$, ε and μ being, respectively, the permittivity and permittivity of the medium
   J and ρ are the current and charge densities, respectively

The electric and magnetic fields are expressed in terms of the vector potential **A** and the scalar potential φ as follows:

$$\mathbf{F} = -\frac{\partial \mathbf{A}}{\partial t} - \nabla \phi \tag{6.2}$$

$$\mathbf{B} = \nabla \times \mathbf{A} \tag{6.3}$$

Taking the curl of the second Maxwell equation and using the third one, we may write

$$\frac{1}{\mu_0} \nabla \times \nabla \times \mathbf{A} + \varepsilon \frac{\partial^2 \mathbf{A}}{\partial t^2} + \varepsilon \nabla \frac{\partial \phi}{\partial t} = \mathbf{J} \tag{6.4}$$

We assume $\mu = \mu_0$: the permeability for free space. We note that $\nabla \times \nabla \times \mathbf{A} = \nabla(\nabla \cdot \mathbf{A}) - \nabla^2 \mathbf{A}$. It is usual to assume Coulomb gauge, in which $\varphi = 0$ and $\nabla \cdot \mathbf{A} = 0$. Therefore,

$$\frac{1}{\mu_0} \nabla^2 \mathbf{A} - \varepsilon \frac{\partial^2 \mathbf{A}}{\partial t^2} = \mathbf{J} \tag{6.5}$$

If $\mathbf{J} = 0$, the solutions for **A** are plane waves, expressed as

$$\mathbf{A}(\mathbf{r},t) = \mathbf{A}_0 \left\{ \exp[j(\mathbf{q.r} - \omega t)] + \text{c.c.} \right\} \tag{6.6}$$

where c.c. stands for complex conjugate, $q^2 = \varepsilon \mu_0 \omega^2$, and the velocity of light in free space is $c = (\varepsilon_0 \mu_0)^{1/2}$. The electric and magnetic fields are given by

$$\mathbf{F} = \frac{\partial \mathbf{A}}{\partial t} = -2\omega \mathbf{A}_0 \sin(\mathbf{q} \cdot \mathbf{r} - \omega t) \tag{6.7}$$

$$\mathbf{B} = \nabla \times \mathbf{A} = -2\mathbf{q} \times \mathbf{A}_0 \sin(\mathbf{q} \cdot \mathbf{r} - \omega t) \tag{6.8}$$

The Poynting vector $\mathbf{S}$ representing the optical power is expressed as

$$\mathbf{S} = (\mathbf{F} \times \mathbf{H}) = \frac{4}{\mu_0} v q^2 |\mathbf{A}_0|^2 \sin^2(\mathbf{q} \cdot \mathbf{r} - \omega t) \hat{\mathbf{q}} \tag{6.9}$$

where:
  $\hat{\mathbf{q}}$ is a unit vector in the direction of $\mathbf{q}$
  $v = \omega/|\mathbf{q}|$ is the velocity of light in the medium having permittivity $\varepsilon = \varepsilon_0 \varepsilon_r$, ε_r being the relative permittivity

Taking the time-averaged value of $\mathbf{S}$, we obtain the expression for the energy density as

$$I = \left| \frac{\mathbf{S}}{v} \right| = \frac{2\varepsilon \omega^2 |\mathbf{A}_0|^2}{c^2} \tag{6.10}$$

We now consider a conducting material in which the current density $\mathbf{J} = \sigma \mathbf{F}$, where σ is the conductivity. The wave equation for the electric field is obtained as

$$\nabla^2 \mathbf{F} = \varepsilon \mu_0 \frac{\partial^2 \mathbf{F}}{\partial t^2} + \sigma \mu_0 \frac{\partial \mathbf{F}}{\partial t} \tag{6.11}$$

The general solution for the field is

$$\mathbf{F} = \mathbf{F}_0 \exp\{j(\mathbf{q} \cdot \mathbf{r} - \omega t)\} \tag{6.12}$$

Putting this form in Equation 6.11, we obtain

$$-q^2 = -\varepsilon \mu_0 \omega^2 - j\sigma \mu_0 \omega \tag{6.13}$$

Since q is a complex number, we may introduce a complex refractive index n to account for the change in phase velocity from the free-space velocity. Thus,

$$n = n_r + jn_i = \left(\varepsilon_r + \frac{j\sigma \mu_0}{\omega} \right)^2 \tag{6.14}$$

so that

$$q = \frac{n_r \omega}{c} + j n_i \frac{\omega}{c} \tag{6.15}$$

Here the subscripts r and i of n represent the real and the imaginary parts, respectively. The electric field, assuming that the wave propagates along the z-direction, may now be expressed as

$$\mathbf{F} = \mathbf{F}_0 \exp\left\{ j\omega\left(\frac{n_r z}{c} - t \right) \right\} \exp\left(\frac{-n_i \omega z}{c} \right) \tag{6.16}$$

The field amplitude is damped exponentially as it propagates along the z-direction. The intensity of light decreases exponentially as

$$I(z) = I_0 \exp(-\alpha z) \tag{6.17a}$$

where α is the absorption coefficient. Since the intensity is proportional to F^2, the absorption coefficient is given by

$$\alpha = \frac{2 n_i \omega}{c} \tag{6.18}$$

The relation between the wave vector q and the angular frequency ω can also be expressed as

$$c^2 q^2 = \omega^2 \varepsilon_r(\omega) \tag{6.19}$$

The quantity $\varepsilon_r(\omega)$ is called the dielectric function. Using Equation 6.19, we may rewrite Equation 6.12 in the following form:

$$\mathbf{F}(\mathbf{r}, t) = \mathbf{F}_0 \exp\left[j\omega\left(\frac{n(\omega)}{c} \mathbf{q} \cdot \mathbf{r} - t \right) \right] \hat{\mathbf{q}} \tag{6.20}$$

where the refractive index $n(\omega)$ is given by

$$n(\omega) = \left[\varepsilon_r(\omega) \right]^{1/2} \tag{6.21}$$

When a conductor is considered, the complex conductivity can be related to the permittivity by the following expressions:

$$\bar{\sigma} = \sigma - j\omega\varepsilon_r\varepsilon_0 \quad \text{and} \quad \bar{\varepsilon} = \varepsilon_r + \left(\frac{j}{\varepsilon_0\omega}\right)\sigma = \left(\frac{j}{\varepsilon_0\omega}\right)\bar{\sigma} \tag{6.22}$$

The dielectric function $\varepsilon(\omega)$ and the refractive index $n(\omega)$ are complex quantities expressed as follows:

$$\bar{\varepsilon} = \varepsilon(\omega) = \varepsilon'_r(\omega) + j\varepsilon''_r(\omega) \quad \text{and} \quad N(\omega) = n(\omega) + jK(\omega)$$

The real part of the RI, $n(\omega)$, and the imaginary part of the RI, $K(\omega)$, called the *extinction coefficient*, are related to the real and imaginary parts of the dielectric function by

$$\varepsilon'_r(\omega) = n^2 - K^2 \tag{6.23a}$$

$$\varepsilon''_r(\omega) = 2nK \tag{6.23b}$$

We now use Equation 6.17, assuming that the absorption coefficient of the material, $\alpha(\omega)$, depends on frequency, and obtain the following relation:

$$\alpha(\omega) = \frac{2\omega K(\omega)}{c} = \frac{\omega\varepsilon''_r(\omega)}{cn(\omega)} \tag{6.24}$$

To complete this section, one should note that the real and imaginary parts of the dielectric function are not independent, but are related to each other by the Kramers–Kronig relation:

$$\varepsilon'(\omega) = \frac{1}{\pi}P\int_{-\infty}^{\infty}\frac{\varepsilon''(\omega')d\omega'}{\omega' - \omega} \tag{6.25a}$$

$$\varepsilon''(\omega) = -\frac{1}{\pi}P\int_{-\infty}^{\infty}\frac{\varepsilon'(\omega')d\omega'}{\omega' - \omega} \tag{6.25b}$$

where P stands for the Cauchy principal value of the integral that follows.

6.3 Absorption Processes in Semiconductors

The absorption processes in semiconductors can best be understood with the help of quantum mechanical theory. This theory treats the radiation as

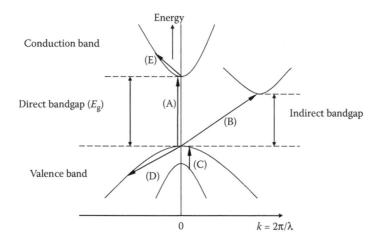

FIGURE 6.1
Illustration of the different absorption processes in semiconductors.

a bunch of photons. When incident on a semiconductor, this bunch induces transitions among different energy levels in different bands in the material. A few representative transitions occurring in a semiconductor are illustrated in Figure 6.1 using the *E*–**k** diagram.

The process indicated by (A) in Figure 6.1 represents direct valence-to-conduction-band transitions (constant *k*-vector) that occur in a direct-bandgap semiconductor such as GaAs. Process (B) illustrates transition from the valence band (VB) to the indirect conduction band (CB) involving a photon and a scattering agent such as a phonon. Such transitions occur in indirect-gap semiconductors such as Ge and Si. Process (C) represents intervalence band transitions; in the figure, transition between the HH and the SO bands has been illustrated, though other types, such as HH to LH transitions, are also possible. Process (D) depicts free-carrier transitions in a VB involving scattering by impurities, phonons, or other mechanisms. Process (E) is for free-carrier transitions in the CB aided by impurities or photon–phonon interactions. Depending on the photon energy, a single type or a combination of a few types of all the absorption processes indicated in the figure may be important. We shall discuss the origin and characteristics of these processes one by one. Indirect band-to-band transition (Process B) is not of interest in the present context.

6.4 Fundamental Absorption in Direct Gap

The first and foremost requirement for a semiconductor material to be useful for lasers is that it should have a direct bandgap. The knowledge of the absorption or the gain coefficient of a direct-gap semiconductor is therefore of extreme importance for the design of a semiconductor laser. In this section, we present the basic conservation laws in direct-gap materials and then proceed to obtain the expression for the absorption coefficient for direct-gap semiconductors.

6.4.1 Conservation Laws

When photons of energy greater than the fundamental bandgap are incident on a semiconductor sample, electrons from the VB absorb the photons and move to the CB. As a consequence, an empty state or a hole is created in the VB. The absorption of photons thus leads to the generation of excess electron–hole pairs (EHPs). The absorption coefficient rises rapidly as the photon energy $\hbar\omega \geq E_g$, the bandgap energy. This process is shown schematically in Figure 6.2.

Two conditions must be fulfilled for this interband transition. The first is energy conservation, that is, the energy of the electron should equal the sum of the energies of the hole and the photon. This leads to the expression

$$E_c(\mathbf{k}_c) = E_v(\mathbf{k}_v) + \hbar\omega \qquad (6.26)$$

where **k**s denote the wave vectors for the respective particles.

The absorption threshold occurs when the photon energy equals the bandgap energy. We may define a threshold or cutoff wavelength from Equation 6.26 as follows:

$$\lambda_c = \frac{hc}{E_g} = \frac{1.24}{E_g \, (\text{eV})} \, (\mu\text{m})$$

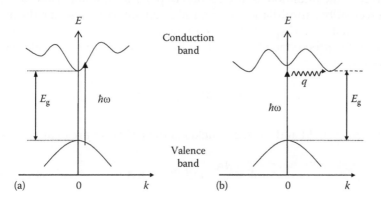

FIGURE 6.2

Fundamental absorption process in a direct-gap semiconductor (a) and in an indirect-gap semiconductor (b).

The last ratio indicates that the value of the cutoff wavelength is in micrometers when the energy is in electronvolts.

The second condition to be satisfied is momentum conservation, stated as follows:

$$\mathbf{k_c} = \mathbf{k_v} + \mathbf{q} \tag{6.27}$$

where $\mathbf{k_c}$ and $\mathbf{k_v}$ are, respectively, the electron and the hole wave vectors. However, in most situations, the photon wave vector $\mathbf{q} \ll \mathbf{k_c}$ and $\mathbf{k_v}$. Therefore, we obtain

$$\mathbf{k_c} = \mathbf{k_v} \tag{6.28}$$

This **k**-conservation is indicated by a vertical line in the E–**k** diagrams in Figures 6.1 and 6.2.

> **Example 6.1:** If the wavelength of the photon is 1 μm, the wave vector is $q = 2\pi/\lambda = 2\pi \times 10^6\,\mathrm{m}^{-1}$. Let the electron energy be 25 meV, which leads to a value for the wave vector of $k_c = (2m_e E/\hbar^2)^{1/2} = 2.56 \times 10^8\,\mathrm{m}^{-1}$ for $m_e = 0.1m_0$. Thus, $q \ll k_e$.

6.4.2 Calculation of Absorption Coefficient

The calculation of the absorption coefficient is performed by quantum-mechanically calculating the probability of transition from a VB state to a CB state. Here, we shall use the semiclassical approach by treating electromagnetic (EM) waves classically but describing the electrons as quantum-mechanical (Bloch) waves (please see Reading List). An alternative way is to consider the radiation as a collection of particles (photons) and to use the photon creation, annihilation, and number operators to describe the absorption, stimulated, and spontaneous emissions.

The unperturbed one-electron Hamiltonian in the semiclassical treatment is given by

$$H_0 = p^2/2m_0 + V(\mathbf{r}) \tag{6.29}$$

The electric field and magnetic fields of the EM wave are expressed as

$$\mathbf{F} = \frac{\partial \mathbf{A}}{\partial t} \quad \text{and} \quad \mathbf{B} = \nabla \times \mathbf{A} \tag{6.30}$$

$$H = \frac{1}{2m_0}(\mathbf{p} + e\mathbf{A})^2 + V(r) \tag{6.31}$$

Writing the term $(\mathbf{p}+e\mathbf{A})^2 = (e\mathbf{A}\cdot\mathbf{p})+(e\mathbf{p}\cdot\mathbf{A})$ and noting that $\mathbf{p}=(\hbar/j)\nabla$, we may write

$$(\mathbf{p}\cdot\mathbf{A})f(r) = \mathbf{A}\cdot\left(\frac{\hbar}{j}\nabla f\right)+\left(\frac{\hbar}{j}\nabla\cdot\mathbf{A}\right)f \tag{6.32}$$

It is safe to neglect the term involving A^2 in Equation 6.31 (see Example 6.2) so that first-order perturbation theory is valid.

Since $\nabla\cdot\mathbf{A}=0$, the Hamiltonian given by Equation 6.31 can now be approximated as

$$H = H_0 + \frac{e}{m_0}\mathbf{A}\cdot\mathbf{p} \tag{6.33}$$

The second term on the right-hand side of Equation 6.33 represents the interaction between radiation and Bloch electrons and is referred to as the electron–radiation interaction Hamiltonian H_{eR}. It is written in two different forms as

$$H_{eR} = \frac{e}{m_0}\mathbf{A}\cdot\mathbf{p} \text{ or } H_{eR} = (-e)\mathbf{r}\cdot\mathbf{F} \tag{6.34}$$

The equivalence of the two can be established when the wave vector of the EM field is small, the so-called electric dipole approximation. The first form of H_{eR} is more frequently used.

> **Example 6.2:** In arriving at Equation 6.32, the term containing A^2 has been neglected in comparison to the $\mathbf{A}\cdot\mathbf{p}$ term. To illustrate this omission, we note that $eA/2p = e|F_0|/2\omega p \approx (e/\omega p)(2I/\varepsilon_0 c n_r)^2$, where we use $\mathbf{F}=-\partial\mathbf{A}/\partial t$ and the relation between electric field and average intensity. Let $I=10^4$ W m^{-2}, $\omega=10^{14}$ s^{-1}, and $p=(3k_B T/m_0)^{1/2}=1\times10^{-25}$ kg m s^{-1}. The above ratio is $\sim 0.5\times10^{-4}$. Hence, the neglect of the A^2 term is justified.

Our next task is to evaluate the rate of transition, which involves the following squared matrix element for transition from a VB state $|v\rangle$ with energy E_v and wave vector $\mathbf{k}_v$ to a CB state $|c\rangle$ with energy E_c and wave vector $\mathbf{k}_c$:

$$\left|\langle c|H_{eR}|v\rangle\right|^2 = (e/m_0)^2\left|\langle c|\mathbf{A}\cdot\mathbf{p}|v\rangle\right|^2 \tag{6.35}$$

The matrix element involves integration in space. The time integration involving $\exp(j\omega t)$ and corresponding time dependencies in Bloch functions may be written as

$$\int \exp\left(\frac{jE_c t}{\hbar}\right) \exp(-j\omega t) \exp\left(\frac{-jE_v t}{\hbar}\right) \propto \delta[E_c(\mathbf{k}_c) - E_v(\mathbf{k}_v) - \hbar\omega] \quad (6.36)$$

The δ-function indicates energy conservation in the absorption process, in which an electron in the VB with energy E_v absorbs a photon of energy $\hbar\omega$ and is excited to the CB state of energy E_c. Similarly, the matrix element involving the complex conjugate term, that is, $\langle c|\exp(j\omega t)|v\rangle$ of the vector potential in Equation 6.35, gives rise to $\delta[E_c(\mathbf{k}_c) - E_v(\mathbf{k}_v) + \hbar\omega]$ and describes stimulated emission of a photon due to transition from state E_c to state E_v. Writing the Bloch functions of electrons and holes, respectively, as

$$|c\rangle = U_{c,k_c}(\mathbf{r}) \exp(j(\mathbf{k}_c \cdot \mathbf{r}) \quad (6.37)$$

and

$$|v\rangle = U_{v,k_v}(\mathbf{r}) \exp(j(\mathbf{k}_v \cdot \mathbf{r}) \quad (6.38)$$

and using the form of vector potential as given by Equation 6.6, we obtain

$$\left|\langle c|\mathbf{A} \cdot \mathbf{p}|v\rangle\right|^2 = |\mathbf{A}_0|^2 \left|\int U_{c,k_c}^* \exp[j(\mathbf{q} - \mathbf{k}_c) \cdot \mathbf{r}](\hat{\mathbf{q}} \cdot \mathbf{p})U_{v,k_v} \exp(j\mathbf{k}_v \cdot \mathbf{r})d\mathbf{r}\right|^2$$

$$(6.39)$$

Since **p** involves the differential operator, we may write

$$\mathbf{p}U_{v,k_v} \exp(j\mathbf{k}_v \cdot \mathbf{r}) = \exp(j\mathbf{k}_v \cdot \mathbf{r})\mathbf{p}U_{v,k_v} + \hbar\mathbf{k}_v U_{v,k_v} \exp(j\mathbf{k}_v \cdot \mathbf{r}) \quad (6.40)$$

The second term above, when multiplied by U_{c,k_c}^* and integrated over space, vanishes due to orthogonality of U_{c,k_c}^* and U_{v,k_v}. We are now left with an integral containing the first term, which can be broken into two parts as follows in view of the periodicity of functions U_c and U_v:

$$\int U_{c,k_c}^* \exp[j(\mathbf{q} - \mathbf{k}_c + \mathbf{k}_v) \cdot \mathbf{r}]\mathbf{p}U_{v,k_v}d\mathbf{r}$$

$$= \left(\sum_l \exp[j(\mathbf{q} - \mathbf{k}_c + \mathbf{k}_v) \cdot \mathbf{R}_l]\right) \int_{\text{unit cell}} U_{c,k_c}^* \exp[j(\mathbf{q} - \mathbf{k}_c + \mathbf{k}_v) \cdot \mathbf{r}']\mathbf{p}U_{v,k_v}d\mathbf{r}'$$

$$(6.41)$$

In the above, we have written $\mathbf{r}=\mathbf{R_l}+\mathbf{r'}$, where $\mathbf{R_l}$ is a lattice vector and $\mathbf{r'}$ lies within one unit cell. One notes that

$$\sum_l \exp\left[j(\mathbf{q}-\mathbf{k_c}+\mathbf{k_v})\cdot\mathbf{R_l}\right]=\delta(\mathbf{q}-\mathbf{k_c}+\mathbf{k_v})$$

The delta function ensures wave vector (momentum) conservation in the absorption process. Using the conservation condition in the integral over unit cell in Equation 6.41 yields the following simplified result:

$$\int_{\text{unit cell}} U^*_{c,k_c}\exp[j(\mathbf{q}-\mathbf{k_c}+\mathbf{k_v})\cdot\mathbf{r'}]\mathbf{p}U_{v,k_v}\,d\mathbf{r'} = \int_{\text{unit cell}} U^*_{c,k_v+q}\mathbf{p}U_{v,k_v}\,d\mathbf{r'} \tag{6.42}$$

This equation is further simplified by assuming $\mathbf{q}\ll\mathbf{k_v}$. Expanding $U_{k+q}\approx U_{k+q}\cdot\nabla_k U_k$, and keeping only the first term, the matrix element may be expressed as

$$\left|\langle c|\hat{\mathbf{e}}\cdot\mathbf{p}|v\rangle\right|^2 = \left(\int_{\text{unit cell}} U^*_{c,k}(\hat{\mathbf{e}}\cdot\mathbf{p})U_{v,k}\,d\mathbf{r'}\right)^2 \tag{6.43}$$

This approximation is known as dipole approximation. In most cases, the momentum matrix element defined above does not depend on $\mathbf{k}$. We therefore write

$$\left|\langle c|H_{eR}|v\rangle\right|^2 = \left(\frac{e}{m_0}\right)^2|A_0|^2|(\mathbf{e}\cdot\mathbf{p})_{cv}|^2 \tag{6.44}$$

where $|(\hat{\mathbf{e}}\cdot\mathbf{p})_{cv}|$ denotes the integral over the unit cell. The transition probability R for photon absorption per unit time by using Fermi's Golden Rule is expressed as

$$R=\left(\frac{2\pi}{\hbar}\right)\sum_{k_c,k_v}\left|\langle c|H_{eR}|v\rangle\right|^2\delta[E_c(\mathbf{k_c})-E_v(\mathbf{k_v})-\hbar\omega] \tag{6.45}$$

Considering unit volume of the crystal, the above equation gives the rate of transition per unit volume due to the absorption process, and $R\hbar\omega$ gives the rate of loss of power per unit volume due to absorption of photons. The power loss can also be expressed in terms of the absorption coefficient α. If I denotes the intensity, then the rate of decrease of I may be expressed as

$$-\frac{dI}{dt} = -\left(\frac{dI}{dz}\right)\left(\frac{dz}{dt}\right) = \frac{c}{n}\alpha I \qquad (6.46)$$

Thus,

$$\alpha = \frac{1}{I}\frac{n}{c}\left(\frac{dI}{dt}\right) = \frac{1}{I}\frac{n}{c}R\hbar\omega \qquad (6.47)$$

Both R and I contain $|\mathbf{A}_0|^2$, and therefore α is independent of intensity of light.

Evaluation of the transition rate R involves summation over $\mathbf{k}_c$ and $\mathbf{k}_v$ as given in Equation 6.45. We write $\mathbf{k}_c = \mathbf{k}_v = \mathbf{k}$ and convert summation into integration by noting $\Sigma_\mathbf{k} \to \left(2/(2\pi)^3\right)\int 4\pi k^2 dk$. The integral is therefore

$$I_k = \sum_\mathbf{k} \delta[E_c(\mathbf{k}) - E_v(\mathbf{k}) - \hbar\omega] = \frac{2}{8\pi^3}\int 4\pi k^2 dk \delta[E_c(\mathbf{k}) - E_v(\mathbf{k}) - \hbar\omega] \quad (6.48)$$

Assuming parabolic E–$\mathbf{k}$ relationships, we have

$$E_c(\mathbf{k}) = E_g + \frac{\hbar^2 k^2}{2m_e} \quad \text{and} \quad E_v(\mathbf{k}) = -\frac{\hbar^2 k^2}{2m_h} \qquad (6.49)$$

where m_e and m_h are the electron and hole effective masses, respectively. The argument of the δ-function becomes $E_g + (\hbar^2 k^2/2m_r) - \hbar\omega$, where the reduced mass is given by

$$m_r^{-1} = m_e^{-1} + m_h^{-1} \qquad (6.50)$$

The integration is performed by transforming k to $E(k)$ and then using the δ-function. Using the expressions for R and I, the final expression for absorption coefficient becomes

$$\alpha(\hbar\omega) = \frac{e^2(2m_r)^{3/2}}{2\pi\varepsilon_0 cnm_0^2\hbar^3\omega}\left(\hbar\omega - E_g\right)^{1/2}\left\langle|p_{cv}^2|\right\rangle, \quad \hbar\omega \geq E_g \qquad (6.51)$$

The term $\langle|p_{cv}^2|\rangle$ is the average of squared momentum matrix element for unpolarized light. The average is obtained by considering the polarization dependence of the term $|(\hat{\mathbf{e}} \cdot \mathbf{p})_{cv}|$.

For transitions involving states near the band edges, both $U_{c,k}$ and $U_{v,k}$ are given by their zone center values. In this approximation,

$$\text{Conduction band: } U_{c,0} = |s\rangle \tag{6.52}$$

$$\text{Valence band: heavy-hole states } |3/2, 3/2\rangle = \left(-1/\sqrt{2}\right)\left(|p_x\rangle + j|p_y\rangle\right)|\alpha\rangle$$

$$|3/2, -3/2\rangle = \left(1/\sqrt{2}\right)\left(|p_x\rangle - j|p_y\rangle\right)|\beta\rangle \tag{6.53a}$$

$$\text{Light-hole states: } |3/2, 1/2\rangle = (-1/6)\left(|p_x\rangle + j|p_y\rangle\right)|\beta\rangle - 2|p_z\rangle|\alpha\rangle \tag{6.53b}$$

Symmetry allows only the following matrix elements to be nonzero:

$$\langle p_x|p_x|s\rangle = \langle p_y|p_y|s\rangle = \langle p_z|p_z|s\rangle = p_{cv} \tag{6.54}$$

This identifies the following allowed transitions with the corresponding matrix elements:

$$\langle HH|p_x|s\rangle = \langle HH|p_y|s\rangle = \left(1/\sqrt{2}\right)\langle p_x|p_x|s\rangle \tag{6.55a}$$

$$\langle LH|p_x|s\rangle = \langle LH|p_y|s\rangle = \left(1/\sqrt{6}\right)\langle p_x|p_x|s\rangle \tag{6.55b}$$

It is also to be noted that

$$\langle HH|p_z|s\rangle = 0 \tag{6.56}$$

The squared matrix element for light polarized along different directions may now be listed. This dependence on orientation is useful in quantum wells (QWs), where heavy-hole (HH) and light-hole (LH) states are no longer degenerate.

z-polarized light: HH→CB: No coupling

$$\text{LH} \rightarrow \text{CB: } |p_{cv}|^2 = (2/3)\left|\langle p_x|p_x|s\rangle\right|^2 \tag{6.57}$$

x-polarized light: HH→CB: $|p_{cv}|^2 = (1/2)|\langle p_x|p_x|s\rangle|^2$

$$\text{LH} \to \text{CB:} \quad \left|p_{cv}\right|^2 = (1/6)\left|\langle p_x|p_x|s\rangle\right|^2 \tag{6.58}$$

y-polarized light: HH→CB: $|p_{cv}|^2 = (1/2)|\langle p_x|p_x|s\rangle|^2$

$$\text{LH} \to \text{CB:} \quad \left|p_{cv}\right|^2 = (1/6)\left|\langle p_x|p_x|s\rangle\right|^2 \tag{6.59}$$

For finite values of k, there is some mixing of HH and LH states. As may be seen for x–y-polarized light, HH–CB coupling is three times stronger than LH–CB coupling.

In the present context, we are considering unpolarized light. Therefore, using Equations 6.57–6.59 we may write

$$\left\langle \left|p_{cv}\right|^2 \right\rangle = \frac{2m_0^2 E_g(E_g + \Delta)}{3m_e(E_g + 2\Delta/3)} \tag{6.60}$$

In the literature, a quantity E_P, defined in Equation 6.61, is used to describe the momentum matrix element:

$$E_P = \frac{2}{m_0}\left|\langle p_x|p_x|s\rangle\right|^2 \tag{6.61}$$

The values of E_p for most semiconductors lie in the range 20–25 eV. The final expression for the absorption coefficient is therefore

$$\alpha(\hbar\omega) = \frac{e^2(2m_r)^{3/2}}{2\pi\varepsilon_0 cnm_0^2\hbar^3\omega}\left(\hbar\omega - E_g\right)^{1/2}\left\langle \left|p_{cv}\right|^2 \right\rangle \tag{6.62}$$

The absorption coefficients for a few common semiconductors as a function of wavelength are presented in Figure 6.3. The presence of a cutoff wavelength for direct-gap materials such as GaAs, InP, and the ternary and quaternary alloys can easily be identified in the diagram. The rise in absorption below λ_c (above E_g) is also apparent.

The absorption coefficient is sometimes expressed in terms of a dimensionless quantity called the *oscillator strength*, denoted as f_{vc} and defined as

$$f_{vc} = \frac{2P^2}{m_0(E_{kc} - E_{kv})} \tag{6.63}$$

The absorption coefficient may be expressed as follows in terms of the oscillator strength:

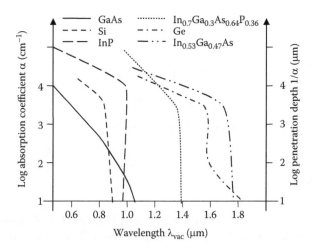

FIGURE 6.3
Absorption coefficients of a few common semiconductors as a function of wavelength. The penetration depths in micrometers are shown on the right.

$$\alpha(\omega) = \frac{e^2 (2m_r)^{3/2}}{4\pi\varepsilon_0 c m_0 n(\omega) \hbar^2} f_{vc} \left(\hbar\omega - E_g\right)^{1/2} \tag{6.64}$$

The absorption coefficient expressed by Equation 6.62 applies to the ideal situation for 0 K, when all states in the VB are full and all states in the CB are empty. The expression becomes modified at finite temperature by including the Fermi functions for both electrons and holes, as will be discussed in Chapter 7.

> **Example 6.3:** We use Equation 6.62 to calculate the absorption coefficient in GaAs. We take $m_r = 0.065 m_0$, $n = 3.6$, a factor of 2/3 for unpolarized light, and $2|p_{cv}|^2/m_0 = E_p = 23$ eV. The absorption coefficient is $\alpha = 5.25 \times 10^4 [(\hbar\omega - E_g)^{1/2}/\hbar\omega]$ cm^{-1}, and the energies are in electronvolts. Take $\hbar\omega = 1.5$ eV and $E_g = 1.43$ eV. The value is 0.97×10^4 cm^{-1} at 0 K. The value is reduced at finite temperatures.

6.5 Intervalence Band Absorption (IVBA)

As noted already, the VBs in typical semiconductors comprise LH, HH, and split-off (SO) bands. At elevated temperatures or in heavily doped p-type materials having Fermi energy below the VB edge, transitions are possible from the LH band to the HH band or from the SO band to the HH or LH band.

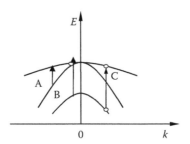

FIGURE 6.4
Schematic of IVBA.

An example of such transitions is given in Figure 6.1 as Process (C), indicating a transition from the SO to the HH band. The different transitions are further illustrated in Figure 6.4, in which processes A, B, and C denote, respectively, the LH–HH, SO–LH, and SO–HH transitions. The inter-VB transitions are forbidden at $\mathbf{k}=0$ due to quantum mechanical selection rules. Direct (vertical) transitions become possible at $\mathbf{k}\neq0$. The momentum matrix elements for this transitions are proportional to k. The absorption bands are broad.

In the following, we shall outline the method of obtaining the absorption coefficient for IVBA by considering transition between the SO and HH bands.

The absorption coefficient for transition from a state of energy $E(\mathbf{k_s})$ in the SO band to a state of energy $E(\mathbf{k_h})$ in the HH band is given as

$$\alpha(\hbar\omega) = \frac{2\pi}{\hbar}\frac{1}{(c/n_r)}\left(\frac{eA_0}{m_0}\right)^2\sum_{\mathbf{k_s},\mathbf{k_h}}\left\langle\left|M_{\mathbf{k_s},\mathbf{k_h}}\right|^2\right\rangle(f_h - f_s)\delta\left|E(\mathbf{k_s} - \mathbf{k_h}) + \hbar\omega\right|\delta_{\mathbf{k_s},\mathbf{k_h}}$$

(6.65)

The energies are expressed as

$$E(\mathbf{k_h}) = -\frac{\hbar^2 k_h^2}{2m_h} \text{ and } E(\mathbf{k_s}) = \Delta_{so} - \frac{\hbar^2 k_h^2}{2ms}$$

(6.66)

The argument of the energy-conserving δ-function becomes

$$E(\mathbf{k_s}) - E(\mathbf{k_h}) + \hbar\omega = -\Delta_{so} - \frac{\hbar^2 k_h^2}{2ms} + \frac{\hbar^2 k_h^2}{2m_h} + \hbar\omega$$

(6.67)

The double summation in Equation 6.65 can be converted to a single integral by using the Kronecker delta specifying $\mathbf{k}$-conservation, that is, $\mathbf{k_s}=\mathbf{k_h}$, so that

$$E(\mathbf{k_s}) - E(\mathbf{k_h}) + \hbar\omega = \hbar\omega - \Delta_{so} + \frac{\hbar^2 k_h^2}{2m_r} \tag{6.68}$$

$$\frac{1}{m_r} = \frac{1}{m_h} - \frac{1}{m_s} \tag{6.69}$$

where m_r is the reduced mass.

The total summation in Equation 6.65 reduces to

$$\sum_{k_h} \rightarrow \frac{2V}{(2\pi)^3} \int k_h^2 \sin\theta d\theta d\varphi dk_h = \frac{2V}{(2\pi)^3} \int 4\pi k_h^2 dk_h \tag{6.70}$$

The square of the absolute value of the matrix element averaged over all directions is

$$\left| M(k_s, k_h) \right|^2 = \hbar^2 A_{sh} k_h^2 \tag{6.71}$$

where A_{sh} is a dimensionless constant, not to be confused with the Einstein A-coefficient.

Putting Equations 6.70 and 6.71 in Equation 6.65, the integration may be easily performed by using the energy-conserving δ-function. Using the expression for A_0, the expression for the IVBA coefficient becomes

$$\alpha(\hbar\omega) = \frac{e^2 A_{sh} (2m_r)^{5/2}}{\pi\varepsilon_0 n_r c\hbar^2 m_0^2} \frac{(\hbar\omega - \Delta)^{3/2}}{(\hbar\omega)} (f_h - f_s) \tag{6.72}$$

6.6 Free-Carrier Absorption

In actual device structures, different semiconductor regions are doped with impurities, and as a result there exist a substantial number of free electrons or holes in different regions. These carriers may absorb photons and make a transition from a state of wave vector $\mathbf{k}$ to another state of wave vector $\mathbf{k'}$ in the same band. These transitions are shown in Figure 6.1 as process (D) and process (E) in the VB and the CB, respectively. These transitions are termed free-hole (D) and free-electron (E) absorption or, in general, free-carrier absorption.

A change in wave vector occurs in the absorption process, and to conserve the wave vector, one must have $|\mathbf{k}' - \mathbf{k}| = \kappa$. The photon wave vector is related to the frequency of radiation by $\kappa = \omega n(\omega)/c$. In order to satisfy both energy and momentum conservation, the energy of the photon should be prohibitively large: larger than the typical bandgap. Therefore, the radiative transitions just discussed cannot account for intraband FCA.

Intraband processes occur with the simultaneous action of a photon and another momentum-conserving agency, that is, scatterers such as phonons, impurities, alloy disorder, or other imperfections. Again, the process is a second-order transition, and the energy of the photon is quite small compared with that needed for interband transitions. The photons introduce negligible change in the $\mathbf{k}$-value of the carrier; the main contribution to wave-vector change comes from the scattering agencies. Quantum-mechanical calculation of the FCA has been described in a number of textbooks and monographs (please see Reading List).

The expressions for the FCA coefficient may also be obtained by using classical electromagnetic theory and the Drude model for conduction (please see Reading List). There is not much difference between the expressions obtained by classical and quantum methods. Due to its simplicity, the classical method looks attractive. Since the expression for the carrier-induced change in refractive index (RI) is also obtained from the theory and will be utilized in subsequent discussions, the derivation seems useful.

Let us consider the motion of a free electron in the CB of a semiconductor under the influence of a sinusoidal electric field. The equation of motion takes the following form:

$$m_e \frac{d^2 x}{dt^2} + m_e g \frac{dx}{dt} = -eF_0 \exp(j\omega t) \tag{6.73}$$

where:

x is the displacement of the electron having effective mass m_e
g is a damping coefficient
F_0 is the amplitude of the impressed electric field varying with angular frequency ω

The first term is the force term, the second term represents damping of electron motion by scattering with lattice vibrations (phonons), impurities, and so on, and the right-hand side represents the applied force. The steady-state solution of the above may easily be written as

$$x = \frac{(eF_0)/m_e}{\omega^2 - j\omega g} \exp(j\omega t) \tag{6.74}$$

If n denotes the free-carrier concentration per unit volume, then the displacement of carriers, x, will produce additional polarization, P_1, given by

$$P_1 = -nex \tag{6.75}$$

so that the total polarization is $P = P_0 + P_1$, where P_0 is the polarization present in the material without free carriers. The relative permittivity is now given as

$$\varepsilon_r = \frac{\varepsilon}{\varepsilon_0} = 1 + \frac{P}{\varepsilon_0 F} = 1 + \frac{P_0}{\varepsilon_0 F} + \frac{P_1}{\varepsilon_0 F} = n_0^2 + \frac{P_1}{\varepsilon_0 F} \tag{6.76}$$

where n_0 is the index of refraction without free carriers. Using Equations 6.74, and 6.75 in Equation 6.76, we may obtain

$$\varepsilon_r = n_0^2 - \frac{(ne^2)/(m_e \varepsilon_0)}{\omega^2 - j\omega g} \tag{6.77}$$

The real and imaginary parts of the relative permittivity may now be expressed as

$$\varepsilon_{rr} = n_0^2 - \frac{(ne^2)/(m_e \varepsilon_0)}{\omega^2 + g^2} \tag{6.78}$$

$$\varepsilon_{ri} = \frac{(ne^2 g)/(m_e \omega \varepsilon_0)}{\omega^2 + g^2} \tag{6.79}$$

At steady state, $d^2x/dt^2 = 0$, and from Equation 6.73

$$m_e g \frac{dx}{dt} = eF \tag{6.80}$$

Since the drift velocity is related to the mobility, μ_e, by the relation

$$\frac{dx}{dt} = \mu_e F \tag{6.81}$$

Equations 6.80 and 6.81 yield

$$g = \frac{e}{\mu_e m_e} = \frac{1}{\tau} \tag{6.82}$$

The electron mobility is related to the momentum relaxation time by the well-known expression $\mu_e = e\tau/m_e$.

Example 6.4: Assume an electron mobility $= 8000$ cm^2 (V·s)$^{-1}$ for GaAs; the effective mass is $0.067m_0$, where m_0 is the free electron mass. The value of g is 3.26×10^{12} s^{-1} and that of τ is 0.3 ps.

Since typical values for $\omega \approx 10^{15}$ s^{-1}, one may neglect g in the denominators of Equations 6.78 and 6.79, which are modified as follows by using Equation 6.82:

$$\varepsilon_{rr} = n_0^2 - \frac{ne^2}{m_e \varepsilon_0 \omega^2} \tag{6.83}$$

$$\varepsilon_{ri} = \frac{ne^3}{m_e^2 \varepsilon_0 \omega^2 \mu_e} \tag{6.84}$$

The exponential loss coefficient α is related to ε_{ri} by $\alpha = k\varepsilon_{ri}/n_r$, where n_r is the refractive index and k is the light wave vector given by $k = \omega/c$. We may write, therefore,

$$\alpha_{fc} = \frac{ne^3 \lambda_0^2}{4\pi^2 n_r m_e^2 \mu_e \varepsilon_0 c^3} \tag{6.85}$$

Equation 6.83 indicates that there is a change of RI also due to free carriers. Writing the small change as Δn_e, the corresponding expression may be written as

$$\Delta n_e = -\frac{ne^2 \lambda_0^2}{8\pi^2 c^2 n_r \varepsilon_0 m_e} \tag{6.86}$$

The change in RI due to free carriers in GaAs, InP, and InGaAsP is given by Bennett et al. [1] in the form

$$\Delta n_e = -\frac{6.9 \times 10^{-22}}{n_r (\hbar\omega)^2} \left\{ \frac{n}{m_e} + p \left(\frac{m_{hh}^{1/2} + m_{lh}^{1/2}}{m_{hh}^{3/2} + m_{lh}^{3/2}} \right) \right\} \tag{6.87}$$

where the photon energy is in electronvolts, effective masses are ratios, and the carrier densities are in cubic centimeters. The prefactor may easily be calculated by using Equation 6.86.

Example 6.5: The FCA in GaAs is calculated by using Equation 6.85 with the following values of parameters: $\lambda_0 = 0.85$ μm, $\mu_e = 7000$ cm^2 V^{-1} s^{-1}, $n_r = 3.6$, $m_e = 0.067\, m_0$, and $n = 10^{18}$ cm^{-3}. We find $\alpha_{fc} = 0.86$ cm^{-1}.

Example 6.6: The quaternary $In_{0.82}Ga_{0.18}As_{0.6}P_{04}$ is lattice matched to InP and has a bandgap of 0.954 eV, corresponding to $\lambda_0 = 1.3$ μm. Here $y = 0.6$. The mass values may be obtained from $m_e = 0.080 - 0.039y = 0.0566$, $m_{hh} = 0.46$, and $m_{lh} = 0.12 - 0.099y + 0.030y^2 = 0.0714$. Taking $n_r = 3.4$, $n = p = 10^{18}$ cm^{-3}, $\Delta n_e = -0.42 \times 10^{-2}$.

6.7 Recombination and Luminescence

The process opposite to absorption is emission. Since in semiconductors the excess electrons and holes recombine, the excess energy is given up in the form of photons. The general name *luminescence* is given to the phenomenon of light emission. In this section, the recombination and luminescence processes will be discussed.

6.7.1 Luminescence Lifetime

Band-to-band absorption leads to the creation of an excess EHP. These excess carriers have a relatively short lifetime and tend to recombine with the emission of a photon. The recombination processes for direct- and indirect-gap semiconductors are shown in Figure 6.5. Just like absorption, a radiative transition can be direct or indirect depending on the type of semiconductor.

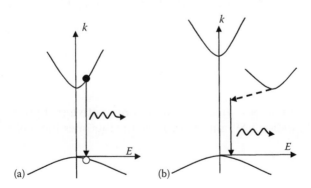

FIGURE 6.5
Recombination processes in a direct gap (a) and an indirect gap (b) semiconductor.

It has been shown in Chapter 2 that the excess carriers decay exponentially with time as

$$\Delta n(t) = \Delta n(0)e^{-t/\tau} \tag{6.88}$$

where $\Delta n(0)$ is the initial excess electron density. It is convenient to introduce a recombination rate, defined as follows:

$$R = -\frac{d\Delta n}{dt} = \frac{\Delta n}{\tau} \tag{6.89}$$

The recombination lifetime is a measure of the average time an excess carrier pair spends in the sample before being lost by recombination. The recombination in a semiconductor may be both intrinsic and extrinsic. Defects in the material as well as the surface of the sample provide recombination channels.

In addition to radiative recombination, nonradiative recombination also occurs, in which the excess energy is given up in the form of phonons or heat waves. These and other defect- and surface-related processes will be discussed in Section 6.8. We now introduce a quantity termed the *internal quantum efficiency* (IQE), which is an important parameter characterizing the efficiency of light-emitting devices.

Let the radiative recombination rate in a material per unit volume be R_r and the rate for nonradiative processes R_{nr}. The total recombination rate for the spontaneous process becomes

$$R_{sr} = R_r + R_{nr} \tag{6.90}$$

The IQE is the ratio of radiative to total recombination rate and is therefore

$$\eta_i = \frac{R_r}{R_r + R_{nr}} \tag{6.91}$$

Now the lifetime of radiative (τ_r) and nonradiative (τ_{nr}) processes may be defined by Equation 6.89, and in terms of the lifetimes the IQE may be written as

$$\eta_i = \frac{\tau_r^{-1}}{\tau_r^{-1} + \tau_{nr}^{-1}} = \left(1 + \frac{\tau_r}{\tau_{nr}}\right)^{-1} \tag{6.92}$$

6.7.2 Carrier Lifetime: Dependence on Carrier Density

We first discuss the dependence of spontaneous recombination lifetime on carrier density for nondegenerate semiconductors. The spontaneous recombination rate is directly proportional to the product of the electron and hole concentrations, as follows:

$$R_{sr} = B_r np \tag{6.93}$$

where n and p are the electron and hole concentrations, respectively, and B_r is a recombination coefficient that may be derived quantum mechanically.

The electron and hole concentrations at thermal equilibrium, denoted by n_0 and p_0, respectively, are related by $n_0 p_0 = n_i^2$, where n_i is the intrinsic carrier concentration. In the presence of excess carriers, the spontaneous recombination rate becomes

$$R_{sr} = B_r(n_0 + \Delta n)(p_0 + \Delta p) \tag{6.94}$$

Since $\Delta n = \Delta p$, and the spontaneous recombination rate $R_{sr}^0 = B_r n_0 p_0$, the recombination rate for the excited carriers R_{sr}^{exc} may be written as

$$R_{sr}^{exc} = R_{sr} - R_{sr}^0 = B_r \Delta n(n_0 + p_0 + \Delta n) \tag{6.95}$$

and therefore the excess carrier lifetime becomes

$$\tau_r = \left[B_r(n_0 + p_0 + \Delta n) \right]^{-1} \tag{6.96}$$

Equation 6.89 has been used to express lifetime in terms of recombination rate.

In the high-injection case, $\Delta n \gg n_0$ or p_0 and $\tau_r = [B_r(\Delta n)]^{-1}$, while in the low-injection case, $\Delta n \ll n_0$ or p_0 and $\tau_r = [B_r(n_0 + p_0)]^{-1}$.

> **Example 6.7:** The carrier lifetime in GaAs is calculated for $\Delta n = 10^{18}\,\text{cm}^{-3}$. Using $B_r = 10^{-10}$, the lifetime is $10^{-8}\,\text{s}$.

6.7.3 Absorption and Recombination

The coefficient B_r depends on the band structure of the material. The value is large in direct-gap material, and hence the recombination lifetime is small. On the other hand, for indirect-gap material, the value of B_r is extremely low, and the recombination lifetime is extremely large. The value of the recombination coefficient may be obtained from the absorption data or from microscopic calculation. It is well known that good absorbers are good emitters. Since the absorption coefficient is small in silicon, its emissive characteristics are also very poor. We present here the treatment of van Roosbroeck and Shockley [2] relating the lifetime with the absorption coefficient.

The principle of detailed balance states that under equilibrium the rate of photoexcitation of carriers across the gap at frequency interval dv should be equal to the rate of generation of photons in the same frequency interval by electron–hole recombination. The rate of photoexcitation is given by

$$R_{pe}^0 = \int \frac{c}{n(v)} \rho(v)\alpha(v)dv \tag{6.97}$$

where:

  $n(v)$ is the RI of the material
  $\rho(v)$ is the photon density
  $\alpha(v)$ is the absorption coefficient

The photon density increases rapidly with wavelength. The main contribution to the integral therefore comes from the vicinity of the absorption edge, where the absorption is weak and the dispersion is small. One may therefore assume the refractive index to be constant and thus write

$$\rho(v) = \frac{8\pi v^2 n^3}{c^3(e^{hv/k_B T} - 1)} \tag{6.98}$$

Using the relation between Planck function for surface emission, $D(v)$, and $\rho(v)$,

$$\rho(v) = 4v^{-2}n^3 D(v) \tag{6.99}$$

Replacing $\rho(v)$ in Equation 6.97 by $D(v)$, and also replacing the integration variable v by wavelength, one obtains

$$R_{pe}^0 = 4n^2 \int_0^\infty D(\lambda)\alpha(\lambda)d\lambda \tag{6.100}$$

The principle of detailed balance demands that

$$R_{sr}^0 = R_{pe}^0 \tag{6.101}$$

Using Equation 6.93, one may write

$$B_r = \frac{R_{pe}^0}{n_i^2} \tag{6.102}$$

In the above equation, valid for the thermal equilibrium condition, the relationship $n_0 p_0 = n_i^2$ has been used. The longest possible lifetime corresponds to the case when $n_0 = n_i$, and it may be proved that

$$\tau_r(\max) = \frac{n_i}{2R_{pe}^0} \qquad (6.103)$$

6.7.4 Microscopic Theory of Recombination

We now present the theory of band-to-band recombination in a direct-gap semiconductor in the same way as the expression for the absorption coefficient has been derived. The spontaneous emission rate for transition between two states $|2\rangle$ in the CB and $|1\rangle$ in the VB may be expressed as

$$R_{sp}(\hbar\omega) = \frac{2\pi}{\hbar} \left(\frac{eA_0}{m_0}\right)^2 \sum_{1,2} \langle |p_{12}|^2 \rangle G(\hbar\omega) f_2 (1 - f_1) \delta(E_{12} - \hbar\omega) \qquad (6.104)$$

where:
    p_{12} is the matrix element between the two states
    G is the photon density of states

The explicit expressions for $|2\rangle$ and $|1\rangle$ are as follows:

$$|2\rangle = |c, k_e\rangle = U_c(\mathbf{r}) \exp(jk_e \cdot \mathbf{r}) \text{ and } |1\rangle = |v, k_h\rangle$$

$$= U_v(\mathbf{r}) \exp(jk_h \cdot \mathbf{r}) \qquad (6.105)$$

As found earlier,

$$\langle c, k_e | e^{jq \cdot r} \hat{q} \cdot \mathbf{p} | v, k_h \rangle = p_{cv} \qquad (6.106)$$

Therefore, using Equation 6.104, the spontaneous emission rate may be reexpressed as

$$R_{sp}(\hbar\omega) = \frac{2\pi}{\hbar} \left(\frac{eA_0}{m_0}\right)^2 \sum_{k_e, k_h} \langle |p_{cv}|^2 \rangle \delta_{k_e, k_h} G(\hbar\omega) f_e(k_e) f_h'(k_h) \delta\big[E(k_e)$$

$$- E(k_h) - \hbar\omega\big] \qquad (6.107)$$

First of all, let us assume that the probability factors are unity. Using the expressions for A_0, G, and the squared matrix element averaged over all polarizations, we obtain

$$R_{sp} = \frac{e^2 n_r}{3\pi\varepsilon_0 m_0 c^3 \hbar^2} \frac{\langle |p_{cv}|^2 \rangle}{m_0} \hbar\omega = \frac{1}{\tau_{sp}} \tag{6.108}$$

In this case, since we are considering two discrete states, the spontaneous emission lifetime is related to Einstein's A-coefficient by $\tau_{sp} = A_{21}^{-1}$.

> **Example 6.8:** For GaAs, $2\langle |p_{cv}|^2 \rangle / m_0 = 23$ eV and $n_r = 3.6$. Let us take $\hbar\omega = 1.43$ eV. We obtain $R_{sp} = 1.14 \times 10^9\,\text{s}^{-1}$ and $\tau_{sp} = 0.58$ ns.

The above result is applicable when both the electron and hole densities are high enough to make f_e and f_h' unity. However, when this approximation is not valid, we encounter a summation of the following form:

$$I = \sum_{k_e, k_h} \delta_{k_e, k_h} f_e(k_e) f_h'(k_h) \delta \big[E(k_e) - E(k_h) - \hbar\omega \big] \tag{6.109}$$

Using **k**-conservation, the argument of the energy-conserving δ-function may be written in terms of the reduced mass as in Equations 6.49 and 6.50. Using a similar procedure, the recombination rate may be expressed as

$$R_{sp}(\hbar\omega) = \frac{e^2 n_r \hbar\omega \langle |p_{cv}|^2 \rangle}{2\pi^3 c^3 m_0^2 \varepsilon_0 \hbar^2} \left(\frac{2m_r}{\hbar^2} \right)^{3/2} (\hbar\omega - E_g)^{1/2} f_e(k_h) f_h'(k_h) \tag{6.110}$$

We assume that the quasi-Fermi levels for both electrons and holes are a few $k_B T$ away from the respective band edges, so that the distribution functions may be approximated as Boltzmann distributions, expressed as

$$f_e(k_h) = \exp\left\{ -\frac{[E_c(k_h) - F_e]}{k_B T} \right\} \tag{6.111a}$$

$$f_h'(k_h) = \exp\left\{ \frac{[E_v(k_h) - F_h]}{k_B T} \right\} \tag{6.111b}$$

Writing $\Delta F = F_e - F_h$ and noting that $E_c(k_h) - E_v(k_h) = \hbar\omega$, the product may be written as

$$f_e(k_h) f'(k_h) = \exp\left\{-\frac{\left[E_g - \Delta F\right]}{k_B T}\right\} \times \exp\left[-\frac{\left(\hbar\omega - E_g\right)}{k_B T}\right] \tag{6.112}$$

Putting this in Equation 6.110, the recombination rate may be expressed as

$$R_{sp}(\hbar\omega) = C'(\hbar\omega, T)(m_r)^{3/2} \left(\hbar\omega - E_g\right)^{1/2} \exp\left\{-\frac{\left[E_g - \Delta F\right]}{k_B T}\right\}$$

$$\times \exp\left[-\frac{\left(\hbar\omega - E_g\right)}{k_B T}\right]^{1/2} \tag{6.113}$$

where the prefactor C' contains the material parameters and the fundamental constants. The quasi-Fermi levels are related to the electron density n and the hole density p by

$$n = 2\left(\frac{m_e k_B T}{2\pi\hbar^2}\right)^{3/2} \exp\left[-\frac{(E_c - F_e)}{k_B T}\right] \tag{6.114a}$$

$$p = 2\left(\frac{m_h k_B T}{2\pi\hbar^2}\right)^{3/2} \exp\left[\frac{(E_v - F_h)}{k_B T}\right] \tag{6.114b}$$

The recombination rate may now be expressed in terms of the np product as follows:

$$R_{sp}(\hbar\omega) = \left[\frac{2\pi}{(\pi k_B T)^{3/2}}\right] C(\hbar\omega, T) np \left(\frac{m_r}{m_e m_h}\right)^{3/2} \left(\hbar\omega - E_g\right)^{1/2} \exp\left[-\frac{\left(\hbar\omega - E_g\right)}{k_B T}\right] \tag{6.115}$$

where:

$$C(\hbar\omega, T) = \frac{e^2 n_r \hbar\omega}{2\pi\varepsilon_0 \hbar^2 m_0^2 c^3} \left(\frac{2\pi\hbar^2}{k_B T}\right)^{3/2} \left\langle |p_{cv}|^2 \right\rangle \tag{6.116}$$

The total spontaneous emission rate is obtained by integrating $R_{sp}(\hbar\omega)$ over all photon energies. It is assumed that $C(\hbar\omega, T)$ is slowly varying, so that it may be treated as constant in the range over which $(\hbar\omega - E_g)^{1/2}\exp[-(\hbar\omega - E_g)/$

$k_BT]$ is appreciable. The integral is of the form $\int\limits_0^\infty x^{1/2}\exp(-x)dx = \sqrt{\pi}/2$, and thus

$$R_{sp} = \int\limits_0^\infty R_{sp}(\hbar\omega)d(\hbar\omega) = npC(\hbar\omega_{max},T)\left(\frac{1}{m_e + m_h}\right)^{3/2} \qquad (6.117)$$

The photon energy at which the integrand is maximum is given by $\hbar\omega_{max}=E_g+k_BT/2$. From Equation 6.93, we may obtain the expression for $B=R_{sp}/np$. The recombination lifetime is now expressed as

$$\frac{1}{\tau_r} = (n_0 + p_0 + \delta n)\frac{(2\pi)^{1/2}e^2\hbar}{\varepsilon_0 c^3 m_0^2(k_BT)^{3/2}}(n_r\hbar\omega)_{\hbar\omega_{max}}\frac{\left\langle|p_{cv}|^2\right\rangle}{(m_e+m_h)^{3/2}} \qquad (6.118)$$

In arriving at the above equation, only one type of hole is considered. It is straightforward to extend the calculation by including both the light and heavy holes, and the expression is

$$\frac{1}{\tau_r} = (n_0 + p_0 + \delta n)\frac{(2\pi)^{1/2}e^2\hbar}{\varepsilon_0 c^3 m_0^2(k_BT)^{3/2}}(n_r\hbar\omega)_{\hbar\omega_{max}}\left\langle|p_{cv}|^2\right\rangle\sum_{v=h,l}\frac{\left[m_v/(m_e+m_v)\right]^{3/2}}{m_{hh}^{3/2}+m_{lh}^{3/2}}$$

$$(6.119)$$

It is easy to conclude that the excess carrier lifetime varies as $T^{3/2}$ and is inversely proportional to the bandgap, as for most III–V semiconductors $(2/m_0)\langle|p_{cv}|^2\rangle \approx 20$ eV.

> **Example 6.9:** The recombination lifetime for GaAs is calculated by using $m_e = 0.067m_0$, $m_{hh} = 0.48m_0$, $m_{lh} = 0.09m_0$, $n_r = 3.6$, $E_p = 23$ eV, $E_g = 1.424$ eV, $\hbar\omega_{max} = 1.437$ eV at 300 K, and injected carrier density $\Delta p = 10^{18}$ cm^{-3}. The calculated values are $B_r = 4.62 \times 10^{-10}$ cm^3 s^{-1} and $\tau_r - (B_r\Delta p)^{-1} = 2.2$ ns.

6.8 Nonradiative Recombination

The radiative recombination in indirect-gap semiconductors, in particular Si, is very inefficient. The excess carriers in silicon decay mostly by nonradiative processes involving trap levels or by the Auger recombination process. In the following, these two processes are discussed.

6.8.1 Recombination via Traps

The dominant recombination mechanism in silicon is via traps. The theory of this process was developed by Shockley and Read [3] and modified by Hall [4]. The model is referred to as the SRH model. The mechanism involves four electron and hole transitions, as shown in Figure 6.6. A trap level E_t first captures an electron (Process 1), and then a hole is captured by the trap filled by the electron (Process 3: the direction of the arrow indicates the transition of an electron, and the hole moves in the opposite direction). The combined result of electron and hole capture is a recombination of the pair. As shown in the right-hand part of the diagram, the trap is free to capture another EHP after this. The inverse processes are the emission of an electron from the filled trap into the CB (Process 2) and the emission of a hole from an empty trap into the VB (Process 4).

The rate of electron capture by the traps, R_{nc}, is proportional to the number of electrons and to the number of empty traps, so that

$$R_{nc} = C_n n (1 - f_t) N_t \qquad (6.120)$$

where:
- N_t is the trap density
- f_t is the occupancy function of the trap level
- C_n is the capture coefficient of the electrons expressed in terms of a capture cross section, σ_n, for electrons and the electron thermal velocity, $v_{thn} = (3k_B T / m_e)^{1/2}$, as shown in Figure 6.6

$$C_n = \sigma_n v_{thn} \qquad (6.121)$$

The rate of electron emission from the traps, R_{ne}, is proportional to the number of filled traps, $f_t N_t$, and is given by

$$R_{ne} = e_n f_t N_t \qquad (6.122)$$

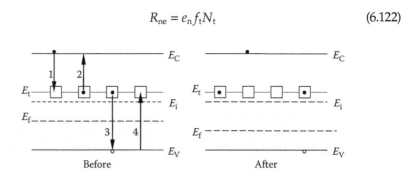

FIGURE 6.6
Four basic steps in trap-assisted recombination: (1) electron capture, (2) electron emission, (3) hole capture, and (4) hole emission.

where e_n is the emission probability of the electron. Under thermal equilibrium, $R_{nc} = R_{ne}$, and hence from Equations 6.120 and 6.122,

$$C_n n_0 = \frac{e_n f_{t0}}{(1 - f_{t0})} \tag{6.123}$$

The equilibrium electron concentration, n_0, is expressed as

$$n_0 = N_c \exp\left[\frac{(E_F - E_c)}{k_B T} \right] \tag{6.124}$$

and the ratio $f_{t0}/(1-f_{t0})$, where f_{t0} is the equilibrium occupancy of the trap level, is expressed using the Fermi–Dirac occupation function as

$$\frac{f_{t0}}{(1 - f_{t0})} = \exp\left[-\frac{(E_t - E_F)}{k_B T} \right] \tag{6.125}$$

where E_F, E_c, and E_t, denote, respectively, the Fermi level, the CB edge, and the energy of the trap level. Using Equations 6.123 and 6.125, we may write

$$e_n = N_c \exp\left[\frac{(E_t - E_c)}{k_B T} \right] = n_t C_n \tag{6.126}$$

The difference between the electron-capture and electron-emission rates is given by

$$R_n = R_{nc} - R_{ne} = C_n N_t [(1 - f_t)n - f_t n_t] \tag{6.127}$$

Proceeding in a similar fashion, the difference between the hole-capture and hole-emission rates may be written as

$$R_p = R_{pc} - R_{pe} = C_p N_t [f_t p - (1 - f_t)p_t] \tag{6.128}$$

The capture coefficient for holes may be defined in terms of capture cross section and thermal velocity for holes by replacing σ_n in Equation 6.121 by σ_p, and by expressing the thermal velocity in terms of hole effective mass, m_h. Under steady state, there is no accumulation of charge, and hence electrons and holes must recombine in pairs. Thus,

$$R_p = R_n = R \tag{6.129}$$

where R is the recombination rate. Equating Equations 6.127 and 6.128, the occupation function may be expressed as

$$f_t = \frac{nC_n + p_t C_p}{C_n(n + n_t) + C_p(p + p_t)}$$ (6.130)

Substituting this in Equation 6.127, one finds

$$R = \frac{pn - n_i^2}{\tau_{pl}(n + n_t) + \tau_{nl}(p + p_t)}$$ (6.131)

The electron and hole lifetimes are expressed as

$$\tau_{nl} = \frac{1}{(v_{thn}\sigma_n N_t)}$$ (6.132a)

$$\tau_{pl} = \frac{1}{(v_{thp}\sigma_p N_t)}$$ (6.132b)

The recombination rates under special conditions may easily be obtained from Equation 6.131. Consider p-type materials, in which electrons are minority carriers. Then $n \ll p \approx N_A$: the acceptor density. In this case, $p \gg p_t$ and also $p \gg n_t$. Equation 6.131 now reduces to

$$R = \frac{n - (n_i^2/N_A)}{\tau_{nl}} = \frac{n - n_0}{\tau_{nl}}$$ (6.133a)

Similarly, for n-type materials

$$R = \frac{p - (n_i^2/N_D)}{\tau_{pl}} = \frac{p - p_0}{\tau_{pl}}$$ (6.133b)

It is noted that the SRH model presented in this section is applicable for describing the nonradiative recombination process via a single deep-level recombination center in the forbidden gap of a semiconductor. Treatment of the nonradiative recombination process via multiple deep-level centers in the forbidden gap of the semiconductor can be found in a classical paper by Sah and Shockley [5].

6.8.2 Auger Recombination

The Auger band-to-band recombination is usually the predominant recombination process occurring in degenerate semiconductors and small-bandgap semiconductors, such as InSb and HgCdTe materials. The Auger recombination process can also become the predominant recombination mechanism under high-injection conditions. There are five different ways by which Auger recombination may occur: (1) direct band-to-band, (2) phonon assisted, (3) trap-assisted electronic transition, (4) trap-assisted hole transition, and (5) donor–acceptor-pair related. The Auger recombination rate R_a is approximately written as

$$R_a = Cn^3 \qquad (6.134)$$

where:
  C is the Auger coefficient
  n is the injected carrier density

The corresponding recombination lifetime is then expressed as

$$\tau_A = \frac{n}{R_a} = (Cn^2)^{-1} \qquad (6.135)$$

In most cases, the effect of Auger recombination is taken into account by using experimentally obtained values of C. In the following, we shall give a simplified picture of the band-to-band Auger process.

6.8.2.1 Band-to-Band Auger Effect

The band-to-band Auger process in a direct-gap semiconductor may be classified into CCCH, CHHS, and CHHL processes, as shown in Figure 6.7a, b, and c, respectively. The symbols C, H, L, and S stand for CB, HH band, LH band, and SO band, respectively. Consider the CCCH process; two electrons, 1 and 2, in the CB and one hole, 1′, are involved in the process. Here, electron 1 recombines with hole 1′, and the emitted photon is absorbed by electron 2, which moves up to a higher state in the CB. The excited electron loses energy by phonon emission and eventually returns to a lower state in equilibrium with other electrons. The other processes may be understood in a similar way with reference to the figures. The recombination process is the inverse of impact ionization, in which an energetic carrier generates an EHP.

6.8.2.1.1 CCCH Process

We present a simple theory of calculating the Auger rate for the CCCH process. The rate should contain the probabilities that the states k_1 and k_2 are

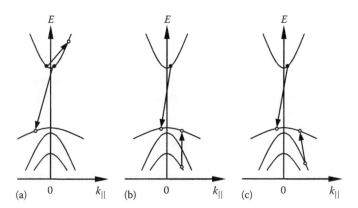

FIGURE 6.7
(a) Auger recombination (CCCH) process in n-type sample. (b) CHHS process occurring in p-type sample. (c) CHHL process in p-type material.

occupied, state k_1' is empty, and state k_2' is empty. The occupation factor may be written as

$$P\left(k_1, k_2, k_1'\right) = f(k_1)f(k_2)[1 - f(k_1')] \qquad (6.136)$$

We assume nondegenerate statistics, and therefore the probability functions are

$$f(k_1) = (n/N_c) \exp(-E_{ck_1} / k_B T)$$
$$f(k_2) = (n/N_c) \exp(-E_{ck_2} / k_B T)$$
$$1 - f(k_1') = (p/N_v) \exp(-E_{vk_1'} / k_B T)$$

Here n and p are, respectively, the electron and hole densities, and N_c and N_v are the effective density of states for the CB and VB. The probability factor may now be written as

$$P\left(k_1, k_2, k_1'\right) = \frac{np}{N_c N_v} \frac{n}{N_c} \exp\left(-\frac{E_{ck_2} + E_{vk_1'} + E_{ck_1}}{k_B T}\right)$$

$$\approx \frac{n}{N_c} \exp\left(-\frac{E_g + E_{ck_2} + E_{vk_1'} + E_{ck_1}}{k_B T}\right) \qquad (6.137)$$

We now aim at finding the energy for which the exponent in Equation 6.137 is maximum. Writing the energies in the exponent of the first term on the right-hand side in terms of wave vectors, we may write

$$\frac{k_1^2}{2m_e} + \frac{k_1^2}{2m_e} + \frac{k_1'^2}{2m_h} = \frac{1}{2m_e}\left[k_1^2 + k_2^2 + \mu k_1'^2\right]; \mu = \frac{m_e}{m_h} \tag{6.138}$$

It is easy to conclude that the probability factor will be maximum for the lowest energy values of the wave vectors $\mathbf{k_1}$, $\mathbf{k_1'}$, and $\mathbf{k_2}$. Since $\mathbf{k_2'}$ is the largest wave vector, we choose

$$\mathbf{k_1} + \mathbf{k_1'} + \mathbf{k_2} = -\mathbf{k_2'} \tag{6.139}$$

We also write $\mathbf{k_1} = a\mathbf{k_1'}$ and $\mathbf{k_2} = b\mathbf{k_1'}$ obtain from Equation 6.138

$$k_2'^2 = (a^2 + b^2 + \mu)k_1'^2 + k_g^2; \quad \hbar^2 k_g^2/2m_e = E_g \tag{6.140}$$

Equation 6.140 leads to the expression

$$k_2' = (a + b + 1)k_1' \tag{6.141}$$

Squaring this equation and eliminating $k_2'^2$ from Equations 6.139 and 6.140, we obtain

$$k_1'^2(1 + 2ab + 2a + 2b - \mu) = k_g^2 \tag{6.142}$$

We need to minimize the following in order to maximize the Auger rate:

$$k_1^2 + k_2^2 + \mu k_1'^2 = k_1'^2(a^2 + b^2 + \mu) \tag{6.143}$$

Using Equation 6.140, we write

$$(a^2 + b^2 + \mu)k_1'^2 = \frac{(a^2 + b^2 + \mu)}{(1 + 2ab + 2a + 2b - \mu)}k_g^2 \tag{6.144}$$

The quantity minimizes when $a = b = \mu$. The energy value for the initial-state electron is then expressed as

$$E_{ck_1} = E_{ck_2} = \mu E_{vk_1'} = \left(\frac{\mu^2}{1 + 3\mu + 2\mu^2}\right)E_g \tag{6.145}$$

The maximum probability function is now written as

$$P(\mathbf{k_1}, \mathbf{k_2}, \mathbf{k_1'}) = \frac{n}{N_c}\exp\left(-\frac{1 + 2\mu}{1 + \mu}\frac{E_g}{k_B T}\right) \tag{6.146}$$

The final-state energy of the electron is

$$E_{ck_2'} = \frac{1+2\mu}{1+\mu} E_g \tag{6.147}$$

When $\mu \ll 1$, the above energy may be approximately expressed as

$$E_{ck_2'} \approx (1+\mu)E_g \tag{6.148}$$

The Auger rate decreases as this threshold energy increases, since the occupation factor decreases with increasing energy. The threshold depends on the bandgap and the ratio of electron and hole masses. Since strain changes the hole mass, the Auger rate is affected also.

The general expression for the Auger rate for a fixed initial state k_i is given in Equation 6.149:

$$W_{Auger}(k_1) = 2\left(\frac{2\pi}{\hbar}\right)\left(\frac{e^2}{\varepsilon}\right)\left(\frac{1}{2\pi}\right)^6$$

$$\times \int dk_2 \int dk_1' \int dk_2' |M|^2 P(k_1, k_2, k_1')\delta\left(E_{ck_1} + E_{ck_2} - E_{vk_1'} - E_{ck_2'}\right) \tag{6.149}$$

The matrix element M is the screened Coulomb matrix element. The evaluation of the multiple integral is cumbersome and involves a knowledge of the accurate band structure. As mentioned already, the Auger rate is proportional to cube of the carrier density.

Other processes, such as CHHS, CHHL, and phonon-assisted Auger recombination, may be calculated by using proper band structures and matrix elements.

In most situations, the cubic dependence of the Auger rate on the carrier density is employed, and the value of the proportionality constant is obtained from the experimental data.

> **Example 6.10:** The Auger coefficient $C = 10^{-28}$ cm^6 s^{-1}. Taking the injected electron density $n = 10^{18}$ cm^{-3}, the Auger recombination lifetime becomes 10 ns.

6.8.3 Other Recombination

6.8.3.1 Surface Recombination

The presence of extra defects and trap levels in the surface of semiconductors increases the recombination rates drastically. Surfaces and interfaces typically contain a large number of recombination centers due to abrupt termination of the crystal. A large number of electrically active states are then created at the surface. In addition, a large number of impurities may be present at the surfaces and interfaces, since they are exposed during the device fabrication process. The net recombination rate due to trap-assisted recombination and generation is given by

$$R_s = \frac{pn - n_i^2}{p + n + 2n_i \cosh\left(\dfrac{E_i - E_{st}}{k_B T}\right)} N_{st} v_{th} \sigma_s \qquad (6.150)$$

This expression is almost identical to that for SRH recombination. In this case, however, the two-dimensional (2-D) density of traps, N_{st}, appears in the expression, as the traps exist only at the surface.

Further simplification of Equation 6.150 is possible. For example, for electrons in the quasi-neutral p-type region, $p \gg n$ and $p \gg n_i$, so that for $E_i = E_{st}$, it can be simplified to

$$R_{s,n} = R_{s,nc} - R_{s,ne} = v_s(n - n_0) \qquad (6.151)$$

where the surface recombination velocity is expressed as

$$v_s = N_{st} v_{th} \sigma_s \qquad (6.152)$$

6.8.3.2 Recombination of Complexes

It has been noticed in connection with absorption processes that bound electron–hole pairs or excitons play a role in low–temperature absorption spectra. An excitonic signature is also observed in the photoluminescence spectra of semiconductors, especially in direct-gap semiconductors. Excitonic effects are also found in the recombination in indirect-gap materials. In particular, the radiative decay of excitons bound to isoelectronic impurities is an important physical process in indirect-bandgap semiconductors such as GaP and silicon, in which band-to-band radiative transitions are forbidden by the k-conservation rule. The introduction of isoelectronic impurities into indirect-bandgap semiconductors such as GaP and its alloys improves the quantum efficiency of optical emission. Emission due to excitons bound to an isoelectronic center is characterized by a long radiative decay time, which indicates stability against nonradiative Auger processes, and high radiative quantum efficiency. The lifetime can be $>10^{-5}$ s compared with $<10^{-7}$ s for bound excitons (BEs) at single donors or acceptors. The binding of both electron and hole is relatively weak, and hence the luminescence from isoelectronic centers usually has photon energy close to the electronic gap, but considerably less than for an exciton bound to shallow donors or acceptors.

Isoelectronic-bound-exciton (IBE) emission from silicon has attracted considerable interest in the last few years and has now been observed for S- or Be-doped Si.

6.9 Carrier Effect on Absorption and Refractive Index

In early semiconductor lasers, heavily doped p-n homojunctions were used to inject a sufficient number of electrons and holes to the junction to establish the condition for gain. In heterojunction lasers, the injected electron and hole density exceeds 10^{18} cm^{-3}. The large carrier density induces changes in the band structure as well as in the absorption coefficient and the RI of the semiconductor. We shall discuss briefly the reasons for the changes and present expressions for the relevant parameters.

6.9.1 Band Filling

When a semiconductor is heavily doped or contains a sufficient number of injected carriers, a decrease in the absorption coefficient above the fundamental gap is observed. The effect, known as the Burstein–Moss effect, is more pronounced for lower-gap materials such as InSb. Electrons either injected or provided by the donors occupy the lower energy states in the CB, and the electrons from the VB do not find empty states there. The same argument applies to absorption in p-type material, but due to larger density of states, a greater number of holes are needed to block the absorption. The effect is alternatively called the band-filling effect.

Assuming parabolic bands and **k**-conservation, the change in absorption at photon energy E due to band filling may be expressed as

$$\Delta\alpha(n,p,E) = \alpha(n,p,E) - \alpha_0(E)$$

$$= \sum_{i=\text{hh,lh}} (C_i/E)(E-E_g)^{1/2}[f_v(E_{ai}) - f_c(E_{bi})1] \qquad (6.153)$$

where:
  α_0 is the absorption coefficient without injection
  C is the prefactor in Equation 6.153
  E_a is the energy level in the VB
  E_b is the energy level in the CB satisfying energy and momentum conservation
  n and p are, respectively, the injected electron and hole density

The change in RI is similarly defined and is expressed in terms of $\Delta\alpha$ by using the Kramers–Kronig relation

$$\Delta n_r(n,p,E) = \frac{2c\hbar}{e^2} P \int_0^\infty \frac{\Delta\alpha(n,p,E')}{E'^2 - E^2} dE'$$ (6.154)

where P stands for the principal value. The change in absorption for $n=p$ may be calculated by first calculating the Fermi levels by numerical methods or by using an approximate formula, for example, the Jouce–Dixon formula, and then calculating the Fermi factors in Equation 6.153. The calculated values of RI change show both negative and positive values around the bandgap. For InP below the bandgap, the linear relation $\Delta n_r = -A \cdot n$ holds good: $A = 1.4 \times 10^{-20}$ (1.2 eV), 7.7×10^{-21} (1.0 eV), and 5.6×10^{-21} (0.8 eV), where the density is in cubic centimeters and the energy is given in the parentheses.

6.9.2 Bandgap Shrinkage

In the presence of high carrier density, the lattice potential that an electron faces is affected by the presence of other electrons, and thus the potential is screened. When the electron gas is dense, the wave functions of adjacent electrons overlap, forming a gas of interacting particles. The electrons will repel one another by Coulomb forces, and electrons with the same spin will avoid each other. The behavior of an electron is thus controlled by other electrons, and the effect is known as correlation. The result of these two effects is a lowering of the CB edge. A similar correlation effect for holes raises the VB edge. The net effect is a bandgap shrinkage, also known as the bandgap renormalization.

The bandgap shrinkage is expressed by the following relation:

$$\Delta E_g = -\left(\frac{e^2}{2\pi\varepsilon_0\varepsilon_s}\right)\left(\frac{3n}{\pi}\right)^{1/3}$$ (6.155)

where ε_s is the relative static permittivity of the semiconductor. The cube root dependence suggests that the shrinkage is proportional to the average interparticle spacing. The following modification by Bennett et al. has been found to give good agreement with experiment:

$$\Delta E_g(n) = \frac{\kappa}{\varepsilon_s}\left(1 - \frac{n}{n_{cr}}\right)^{1/3} ; \quad n_{cr} = 1.6 \times 10^{24}\left(\frac{m_e}{1.4\varepsilon_s}\right)^3$$ (6.156)

The critical carrier density n_{cr} is in cubic centimeters. The fitting parameter κ is 0.11 for holes, 0.125 for electrons, and 0.14 when both electrons and holes are present in equal numbers.

> **Example 6.11:** Taking $\varepsilon_s = 13\varepsilon_0$ and $n = (\pi/3) \times 10^{24}\,\text{m}^{-3}$ for GaAs, the band-gap shrinkage is 22.2 meV from Equation 6.155. $n_{cr} = 5.36 \times 10^{16}\,\text{cm}^{-3}$ for n-GaAs.

Assuming that the bandgap shrinkage causes a rigid shift of the absorption curve, the change in absorption is given by

$$\Delta\alpha(n, E) = (C/E)\left[(E - E_g - \Delta E_g(n))^{1/2} - (E - E_g)^{1/2} \right] \tag{6.157}$$

The change in RI may then be calculated from the Kramers–Kronig relation.

6.9.3 Free-Carrier Absorption

The effect of free carriers on absorption and RI change has already been discussed in Section 6.6. The expression derived there should be modified to include the contributions from the heavy and light holes. Since the concentrations of heavy and light holes are proportional to their effective masses to the 3/2 power, the following expression is obtained by inserting the values of the fundamental constants [1,6,7]:

$$\Delta n_r = \frac{-6.9 \times 10^{-22}}{n_r E^2}\left[\frac{n}{m_e} + p\left(\frac{m_{hh}^{1/2} + m_{lh}^{1/2}}{m_{hh}^{3/2} + m_{lh}^{3/2}} \right) \right] \tag{6.158}$$

6.10 Excitons

As already discussed, the ground state of a semiconductor material consists of a fully occupied VB and a completely empty CB, well above the top of the VB and separated by a forbidden energy gap. When the material is excited, an electron from the VB is transferred to one of the unoccupied states in the CB. Thus, EHPs are generated. These electrons and holes are normally thought to be uncorrelated. But in the actual situation, as the electrons and holes are oppositely charged, an attractive Coulomb interaction is assumed to occur between them. As a result, a hydrogen-like bound pair is created in place of a free EHP. These complex elementary excitations in solids that

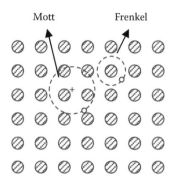

FIGURE 6.8
A conceptual picture of Frenkel- and Mott–Wannier-type excitons.

arise due to Coulomb interaction between electrons and holes are known as *excitons*. Their lowest energy level is above the ground-state energy of the unexcited crystal, the energy separation being slightly less than the band-gap E_g.

6.10.1 Classification

The excitons in solids are classified into one of two types: a Frenkel type or a Wannier type. In highly ionic or molecular crystals, the atoms interact very weakly. When an elementary excitation takes place, it may be thought of as occurring in a single atom or molecule. The excitation is more or less localized in space, extending over at most a few atomic sites. The corresponding exciton is known as a Frenkel-type exciton. It is, however, free to move in the crystal and to provide a means for transfer of energy from one point to another. On the other hand, in semiconductors having covalent bonding, the electrons are shared by many atoms. The excitons created in these materials are called Wannier-type excitons, and they extend over many atomic sites in the crystal. These are usually formed in covalently bonded Gr IV materials as well as in weakly ionic III–V compound semiconductors. Wannier excitons are also called Mott–Wannier or simply Mott excitons. A conceptual picture of the two types is given in Figure 6.8. Excitons have a very strong effect on the absorption process in direct as well as indirect semiconductors.

6.10.2 Exciton Binding Energy in Bulk

Since two particles, an electron and a hole, are involved, the problem is solved by using the two-particle Schrödinger equation

$$\left[-\frac{\hbar^2}{2m_e} \nabla_e^2 - \frac{\hbar^2}{2m_h} \nabla_h^2 - \frac{e^2}{4\pi\varepsilon |\mathbf{r}_e - \mathbf{r}_h|} \right] \psi_{ex} = E\psi_{ex} \tag{6.159}$$

where:

m_e (m_h) is the electron (hole) effective mass
$|r_e - r_h|$ is the separation between the positions of the two particles

The standard two-body problem is transformed into a one-body problem by introducing the following transformations:

$$\mathbf{r} = \mathbf{r}_e - \mathbf{r}_h; \quad \mathbf{k} = \frac{m_e \mathbf{k}_e + m_h \mathbf{k}_h}{m_e + m_h}; \quad \mathbf{R} = \frac{m_e \mathbf{r}_e + m_h \mathbf{r}_h}{m_e + m_h}; \quad \mathbf{K} = \mathbf{k}_e - \mathbf{k}_h \quad (6.160)$$

The transformed Hamiltonian reads

$$H = \frac{\hbar^2 K^2}{2(m_e + m_h)} + \left[\frac{\hbar^2 k^2}{2m_r} - \frac{e^2}{4\pi\varepsilon |r|} \right] \quad (6.161)$$

where m_r is the reduced mass. The first term of the above Hamiltonian is due to center-of-mass motion of the electron–hole system, while the second term is related to their relative motion. The first term gives the plane wave solution

$$\psi_{cm} = \exp(j\mathbf{K} \cdot \mathbf{R}) \quad (6.162)$$

The envelope function $F(\mathbf{r})$ related to the second part satisfies the relation

$$\left(\frac{\hbar^2 k^2}{2m_r} - \frac{e^2}{4\pi\varepsilon |r|} \right) F(\mathbf{r}) = EF(\mathbf{r}) \quad (6.163)$$

which is the equation for the H-atom problem, and the eigenvalues are given by

$$E_n = -\frac{m_r e^4}{2(4\pi\varepsilon)^2 \hbar^2} \frac{1}{n^2} = \left(\frac{m_r}{m_0} \right) \left(\frac{\varepsilon_0}{\varepsilon} \right)^2 \frac{1}{n^2} \times 13.6 \text{ eV} \quad (6.164)$$

The complete wave function may be written as

$$\psi_{nK} = e^{j\mathbf{K} \cdot \mathbf{R}} F_n(\mathbf{r}) \varphi_e(\mathbf{r}_e) \varphi_h(\mathbf{r}_h) \quad (6.165)$$

Example 6.12: For GaAs, $m_e = 0.067m_0$ and $m_h = 0.5m_0$, and thus $m_r = 0.059m_0$. Taking $\varepsilon = 13.1\varepsilon_0$, the binding energy for the 1s ($n = 1$) exciton is 4.7 meV.

The method of calculation of the excitonic absorption coefficient has been given in many references and textbooks, and will not be presented here. The calculated absorption spectra have the following features:

1. There are distinct absorption peaks corresponding to discrete excitonic states, such as 1s, 2s, 2p, and so on. Ideally, these peaks are sharp. However, excitons undergo collisions with phonons, impurities, and other defects, which contributes to the broadening of the excitonic peaks.

2. The higher-lying excitonic states are closely spaced to make a continuum. The calculated absorption spectra show a continuous rise with energy, and the variation follows the $(\hbar\omega - E_g)^{1/2}$ law given by Equation 6.51 for band-to-band transition in the direct-gap semiconductor. For experimental absorption spectra in GaAs, the reader is referred to the work of Sturge [8]; see also Figure 6.10.

6.10.3 Excitonic Processes in QWs

In the following, the theory of excitonic processes in direct-gap materials is briefly outlined.

6.10.3.1 Excitons in 2-D: Preliminary Concepts

In bulk semiconductors, mutual Coulomb interaction between electrons and holes becomes prominent at low temperatures. The excitons dissociate at higher temperatures due to their small binding energy. In a QW or other low-dimensional system, the situation is altogether different. Take, for example, the values of binding energy and Bohr radius in bulk GaAs, which are, respectively, 4.2 meV and 15 nm. As shown in Figure 6.9, the excitonic orbit is spherical in bulk GaAs. However, if the pair is created in a QW of width less than 15 nm, the Bohr radius, the orbit becomes squeezed, as shown in Figure 6.9. The confinement of electrons and holes reduces the physical separation between the particles along the z-direction, as a result of which the Coulombic interaction and hence the binding energy increase. Shinada and Sugano [9] first showed that the binding energy for

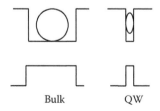

Bulk   QW

FIGURE 6.9
Exciton orbits in bulk and QWs.

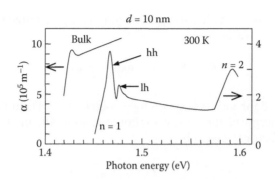

FIGURE 6.10
Excitonic absorption spectra in bulk GaAs and GaAs–AlGaAs MQWs.

pure 2-D systems increases fourfold. The measured optical absorption spectra, shown in Figure 6.10, indicate that sharp excitonic peaks occur for different subbands (denoted by the symbol n). In addition, as the light and heavy holes get separated in QWs, there are two excitonic peaks. These peaks are barely resolved in the spectra for a 14 nm wide sample.

6.10.3.2 Excitons in Purely 2-D Systems

The hydrogenic levels in a purely 2-D system were calculated by Shinada and Sugano [9]. Although a real system differs in many ways from the ideal system they considered, we prefer to give a brief sketch of their work. The equation describing the relative motion in a purely 2-D system is

$$-\frac{\hbar^2}{2m_r}\left(\frac{\partial^2}{\partial x^2}+\frac{\partial^2}{\partial y^2}\right)\phi-\frac{e^2}{4\pi\varepsilon(x^2+y^2)^{1/2}}\phi=E\phi \tag{6.166}$$

The wave function satisfying Equation 5.49 is written as

$$\phi=(2\pi)^{-1/2}R(\mathbf{r})\exp(jm\vartheta) \tag{6.167}$$

The radial function $R(\mathbf{r})$ may be substituted in Equation 6.167. The complete treatment will not be given here. It is found that there are different solutions giving rise to bound-state and discrete (quasi-continuum) and continuum-state eigenvalues. The bound-state energies are expressed as

$$E_n=-\frac{R_x}{(n+1/2)^2},\quad n=0,1,2,\dots\quad R_x=\frac{m_r e^4}{32\pi^2\hbar^2\varepsilon^2} \tag{6.168}$$

where R_x is the effective Rydberg constant. The normalized ground-state wave function is written as

$$R_{0,0}(r) = (4/a_0)\exp(-2r/a_0) \tag{6.169}$$

The effective Bohr radius is given by $a_0 = 4\pi\varepsilon\hbar^2/m_r e^2$. From Equation 6.168, the binding energy of the lowest excitonic state is $4R_x$ $(n=0)$. For discrete states, the oscillator strength is

$$f_{osc} = \frac{2}{Am_0\hbar\omega}\left|\langle f|e^{jk_p\cdot r}||i\rangle\right|^2 \tag{6.170}$$

while for continuous and quasi-continuous states, the absorption coefficient is given by

$$\alpha(\hbar\omega) = \frac{\pi\varepsilon}{m_0 c n_r \omega a_0}\left|\langle f|e^{jk_p\cdot r}||i\rangle\right|^2 D(\hbar\omega) \tag{6.171}$$

where D denotes the joint density of states function. As noted in this chapter, the matrix element connecting the initial (i) and final (f) states is proportional to the value of the envelope function at $\mathbf{r}=0$, that is, $\phi(\mathbf{r}=0)$. Thus, in 2-D also

$$\left|\langle f|e^{jk_p\cdot r}\mathbf{a}\cdot\mathbf{p}|i\rangle\right|^2 = A\beta\phi_{n(or)k,m}(0)\left|\langle c|\mathbf{a}\cdot\mathbf{p}|v\rangle\right|^2 \tag{6.172}$$

Since ϕ_{nm} and ϕ_{km} are nonvanishing only for $m=0$, only s-states need be considered for calculating absorption. For transitions to the discrete energy states

$$|\phi_{n0}(0)|^2 = \left[\pi a_0^2(n+1/2)^3\right]^{-1} \tag{6.173}$$

which can be derived easily by using Equation 6.167. The oscillator strength for transition to the nth state is

$$f_0 = \frac{2}{m_0\pi a_0^2\hbar\omega}(n+1/2)^{-3}\left|\langle|p_{cv}|^2\rangle\right| \tag{6.174}$$

taking into account the spin degeneracy factor 2. When n is large, the states may be assumed to be distributed quasi-continuously, and the density-of-states function may be defined as

$$D(E) = 2 \left| \left(A \frac{\partial E}{\partial n} \right)^{-1} \right| = \frac{(n+1/2)^3}{AR_x} \tag{6.175}$$

with a factor 2 for spin degeneracy included. Using Equations 6.171 and 6.175, the absorption coefficient is expressed as

$$\alpha_{qc}(\hbar\omega) = \frac{8\pi\varepsilon}{m_0 c n_r a_0 \omega} \left\langle \left| p_{cv} \right|^2 \right\rangle \tag{6.176}$$

The quantum number n does not appear in the expression. The unbound states form a continuum of energy, and for these states one obtains from Equation 6.167

$$\left| \phi_{k0}(0) \right|^2 = \frac{\exp(\pi\gamma)}{A \cosh(\pi\gamma)} \tag{6.177}$$

and the joint density of states function is a step function given by

$$D(E) = \frac{m_r}{\pi\hbar^2} \quad (E > 0) \tag{6.178}$$

By using Equations 6.171 and 6.178, the absorption coefficient may be expressed as

$$\alpha_{cont}(\hbar\omega) = \frac{4\pi\varepsilon}{m_0^2 c n_r a_0 \omega} \frac{\exp(\pi\gamma)}{\cosh(\pi\gamma)} \left\langle \left| p_{cv} \right|^2 \right\rangle \tag{6.179}$$

This may be compared with the expression for the absorption coefficient for a purely 2-D system for band-to-band transitions, which is readily obtained as

$$\alpha_{bb}(\hbar\omega) = \frac{4\pi\varepsilon}{m_0^2 c n_r a_0 \omega} \left\langle \left| p_{cv} \right|^2 \right\rangle = \alpha_{cont}(\hbar\omega) \left[\frac{\cosh(\pi\gamma)}{\exp(\pi\gamma)} \right] \tag{6.180}$$

In the limit $\gamma = (R_x/E)^{1/2} \to 0$, the factor $\left[\cosh(\pi\gamma) / \exp(\pi\gamma) \right] = 1$.

Thus, in the limit $R_x \to 0$ or $E \gg R_x$, one obtains $\lim_{\gamma \to 0} (\alpha_{cont} - \alpha_{bb}) = 0$, which should be expected from physical considerations. This conclusion was also drawn for bulk materials, for which the absorption spectra for the continuum states merge with band-to-band spectra obeying the $(\hbar\omega - E_g)^{1/2}$ law at high values of photon energies.

Example 6.13: The binding energy for 2-D excitons in GaAs is $4 \times 4.2 = 16$ meV. The value in real QWs is ~11 meV.

Example 6.14: The excitonic Bohr radius and binding energy for a Ge QW with $d = 5$ nm are calculated using $m_e = 0.042\, m_0$ and $m_{hh} = 0.284\, m_0$, so that the reduced mass is $m_r = m_e m_{hh}/(m_e + m_{hh}) = 0.036\, m_0$. The effective Bohr radius is $a_0 = 4\pi\varepsilon\hbar^2/m_r e^2 = 23.5$ nm, using $\varepsilon = 16\varepsilon_0$ for Ge. The exciton binding energy as expressed by Equation 6.168 is $E_0 = 4m_r e^4/32\pi^2\hbar^2\varepsilon^2 = 7.64$ meV.

6.10.3.3 Excitonic Absorption in Direct-Gap QWs

The simplified picture for purely 2-D systems given in Section 6.12.3.2 is modified in a real QW. First, the binding energy does not increase fourfold. However, the increase is high enough that excitons survive at room temperature. The excitonic absorption spectra for a GaAs–AlGaAs MQW are shown in Figure 6.10, and sharp excitonic absorption peaks are clearly visible [10]. In the absorption spectra for bulk GaAs, included in Figure 6.10, the excitonic peak is barely visible at room temperature. The spectrum for QW for the first subband ($n = 1$) shows two peaks for HH and LH absorption. However, the peaks are not separated for the $n = 2$ level.

The theory of absorption developed for a purely 2-D system is modified for a real QW, since the excitonic envelope functions are modified due to finite width of the QW and finite barrier height, and in addition the binding energy of excitons should be calculated as a function of separation of electrons and holes in both the xy plane and the z-direction. We refer the interested reader to the existing literature.

PROBLEMS

6.1. Show that the absorption coefficient for a material with finite conductivity σ may be related with the absorption coefficient by $\alpha = \sigma/nc\varepsilon_0$, where n is the real part of RI.

6.2. Using the expressions for the real and imaginary parts of the susceptibility derived in Problem 6.1, show that the two are related by Kramers–Kronig relationships. Choose a proper contour in the complex ω-plane to arrive at these relationships.

6.3. Express Einstein's A- and B-coefficients in terms of momentum matrix elements.

6.4. Establish the equivalence of the two forms of perturbation Hamiltonian as given in Equation 6.34.

6.5. Using energy and momentum conservations, prove that absorption from the VB to the degenerate CB starts at absolute zero at a photon energy $\hbar\omega = E_g + (E_f - E_c)(1 + m_e/m_h)$.

6.6. Calculate the value of squared momentum matrix element for GaAs using Equation 6.60. Hence, find the value of E_p.

6.7. Derive the general expression for the absorption coefficient, taking into account the density of states functions and occupational probabilities of both the bands.

6.8. Assume that the occupational probabilities in Equation 6.72 are governed by a Maxwellian distribution. Show that the absorption coefficient is proportional to $\exp[-(\hbar\omega - \Delta/k_BT)(m_s/m_s - m_h)]$ for the HH–SO transition.

6.9. Give reasons why the absorption spectrum for IVBA is peaked. Prove that the peak occurs at a photon energy $(\hbar\omega)_{max} = \Delta + (3/2)k_BT[(m_s - m_h)/m_h]$.

6.10. Prove that in an intraband FCA process, electrons must absorb or emit phonons in order to conserve momentum.

6.11. Consider Equation 6.85 for FCA. Assuming that the electron mobility is limited by deformation-potential acoustic-phonon scattering, comment on how the free-carrier absorption depends on temperature.

6.12. Consider intraband FCA. Assume that the angle between the initial wave vector **k** and the light wave vector **κ** is θ. Show that the energy of the photon to satisfy energy and momentum conservation conditions may be expressed as

$$\hbar\omega = 2\left[m^*\left(\frac{c}{n(\omega)}\right)^2 - \hbar k\left(\frac{c}{n(\omega)}\right)\cos\theta\right]$$

Take $m^* = 0.25m_0$, $n(\omega) = 3.5$, and $k = 0$. Calculate the value of photon energy needed.

6.13. Starting from Equation 6.86, valid for free electrons, obtain Equation 6.87, the complete equation for FCA considering both electrons and holes.

6.14. Assume that the electron and hole densities at energies E_2 in CB and E_1 in VB are given as $n(E_2) = A\exp[-(E_2 - E_c/k_BT)]$ and $p(E_1) = B\exp[-(E_v - E_1/k_BT)]$. Show that the spectral distribution of photon energy $\hbar\omega = E_2 - E_1$ is expressed as $P(\hbar\omega) = K\exp(\hbar\omega - E_g)\exp[-(\hbar\omega - E_g)/k_BT]$. A, B, and K are constants of proportionality.

6.15. Prove that the maximum recombination lifetime may be expressed as $\tau_r(max) = (2B_rn_i)^{-1}$, appropriate for an intrinsic semiconductor, and that the radiative lifetime is reduced if the material is made n- or p-type.

6.16. Prove that for a p-type material the lifetime under low injection is given by $\tau_r = [B_rN_A]^{-1}$, where N_A is the acceptor concentration.

6.17. Assume that the reduced mass in semiconductors is approximately equal to the electron effective mass. Hence, show that the exciton binding energy scales roughly with the bandgap.

6.18. Calculate the binding energy of excitons in $Cd_xZn_{1-x}Se$ for $x = 0, 0.2$, 0.5, and 1.0 using the following parameters: $m_e/m_0 = (0.16 - 0.03x)$, $\gamma_2(x) = 1.14 + 0.22x$, $\gamma_1(x) = 4.30 + 0.65x$, $\gamma_3(x) = 1.84$, $n_r(x) = 3.022 + 0.07x$.

6.19. The linewidth of excitonic absorption may be expressed as $\gamma = \gamma_{in} + \gamma_0 T + \gamma_{LO} / [\exp(\hbar\omega_{LO} / k_B T) - 1]$. The inhomogeneous part is due to fluctuation of QW width, the second term is due to exciton–acoustic-phonon scattering, and the last term is due to exciton–LO phonon scattering. Calculate the linewidth for 10 nm GaAs QW having width fluctuation of 0.3 nm. Assume $\gamma_0 = 5$ μeV K^{-1} and $\gamma_{LO} = 20$ meV.

Reading List

Agrawal, G. P. and N. K. Dutta, *Long Wavelength Semiconductor Lasers*. van Nostrand, New York, 1986.

Anselm, A. I., *Introduction to Semiconductor Theory*, trans. M. M. Samokhvalov. Mir, Moscow, 1981.

Basu, P. K., *Theory of Optical Processes in Semiconductors: Bulk and Microstructures*. Clarendon Press, Oxford, 2003.

Bebb, H. B. and E. W. Williams. In *Semiconductors and Semimetals*, vol. 8, ed. Willardson, R. K. and Beer, A. C., Academic Press, New York, pp. 181–320, 1971.

Casey, Jr., H. C. and M. B. Panish, *Heterostructure Lasers, Part A, Fundamental Principles*. Academic, New York, 1978.

Chuang, S. L., *Physics of Optoelectronic Devices*. Wiley, New York, 1995.

Ghatak, A. K. and K. Thyagarajan, *Optical Electronics*. Cambridge University Press, Cambridge, 1989.

Loudon, R., *The Quantum Theory of Light*, 2nd edition. Oxford University Press, Oxford, 1983.

Manasreh, O., *Semiconductor Heterojunctions and Nanostructures*. McGraw Hill, New York, 2005.

Nag, B. R., *Electron Transport in Compound Semiconductors*. Springer, Berlin, 1980.

Ridley, B. K., *Quantum Processes in Semiconductors*, 5th edition. Clarendon, Oxford, 2000.

Shur, M., *Physics of Semiconductor Devices*. Prentice Hall, Englewood Cliffs, NJ, 1990.

Singh, J., *Electronic and Optoelectronic Properties of Semiconductor Structures*. Cambridge University Press, Cambridge, 2003.

Wang, S., *Fundamentals of Semiconductor Theory and Device Physics*. Prentice Hall, Englewood Cliffs, NJ, 1989.

Yariv, A., *Quantum Electronics*, 3rd edition. Wiley, New York, 1989.

Yu, P. and M. Cardona, *Fundamentals of Semiconductors*. Springer, Berlin, 1995.

References

1. Bennett, B. R., R. A. Soref, and J. A. Del Alamo, Carrier induced change in refractive index of InP, GaAs, and InGaAsP, *IEEE J. Quantum Electron.* 26, 113–122, 1990.
2. van Roosbroeck, W. and W. Shockley, Photon radiative recombination of electrons and holes in Ge, *Phys. Rev.* 94, 1558–1560, 1954.
3. Shockley, W. and W. T. Read, Statistics of recombination of electrons and holes, *Phys. Rev.* 87, 835–842, 1952.
4. Hall, R. N. Electron-hole recombination in germanium, *Phys. Rev.* 152, 387, 1952.
5. Sah, C.-T. and W. Shockley, Electron-hole recombination statistics in semiconductors through flaws with many charge conditions, *Phys. Rev.* 109, 1103–1115, 1958.
6. Bandyopadhyay, A. and Basu, P. K., A comparative study of phase modulation in InGaAsP/InP and GaAs/AlGaAs based P-i-N and P-p-n-N structures, *J. Lightw. Technol.* 10, 1438–1442, 1992.
7. Bottledooren, D. and R. Baets, Influence of bandgap shrinkage on the carrier induced refractive index change in InGaAsP, *Appl. Phys. Lett.* 54, 1989–1991, 1989.
8. Sturge, M. D., Optical absorption in gallium arsenide between 0.6 and 2.75 eV, *Phys. Rev.* 127, 768, 1962.
9. Shinada, M. and S. Sugano, Interband optical transitions in extremely anisotropic semiconductors, *J. Phys. Soc. Jpn.* 21, 1936–1946, 1966.
10. Schmitt-Rink, S., D. S. Chemla, and D. A. B. Miller, Linear and nonlinear optical properties of semiconductor quantum wells, *Adv. Phys.* 38, 89–188, 1989.

7

Models for DH Lasers

7.1 Introduction

The structure and the elementary operating principle of semiconductor lasers have been introduced in Chapter 1. The basic semiconductor physics and principles of the semiconductor p-n junction, essential for understanding the operation of lasers, are discussed in Chapter 2. Following the discussion in Chapter 3 regarding the modification of band structure by alloying, or by using strained or unstrained quantum nanostructures, the band structure calculation using the **k.p** method has been introduced in Chapter 4. The role of passive optical devices in lasers is mentioned in Chapter 5. The optical processes occurring in semiconductors are discussed in Chapter 6.

We are now in a position to develop the models for semiconductor double heterostructure (DH) lasers by utilizing the theory and equations presented in earlier chapters. There are two approaches to understanding the operation of semiconductor lasers. The first one is based on first principles; in its most elementary form, it relies on the semiclassical theory of electron–photon interaction and band structure using **k.p** theory. There are more sophisticated methods, but these models would go beyond the scope of this book. The model thus relies on the intrinsic optoelectronic processes in bulk semiconductors that form the active layer in a DH. The second model is based on rate equations for carriers and photons. The rate equation model is used to predict the modulation characteristics and transient behavior of diode lasers.

In the current chapter, we will develop these two models. It should be mentioned that DH lasers are not much in use today. However, the models developed to account for the behavior of DH lasers are important as they are suitably modified to form the models for more recent devices based on quantum nanostructures. The models for DH lasers therefore serve as a suitable introduction to a theoretical understanding of the operation and characteristics of semiconductor lasers.

7.2 Gain in DH Lasers

In Section 6.4, the transition rate from a valence band state to a conduction band state conserving momentum has been calculated and from that rate the expression for the absorption coefficient has been obtained. The same transition rate can be used to obtain the expression for Einstein's B_{12} coefficient and from it the coefficients A and B_{21}.

Under usual circumstances, a light beam falling on a semiconductor sample is absorbed. However, if a condition for population inversion is created in the semiconductor, then instead of absorption, the electromagnetic (EM) radiation is amplified. In this section, we shall establish the condition for population inversion in a semiconductor and then obtain the expression for the gain coefficient in terms of the B-coefficient. It is straightforward to relate the gain coefficient to the absorption coefficient.

7.2.1 Absorption and Gain

The calculation of the emission and absorption rates in a semiconductor must take into consideration the band picture as well as the Fermi occupational probabilities in the conduction and valence bands. Let us consider an energy level E_2 in the conduction band and an energy level E_1 in the valence band of a direct-gap semiconductor. Spontaneous emission from E_2 to E_1 can occur only when the upper state is occupied and the lower state is empty (i.e., occupied by a hole). The occupational probabilities for the electrons in the conduction and valence bands are expressed by Fermi–Dirac statistics and are given by

$$f_c(E_2) = \left\{ 1 + \exp\left[\frac{(E_2 - F_e)}{k_B T} \right] \right\}^{-1} \tag{7.1}$$

and

$$f_v(E_1) = \left\{ 1 + \exp\left[\frac{(E_1 - F_h)}{k_B T} \right] \right\}^{-1} \tag{7.2}$$

Here F_e and F_h are the quasi-Fermi levels (under nonequilibrium situations), respectively, for electrons and holes. Now consider the small energy intervals dE_2 and dE_1 around levels E_2 and E_1, respectively, in the conduction and valence bands. The number of states in these intervals is, respectively, $S_c(E_2)dE_2$ and $S_v(E_1)dE_1$, where S_c and S_v denote, respectively, the density of states (DOS) in the respective bands. Assuming parabolic DOS functions, one may write their expressions as

$$S_c(E_2) = \frac{(2m_e)^{3/2}}{2\pi^2 \hbar^3} (E_2 - E_c)^{1/2} \tag{7.3a}$$

$$S_v(E_1) = \frac{(2m_h)^{3/2}}{2\pi^2 \hbar^3} (E_v - E_1)^{1/2} \tag{7.3b}$$

The rate of stimulated emission from state E_2 in the conduction band to state E_1 in the valence band depends on four factors: (1) the transition probability per unit time expressed in terms of Einstein's coefficient B_{21}; (2) the probability that E_2 is occupied; (3) the probability that E_1 is unoccupied; and (4) the photon density, $n_{ph}(\hbar\omega)$. Thus

$$r_{21}(st) = B_{21}n_{ph}(\hbar\omega)S_c(E_2)S_v(E_1)\{f_c(E_2)[1-f_v(E_1)]\} \tag{7.4}$$

Using similar arguments, the rate of absorption from E_1 to E_2 may be expressed as

$$r_{12}(abs) = B_{12}n_{ph}(\hbar\omega)S_c(E_2)S_v(E_1)\{f_v(E_1)[1-f_c(E_2)]\} \tag{7.5}$$

The spontaneous emission rate involves Einstein's A-coefficient and may be written for the present case as

$$r_{21}(sp) = A_{21}S_c(E_2)S_v(E_1)\{f_c(E_2)[1-f_v(E_1)]\} \tag{7.6}$$

In thermal equilibrium, the upward transition rate ($r_{12}(abs)$) equals the total downward transition rate [$r_{21}(sp) + r_{21}(st)$] and both of the quasi-Fermi levels merge with the equilibrium Fermi level. The calculation is similar to that given in Section 1.2.2. We obtain for the photon density:

$$n_{ph}(\hbar\omega) = \frac{n_r^3(\hbar\omega)^2}{\pi^2\hbar^3c^3\left[\exp(\hbar\omega/k_BT)-1\right]} \tag{7.7}$$

It is assumed that $B_{12} = B_{21} = B$, and A_{21} is related to B by

$$A_{21} = \frac{n_r^3(\hbar\omega)^2}{\pi^2\hbar^3c^3}B \tag{7.8}$$

It has already been shown that the ratio A_{21}/B is equal to the number of EM modes per unit volume or the mode density $m(E)dE$ such that

$$m(E)dE = \frac{n_r^3(E)^2}{\pi^2\hbar^3c^3}dE \tag{7.9}$$

Under conditions away from thermal equilibrium, the rates of radiative transitions do not balance the rate of absorption. The net stimulated emission rate is then expressed as

$$r^0(st) = r_{21}(st) - r_{12}(abs) = Bn_{\text{ph}}(E_{21})S_c(E_2)S_v(E_1)\{f_c(E_2) - f_v(E_1)\} \quad (7.10)$$

In order for the net stimulated emission rate to be positive, that is, $r^0(st) > 0$, it is required that $f_c(E_2) > f_v(E_1)$. From Equation 7.10, this condition is satisfied when

$$F_e - F_h \geq E_2 - E_1 > E_g \quad (7.11)$$

The above condition is known as the Bernard–Duraffourg condition [1]. The separation of the quasi-Fermi levels must therefore exceed the bandgap so that the rate of stimulated emission exceeds the rate of absorption. In equilibrium, the Fermi levels in the p and n layers align with each other, that is, $F_e = F_h$. In order to separate them, a pumping scheme is needed. For semiconductors, the most convenient way to effect this is to use a p-n junction, and to apply a forward bias to allow the injection of both electrons and holes in sufficient numbers. The condition of population inversion is then established in the junction.

The optical power gain coefficient per unit length, g, is defined as

$$\frac{dI}{dz} = gI \quad (7.12)$$

where:
    I is the light intensity
    z is the direction of propagation of the light wave

The intensity is the energy crossing per unit area per unit time and is expressed as

$$I = v_g \hbar \omega n_{\text{ph}}(\hbar \omega) \quad (7.13a)$$

where $v_g = d\omega/dk = c/n_r$ is the group velocity of light in the material. We may write from Equation 7.13a

$$\frac{dI}{dt} = v_g \hbar \omega \frac{d n_{\text{ph}}(\hbar \omega)}{dt} = v_g \hbar \omega r^0(st) \quad (7.13b)$$

Using the relation $dI/dz = (dI/dt)(dz/dt)^{-1} - (1/v_g)(dI/dt)$ and Equations 7.12 and 7.13a, we may rewrite Equation 7.13b as

$$\frac{dI}{dz} = gI = gv_g \hbar \omega n_{\text{ph}}(\hbar \omega) = \frac{1}{v_g}\frac{dI}{dt} \quad (7.14)$$

Comparing Equation 7.13b with Equation 7.14, we obtain

$$r^0(st) = v_g g n_{ph}(\hbar\omega) \tag{7.15}$$

Thus, using Equation 7.10, the gain coefficient is expressed as

$$g = \frac{r^0(st)}{n_{ph}(\hbar\omega)} \frac{n_r}{c} = \frac{n_r}{c} B S_c(E_2) S_v(E_1) \{ f_c(E_2) - f_v(E_1) \} \tag{7.16}$$

Einstein's coefficient B, which is the transition rate due to the stimulated emission of photons from State 2 to State 1, is expressed from time-dependent perturbation theory as

$$B = \frac{\pi e^2}{m_0^2 \varepsilon_0 n_r^2 \omega} \langle 1|\mathbf{p}|2\rangle^2 \tag{7.17}$$

As already noted in Chapter 6, $\mathbf{p}$ is the momentum operator and $\langle 1|\mathbf{p}|2\rangle$ denotes the momentum matrix element connecting States $|1\rangle$ and $|2\rangle$ in the valence and conduction bands, respectively.

Since the momentum operator is related to the position operator $\mathbf{r}$ by $\mathbf{p} = m_0(d\mathbf{r}/dt)$, one obtains

$$\langle 1|\mathbf{p}|2\rangle = m_0 \frac{d}{dt}\langle 1|\mathbf{r}|2\rangle = i\omega m_0 \langle 1|\mathbf{r}|2\rangle,$$

assuming $r \propto \exp(i\omega t)$. The B-coefficient may therefore be written as

$$B = \frac{\pi e^2 \omega}{\varepsilon_0 n_r^2} \langle 1|\mathbf{r}|2\rangle^2 = \frac{\pi\omega}{\varepsilon_0 n_r^2} \mu^2, \quad \mu^2 = \langle 1|e\mathbf{r}|2\rangle^2 \tag{7.18}$$

where μ is called the dipole moment. It is straightforward to show that

$$A_{21} = \frac{e^2 n_r \omega}{\pi m_0^2 \varepsilon_0 \hbar c^3} \langle 1|\mathbf{p}|2\rangle^2 = \frac{n_r \omega^3}{\pi \varepsilon_0 \hbar c^3} \mu^2 \tag{7.19}$$

Equation 7.10 has been derived by considering a pair of states E_2 and E_1 maintaining the relation $E_2 - E_1 = \hbar\omega = E_{ph}$. As these energy states form bands in a semiconductor, there may be a number of such pairs separated by the photon energy E_{ph}. To derive the expression for the overall net stimulated emission rate, we have to sum over such pairs of states. In the following, we denote the conduction band states by E and denote the energy of the matching valence band state by $E_1 = E - E_{ph}$. Using Equations 7.10 and 7.17, the net stimulated emission rate is therefore expressed as

$$R_{st}(E_{ph}) = \frac{\pi e^2 \hbar n_{ph}(E_{ph})}{m_0^2 \varepsilon_0 n_r^2 E_{ph}} \int_0^\infty \langle 1|\mathbf{p}|2\rangle^2 S_c(E) S_v(E - E_{ph}) \left[f_c(E) - f_v(E - E_{ph}) \right] dE$$

(7.20)

Similarly, the optical power gain coefficient is given by

$$g(E_{ph}) = \frac{\pi e^2 \hbar}{m_0^2 \varepsilon_0 n_r c E_{ph}} \int_0^\infty \langle 1|\mathbf{p}|2\rangle^2 S_c(E) S_v(E - E_{ph}) \left[f_c(E) - f_v(E - E_{ph}) \right] dE$$

(7.21)

Using Equations 7.6, 7.8, and 7.17, the spontaneous recombination rate is expressed as

$$R_{sp}(E_{ph}) = \frac{e^2 n_r E_{ph}}{\pi m_0^2 \varepsilon_0 n_r \hbar^2 c^3} \int_0^\infty \langle 1|\mathbf{p}|2\rangle^2 S_c(E) S_v(E - E_{ph}) f_c(E) \left[1 - f_v(E - E_{ph}) \right] dE$$

(7.22)

It is interesting to note that if we put $f_v = 1$ and $f_c = 0$ in Equation 7.21 and apply the **k**-conservation condition, we may recover Equation 6.62 for the absorption coefficient $\alpha(\hbar\omega)$.

The calculation of the gain coefficient in general involves numerical integration. We may, nevertheless, obtain an idea of the variation of gain with photon energy by using some approximations. Assume that the quasi-Fermi levels satisfy the following conditions:

$$F_e - E_c < 4k_B T \quad \text{and} \quad E_v - F_h < 4k_B T$$

(7.23)

Then the Fermi functions may be approximated as straight lines such that

$$f_v(E) = \frac{1}{2} + \frac{F_h - E}{4k_B T} \quad \text{and} \quad f_c(E) = \frac{1}{2} - \frac{E - F_e}{4k_B T}$$

(7.24)

The integration may be performed analytically by considering the parabolic nature of the DOS functions. The expression for the gain becomes

$$g(\hbar\omega) = K(\Delta F - \hbar\omega)(\hbar\omega - E_g)^2$$

(7.25)

where all the constants are lumped into the prefactor K, and ΔF is the difference between the quasi-Fermi levels. The gain curve therefore covers the range from E_g to ΔF, showing a maxima at $\hbar\omega = (1/3)(E_g + 2\Delta F)$ and then becoming negative (absorption) when $\hbar\omega > \Delta F$. With increasing injection, the values of ΔF increase and the gain curve covers a wider range of photon energy. Figure 7.1a shows how the band states are filled under heavy injection. The qualitative variation of the gain spectra is shown in Figure 7.1b for different injected carrier densities as the parameter. The maximum value of gain, g_{max}, for each injected carrier density is plotted in Figure 7.1c against the corresponding carrier density. The plot shown is approximately linear; however, the actual variation is nonlinear, which will be discussed in Section 7.3.2. It is found that below the transparency carrier density, n_{tr}, the gain is zero.

The position of the Fermi levels for a given carrier density is usually calculated numerically. Various approximate formulas are, however, available in the literature [2,3], which may be used for this purpose. The use of one such formula is illustrated in the examples given in this chapter.

The expression for gain, Equation 7.21, has been derived by assuming that the states E_1 and E_2 are sharp levels. The levels are usually broadened due to scattering and other effects. Here, we assume that each state is broadened, that the energy states related to the transitions have an energy width $\hbar/\tau_{in}$, and that the spectral shape is Lorentzian and is given by

$$L(E) = \frac{1}{\pi} \frac{\hbar/\tau_{in}}{(E - E_{ph})^2 + (\hbar/\tau_{in})^2} \tag{7.26}$$

where τ_{in} is the relaxation time due to scattering and transitions. The gain coefficient is now modified to the following form:

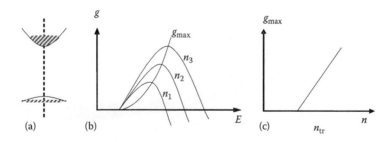

FIGURE 7.1
(a) Band state filling under heavy injection. Shaded regions represent occupied states. (b) Variation of gain with photon energy with injected electron density as the parameter. (c) Plot of maximum gain with injected carrier density; n_{tr} is the transparency carrier density.

$$g(E_{ph}) = \frac{\pi e^2 \hbar}{m_0^2 \varepsilon_0 n_r c E_{ph}} \int_0^\infty \langle 1|\mathbf{p}|2 \rangle^2 S_{red}(E) \left[f_c(E) - f_v(E - E_{ph}) \right] L(E) dE \quad (7.27)$$

where S_{red} stands for reduced DOS, which may be defined by replacing m_e by the reduced mass m_r in Equation 7.3a,b.

Example 7.1: Let us assume that the electron and the hole densities injected into the active GaAs layer in a laser are $n=p=10^{24}$ m^{-3}. With $m_e=0.067m_0$ and $m_h=0.5m_0$, the effective DOSs at $T=300$ K are $N_c=4.35\times10^{23}$ m^{-3} and $N_v=8.87\times10^{24}$ m^{-3}. The quasi-Fermi levels are $F_e-E_c=41.8$ meV and $E_v-F_h=55.4$ meV. Thus, the range of photon energies for which laser action takes place is from 1.424 to 1.521 eV.

Example 7.2: The maximum value of gain in an InP/In$_{0.53}$Ga$_{0.47}$As/InP DH laser for an injected carrier density of 4×10^{18} cm^{-3} is about 85 m^{-1}. The values of the parameters used in the calculation are given in Table 7.1. The value is calculated with $\tau_{in}=0.1$ ps.

Figure 7.2 shows plots of the gain spectra in an In$_{0.53}$Ga$_{0.47}$As layer clad between two InP layers for different values of injected carrier densities. The peak gain occurs at photon energies away from the gap energy and the peak gain shifts to higher photon energies with an increased injection level. The

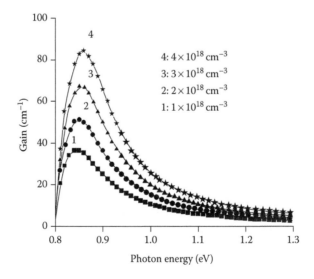

FIGURE 7.2
Calculated gain spectra for InGaAs/InP for different injection carrier densities.

TABLE 7.1

Effective Mass, Permittivity, Bandgap and Spin-Orbit Splitting of InAs and GaAs

Material	m_e/m_0	m_h/m_0	m_{lh}/m_0	ε_s (F m^{-1})	E_g (eV)	Δ (eV)
InAs	0.023	0.4	0.026	15.15 ε_0	0.354	0.38
GaAs	0.0663	0.5	0.087	13.1 ε_0	1.424	0.34

values of the different parameters are given in Table 7.1. The values for the alloy are calculated by linear interpolation.

7.3 Threshold Current

An important parameter for all lasers is the threshold current, which denotes the onset of self-sustained oscillation. The expression for the gain coefficient obtained in Section 7.2 may be used to derive the expression for the threshold current. In this section, we consider two models used; in the first one, the gain increases linearly with the rise in the current above its threshold value. In the second model, the gain coefficient shows nonlinear behavior.

7.3.1 Linear Gain Model

The current density J in a junction laser is expressed in terms of the total recombination rate, R_{sp}, given by Equation 7.22 and the width of the active region, d, by

$$J = edR_{sp} \tag{7.28a}$$

Under steady state, the current density is also proportional to the injected carrier density by the following relation:

$$J = \frac{end}{\tau_r} = \frac{end}{\eta_i \tau_{sp}} \tag{7.28b}$$

where τ_r is the total recombination time and is related to the radiative recombination time, τ_{sp}, by $\tau_r = \eta_i \tau_{sp}$, η_i being the internal quantum efficiency.

We assume that the peak gain varies linearly with the injected carrier density, n, above the transparency carrier density, n_{tr}, as shown qualitatively in Figure 7.1c. The peak gain may be expressed as

$$g_{max} = a(n - n_{tr}) \tag{7.29a}$$

where $a = \partial g_{max}/\partial n$ is the proportionality constant, called the differential gain. Using Equation 7.28b, we may express the peak gain in terms of current density as

$$g_{max} = \frac{a\eta_i \tau_{sp}}{ed}(J - J_{tr}) \qquad (7.29b)$$

The transparency current density may therefore be written as

$$J_{tr} = \frac{ed}{\eta_i \tau_{sp}} n_{tr} \qquad (7.29c)$$

At threshold, $\Gamma g_{max} = \Gamma g_{th} = \alpha + \alpha_m$, where Γ is the optical confinement factor, α is the material loss, and α_m is the mirror loss. Using Equation 7.29b, the expression for the threshold current density becomes

$$J_{th} = J_{tr} + \frac{ed}{\Gamma\eta_i \tau_{sp} a}\left[\alpha + \frac{1}{2L}\ln\left(\frac{1}{R_1 R_2}\right)\right] \qquad (7.30)$$

In Equation 7.30, we have included the explicit expression for the mirror loss α_m.

> **Example 7.3:** An estimate of the threshold current density is given for a GaAs DH laser using the following values for the parameters: $L = 200\ \mu m$, $d = 0.2\ \mu m$, $R_1 = R_2 = 0.32$, $\alpha = 10^3\ m^{-1}$, $n_{tr} = 1.8 \times 10^{24}\ m^{-3}$, $\Gamma = 0.02$, $\tau_{sp} = 1$ ns, $\eta_i = 0.8$, and $a = 7 \times 10^{-20}\ m^2$.
>
> The transparency current density is $0.72 \times 10^8\ A\,m^{-2}$. The mirror loss is $0.57 \times 10^4\ m^{-1}$. The second term on the right-hand side of Equation 7.30 is $16.75 \times 10^7\ A\,m^{-2}$. Therefore, the threshold current density $J_{th} = 23.95 \times 10^7\ A\,m^{-2}$. If the width of the laser is $w = 10\ \mu m$, so that the area is $10 \times 200\ \mu m^2$, the threshold current is 0.48 A.

We use the values of the parameters given in Example 7.3 to calculate J_{th} versus L, the length of the cavity. Figure 7.3 shows the variation of the threshold current density. The steep rise of J_{th} with the decreasing cavity length is due to increased mirror loss. The threshold current, on the other hand, increases with the increase in L. The results shown in Figure 7.3 are based on the constant value of the mode confinement factor.

The expressions for the light power output, the external quantum efficiency, and the power conversion efficiency of a semiconductor laser have been given in Section 1.3.3. The power output can be expressed as

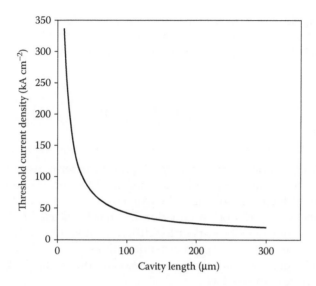

FIGURE 7.3
Variation of threshold current density with cavity length.

$$P_0 = \eta_i \left(\frac{\alpha_m}{\alpha_i + \alpha_m} \right) \frac{\hbar\omega}{e} (I - I_{th}) = \eta_d \frac{\hbar\omega}{e} (I - I_{th}) \qquad (7.31)$$

where η_d is called the differential quantum efficiency. Equation 7.31 represents the total power coming out of both mirrors. For unequal reflectivities, the fraction coming out of each mirror is different. For example, if $R_2 = 1$, all power will come out of the front mirror.

> **Example 7.4:** The calculated value of α_m in Example 7.3 is 5.7×10^3 m^{-1}. Using $\alpha = 10^3$ m^{-1} and $\eta_i = 0.8$, the differential quantum efficiency is 0.68.

7.3.2 Nonlinear Gain Model: Effect of Cavity Length and Reflectivity of Mirrors

For a given output power, it is necessary to design the laser structure for minimum current. For this purpose, an analytical expression for gain versus carrier density is needed. This relationship is by no means linear as has been assumed earlier. Some authors have used the following three-parameter logarithmic formula to relate gain with injected carrier density:

$$g = g_{0s} \ln \frac{n + n_s}{n_{tr} + n_s} \qquad (7.32)$$

In this expression, g_{0s} is an empirical gain coefficient, n_{tr} is the transparency carrier density, and n_s is a parameter that is used to make ln finite at $n=0$, so that the gain equals the absorption. Equation 7.32 may be further simplified as

$$g = g_0 \ln \frac{n}{n_{tr}} \quad (g \geq 0) \tag{7.33}$$

A plot of the modal gain versus the injected carrier density is shown in Figure 7.4. The various quantities defined in Equation 7.33 are shown in Figure 7.4 by drawing a tangent to the gain curve.

The logarithmic variation in Equation 7.32 has been used to fit the calculated gain–carrier density curves for different quantum well (QW) structures to give the values for g_0 and other quantities.

Now using Equation 7.33 and the expression for the modal gain, the threshold gain may be expressed as

$$n_{th} = n_{tr} e^{g_{th}/g_0} = n_{tr} e^{(\alpha_i + \alpha_m)/\Gamma g_0} \tag{7.34}$$

The threshold current may be expressed as

$$\frac{\eta_i I_{th}}{eV} = \left(R_{nr} + R_{sp} + R_t \right)_{th} = \frac{n_{th}}{\tau_r} = \left(An + Bn^2 + Cn^3 \right) \tag{7.35}$$

where V is the volume of the active region and the other symbols have been defined already. We now consider two different dominant recombination processes.

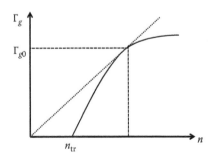

FIGURE 7.4
Modal gain vs. injected carrier density.

7.3.2.1 Dominant Radiative Recombination

It is to be noted that for the best laser material, the dominant recombination at threshold is due to spontaneous emission. Therefore, $I_{th} \cong Bn_{th}^2 eV/\eta_i$ and Equation 7.35 reduces to

$$I_{th} \cong \frac{eVBn_{tr}^2}{\eta_i} e^{2(\alpha_i + \alpha_m)/\Gamma g_0} \tag{7.36}$$

Equations 7.31 and 7.36 are now used to obtain a closed-form expression for the power output as a function of the applied current. Our interest is, however, on the power output P_{01} from front mirror 1. Solving, we get

$$I \cong \frac{eP_{01}(\alpha_i + \alpha_m)}{F_1 \eta_i \hbar \omega \alpha_m} + \frac{eVBn_{tr}^2}{\eta_i} e^{2(\alpha_i + \alpha_m)/\Gamma g_0} \tag{7.37}$$

The first term represents the additional current required above threshold to obtain power P_{01} and the second term is the threshold current. The factor F_1 is the fraction of the total output power coming out of mirror 1.

The following points should be taken into consideration to reduce the drive current:

1. Reduce the transparency value and the differential gain of the active material. The need for a larger differential gain is met by using QW, strained QW, quantum wire (QWR), and even quantum dot (QD) lasers.
2. Reduce cavity loss ($\alpha_i + \alpha_m$) and volume V subject to retaining a large confinement factor.
3. Use of high-reflectivity mirrors or short cavity length will reduce mirror loss. This advocates for the use of vertical-cavity surface-emitting lasers (VCSELs) or short-length in-plane lasers.
4. Use of multiple quantum well (MQW) lasers.

7.3.2.2 Dominant Nonradiative Recombination

For long-wavelength emission, the nonradiative recombination, in particular the Auger process, is more important. This adds to the threshold current. This is the reason why, despite a higher gain coefficient, InGaAsP/InP lasers show a higher threshold than GaAs/AlGaAs lasers. In the presence of higher-order nonradiative recombination, the rate of which is given by Cn_{th}^3, Equation 7.36 and the second term in Equation 7.37 should be increased by

$$I_{nr,th} \cong \frac{eVCn_{tr}^3}{\eta_i} e^{3(\alpha_i + \alpha_m)/\Gamma g_0} \tag{7.38}$$

Example 7.5: For InGaAsP, $C \sim 3 \times 10^{-29}$ cm³s⁻¹ at 1.3 μm and $\sim 6 \times 10^{-29}$ cm³s⁻¹ at 1.55 μm. With $B \sim 10^{-10}$ cm³ s⁻¹, the nonradiative terms dominate for $N_{th} > 3 \times 10^{18} (1.5 \times 10^{18})$ at 1.3 (1.55) μm.

7.3.3 Leakage Current

The carriers injected into the active layer of a DH laser have a distribution of energy as illustrated in Figure 7.5 for electrons in the conduction band. Once the energy of the injected electrons reaches the barrier height between the active and the cladding layers, that is, $E \geq \Delta E_c$ for conduction band electrons, they are free to diffuse into the cladding layer, giving rise to a leakage component of current. In Figure 7.5, the high-energy tail of the electron distribution crosses the band offset between the active n-layer and the p-type cladding layer. It is easy to conclude that the higher the temperature, the greater the number of carriers with energy in excess of the band offset.

The density of electrons above the barrier, n_{lk}, may be calculated from the following expression:

$$n_{lk} = \int_{E_B}^{\infty} S_c(E) f(E) dE \tag{7.39}$$

where:
  E_B is the barrier height in the heterobarrier
  $S_c(E)$ is the DOS function
  $f(E)$ is the Fermi distribution

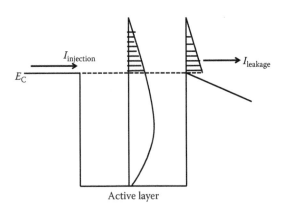

FIGURE 7.5
Carrier distribution in the active layer. The shaded area indicates the number of electrons with energy exceeding the band offset.

The expression may be written explicitly for a parabolic DOS function as

$$n_{lk} = \frac{1}{2\pi^2}\left(\frac{2m_e}{\hbar^2}\right)^{3/2}\int_{E_B}^{\infty}\frac{(E-E_c)^{1/2}}{1+\exp\left[(E-F_e)/k_BT\right]}dE \tag{7.40}$$

A fraction of the electrons will suffer reflection at the heterobarrier. Also, the electrons have random thermal velocity and thus a fraction of the electrons will actually move into the cladding layer. Some fraction of the electrons coming to the p-type cladding will diffuse in the opposite direction to come back into the active region. All these factors should be taken into account to know the actual number of electrons that succeed to diffuse along the p-cladding layer and are collected by the terminal.

In an alternative approach, the electron populations in both the active and the cladding layers are assumed to be in thermal equilibrium such that the quasi-Fermi levels in both the layers are aligned. With this assumption and using the Boltzmann distribution, the electron population at the edge of the heterobarrier may be written as

$$n_{p0} = N_c \exp\left[\frac{(F_e - E_B)}{k_BT}\right] \tag{7.41}$$

The distribution of the minority electrons in the p-cladding layer is obtained by solving a continuity equation in the presence of diffusion and recombination. The solution takes the following form:

$$n(x) = n_{p0}\exp\left[\frac{-x}{L_n}\right] \tag{7.42}$$

where the diffusion length of the minority electron is $L_n = \sqrt{D_n\tau_n}$, D_n and τ_n being the diffusion coefficient and minority carrier lifetime, respectively.

It is assumed first that there is no electric field in the p-cladding and so the current is entirely due to diffusion. The current density is then obtained by the standard method and may be written as

$$J_n\big|_{x=0} = eD_n(dn/dx)\big|_{x=0} = eD_n n_{p0}/L_n \tag{7.43}$$

Example 7.6: Consider an $Al_{0.3}Ga_{0.7}As/GaAs/Al_{0.3}Ga_{0.7}As$ DH laser. The barrier height in the conduction band is obtained by assuming the 65/35 rule as 250.6 meV. If $10^{24}\,m^{-3}$ electrons are injected into the laser, the leakage current, $J_{leakage} = eD_n n_{p0}/L_n$, will be 10.19 A cm^{-2}, with $D_n = 220$ cm^2 s^{-1} and $\tau_n = 1$ ns. Using Equation 7.41, n_{p0} can be calculated as $1.357 \times 10^{20}\,m^{-3}$.

Example 7.6 shows that the leakage current density in the GaAs–AlGaAs system is a small fraction of the threshold current density. However, in systems having small values of heterobarrier, for example, AlInGaP, carrier leakage may pose a problem. Carrier leakage also increases with a rise in temperature. If, in addition, the cladding layer has a high defect density, the reduced minority carrier lifetime can lead to higher carrier leakage currents.

7.4 Effect of Electric Field in Cladding on Leakage Current

An electric field may exist in the p-cladding layer if the resistance of the cladding is high. This electric field will assist the diffusion of injected carriers away from the active region, thereby increasing the leakage current. The current density including the electric field as well as a contact layer at a distance of x_p from the heterobarrier has been calculated and is expressed as [4]

$$J_n = eD_nN_{p0}\left[\sqrt{\frac{1}{L_n^2}+\frac{1}{L_{nf}^2}}\coth\sqrt{\frac{1}{L_n^2}+\frac{1}{L_{nf}^2}}x_p+\frac{1}{L_{nf}}\right] \tag{7.44a}$$

where

$$L_{nf} = \frac{2k_BT}{e}\frac{\sigma_p}{J_{tot}} \tag{7.44b}$$

Here, σ_p is the conductivity of the p-cladding layer, J_{tot} is the total diode current density, and L_{nf} is the drift length in analogy with the diffusion length. For low current and high p-doping, $L_{nf}\gg L_n$ and the diffusion current dominates. However, for high current and low p-doping, it is possible that $L_{nf}\ll L_n$ and the drift component exceeds the diffusion component. If $L_{nf}\ll L_n$, and x_p, then $J_n\to e\mu_n n_{p0}J_{tot}/\sigma_p$. This indicates that the leakage current becomes prohibitively large for a low heterobarrier leading to a higher sensitivity to temperature.

7.5 Gain Saturation

An important point to note about all lasers is that, in the steady state, the gain above threshold is equal to the gain at threshold, that is,

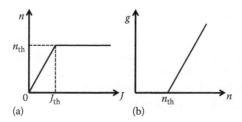

FIGURE 7.6
Plots illustrating gain saturation. (a) Plot of carrier density vs. drive current and (b) plot of gain vs. carrier density.

$$g(I > I_{th}) = g_{th} \tag{7.45}$$

If the gain above threshold exceeds the threshold gain, then the EM field amplitude will continue to increase without limit, and this certainly cannot occur at steady state. Furthermore, as the gain increases monotonically with the carrier density, the latter must also clamp to its threshold value. This means that under steady state,

$$n(I > I_{th}) = n_{th} \tag{7.46}$$

The inherent mechanism of gain saturation above threshold may be understood as follows. As the current exceeds the threshold value, the carrier density and the gain increase initially above their threshold values over a period of about a few nanoseconds. At the same time, the stimulated recombination increases, reducing both the carrier density and the gain, so that a new dynamic equilibrium is reached and both the carrier density and the gain return to their threshold values. Thus, the stimulated emission process uses all the extra electrons above its threshold value. Figure 7.6a shows a plot of the carrier density versus the drive current density, while Figure 7.6b shows a plot of the gain versus the carrier density The carrier density is clamped to its threshold value when $I > I_{th}$.

7.6 Rate Equation Model

We now introduce the rate equation model, which is useful in relating the light power output with the bias current, in predicting the modulation (ac) response of the laser diode (LD), and obtaining the transient characteristics when a step increase in current is applied.

7.6.1 Rate Equations

The behavior of diode lasers may be studied by considering two coupled equations involving the carrier densities and the photon densities. At the outset, it is assumed that the electron density, n, equals the hole density, p. Therefore, the number of equations reduces to two.

The rate equation for n is written as

$$\frac{dn}{dt} = \frac{J}{ed} - G(n)n_{ph} - \frac{n}{\tau_n} \tag{7.47}$$

where:

- J is the current density
- d is the thickness of the active layer
- $G(n)$ is the rate of amplification due to stimulated emission
- n_{ph} is the photon density
- τ_n is the carrier lifetime

The first term on the right-hand side of Equation 7.47 represents the rate of increase of the electron density due to the injected current and the other two terms represent the rate of loss of the carriers. The second term occurs due to stimulated emission and is therefore proportional to the photon density. The third term expresses the decay rate of carriers via the spontaneous emission process.

The rate equation for photon density may similarly be developed considering gain and loss terms in the following way:

$$\frac{dn_{ph}}{dt} = G(n)n_{ph} - \frac{n_{ph}}{\tau_{ph}} + \beta\frac{n}{\tau_r} \tag{7.48}$$

The first term, the gain term for photon density, obviously occurs due to the stimulated process; the second term is the rate of decrease of the photon density in the cavity due to absorption and mirror loss, the time constant for which is the photon lifetime, τ_{ph}; and the last term represents the coupling rate of spontaneously emitted photons into the lasing mode, β being the spontaneous emission coupling factor.

Assuming that n_{tr} denotes the transparency carrier density, we may express the stimulated emission amplification factor by

$$G(n) = \Gamma_a g_0(n - n_{tr}) \tag{7.49}$$

where:

- g_0 is the differential amplification rate
- Γ_a is the optical confinement factor of the active layer

The carrier lifetime τ_n in Equation 7.47 contains both the radiative and non-radiative parts and may be expressed as

$$\frac{1}{\tau_n} = \frac{1}{\tau_r} + \frac{1}{\tau_{nr}} \tag{7.50}$$

where the subscripts r and nr refer, respectively, to radiative and nonradiative lifetimes. The photon lifetime is the average time that the photons spend in the cavity and is expressed as

$$\frac{1}{\tau_{ph}} = \frac{c}{n_r}\left(\alpha + \frac{1}{2L}\ln\frac{1}{R_1 R_2}\right) \tag{7.51}$$

The spontaneous emission coupling factor β is defined as

$$\beta = \frac{\text{spontaneous emission coupling rate to the lasing mode}}{\text{total spontaneous emission rate}} \tag{7.52}$$

Assume that the spontaneous emission spectrum has a Lorentzian lineshape of full-width at half-maximum (FWHM) $\Delta\omega$ around the central frequency of ω_0 expressed as

$$r_{sp} = r_{sp0}\frac{(\Delta\omega/2)^2}{(\omega - \omega_0)^2 + (\Delta\omega/2)^2} \tag{7.53}$$

where r_{sp0} is a prefactor.

The total spontaneous emission rate, R_{sp}, is calculated by considering the number of modes dN with two polarizations within the volume V, a solid angle $d\Omega$, and an angular frequency range. For the continuous distribution of modes, dN is given by

$$dN = Vm(\omega)\,d\omega\frac{d\Omega}{4\pi} = V\frac{n_r^3\omega^2}{\pi^2 c^3}\,d\omega\frac{d\Omega}{4\pi} \tag{7.54}$$

The total spontaneous emission rate is then

$$R_{sp} = \int r_{sp}\,dN = r_{sp0}\frac{V}{2\pi}\left(\frac{n_r}{c}\right)^3\omega_0^2\Delta\omega \tag{7.55}$$

The spontaneous emission coupling factor is thus expressed by using Equations 7.52 through 7.55 in terms of the free-space wavelength λ_0 and spectral linewidth $\Delta\lambda$ as

$$\beta = \Gamma_a \frac{r_{sp}}{R_{sp}} = \Gamma_a \frac{2\pi}{V} \left(\frac{c}{n_r}\right)^3 \frac{1}{\omega_0^2 \Delta\omega} = \frac{\Gamma_a}{4\pi^2 n_r^3 V} \frac{\lambda_0^4}{\Delta\lambda} \tag{7.56}$$

To obtain Equation 7.56 from Equation 7.55, the relation $\Delta f/f_0 = \Delta\lambda/\lambda_0$ is used.

Example 7.7: An estimate of β is made using the following parameter values: $\Gamma_a = 0.02$, $\lambda_0 = 1.55\ \mu m$, $\Delta\lambda = 10\ nm$, $L = 200\ \mu m$, $w = 90\ \mu m$, $d = 0.2\ \mu m$, and $n_r = 3.5$. The calculated value is 0.59×10^{-5}.

7.6.2 Steady-State Solutions

The steady-state solutions to the rate equations are obtained by putting $d/dt = 0$. Also, below threshold the net stimulated emission is negligible and $n_{ph} = 0$. Putting these in Equation 7.47 leads to $n = J\tau_n/ed$. Assuming this equality to be valid up to the threshold, one may relate the threshold current density, J_{th}, to the threshold electron density, n_{th}, as

$$J_{th} = \frac{ed}{\tau_n} n_{th} \tag{7.57}$$

Since $\beta \approx 10^{-5}$, we neglect this term in Equation 7.49 and write for the steady state:

$$G(n) = \Gamma_a g_0 (n - n_{tr}) = \frac{1}{\tau_{ph}} \tag{7.58}$$

Writing the modal gain $\Gamma_a g_0 = \alpha + (1/2L)\ \ln(1/R_1 R_2)$ and relating this to the photon lifetime, one may express the threshold carrier concentration by using the above relation for $G(n)$ as

$$n_{th} = n_{tr} + \frac{1}{\Gamma_a g_0 \tau_{ph}} \tag{7.59}$$

In diode lasers, changes in the length, facet reflectivities, and refractive index (RI) during laser operation are small, so that the right-hand side of Equation 7.59 may be assumed constant above threshold. This leads to the following relation for the threshold current density:

$$J_{th} = \frac{ed}{\tau_n} n_{th} = \frac{ed}{\tau_n} \left(n_{tr} + \frac{1}{\Gamma_a g_0 \tau_{ph}} \right) \tag{7.60}$$

Equation 7.60 is identical to Equation 7.30, since $g_0 = a(dz/dt) = a(c/n_r)$.

Example 7.8: The variation of the threshold current density with thickness d of $In_{0.53}Ga_{0.47}As$ clad between two InP layers is calculated using the following parameter values: $L = 200$ μm, $n_{tr} = 2 \times 10^{18}$ cm^{-3}, $\tau_n = 1$ ns, $a = 10^{-15}$ cm^2, $\alpha = 10$ cm^{-1}, $R = 0.35$, $\Gamma = \beta^2/2 + \beta^2$, $\beta = (2\pi d / \lambda_0)\sqrt{n_a^2 - n_{cl}^2}$, n_a (InGaAs) $= 3.77$, and n_{cl} (InP) $= 3.544$. The value for $d = 0.1$ μm is $J_{th} = 3$ kA cm^{-2}.

Figure 7.7 shows a plot of J_{th} versus d, using the aforementioned values. For large values of d, the current density increases linearly. However, for small values of d, the mode confinement factor decreases, thereby increasing J_{th}. There is an optimum value for d for which J_{th} is minimum.

7.6.3 Light Output versus Injected Current

We first neglect the coupling of the spontaneous emission to the lasing mode, that is, take $\beta = 0$. Below threshold, it is assumed that there is no photon and therefore $n_{ph} = 0$. Above threshold, the carrier concentration, n, is clamped to

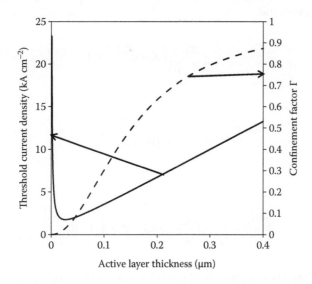

FIGURE 7.7
J_{th} and mode confinement factor vs. active layer thickness d for InP–InGaAs–InP DH.

its threshold value, n_{th}. Now from the first rate equation (Equation 7.47), the photon density is written as

$$n_{ph} = \frac{1}{G(n)}\left(\frac{J}{ed} - \frac{n_{th}}{\tau_n}\right) = \frac{\tau_{ph}}{ed}(J - J_{th}) \tag{7.61}$$

where use has been made of Equations 7.57 and 7.58 in relating $G(n)$ with τ_{ph} and n_{th} with J_{th}.

The light power, $P = \hbar \omega n_{ph}$, therefore increases linearly as the current density exceeds its threshold value.

When coupling of spontaneous emission into the lasing mode is considered, the steady-state rate equations are written in the following forms using Equations 7.47 through 7.49:

$$\frac{J}{ed} = \Gamma_a g_0 (n - n_{tr}) n_{ph} + \frac{n}{\tau_n} \tag{7.62}$$

$$\frac{n_{ph}}{\tau_{ph}} = \Gamma_a g_0 (n - n_{tr}) n_{ph} + \beta \frac{n}{\tau_n} \tag{7.63}$$

where the assumption that nonradiative recombination is negligible leading to $\tau_r \approx \tau n$ has been introduced. Solving these two equations, the expressions for n and n_{ph} become

$$n = \frac{n_{th}}{2(1-\beta)}\left(X - \sqrt{X^2 - Y}\right) \tag{7.64a}$$

$$n_{ph} = \frac{\beta}{\Gamma_a g_0 \tau_n} \frac{X - \sqrt{X^2 - Y}}{2(1-\beta) - \left(X - \sqrt{X^2 - Y}\right)} \tag{7.64b}$$

where

$$X = 1 + \frac{J}{J_{th}} - \beta\frac{n_0}{n_{th}} \quad \text{and} \quad Y = 4(1-\beta)\frac{J}{J_{th}} \tag{7.64c}$$

The calculated results for the carrier density and the light power output for $\beta > 1$ are also included in Figure 7.8. The threshold is not well defined when $\beta \neq 0$.

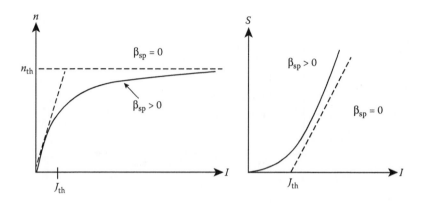

FIGURE 7.8
Carrier concentration and light power output vs. injected current for $\beta=0$ and $\beta\neq0$.

7.7 Rate Equations: Solution of Time-Dependent Problems

In the previous section, the rate equations are employed to obtain the steady-state behavior of the LD. In this section, a few time-dependent problems will be addressed.

7.7.1 Transient Characteristics of a Fabry–Perot LD

We now examine the response of an LD when a current pulse is applied to the laser. The diode is biased with a steady bias current density $J_B<J_{th}$, and the amplitude of the pulse is sufficient to reach the threshold current density. The situation is shown in Figure 7.9.

We use rate equations neglecting the coupling of spontaneous emission. Also below threshold, there is no emission and $n_{ph}=0$. As a result, Equation 7.47 reduces to

$$\frac{dn}{dt} = \frac{J}{ed} - \frac{n}{\tau_n} \tag{7.65a}$$

The current density has a steady bias J_B to which a step pulse is applied at $t=0$ and we therefore write

$$J = J_B + J_1 u(t) \tag{7.65b}$$

where $u(t)$ is a step function of unity amplitude defined as $u(t)=0$ for $t<0$ and $u(t)=1$ for $t\geq0$. We substitute Equation 7.65b into Equation 7.65a and take

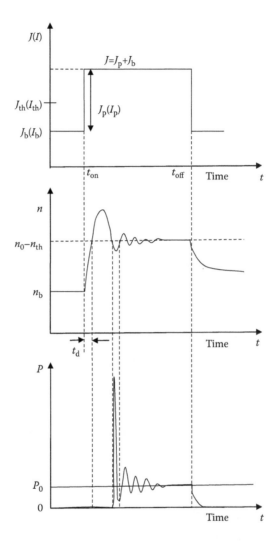

FIGURE 7.9
Turn-on delay time and relaxation oscillation for a current pulse applied to the LD in addition to a bias current. P is the power proportional to the photon density n_{ph}.

the Laplace transform. Expressing the Laplace transform of $n(t)$ by $N(s)$ and denoting $n(0) = n_B = \tau_n J_B/(ed)$, we obtain

$$sN(s) - n(0) = sN(s) - \frac{\tau_n J_B}{ed} = \frac{J_1 + J_B}{ed}\frac{1}{s} - \frac{N(s)}{\tau_n} \qquad (7.66)$$

This leads to the following expression for $N(s)$:

$$N(s) = \left(\frac{1}{s} - \frac{1}{s + \tau_n^{-1}}\right)\frac{\tau_n (J_1 + J_B)}{ed} + \frac{\tau_n J_B}{ed}\frac{1}{s + \tau_n^{-1}} \tag{7.67}$$

We now take the inverse Laplace transform of $N(s)$ and obtain

$$n(t) = \frac{\tau_n (J_1 + J_B)}{ed} u(t) - \frac{\tau_n (J_1 + J_B)}{ed} e^{-t/\tau_n} + \frac{\tau_n J_B}{ed} e^{-t/\tau_n}$$

$$= \frac{\tau_n (J_1 + J_B)}{ed} u(t) - \frac{\tau_n J_1}{ed} e^{-t/\tau_n} \tag{7.68}$$

When $t \geq 0$, $u(t) = 1$, writing the total current as $J = J_1 + J_B$, Equation 7.68 reduces to

$$n(t) = \frac{\tau_n J}{ed} - \frac{\tau_n J_1}{ed} e^{-t/\tau_n} \tag{7.69}$$

The bias current for $t < 0$ is insufficient to initiate laser action for which the carrier density $n_B < n_{th}$. With a current pulse applied at $t = 0$, the threshold condition is reached. However, due to the finite lifetime of carriers, it takes some time, termed *the turn-on delay*, t_d, for carriers to increase from n_B to n_{th}. The expression for t_d is then obtained from

$$n(t_d) = n_{th} = \frac{\tau_n J_{th}}{ed} \tag{7.70}$$

and using Equations 7.69 and 7.70, the turn-on delay time may be expressed as

$$t_d = \tau_n \ln \frac{J - J_B}{J - J_{th}} \tag{7.71}$$

Example 7.9: Let $J = 1.2 J_{th}$ and $\tau_n = 2$ ns. For $J_B = 0.6 J_{th}$, $t_d = 2.2$ ns while for $J_B = 0.8 J_{th}$, $t_d = 1.39$ ns.

In order to generate high-speed optical signals by modulating the injection current, the turn-on delay must be made short. Example 7.9 points out the advantage of using bias current as close as the threshold current. Also, a shorter carrier lifetime and a lower threshold current density may yield high-speed modulation.

7.7.2 Relaxation Oscillation

An important outcome of applying a current pulse to the LD is the appearance of ringing or relaxation oscillation in the output light pulse as illustrated in Figure 7.9. The oscillation frequency and the decay constant may be calculated from the rate equations. However, an exact solution may only be obtained from a large signal analysis, which involves numerical solutions of the coupled differential equations. We provide instead a small-signal analysis that qualitatively explains the observed phenomena and yields approximate expressions for the relaxation frequency.

The analysis begins with the assumption that the carrier concentration, n, the photon density, n_{ph}, and the current density, J, may be expressed as

$$n = n_0 + \delta n, \quad n_{ph} = n_{ph0} + \delta n_{ph},$$

$$J = J_0 + \delta J > J_{th}, \quad n_0 \gg \delta n, \quad n_{ph0} \gg \delta n_{ph}, \quad J_0 \gg \delta J \tag{7.72}$$

In Equation 7.72, the steady-state quantities are denoted using subscript 0, and δn, δn_{ph}, and δJ are small changes from the respective steady-state values. Assuming that $J_b \gg J_p$ and neglecting several initial sharp peaks in the relaxation oscillation, the conditions for the small-signal analysis are satisfied. Hence, we put $J_b = J_0$ and $\delta J = J_p$.

Neglecting the coupling of the spontaneous emission into the lasing mode, Equations 7.47 and 7.48 reduce to

$$\frac{dn}{dt} = \frac{J}{ed} - G(n)n_{ph} - \frac{n}{\tau_n} \tag{7.73}$$

$$\frac{dn_{ph}}{dt} = G(n)n_{ph} - \frac{n_{ph}}{\tau_{ph}} \tag{7.74}$$

We put the steady-state condition $(d/dt = 0)$ in Equations 7.73 and 7.74 and obtain

$$\frac{J_0}{ed} - G(n_0)n_{ph0} - \frac{n_0}{\tau_n} = 0 \tag{7.75}$$

$$G(n_0) = \frac{1}{\tau_{ph}} \tag{7.76}$$

Using the relation $G(n) = \Gamma g_0(n - n_{tr})$, one may write

$$G(n) = G(n_0 + \delta n) = \Gamma g_0 (n_0 + \delta n - n_{tr})$$

$$= \Gamma g_0 (n_0 - n_{tr}) + \Gamma g_0 \delta n = G(n_0) + \frac{\partial G}{\partial n} \delta n \qquad (7.77)$$

The differential gain introduced above is defined as

$$\Gamma g_0 = \frac{\partial G}{\partial n} \qquad (7.78)$$

Inserting Equation 7.78 into the rate Equations 7.73 and 7.74, using the equations for steady state and neglecting the second-order small term $\delta n \cdot \delta n_{ph}$, the rate equations involving small deviations may be written in the form:

$$\frac{d}{dt} \delta n = \frac{\delta J}{ed} - \frac{\delta n_{ph}}{\tau_{ph}} - \frac{\partial G}{\partial n} n_{ph0} \delta n \qquad (7.79)$$

$$\frac{d}{dt} \delta n_{ph} = \frac{\partial G}{\partial n} n_{ph0} \delta n \qquad (7.80)$$

From Equation 7.80, we may write

$$\delta n = \frac{1}{(\partial G / \partial n) n_{ph0}} \frac{d}{dt} \delta n_{ph0} \qquad (7.81)$$

Differentiating Equation 7.80 once again and using Equations 7.79 through 7.81, we may write

$$\frac{d^2}{dt^2} \delta n_{ph} + \left(\frac{\partial G}{\partial n} n_{ph0} + \frac{1}{\tau_n} \right) \frac{d}{dt} \delta n_{ph} + \frac{\partial G}{\partial n} \frac{n_{ph0}}{\tau_{ph}} \delta n_{ph} = \frac{\partial G}{\partial n} \frac{n_{ph0}}{ed} \delta J \qquad (7.82)$$

Assuming that $\delta n_{ph} \propto \exp(-j\omega t)$, the solution may be written as

$$\delta n_{ph} = \frac{K}{\omega^2 - \omega_r^2 - j\omega\gamma_0}$$

where K is related to the right-hand side of Equation 7.82. This equation represents a damped harmonic oscillator. The relaxation oscillation frequency is expressed as

$$f_r = \frac{1}{2\pi} \sqrt{\frac{\partial G}{\partial n} \frac{n_{ph0}}{\tau_{ph}}} \tag{7.83}$$

and the damping constant is given by

$$\gamma_0 = \frac{\partial G}{\partial n} n_{ph0} + \frac{1}{\tau_n} \tag{7.84}$$

For high-speed operation, the optical pulses should return to their steady-state values as quickly as possible as the turn-on current is applied. This requires that both the damping constant and the relaxation oscillation frequency should be large. Therefore, a large differential gain, a large steady-state photon density, a short carrier lifetime, and a short photon lifetime are needed for high-speed operation.

We now express the decay constant and the relaxation oscillation frequency in terms of the current density. The threshold and transparency carrier densities are related to the corresponding current densities by the relation $n_{th} = (\tau_n J_{th}/ed)$ and $n_{tr} = (\tau_n J_{tr}/ed)$. Again, from Equation 7.59, one may write

$$n_{th} - n_{tr} = \frac{1}{\Gamma g_0 \tau_{ph}} \tag{7.85}$$

Substituting Equation 7.85 into Equation 7.78 and using the relation between n and J as given above, we obtain

$$\frac{\partial G}{\partial n} = \frac{ed}{\tau_n \tau_{ph}(J_{th} - J_{tr})} \tag{7.86}$$

Using Equations 7.61, 7.83, 7.84, and 7.86, one may thus write

$$\gamma_0 = \frac{1}{\tau_n} \frac{J - J_{tr}}{J_{th} - J_{tr}} \tag{7.87}$$

$$f_r = \frac{1}{2\pi} \sqrt{\frac{1}{\tau_n \tau_{ph}} \frac{J - J_{th}}{J_{th} - J_{tr}}} \tag{7.88}$$

7.8 Modulation Response

One of the advantages of semiconductor lasers is that they can be directly modulated. The information to be transmitted is applied to the laser in the form of an alternating current (ac) superimposed on the bias current. Due to the linearity of light power output versus the drive current characteristics of LDs, the ac is converted into an alternating optical intensity or power of the same frequency. This scheme is illustrated in Figure 7.10.

If the frequency of the modulating current is small, there is faithful reproduction of light power output. However, for larger modulation frequencies, the output power variation cannot follow the input current variation due to the inherent time lag between the build up of electrons and the build up of photons. The highest frequency by which the LD can be modulated is therefore a parameter of importance in optical fiber communication.

7.8.1 Small-Signal Analysis

To determine the modulation response of an LD, we resort to the same small-signal analysis using the rate equations, which are written below in a slightly different form:

$$\frac{dn}{dt} = \frac{I}{Ve} - \frac{n}{\tau_n} - g_0(n - n_{tr})n_{ph} \tag{7.89}$$

$$\frac{dn_{ph}}{dt} = g_0(n - n_{tr})n_{ph}\Gamma - \frac{n_{ph}}{\tau_{ph}} \tag{7.90}$$

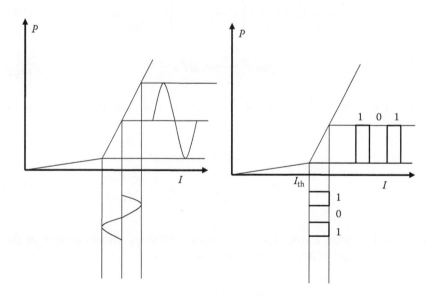

FIGURE 7.10
Direct analog and digital modulation of laser intensity.

In Equations 7.89 and 7.90, I is the injection current, V is the volume of the active region, $g_0(n - n_{tr})$ is the net rate of the induced transition per unit volume, and the constant g_0 is the temporal growth coefficient related to the differential gain coefficient by $g_0 = a(c/n_r)$. The effect of spontaneous emission coupling into the laser mode is neglected here.

If we put $d/dt = 0$, we obtain the following steady-state solutions:

$$0 = \frac{I_0}{Ve} - \frac{n_0}{\tau_n} - g_0(n_0 - n_{tr})n_{ph0} \tag{7.91}$$

$$0 = g_0(n_0 - n_{tr})n_{ph0}\Gamma - \frac{n_{ph0}}{\tau_{ph}} \tag{7.92}$$

Assume now that the current is made up of a direct current (dc) and an ac component such that $I = I_0 + i_1 \exp(j\omega_m t)$, where ω_m is the modulation frequency. The current modulation will bring about changes in the electron and photon densities, so that we may write $n = n_0 + n_1 \exp(j\omega_m t)$ and $n_{ph} = n_{ph0} + p_1 \exp(j\omega_m t)$, where n_0 and n_{ph0} are the steady-state solutions and n_1 and p_1 are small-signal amplitudes. Using these in the rate equations and also noting from Equation 7.92 that $g_0(n_0 - n_{tr}) = (\tau_{ph}\Gamma)^{-1}$, we obtain the following equations for n_1 and p_1:

$$-j\omega_m n_1 = -\frac{i_1}{Ve} - \left(\frac{1}{\tau_n} + g_0 n_{ph0}\right)n_1 + \frac{1}{\tau_{ph}\Gamma}p_1 \tag{7.93}$$

$$j\omega_m p_1 = -g_0 n_{ph0}\Gamma n_1 \tag{7.94}$$

Solving Equations 7.93 and 7.94, the small-signal photon density may be expressed as

$$p_1(\omega_m) = \frac{-(i_1/Ve)g_0 n_{ph0}\Gamma}{\omega_m^2 - j\omega_m\left(\dfrac{1}{\tau_n} + g_0 n_{ph0}\right) - \dfrac{g_0 n_{ph0}}{\tau_{ph}}} \tag{7.95}$$

As may be easily found, the modulation efficiency shows a peak at the relaxation oscillation angular frequency, ω_r, given by

$$\omega_r^2 = \frac{g_0 n_{ph0}}{\tau_{ph}} - \frac{1}{2}\left(\frac{1}{\tau_n} + g_0 n_{ph0}\right)^2 \tag{7.96}$$

The expression for the photon lifetime is $\tau_{ph} = (n_r/c)[\alpha - (1/L)\ln R]^{-1}$. Let $L = 300$ μm, $n_r = 3.4$, $R = 0.32$, and $\alpha = 10$ cm^{-1}. This gives $\tau_{ph} \sim 1$ ps. Taking $g_0 n_{ph0} \sim 10^9$ s^{-1}, the second term in the above expression becomes negligible. Therefore, one may write

$$\omega_r^2 = \frac{g_0 n_{ph0}}{\tau_{ph}} \tag{7.97}$$

This expression agrees with the earlier expression, Equation 7.83, for the relaxation oscillation frequency.

Figure 7.11 shows the modulation efficiency, $M(f_m)$, as a function of the modulation frequency, f_m, with the injected current density as a parameter. The plot shows a maxima at the resonant frequency, f_r, and then a drastic decrease with frequency. The resonance frequency therefore indicates the maximum frequency of the signal by which the laser may be modulated, that is, it sets the modulation bandwidth. As shown in Figure 7.11, the resonance peak shifts to higher values with increasing bias current, thereby enhancing the modulation bandwidth.

Example 7.10: We consider a DH laser of $L = 200$ μm, $w = 3$ μm, and an active layer thickness $d = 0.1$ μm. The power output per facet is 2 mW, loss $= 10$ cm^{-1}, mirror reflectivity $R = 0.32$ and RI $= 3.6$, and photon energy $hf = 1.43$ eV. The photon density is obtained from the relation power/area $= [(1-R)n_{ph0}hf(c/n_r)]$. This gives $n_{ph0} = 5.14 \times 10^{20}$ m^{-3}. The photon lifetime is $\tau_{ph} = 12.5$ ps. The gain $g_0 = B(c/n_r) = 0.833 \times 10^{-12}$ m^3 s^{-1}, taking $B = 1.0 \times 10^{-20}$ m^{-2}. Putting all the values in Equation 7.97, the resonance frequency $f_r = 0.94$ GHz is obtained.

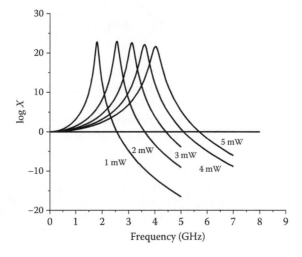

FIGURE 7.11
Modulation response of a DH laser showing resonance peaks for different output powers.

Figure 7.11 plots the normalized modulation response of a DH laser for different output power values. The following values are chosen for the calculation: $L = 300$ μm, $n_r = 3.6$, $\alpha = 10$ cm^{-1}, $R = 0.32$, $B = 1.5 \times 10^{16}$ cm^2, $\tau_{ph} = 2$ ns, and $E_g = 1.43$ eV. The response is flat at low modulation frequencies. However, as the frequency is increased, there is a strong interaction between electrons and photons, which manifests in the appearance of the resonance peak. As expected from Equation 7.97, the resonance peak shifts to higher frequencies as the output power increases.

7.8.2 Gain Suppression and Frequency Chirp

The modulation of current in a DH laser introduces additional effects, which in turn alter the nature of the modulation response and the value of the modulation bandwidth. We now consider the effect of current modulation on the carrier density in the laser. The current is expressed as

$$I(t) = I_0 + i_1(\omega_m) \exp(j\omega_m t) \tag{7.98}$$

The corresponding variation of the photon density and the carrier density in the active region may be expressed, respectively, as $n_{ph}(t) = n_{ph0} + p_1(\omega_m) \exp(j\omega_m t)$ and $n(t) = n_0 + n_1(\omega_m)\exp(j\omega_m t)$. Ideally, it is desired that the modulation response $p_1(\omega_m)/i_1(\omega_m)$ should be constant independent of the modulation frequency and the carrier density $n_1(\omega_m) = 0$ above threshold indicating perfect gain saturation. The expression for carrier modulation becomes

$$n_1(\omega_m) = \frac{-j(i_1 / Ve)\omega_m}{\omega_m^2 - j\omega_m \left(\dfrac{1}{\tau_n} + g_0 n_{ph0} \right) - \dfrac{g_0 n_{ph0}}{\tau_{ph}}} \tag{7.99}$$

This suggests that under dynamic conditions, the carrier density, and hence the gain, is not clamped at the threshold value but has an oscillating component having an amplitude given by the above expression. It is easy to infer from Equation 7.95 that the modulation response $p_1(\omega_m)/i_1(\omega_m)$ increases linearly as ω_m / ω_r^2, then shows a peak at ω_r, and at still higher frequencies decreases as $1/\omega_m$. As is well known, the injected carriers contribute to the magnitude of the RI. Thus, the modulation of the carrier density leads to a modulation of the index of refraction, which in turn produces a frequency modulation (FM) of the output optical field. This undesired FM causes spectral broadening of the optical field. In dispersive fibers, this broadening leads to the spread of the optical pulse with distance affecting the information transmission capacity.

We need to first discuss two new physical concepts: (1) the gain suppression (compression) effect and (2) the amplitude-phase coupling effect.

7.8.2.1 Gain Suppression

As pointed out earlier while solving the coupled rate equations, the gain of a laser saturates at high carrier density or high optical power. There is yet another mechanism that leads to gain compression at high power. The origin of this mechanism is complicated: it may be due to spectral "hole burning" or due to the hot carrier effect (see Yariv and Yeh in Reading list). In the following, this gain suppression will be taken care of by a phenomenological *gain compression factor* ε. The optical gain constant is now expressed as

$$G(n, n_{ph}) = G(n)(1 - \varepsilon n_{ph}) \approx G(n_{th}) + g_0(n - n_{th}) - \varepsilon G(n_{th})n_{ph} \quad (7.100)$$

The gain suppression due to photon density is expressed by the factor $(1 - \varepsilon n_{ph})$. Equation 7.100 is a Taylor series expansion of the gain about the threshold point, $n = n_{th}$ and $n_{ph} = 0$. The rate equations are then rewritten as

$$\frac{dn}{dt} = \frac{I}{Ve} - \frac{n}{\tau_n} - G(n, n_{ph})n_{ph} \quad (7.101)$$

$$\frac{dn_{ph}}{dt} = \Gamma G(n, n_{ph})n_{ph} - \frac{n_{ph}}{\tau_{ph}} \quad (7.102)$$

Using the steady-state condition, $d/dt = 0$, Equations 7.101 and 7.102 reduce to

$$0 = \frac{I_0}{Ve} - \frac{n_0}{\tau_n} - [G(n_{th}) + g_0(n_0 - n_{th}) - \varepsilon G(n_{th})n_{ph0}]n_{ph0} \quad (7.103)$$

$$0 = \Gamma[G(n_{th}) + g_0(n_0 - n_{th}) - \varepsilon G(n_{th})n_{ph0}]n_{ph0} - \frac{n_{ph0}}{\tau_{ph}} \quad (7.104)$$

Using the threshold condition, that is, $n_0 = n_{th}$ and $n_{ph0} = 0$, one obtains from the second equation:

$$G(n_{th}) = \frac{1}{\Gamma \tau_{ph}} \quad (7.105)$$

This may be used to simplify Equations 7.103 and 7.104 for the steady state as follows:

$$0 = \frac{I_0}{Ve} - \frac{n_0}{\tau_n} - \frac{n_{ph0}}{\Gamma \tau_{ph}} \quad (7.106)$$

$$0 = g_0(n_0 - n_{th}) - \frac{\varepsilon n_{ph0}}{\Gamma \tau_{ph}} \qquad (7.107)$$

The small-signal analysis of Equations 7.101 and 7.102 leads to the following:

$$j\omega_m n_1 = \frac{i_1}{Ve} - \left(\frac{1}{\tau_n} + g_0 n_{ph0}\right) n_1 - \frac{(1 - \varepsilon n_{ph0})}{\Gamma \tau_{ph}} p_1 \qquad (7.108)$$

$$j\omega_m p_1 = \Gamma g_0 n_{ph0} n_1 - \frac{\varepsilon n_{ph0}}{\tau_{ph}} p_1 \qquad (7.109)$$

Solving for p_1 from Equations 7.108 and 7.109, one obtains

$$p_1(\omega_m) = \frac{-(i_1 / Ve)g_0 n_{ph0}\Gamma}{\omega_m^2 - j\omega_m\left(\frac{1}{\tau_n} + g_0 n_{ph0} + \frac{\varepsilon n_{ph0}}{\tau_{ph}}\right) - \left(\frac{g_0 n_{ph0}}{\tau_{ph}} + \frac{\varepsilon n_{ph0}}{\tau_n \tau_{ph}}\right)} \qquad (7.110)$$

This equation reduces to Equation 7.95 when $\varepsilon = 0$.

For typical values of $\varepsilon = 10^{-23}\,\text{m}^{-3}$, $\tau_{ph} = 4$ ns, $g_0 = 2 \times 10^{-12}\,\text{m}^3\text{s}^{-1}$, and $\varepsilon/\tau_{ph} \gg g_0$, the peak modulation response occurs at

$$\omega_r = \sqrt{\frac{g_0 n_{ph0}}{\tau_{ph}} - \frac{\varepsilon^2 n_{ph0}^2}{2\tau_{ph}^2}} \qquad (7.111)$$

This equation indicates that in the presence of gain suppression, the modulation resonance frequency does not show a monotonic increase with optical power, but reaches a maximum value for the photon density $(n_{ph0})_{max} = g_0\tau_{ph}/\varepsilon^2$. For this photon density, the maximum value of the resonance frequency takes the form

$$(\omega_r)_{max} = \frac{g_0}{\sqrt{2\varepsilon}} \qquad (7.112)$$

Example 7.11: Using the above-quoted values, $(n_{ph0})_{max} = g_0\tau ph/\varepsilon^2$ $= 2 \times 10^{-12} \times 10^{-12}/10^{-46} = 2 \times 10^{22}$ photons m^{-2}. The corresponding maximum resonant modulation frequency is $f_r = g_0/(2)^{3/2}\pi\varepsilon = 22.4$ GHz.

Gain suppression therefore puts an upper limit on the modulation bandwidth of a DH laser.

7.8.2.2 Amplitude-Phase Coupling

The amplitude of the carrier density variation may be obtained by solving the coupled rate Equations 7.101 and 7.102 and is expressed as

$$n_1(\omega_m) = \frac{-\left(j\omega_m + \dfrac{\varepsilon n_{ph0}}{\tau_{ph}}\right)\left(\dfrac{i_1}{eV}\right)}{\omega_m{}^2 - j\omega_m\left(\dfrac{1}{\tau_n} + g_0 n_{ph0} + \dfrac{\varepsilon n_{ph0}}{\tau_{ph}}\right) - \left(\dfrac{g_0 n_{ph0}}{\tau_{ph}} + \dfrac{\varepsilon n_{ph0}}{\tau_n \tau_{ph}}\right)} \tag{7.113}$$

Using Equation 7.109, the relationship between n_1 and p_1 becomes

$$n_1(\omega_m) = \frac{-\left(j\omega_m + \dfrac{\varepsilon n_{ph0}}{\tau_{ph}}\right)}{\Gamma g_0 n_{ph0}} p_1(\omega_m) \tag{7.114}$$

We write $n(t) = n_0 + \Delta n(t)$, $n_{ph}(t) = n_{ph0} + \Delta p(t)$, and $j\omega = d/dt$ in Equation 7.114 to obtain

$$\Delta n(t) = \frac{1}{\Gamma g_0}\left(\frac{1}{n_{ph0}}\frac{dn_{ph}}{dt} + \frac{\varepsilon}{\tau_{ph}}\Delta p(t)\right) \tag{7.115}$$

which is a more general equation and applies to any arbitrary photon density fluctuation.

The index of refraction in a gain medium is a complex number and is written as

$$n_r(t) = n_r'(t) - jn_r''(t)' \tag{7.116}$$

The imaginary part of the RI accounts for gain or absorption and is related to the spatial variation of the EM field, the expression for which is

$$E(z) \propto E_0 \exp\left(-\frac{j\omega}{c}(n_r' - jn_r'')z\right) = E_0 \exp\left(-\frac{j\omega n_r'}{c}z\right)\exp\left(-\frac{\omega n_r''}{c}z\right) \tag{7.117}$$

The spatial growth parameter is now converted to a temporal parameter to fit into the rate equations. Thus

$$\frac{d|E|}{dt} = \frac{\partial|E|}{\partial z}\frac{dz}{dt} \cong -\left(\frac{\omega n_r''}{c}\right)\frac{c}{n_r'}|E| = -\frac{\omega n_r''}{n_0'}|E| \tag{7.118}$$

The exponential growth constant $G(n, n_{pho})$ for the photon density (optical intensity, i.e., $\propto |E|^2$) is related to the imaginary part of the RI by

$$G = -\frac{2\omega n_r''}{n_r'} = -\frac{4\pi f n_r''}{n_r'} \tag{7.119}$$

From this equation, it follows that

$$\frac{\partial n_r''}{\partial n} = -\frac{n_r'}{4\pi f}\frac{\partial G}{\partial n} = -\frac{n_r'}{4\pi f}g_0 \tag{7.120}$$

It follows that the fluctuation of the carrier density introduces a change in the absorption (or gain) coefficient as follows:

$$\Delta n_r'' = -\frac{n_r'}{4\pi f}g_0\Delta n(t) \tag{7.121}$$

The real and imaginary parts of the RI change are related by a parameter called the α-parameter, also known as the Henry α-factor or the linewidth enhancement factor, defined as

$$\alpha = \frac{\Delta n_r'}{\Delta n_r''} \tag{7.122}$$

From Equations 7.121 and 7.122, one obtains

$$\Delta n_r' = -\frac{\alpha n_r' g_0}{4\pi f}\Delta n(t) \tag{7.123}$$

This change in the RI causes the following change in the laser frequency:

$$\frac{\Delta f}{f} = -\frac{\Delta n_r'}{n_r'}\Gamma = \frac{\alpha \Gamma A}{4\pi f}\Delta n(t) \tag{7.124}$$

where use is made of the resonance condition in a cavity given by $f_m = (mc)/(2n_r L)$ and it is assumed that the RI change occurs only in the active region and therefore the confinement factor accounts for the partial filling of the resonator by the active medium.

7.9 Temperature Dependence of Threshold Current

As already mentioned in Chapter 1, the threshold current of a DH laser varies with temperature in accordance with the following empirical relation:

$$I_{th}(T) = I_0 \exp(T / T_0) \tag{7.125}$$

where:

I_0 is a constant
T_0 is the characteristic temperature

The observed value for T_0 of GaAs lasers is >120 K near room temperature, while that for InGaAsP lasers is in the range of 50–70 K. The high temperature sensitivity of InGaAsP lasers limits their performance under high temperature operation. When these lasers are operated under continuous wave (CW) mode at room temperature, the maximum power output is limited by the thermal runaway process; more current is needed to offset the effect of increased I_{th} in reducing the power, and the increased current increases the temperature.

Much experimental and theoretical work has been performed to understand the reason for the high temperature sensitivity of InGaAsP DH lasers. The experimental studies have included the investigation of the temperature dependence of the threshold current, measurement of the carrier lifetime at threshold and the optical gain, and the sublinearity of spontaneous emission. All these studies indicate a nonradiative carrier loss at high temperatures. One of the important components of the carrier loss is the heterobarrier leakage that increases with temperature, as discussed in Section 7.3.3. The leakage is more severe for emission wavelengths less than 1.1 µm due to small band offset in the InGaAsP–InP heterojunction. However, for emission at 1.3 and 1.55 µm this loss is small, about 10%–30% of the total current and is due to electron leakage. The leakage can be reduced by choosing the doping and thickness of the p-type cladding layer.

Another important mechanism for nonradiative carrier loss at high temperature is Auger recombination. The measured value of the Auger coefficient in InGaAsP is indeed quite high in comparison with the values for the GaAs system. There are suggestions that intervalence band absorption in the active layer due to the transition between the split-off band and the heavy hole band may contribute significantly in limiting the value of T_0. However, in experiments lower values of absorption in InGaAsP are obtained. Typical values are 12 cm^{-1} at 1.3 μm and 25 cm^{-1} at 1.6 μm at a carrier concentration of 10^{18} cm^{-3}. These values are quite low and cannot explain the observed temperature dependence of I_{th}. This leads to the conclusion that Auger recombination is the most dominant nonradiative mechanism and is primarily responsible for low values of T_0. A theoretical calculation has been made of the temperature dependence of the radiative current density and the Auger recombination current density at threshold. When these two are added, the total threshold current density shows more or less the same behavior as observed experimentally. The calculated T_0 from the theoretical curves are 70 K at 1.3 μm and 61 K at 1.55 μm, in agreement with the values obtained experimentally.

PROBLEMS

7.1. Establish Equation 7.11: the Bernard–Durraffourg condition.

7.2. Put $f_c = 1$ and $f_v = 0$ in Equation 7.21 and apply the **k**-conservation condition; show that it is possible to recover Equation 6.62 for the absorption coefficient $\alpha(\hbar\omega)$.

7.3. Calculate $S_{c(v)}$ for In$_{0.53}$Ga$_{0.47}$ As using $m_e = 0.042\, m_0$, $m_{lh} = 0.0503\, m_0$, and $m_{hh} = 0.465\, m_0$. The injected electron and hole densities are $n = p = 10^{18}$ cm^{-3}. Calculate the range of photon energies over which positive gain may be obtained at 0 K.

7.4. Derive the expression for the B-coefficient as given by Equation 7.19.

7.5. At low temperature, the Fermi functions are boxlike. Using this in Equation 7.21, obtain the expression for gain.

7.6. Prove that the material gain coefficient may be expressed as

$$g\left(E_{ph}\right) = \frac{\pi^2 c^2 \hbar^2}{8 n_r^2} \left(\frac{\Delta E}{E}\right)^2$$

$$C \frac{\sinh\left[\left(F_e + F_h - \Delta E\right)/2k_B T\right]}{\cosh\left[\left(F_e + F_h - \Delta E\right)/2k_B T\right] + \cosh\left[\left(F_e + F_h\right)/2k_B T\right]}$$

where C includes all other parameters, $\Delta E = E_{ph} - E_g$ [5].

7.7. Derive Equation 7.25. Show that the maximum gain occurs at photon energy $= (1/3)(E_g + 2\Delta F)$.

7.8. There is absorption instead of gain when the injection level is below the transparency carrier density. Using Equation 7.25, give plots of absorption spectra for different injected carrier densities.

7.9. Work out the steps to show that the reduced DOS appears in Equation 7.27 in place of the $S_c \times S_v$ product in Equation 7.21. Use energy and momentum conservation conditions to arrive at this result.

7.10. Calculate the current needed for 1 mW power output from the front face of an InGaAsP/InP laser. Assume $n_{tr}=1\times10^{18}$ cm^{-3}, $\alpha=5$ cm^{-1}, $L=200$ μm, $d=0.2$ μm, $w=5$ μm, $\Gamma=0.2$, $\eta=0.8$, $R_1=0.3$, $R_2=0.99$, $F_1=0.25$, and $g_0=2200$ cm^{-1}. Use the values of B and C given in Examples 7.4 and 7.5.

7.11. Calculate the heterobarrier leakage current for low band offset and low conductivity.

7.12. Obtain Equation 7.44a for the heterobarrier leakage current in the presence of an electric field.

7.13. Consider Equation 7.60 expressing the variation of J_{th} with thickness d of the active layer. Assume that the confinement factor is given by $\Gamma_t=\beta^2/(2+\beta^2)$, where $\beta=(2\pi d/\lambda_0)\sqrt{n_a^2-n_{cl}^2}$, the subscripts a and cl refer to the active and cladding layers, respectively. Show that the variation of J_{th} is as depicted in Figure 7.7.

7.14. Obtain an expression for the optimum value of d that leads to the lowest threshold current density. Using appropriate values, compare the value with that shown in Figure 7.7.

7.15. The maximum gain in a semiconductor is expressed as $g_{max}=\beta(J-J_0)$, where J is the current density, J_0 is the transparency current density, and d is the thickness of active layer. Prove that the threshold condition in a double heterojunction laser is

$$J_{th} = \frac{J_0 d}{\eta_{int}} + \frac{d}{\Gamma\beta\eta_{int}}\left[\alpha+(1/2L)\ln(1/R_1R_2)\right]$$

7.16. The mode number m in a Fabry–Perot laser satisfies the condition $m\lambda=2nL$. Prove that the relation between the wavelength separation and the mode separation is $\Delta\lambda = (-\lambda^2\Delta m)/(2nL[1-(\lambda/n)(dn/d\lambda)])$. Calculate the wavelength separation between adjacent modes when $\lambda=1.5$ μm, $n=3.4$, $dn/d\lambda=-0.26$ μm^{-1}, and $L=400$ μm.

7.17. Prove that the differential amplification factor g_0 defined in Equation 7.49 is related to the gain coefficient a by $g_0=(c/n_r)a$.

Reading List

Agrawal, G. P. and N. K. Dutta, *Semiconductor Lasers*, 2nd edn. Van Nostrand Reinhold, New York, 1993.

Basu, P. K., *Theory of Optical Processes in Semiconductors: Bulk and Microstructures*. Oxford University Press, Oxford, 1997.

Bhattacharyya, P., *Semiconductor Optoelectronic Devices*, 2nd edn. Prentice Hall, Englewood Cliffs, NJ, 1996.

Casey, H. C. and M. B. Panish, *Heterostructure Lasers: Fundamentals and Principles*. Academic Press, San Diego, 1979.

Chow, W. W., S. W. Koch, and M. Sargent III., *Semiconductor Laser Physics*. Springer, Berlin, 1994.

Chuang, S. L., *Physics of Photonic Devices*. Wiley, New York, 2009.

Coldren, L. A., S. W. Corzine, and M. L. Masanovi'c, *Diode Lasers and Photonic Integrated Circuits*, 2nd edn. Wiley, New York, 2012.

Numai, T., *Fundamentals of Semiconductor Lasers*. Springer, New York, 2004.

Singh, J., *Semiconductor Optoelectronics: Physics and Technology*. McGraw Hill, Singapore, 1995.

Singh, J., *Electronic and Optoelectronic Properties of Semiconductor Structures*. Cambridge University Press, Cambridge, 2003.

Suhara, T., *Semiconductor Laser Fundamentals*. Marcel Dekker, New York, 2004.

Yariv, A. and P. Yeh, *Photonics: Optical Electronics in Modern Communication*, 6th edn. Oxford University Press, Oxford, 2007.

References

1. Bernard, M. G. A. and G. Duraffourg, Laser conditions in semiconductors, *Phys. Stat. Sol.* 1, 699–703, 1961.
2. Joyce, W. B. and R. W. Dixon, Analytic approximations for the Fermi energy of an ideal Fermi gas, *Appl. Phys. Lett.* 31, 354–356, 1977.
3. Nilsson, N. G., An accurate approximation of the generalized Einstein relation for degenerate semiconductors, *Phys. Stat. Sol.* A19, K75–K78, 1973.
4. Bour, D. P., D. W. Treat, R. L. Thornton, R. S. Geels, and D. F. Welsh, Drift leakage current in AlGaInP quantum-well lasers, *IEEE J. Quantum Electron.* QE-29, 1337–1342, 1993.
5. Osinski, M. and M. J. Adam, Gain spectra of quaternary semiconductors, *IEEE Proc. J.* 129, 229–236, 1982.

8

Quantum Well Lasers

8.1 Introduction

The basic structure of a quantum well (QW) laser is not much different from the double heterostructure (DH) laser structure. However, the width of the active layer is approximately a few tens of nanometers, so that quantum confinement of the carriers occurs within the active layer. The laser may contain a single QW or a multiple QW (MQW) in the active region.

As pointed out in Chapter 3, the density of states (DOS) function of a QW is staircase like. For this reason, electrons and holes populate the levels near the subband edges. The gain spectrum is sharper than in a DH laser. This leads to reduced threshold current density, higher quantum efficiency, larger modulation bandwidth, and less frequency chirp.

In this chapter, we first present different QW structures, and then develop the theory of gain in single-QW the MQW structures. The modulation response of QW lasers is studied by using the rate equation model. The special features of strained QW lasers are then introduced and explained. The present chapter ends with some discussions on Type II QW lasers and on tunnel-injection QW lasers.

8.2 Structures

Figure 8.1a shows the band diagram of a single QW structure, in which the thickness of the active QW is a few tens of nanometers. As noted earlier, the small width leads to a lower optical confinement factor, thereby drastically increasing the threshold current density. To increase the optical confinement factor, several improved structures, as shown in Figure 8.1b–e, have been introduced.

The separate confinement heterostructure (SCH) shown in Figure 8.1b enhances the confinement factor. As is well known, the refractive index (RI) of semiconductors increases with a decrease in the bandgap. In this

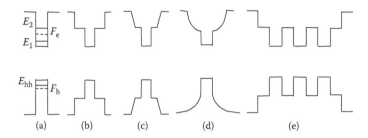

FIGURE 8.1
(a) SQW; (b) SCH-QW; (c) linear GRIN-SCH; (d) parabolic GRIN-SCH; (e) modified MQW.

structure, the bandgap is modified in two steps. The outer potential confines the optical mode in the QW, while the inner potential well confines the carriers. The term SCH arises due to separation of light waves and carriers by two separate wells. Figure 8.1c and d show the band diagrams of graded index (GRIN)-SCH, in which the potential and RI profiles vary in graded fashion. While the variations are linear in Figure 8.1c, variation of both the bandgap and the RI in Figure 8.1d is parabolic.

Use of the MQW structure increases the value of the confinement factor. However, in the presence of multiple barriers, the carriers are not uniformly distributed among the QWs due to longer propagation distance. To circumvent this, the modified MQW structure shown in Figure 8.1e has been developed. In this structure, the energy barriers between QWs are lower than those in the cladding layers, which ensures higher carrier injection efficiency and uniform carrier distribution over all the QWs.

In GRIN-SCH, the optical confinement factor Γ is proportional to the active layer thickness d with a small d, while Γ in the SQW is proportional to d^2. Therefore, when the active layer is thin, a relatively large Γ is obtained in GRIN-SCH.

8.3 Interband Transitions

We first develop the theory of absorption and gain in a rectangular QW having type I band alignment. The most common example of this is the well-studied GaAs/Ga$_{1-x}$Al$_x$As system. The structure and its simplified band diagram are reproduced from Chapter 2 in Figure 8.2. As noted already, electrons and holes injected into the conduction and valence subbands (heavy-hole [HH] or light-hole [LH]) in the GaAs well recombine spontaneously as well as by the stimulated process. When the condition for population inversion between conduction and valence subbands is established, amplification of electromagnetic (EM) radiation occurs.

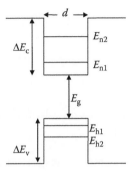

FIGURE 8.2
Simplified band diagram of a rectangular QW showing a few subbands.

In the simplified band diagram shown in Figure 8.2, the band discontinuities ΔE_c and ΔE_v occur abruptly at the heterointerfaces. The particle motions are then quantized along the z-direction. The calculation of absorption coefficient, gain, and spontaneous emission rate in the ideal structure parallels the method used for bulk semiconductors. In the following, the basic theory is developed by considering transition between a conduction subband and a valence subband.

8.3.1 Absorption

The absorption coefficient for transition from a valence band state of energy E_1 to a conduction band (CB) state of energy E_2 in a bulk semiconductor may be written as

$$\alpha(\hbar\omega) = \frac{2\pi}{\hbar(c/\eta)}\left(\frac{eA_0}{m_0}\right)^2 \sum_{1,2}|p_{12}|^2(f_1 - f_2)\delta(E_{12} - \hbar\omega) \tag{8.1}$$

where

$$p_{12} = \langle 1|\exp(-i\mathbf{q}\cdot\mathbf{r})\varepsilon_\lambda.\mathbf{p}|2\rangle \tag{8.2}$$

In Equation 8.1, f_1 and f_2 are the Fermi occupational probabilities of the two states. In the present context, we shall assume that an electron in the mth HH subband with two-dimensional (2-D) wave vector $\mathbf{k}_h$ absorbs a photon and moves to a state of 2-D wave vector $\mathbf{k}_e$ in the nth subband in the CB. To make the notation simple, we do not use any subscript to denote the in-plane wave vectors, but write the position coordinate as $\mathbf{r}=(\rho,z)$, in terms of the 2-D vector ρ and the coordinate z along the direction of quantization. The wave functions are then written as

$$|1\rangle = |h, m, \mathbf{k}_h\rangle = U_h(\rho, z) \exp(i\mathbf{k}_h.\rho)\phi_{hm}(z) \tag{8.3a}$$

$$|2\rangle = |c, n, \mathbf{k}_e\rangle = U_c(\rho, z) \exp(i\mathbf{k}_e.\rho)\phi_{cn}(z) \tag{8.3b}$$

where:
  Us   are the cell periodic parts of the Bloch function
  φs   are envelope functions

Using the components of photon wave vector $\mathbf{q} \equiv (q_\rho, q_z)$, Equation 8.2 may be written as

$$p_{12} = \left|\langle c, n, \mathbf{k}_e | \exp(i\mathbf{q} \cdot \rho + iq_z z)\varepsilon_\lambda.\mathbf{p} | h, m, \mathbf{k}_h\rangle\right| \tag{8.4}$$

The matrix element is evaluated by using the method outlined in Section 6.4.2. Considering the ρ-integration and using the arguments presented earlier, we note that

$$\mathbf{k}_e = \mathbf{k}_h + \mathbf{q}_\rho \approx \mathbf{k}_h \tag{8.5}$$

which represents momentum conservation in the QW layer plane. As usual, we neglect the momentum of the photon. Since the motion of the particles is quantized along the z-direction, there is no momentum conservation (selection rule) for that direction. The matrix element may therefore be written as

$$|p_{12}|^2 = \left\langle |p_{cv}|^2\right\rangle_{QW} \delta_{\mathbf{k}_e, \mathbf{k}_h} C_{mn} \tag{8.6}$$

where C_{mn} is written as

$$C_{mn} = \left|\langle \phi_{hm} | \phi_{cn}\rangle\right|^2 = \left|\int \phi_{hm}\phi_{cn}dz\right|^2 \tag{8.7}$$

In the present case, $\langle |p_{cv}|^2\rangle_{QW}$ is the momentum matrix element for transition between conduction and valence subbands in a QW, and it differs from the momentum matrix element for bulk semiconductors for reasons to be discussed later.

The absence of a selection rule for the z-direction gives rise to a term C_{mn}, which is due to overlap between the envelope functions for electrons and holes in the respective subbands. Considering an infinitely large potential barrier, both the envelope functions are sin or cos functions, and it is easy to prove that $C_{mn} = \delta_{mn}$, where δ is the Kronecker δ-function. In a practical situation, the finite values of ΔE_c and ΔE_v, their inequalities, and the change in effective masses from the well to the barrier lead to a violation of this selection rule. In general, the value of C_{mn} is obtained from numerical method even if the shape of the QW is rectangular.

For wide QWs, it is a good approximation to take $C_{mn} = 1$. For the case of narrow rectangular QWs, the following equation has been derived for C_{mn} [1]:

$$C_{mn} = \int_{-\infty}^{\infty} 2 \sqrt{ \begin{vmatrix} \dfrac{E_{C,str} \cdot E_{V,str}}{\Delta E_C \Delta E_V} & \dfrac{m_{hh} \Delta E_V - m_e \Delta E_C}{m_{hh} \cdot E_{V,str} - m_e \cdot E_{C,str}} \end{vmatrix} } \times \frac{\sqrt{m_E (\Delta E_C - E_{C,str}) m_{hh} (\Delta E_V - E_{V,str})}}{\sqrt{m_e (\Delta E_C - E_{C,str})} + \sqrt{m_{hh} (\Delta E_V - E_{V,str})}}$$

$$\times \left(1 + \sqrt{\frac{m_e (\Delta E_C - E_{C,str}) d^2}{2\hbar^2}} \right)^{-1/2} \times \left(1 + \sqrt{\frac{m_{hh} (\Delta E_V - E_{V,str}) d^2}{2\hbar^2}} \right)^{-1/2}$$

$$(8.7a)$$

where ΔE_C and ΔE_V are CB and valence band offsets at the QW–optical confinement layer (OCL) heteroboundary. The squared momentum matrix element for QW is now expressed in terms of the same for bulk as follows:

$$\left\langle \left| p_{cv} \right|^2 \right\rangle_{QW} = A_{mn} \left\langle \left| p_{cv} \right|^2 \right\rangle_{bulk} \tag{8.8}$$

Using this, we may now express the absorption coefficient for photon energy $\hbar\omega$ as

$$\alpha(\hbar\omega) = \frac{2\pi\eta}{\hbar c} \left(\frac{eA_0}{m_0} \right)^2 A_{mn} C_{mn} \sum_{k_e, k_h} \left\langle \left| p_{cv} \right|^2 \right\rangle_{bulk} \delta_{k_e, k_h} (f_e - f_h) \delta \left[E_e (\mathbf{k}_e) \right.$$

$$\left. - E_h (\mathbf{k}_h) - \hbar\omega \right] \tag{8.9}$$

In Equation 8.9, the proper subscripts (e and h) are used for Fermi functions, and initial and final energies are written instead of subscripts 1 and 2 as in Equation 8.2. Assuming a parabolic E–k relationship, the energies of the final and initial states are expressed as

$$E(\mathbf{k}_e) = E_{cn} + \frac{\hbar^2 k_e^2}{2m_e} \tag{8.10a}$$

$$E(\mathbf{k_h}) = E_{hm} + \frac{\hbar^2 k_h^2}{2m_h} \tag{8.10b}$$

The occupational probabilities are expressed as

$$f_e(E_e) = \left\{1 + \exp\left[\left(E_{cn} + \hbar^2 k_e^2 / 2m_e - F_e\right)/k_BT\right]\right\}^{-1} \tag{8.11a}$$

$$f_h(E_h) = \left\{1 + \exp\left[\left(E_{hm} + \hbar^2 k_h^2 / 2m_h - F_h\right)/k_BT\right]\right\}^{-1} \tag{8.11b}$$

where F_e and F_h are the respective quasi-Fermi levels. The double summation in Equation 8.9 is reduced to a summation over $\mathbf{k_h}$ only due to **k**-conservation, and the argument of the energy-conserving δ-function is then

$$E(\mathbf{k_e}) - E(\mathbf{k_h}) - \hbar\omega = \left(E_g + E_{cn} + E_{hm}\right) - \left(\hbar^2 k_h^2 / 2m_r\right) - \hbar\omega \tag{8.12}$$

where m_r is the reduced mass. The sum over $\mathbf{k_h}$ in Equation 8.9 is now converted into an integral by using the prescription

$$\sum_{\mathbf{k_h}} \rightarrow \frac{2S}{(2\pi)^2} \int \delta\left(E_{gmn} - \hbar^2 k_h^2 / 2m_r\right) 2\pi k_h dk_h \left[f_h(k_h) - f_e(k_h)\right] \tag{8.13}$$

where $E_{gmn} = E_g + E_{cn} + E_{hm}$ is the effective bandgap in the QW. The integration in Equation 8.13 is facilitated by the presence of the δ-function, and the final expression takes the form

$$\alpha(\hbar\omega) = \frac{e^2 \left\langle |p_{cv}|^2 \right\rangle}{\varepsilon_0 m_0^2 c\hbar n_r \hbar\omega d} m_r C_{mn} A_{mn} \left(f_h - f_e\right) H\left(\hbar\omega - E_{gmn}\right) \tag{8.14}$$

where $H(x)$ is the Heaviside step function.

The Fermi function for electrons under **k**-conservation takes the form

$$f_e = \left[1 + \exp\left\{\left(E_{cn} + m_r\hbar^2 k^2 / m_e - F_e\right)/k_BT\right\}\right]^{-1} \tag{8.15}$$

and similarly for holes. Equation 8.14 may be compared with the expression for the bulk material. In both cases, the absorption coefficients are proportional to the respective joint DOS function. Therefore, while in the bulk $\alpha(\hbar\omega) \propto (2m_r/\hbar^2)^{3/2}(\hbar\omega - E_g)^{1/2}$, in QWs α is constant for one pair of subbands.

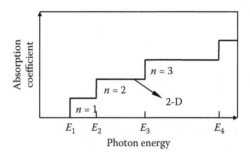

FIGURE 8.3
Ideal step-like absorption spectra of QW.

As more and more pairs of subbands are involved, α changes in a step-like fashion, and its expected variation is shown in Figure 8.3.

8.3.2 Polarization-Dependent Momentum Matrix Elements

The momentum matrix element for optical transitions in a QW depends on how the electric field of the EM wave is oriented with respect to the direction of quantization. To understand this, we first consider that the electron wave vector **k** is parallel to the z-axis, the direction of quantization. We shall consider only HH and LH bands. If α and β denote, respectively, the spin-up and spin-down states, then the wave functions for the CB are written as

$$|s\alpha\rangle \quad \text{and} \quad |s\beta\rangle \tag{8.16a}$$

The corresponding wave functions for the HH are

$$\left|\frac{3}{2},\frac{3}{2}\right\rangle = \frac{1}{\sqrt{2}}\left|(x+iy)\alpha\right\rangle \tag{8.16b}$$

$$\left|\frac{3}{2},-\frac{3}{2}\right\rangle = \frac{1}{\sqrt{2}}\left|(x-iy)\beta\right\rangle \tag{8.16c}$$

Similarly for LH

$$\left|\frac{3}{2},\frac{1}{2}\right\rangle = \frac{1}{\sqrt{6}}\left|2z\alpha+(x+iy)\beta\right\rangle \tag{8.16d}$$

$$\left|\frac{3}{2}, -\frac{1}{2}\right\rangle = \frac{1}{\sqrt{6}}\left|2z\beta - (x - iy)\alpha\right\rangle \tag{8.16e}$$

The momentum matrix elements between the CB and the HH band for different axes are as follows:

$$x\text{-axis:}\frac{1}{\sqrt{2}}\sqrt{3}M; \ y\text{-axis:}\pm i\frac{1}{\sqrt{2}}\sqrt{3M}; \ z\text{-axis:}0 \tag{8.17}$$

M is defined as

$$\sqrt{3}M \equiv \langle s|p_x|x\rangle = \langle s|p_y|y\rangle = \langle s|p_z|z\rangle$$

$$= \left(\frac{m_0}{m_e} - 1\right)^{1/2}\left[\frac{1}{2m_e}\frac{E_g(E_g + \Delta)}{E_g + (2/3)\Delta}\right]^{1/2} \tag{8.18}$$

where:
  m_0 is the free electron mass
  m_e is the electron effective mass
  E_g is the bandgap energy
  Δ is the split-off energy due to spin–orbit interaction

In order for the matrix element averaged over all directions to become M, the factor $\sqrt{3}$ has been introduced.

As assumed earlier, the QW is grown along the z-axis, and xy is the layer plane. For light propagating along the $+x$ axis with its E vector along y, the propagation is with the transverse electric (TE) mode, and if the electric field is along the z-axis, the transverse magnetic (TM) mode propagates. This is illustrated in Figure 8.4a. As the direction of propagation vector $\mathbf{k}$ is arbitrary, we express its direction in a polar coordinate system as shown in Figure 8.4b. The k_z component is discrete, for example, for infinite barrier QW $k_z = (n\pi/L)$; the x- and y-components of $\mathbf{k}$ are, however, continuous.

With the chosen direction of $\mathbf{k}$ shown in Figure 8.4b, the momentum matrix elements between the conduction and HH bands are

$$x\text{-axis:}\frac{1}{\sqrt{2}}\sqrt{3}M(\cos\theta\cos\phi \mp i\sin\phi), \ y\text{-axis:}\frac{1}{\sqrt{2}}\sqrt{3}M(\cos\theta\sin\phi \pm i\cos\phi) \tag{8.19a}$$

$$z\text{-axis:}-\frac{1}{\sqrt{2}}\sqrt{3}M\sin\theta \tag{8.19b}$$

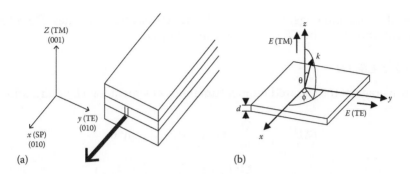

FIGURE 8.4
(a) QW layer and electric field vector for TE polarization; (b) polar coordinate system to obtain polarization dependence of matrix element.

Since the square of optical transition matrix element is proportional to $\langle 1|\mathbf{E}\cdot\mathbf{p}|2\rangle^2$, only the momentum matrix element with a component parallel to $\mathbf{E}$ contributes to the optical transition. We consider a wave vector $\mathbf{k}_n$ for a quantum number n and average the square of the momentum matrix element over all the directions in the xy plane by fixing the z-component k_{nz} of $\mathbf{k}_n$.

8.3.2.1 CB–HH

For TE mode, $\mathbf{E}\|y$, the average of the squared momentum matrix element is obtained by taking the second expression as

$$\langle M^2\rangle_{hh,TE} = \frac{3M^2}{2}\frac{1}{2\pi}\int_0^{2\pi}\left(\cos^2\theta\sin^2\phi+\cos^2\phi\right)d\phi = \frac{3M^2}{4}\left(1+\cos^2\theta\right)$$

$$= \frac{3M^2}{4}\left(1+\frac{k_z^2}{k^2}\right) = \frac{3M^2}{4}\left(1+\frac{E_{z,n}}{E_n}\right) \tag{8.20}$$

where $E_{z,n}$ and E_n are, respectively, the quantized energy and the total energy of the nth subband.

For TM mode, $\mathbf{E}\|z$, and the squared average momentum matrix element is given by

$$\langle M^2\rangle_{hh,TM} = \frac{3M^2}{2}\sin^2\theta = \frac{3M^2}{2}\left(1-\cos^2\theta\right) = \frac{3M^2}{2}\left(1-\frac{k_z^2}{k^2}\right)$$

$$= \frac{3M^2}{2}\left(1-\frac{E_{z,n}}{E_n}\right) \tag{8.21}$$

At the subband edge, $E_{z,n} = E_n$ and the matrix element vanishes. Therefore, optical transitions by TE modes only are allowed.

8.3.2.2 CB–LH

The average of the squared momentum matrix element for TE polarization is

$$\langle M^2 \rangle_{\text{lh,TE}} = \frac{M^2}{4}\left(1 + \cos^2\theta\right) + M^2 \sin^2\theta$$

$$= \frac{M^2}{4}\left(1 + \frac{E_{z,n}}{E_n}\right) + M^2\left(1 - \frac{E_{z,n}}{E_n}\right) \tag{8.22}$$

Similarly, for TM mode

$$\langle M^2 \rangle_{\text{lh,TM}} = \frac{M^2}{2}\sin^2\theta + 2M^2\cos^2\theta = \frac{M^2}{2}\left(1 - \frac{E_{z,n}}{E_n}\right) + 2M^2\frac{E_{z,n}}{E_n} \tag{8.23}$$

8.3.3 Gain in QWs

The expression for the absorption coefficient given by Equation 8.14 indicates that its value is zero when $f_e = f_h$. If the charge neutrality condition, that is, $n = p$, is assumed, then this equality occurs at the transparency carrier density, n_{tr}. Above this density, $f_e > f_h$, $F_e - F_h > \hbar\omega$, and the EM wave is amplified instead of undergoing absorption. It is easy to conclude that the maximum gain occurs at $\hbar\omega = E_{gmn}$, and its expression is

$$g_{\max} = g_0\left[f_e(E_{cn}) - f_h(E_{vm})\right] \tag{8.24}$$

where

$$g_0 = \frac{e^2 \langle |p_{cv}|^2 \rangle}{\varepsilon_0 m_0^2 c\hbar n_r \hbar\omega d} m_r C_{mn} A_{mn} \tag{8.25}$$

Example 8.1: The value for maximum gain in a GaAs–AlGaAs QW is now calculated, assuming $f_e - f_h = 1$ and $C_{mn} = 1$. The parameters chosen are $d = 5$ nm; $\hbar\omega = 1.45$ eV; in-plane masses: $m_e = 0.067m_0$, $m_{hh} = 0.112m_0$; $n_r = 3.6$; $A_{mn}\langle|p_{cv}|^2\rangle = m_0 E_p/4$; $E_p = 25$ eV (this corresponds to the TE polarized case). The reduced mass is $m_r = 0.042m_0$. Putting all the values into Equation 8.25, $g_0 = 9.2 \times 10^5$ m^{-1} or 9200 cm^{-1} is obtained.

The quantities $f_e(E_{cn})$ and $f_h(E_{vm})$ are the Fermi occupational probabilities of electrons at the subband edges. Under charge neutrality, both of these may be expressed in terms of injected carrier density n. It is straightforward to write n in terms of f_e as follows:

$$n = \sum_{l=0}^{\infty} n_l \ln\left[1+\frac{f_e}{1-f_e}\exp(-\varepsilon_{cl})\right]; \quad n_l = \frac{k_B T m_{el}}{\pi\hbar^2 d} \tag{8.26}$$

where:
the lth subband is located at ε_{cl} in units of $k_B T$
$\varepsilon_{c0} = 0$

A similar equation is obtained for holes with f_e replaced by $1-f_h$, ε_{cl} replaced by ε_{vl}, and n_l by p_l.

The expressions for f_e and f_h in terms of injected carrier density may now be expressed as [2]

$$f_e = 1-\exp(-n/N_s) \tag{8.27a}$$

$$f_h = \exp(-n/P_s) \tag{8.27b}$$

$$N_s = \sum_l n_l \exp(-\varepsilon_{cl}) \tag{8.27c}$$

The validity criterion of the above approximation is

$$\sum_{l=1} \frac{p_l e^{-\varepsilon_{vl}}(1-e^{-\varepsilon_{vl}})}{2P_s^2}p \ll 1 \tag{8.28}$$

Assuming that only one conduction and one valence subband are occupied and the charge neutrality condition is valid, and using $m_e = 0.067\, m_0$, $m_{hh} = 0.15\, m_0$, $m_{lh} = 0.23\, m_0$, $d = 10$ nm, and $T = 300$ K, we plot f_e, f_{hh}, and f_{lh} in Figure 8.5 as a function of injected electron density. The following conclusions may be drawn from this figure:

1. $P_s > N_s$ due to larger in-plane hole mass, even when it is assumed that a single subband is occupied. Since hole subbands are closely spaced, the actual value of P_s would be still larger than N_s. Therefore, f_h decreases more slowly than the function f_e rises.

2. Since for transparency $f_e = f_h$, the transparency occurs at a larger value of electron density.

3. Due to larger in-plane mass of hh, f_{hh} decreases more rapidly than f_{lh}, leading to lower transparency carrier density for the hh subband.

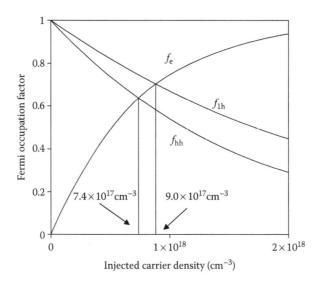

FIGURE 8.5
Plot of Fermi functions for the lowest subbands.

4. If the subband having lower in-plane effective mass is raised well
   above the other bands, f_h will decrease still faster with n, leading
   to a lower value of n_{tr}. This is an efficient way of creating population inversion, and it forms the basis of lower threshold current in
   strained QW lasers.

5. The effect of doping of the active layer may be understood by replacing n/N_s by $(n+N_D)/N_s$ and p/P_s by $(p+N_A)/P_s$, in Equations 8.27a
   through c, where N_D and N_A are the donor and acceptor densities,
   respectively. It is easy to conclude that the value of n_{tr} is reduced, but
   the reduction is more pronounced for p-doping due to larger hole
   mass.

The effect of reduced in-plane mass on the position of the quasi-Fermi level
may be understood from Example 8.2. A smaller in-plane mass pushes the
quasi-Fermi level higher into the subband, and the occupancy factor f_e at the
subband edge is increased.

> **Example 8.2:** The increase of F_e due to smaller in-plane mass may be
> understood if the expression for 2-D electron density at 0 K is considered:
> $n_{2D} = (m_e/\pi\hbar^2)(F_e - E_0)$. Keeping n_{2D} unchanged, if the mass is changed from
> m_{e1} to m_{e2}, $m_{e1} > m_{e2}$, the quasi-Fermi level is increased from F_{e1} to F_{e2}.

The ideal gain spectra for a QW form the basis for comparison between
a DH and a QW laser. The band diagram, gain spectra, and variation of

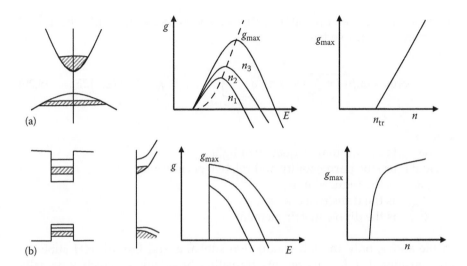

FIGURE 8.6
Band diagram, gain spectra, and g_{max} vs. n in a DH (a) and a QW (b) laser.

maximum gain with injected carrier density for both DH and QW lasers are shown in Figure 8.6. As may be seen, the maximum gain for DH occurs at an energy higher than the bandgap, and its position shifts to higher energy with increasing n. On the other hand, the maximum gain for a QW always occurs at the effective gap energy. The differences are

1. The emission starts at a higher energy in QW.
2. Since the peak gain always occurs at the effective bandgap, the added injected carriers contribute to the peak gain. In DH, added carriers shift the peak gain to higher energy, and the carriers below g_{max} are wasted.
3. In QWs, g_{max} saturates at some value of n, because all the available electron and hole states are fully inverted. This is not the case in DH, in which the peak gain increases monotonically with n.
4. The transparency carrier density n_{tr} and the threshold current density are lower in QW lasers.

8.4 Model Gain Calculation: Analytical Model

We shall first present a completely analytical model for optical gain developed by Makino [3]. Assuming that the E–$\mathbf{k}$ relation in all subbands is parabolic and that all optical transitions obey the k-selection rule, the optical gain may be expressed as

$$g(\omega) = \omega\sqrt{\mu/\varepsilon} \sum_{n=0} \left(m_r / \pi\hbar^2 d\right) \int_{E_g+E_{cn}+E_{vn}}^{\infty} \left\langle R_{cv}^2 \right\rangle \left(f_e - f_h\right) L\left(E_{cv}\right) dE_{cv} \qquad (8.29)$$

where:
- ω is the angular frequency of light
- μ and ε are the permeability and relative permittivity
- m_r is the reduced mass
- E_{cv} is the transition energy
- $\left\langle R_{cv}^2 \right\rangle$ is the dipole matrix element

In the above, only the nth conduction subband and the nth HH subband are considered. $L(E_{cv})$ represents transition broadening, which is usually expressed by the Lorentzian function

$$L\left(E_{cv}\right) = \frac{\hbar / \tau_{in}}{\left(E_{cv} - \hbar\omega\right)^2 + \left(\hbar / \tau_{in}\right)^2} \qquad (8.30)$$

where τ_{in} is the intraband relaxation time. The Fermi functions are given by

$$f_e = \frac{1}{\left[1 + \exp\left(\varepsilon_{cn} - F_e\right)/k_B T\right]} \qquad (8.31a)$$

$$f_h = \frac{1}{\left[1 + \exp\left(\varepsilon_{vn} - F_h\right)/k_B T\right]} \qquad (8.31b)$$

The symbols ε_{cn} and ε_{vn} represent the total energies of electrons and holes in subband n. The quasi-Fermi levels are related to the densities of electrons and holes injected into the well by the expressions

$$n = \frac{m_{ew} k_B T}{\left(\pi\hbar^2 d\right) \sum_n \ln\left[1 + \exp\left(F_e - E_{cn}\right)/k_B T\right]} \qquad (8.32a)$$

$$p = n = \frac{m_{hhw} k_B T}{\left(\pi\hbar^2 d\right) \sum_n \ln\left[1 + \exp\left(F_h - E_{vn}\right)/k_B T\right]} \qquad (8.32b)$$

where the subscript w refers to the well material.

8.4.1 Approximate Fermi Functions

For the ground state ($n = 0$), the Fermi functions may be expressed as

$$f_e = \frac{1}{\left[1 + \exp\left\{(2\Delta E / \Gamma_e) - \chi_e\right\}\right]} \tag{8.33a}$$

$$f_h = \frac{1}{\left[1 + \exp\left\{(2\Delta E / \Gamma_h) - \chi_h\right\}\right]} \tag{8.33b}$$

The symbols in Equations 8.33a and 8.33b are defined as

$$\Gamma_i = 2k_B T m_{iw} / m_r, \quad (i = e, hh) \tag{8.34a}$$

$$\chi_e = (F_e - E_{c0}) / k_B T \tag{8.34b}$$

$$\chi_h = -(E_{v0} + E_g + F_h) / k_B T \tag{8.34c}$$

$$\Delta E = E_{cv} - E_{tr} \tag{8.34d}$$

$$E_{tr} = E_g + E_{c0} + E_{v0} \tag{8.34e}$$

The quantities χ_e and χ_h may be expressed approximately by [2]

$$\chi_e = \ln\left[\exp(n / N_s) - 1\right] \tag{8.35a}$$

$$\chi_h = \ln\left[\exp(n / P_s) - 1\right] \tag{8.35b}$$

where N_s and P_s are given as

$$N_s = \frac{m_{ew} k_B T}{\left(\pi \hbar^2 d\right) \sum_n \ln\left[1 + \exp(F_e - E_{cn}) / k_B T\right]} \tag{8.36a}$$

$$P_s = \frac{m_{hhw} k_B T}{\left(\pi \hbar^2 d\right) \sum_n \ln\left[1 + \exp(F_h - E_{vn}) / k_B T\right]} \tag{8.36b}$$

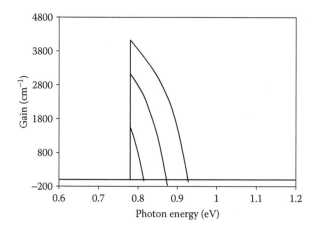

FIGURE 8.7
Gain spectra as a function of photon energy with injected electron density as a parameter.

The analytical expressions for the Fermi functions, given by Equations 8.33a and 8.33b, depend on ΔE, the photon energy, and on carrier densities, given by Equations 8.35a and 8.35b.

8.4.2 Gain without Broadening

In this case,

$$L(E_{cv}) = \pi\delta(E_{cv} - \hbar\omega) \tag{8.37}$$

Considering the lowest subbands, the gain expression given by Equation 8.29 reduces to

$$g(\omega) = K(f_e - f_h) \tag{8.38}$$

where K may be easily obtained from Equation 8.29, and ΔE in Equation 8.33 is given as $\Delta E = \hbar\omega - E_{tr}$.

> **Example 8.3:** The gain spectra of InGaAs QW with InGaAsP barrier for TE modes are shown in Figure 8.6 for $d = 8$ nm. $m_{ew} = 0.041m_0$ and $m_{hhw} = 0.424m_0$. The subband energies are calculated by using the approximate formulas given by Makino (see Chapter 3). The gain is plotted as a function of photon energy with injected carrier density as a parameter. The solid curve represents the numerical results, while the dashed curves are based on approximate formulas.

Figure 8.6 indicates that the maximum gain occurs at $\Delta E = 0$. The Fermi functions at the gain peak are obtained from Equation 8.33 by setting $\Delta E = 0$, and the expressions are

$$f_{e0} = 1 - \exp(-n / N_s) \tag{8.39a}$$

$$f_{h0} = \exp(-n / P_s) \tag{8.39b}$$

These equations were originally developed by Vahala and Zah [2]. Here, N_s may be expressed as

$$N_s = \sum_l n_l \exp(-\varepsilon_{cl}) \tag{8.40}$$

where the *l*th subband is located at ε_{cl} in units of $k_B T$ and $\varepsilon_{c0} = 0$. A similar equation may be written for P_s with f_e replaced by $1 - f_h$, ε_{cl} by ε_{vl}, and n_l by p_l.

Assume now that $f_{e0} - f_{h0} = A\ln(n/N_0)$. The prefactor A is chosen such that at $n = eN_0$, $f_{e0} - f_{h0} = 1 - \exp(-en/N_s) - \exp(-en/P_s)$. With this, we may write

$$g = g_0 \ln(n / N_0) \tag{8.41}$$

where

$$g_0 = K\left[1 - \exp(-eN_0 / N_s) - \exp(-eN_0 / P_s)\right] \tag{8.42}$$

The expression for gain (Equation 8.41) may be written in terms of the current density J, by using the relation $J = AN + BN^2 + CN^3$ and approximating this as $J = B_{eff}N^2$, where B_{eff} is the effective recombination coefficient. It is now easy to show from Equation 8.41 that

$$g = (g_0 / 2)\ln(J / J_0) \tag{8.43}$$

and

$$\Gamma g = \Gamma \xi \ln(J / J_0) \tag{8.44}$$

where:
  Γ is the mode confinement factor
  ξ is a proportionality constant
  the transparency current density $J_{tr} = B_{eff}N_0^2$

This form of modal gain has been used in a number of works.

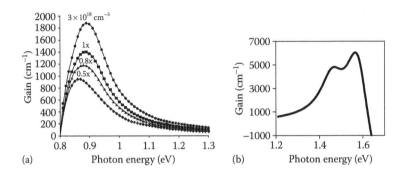

FIGURE 8.8
Gain spectra of (a) InGaAs QW, assuming one subband occupancy $\tau_{in} = 0.1$ ps, and (b) GaAs, considering higher levels.

8.4.3 Effect of Broadening

Analytical approximation for gain with Lorentzian broadening has also been derived by Makino. The interested reader is encouraged to follow the analysis. We do not include such theory. Rather, we show in Figure 8.8 the form of gain spectra by numerically integrating Equation 8.29 using the complete form of the Fermi functions. The curves show that the maximum gain still occurs at the effective bandgap energy, but the peaks shift to higher photon energies with increasing injection.

In Figure 8.8, we show the gain spectra for two different materials. Figure 8.8a plots the gain spectra for the InGaAs/InP system for different values of injection-carrier density, assuming that all the carriers occupy the lowest subband. The spectra show an increase with photon energy, reach a peak, and then decrease.

The variation of gain with photon energy for a GaAs QW with $Al_{0.3}Ga_{0.7}As$ as the barrier for a 2-D injected carrier density of 8×10^{12} cm^{-2} or a volume-carrier density of 8×10^{18} cm^{-3} ($d = 10$ nm) is shown in Figure 8.8b. In this case, detailed calculation shows that the second subband is occupied. The spectra contain two peaks corresponding to two subbands.

8.5 Recombination in QWs

The rate of spontaneous emission from a CB state of energy E_2 to the valence band state of energy E_1 is expressed as

$$R_{sp}(\hbar\omega)d\hbar\omega = \frac{2\pi}{\hbar} \sum |p_{21}|^2 G(\hbar\omega) f_2 (1 - f_1) \delta(E_{21} - \hbar\omega) d\hbar\omega \qquad (8.45)$$

The wave functions for the nth conduction subband and the mth valence subband are given by Equation 8.3. The squared matrix element is given by Equation 8.6. Using the expression for the optical DOS $G(\hbar\omega)$, the spectrally dependent spontaneous emission rate is given by

$$R_{sp}(\hbar\omega) = \frac{e^2 n_r m_r (\hbar\omega)}{\pi^2 m_0^2 \varepsilon_0 c^3 \hbar^4} \sum_{m,n} \langle |p_{cv}|^2 \rangle C_{mn} f_e(E_{cn} - F_e)[1 - f_h(E_{vm} - F_h)] \quad (8.46)$$

where F_e and F_h are the quasi-Fermi levels. The emitted radiation is unpolarized, and hence the factor $A_{mn} = 1$.

The total recombination rate is obtained by integrating Equation 8.46 over all photon energies, and thus

$$R_{sp} = B' \int f_e(E')[1 - f_h(E)]dE \quad (8.47)$$

where B' is the prefactor. The occupational probabilities are expressed in terms of n and p in the following way:

$$f_e = \frac{n\pi\hbar^2}{m_e k_B T} \frac{1}{\left[\exp\{(E - F_e)/k_B T\} + 1\right]} \quad (8.48a)$$

$$f_h = \frac{p\pi\hbar^2}{m_h k_B T} \frac{1}{\left[\exp\{(F_h - E)/k_B T\} + 1\right]} \quad (8.48b)$$

Using Equations 8.48a and b in Equation 8.47, the total recombination rate may be calculated. A useful analytical expression may be obtained if the carriers are assumed to obey nondegenerate statistics. Then,

$$f_e(1 - f_h) \rightarrow \exp\left[-(\hbar\omega - E_{gmn})/k_B T\right]\exp\left[-(E_{gmn} - \Delta F)/k_B T\right]$$

where $\Delta F = F_e - F_h$. Inserting this in Equation 8.47 and noting that the prefactor is a slowly varying function of $\hbar\omega$, the integral takes the form $\int_0^\infty \exp(-x)dx = 1$. It is also noted that the product $np \propto \exp[-(E_{gmn} - \Delta F)/k_B T]$. Therefore, the total spontaneous emission rate may be expressed as

$$R_{sp} = \frac{e^2 n_r \hbar\omega m_r}{\pi^2 m_0^2 \varepsilon_0 dc^3 \hbar^4} \langle |p_{cv}|^2 \rangle C_{mn} \frac{(\pi\hbar^2)^2}{m_e m_h (k_B T)} np = Bnp \quad (8.49)$$

The lifetime of excess carriers in the QW is defined as $\tau = (Bn_0)^{-1}$.

Example 8.4: We consider GaAs QW with $m_e = 0.067m_0$ and $m_{hh} = 0.112m_0$, so that $m_r = 0.042m_0$. We also take $n_r = 3.6$ and $\langle|p_{cv}|\rangle = 2.7m_0E_g$, carrier density $n = 10^{12}$ cm^{-2}, and $C_{mn} = 1$. The calculated value of $B_{3D} = B_{2D} \cdot d = 8.33 \times 10^{-8}$ and $\tau = 1.2$ ns. The measured value is 6 ns, close to this estimate.

Radiative recombination in a QW occurs after excess electron–hole pairs (EHPs) are created by injection or photoexcitation. When an intense photon pump of energy $\hbar\omega_p > E_{gmn}$ is incident, the electrons are lifted from the valence band to a state well above the subband edge in the CB. The electrons and holes are essentially hot, and they give up their excess energy by longitudinal optic phonon emission, and quickly relax to the bottom or top of the respective subbands. The excess EHPs then recombine to give the emission (photoluminescence in the case of photon absorption) spectra. From Equation 8.46, the spectra depend on $f_e(1 - f_h)$. The Fermi functions are described by Equation 8.48 but are characterized by carrier temperatures; for example, we use electron and hole temperatures T_e and T_h in place of T. The recombination spectra may then be expressed as

$$R_{sp}(y) \propto \exp\left(\frac{-y}{k_B T^*}\right) H(y)$$

(8.50)

where:

$$\frac{1}{k_B T^*} = \left(\frac{1}{k_B T_e}\right)\left(\frac{m_h}{M}\right) + \left(\frac{1}{k_B T_h}\right)\left(\frac{m_e}{M}\right)$$

$y = \hbar\omega - E_{gmn}$
$M = m_e + m_h$
$H(y)$ denotes Heaviside step function

If $T_e = T_h = T^*$, the tail of the photoluminescence spectra gives the carrier temperature [4].

8.6 Loss Processes in QW Lasers

The threshold characteristics of a QW laser are determined by various loss processes within the well and in the barriers. Among them, the free-carrier absorption, intervalence band absorption (IVBA), and Auger processes play dominant roles. All these processes occurring in bulk semiconductors and

DH lasers have been discussed in Chapters 2, 6, and 7. In this section, we shall consider the IVBA and the Auger recombination and point out how the restricted motion in the QW modifies the relevant expressions for bulk semiconductors.

8.6.1 Intervalence Band Absorption

This process, which becomes important for small-bandgap semiconductors, is illustrated in Figure 8.9 by considering HH and split-off (SO) bands. The photon emitted by the transition from the conduction subband to the HH subband, indicated by a, is reabsorbed by an electron in the SO band to move into the HH band (b in Figure 8.9). In other words, a hole is transferred from the HH to the SO band.

Conservation of energy and momentum leads to the following expression for the photon energy:

$$E_{IVBA} = m_s \frac{(\hbar\omega - \Delta_n)}{(m_{hh} - m_s)} \tag{8.51}$$

where:

$\Delta_n = \Delta + E_{sn} - E_{hhn}$

Δ is the spin-orbit-splitting energy in the bulk

E_{sn} and E_{hhn} are the nth subband energies in the SO and HH bands, respectively

ms denote the effective masses in the respective bands

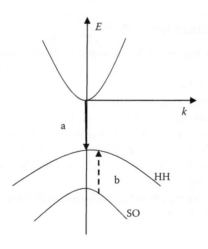

FIGURE 8.9
Intervalence band absorption. The photon emitted through process (a) is reabsorbed in process (b) by SO–HH transition.

The theory of intersubband absorption in a QW may be modified to give the IVBA coefficient in QWs. Assuming that the matrix element is independent of **k**, the IVBA absorption coefficient is expressed as

$$\alpha(\hbar\omega) = \frac{B}{d}\frac{m_s m_{hh}}{m_{hh} - m_s}\left\langle |p_{shh}|^2\right\rangle(f_s - f_{hh}) \tag{8.52}$$

where:

    B is a constant of proportionality

    p_{shh} is the momentum matrix element

    fs denote the Fermi occupation probabilities

Assuming that $f_s \sim 1$ and $(1 - f_{hh})$ is described by the Boltzmann distribution, we may write

$$\alpha_{shh}(QW) = \frac{e^2\omega}{cn_r\varepsilon_0 d\hbar^2}\frac{m_s m_{hh}}{m_{hh} - m_s}p\frac{\sum_n \langle z\rangle_n^2 \exp\left\{[-E_{hhn} - E(\mathbf{k})]/k_BT\right\}}{\sum_n \exp(-E_{hhn}/k_BT)} \tag{8.53}$$

Each term in the summation is a product of the dipole matrix element for a pair of subbands and the Boltzmann factor, and p is the total hole density.

Calculations performed by Asada et al. [5] for InGaAs/InP QWs indicate that the IVBA is of the same order as in bulk.

The effect of reduction of in-plane hole mass on IVBA will be discussed in Section 8.9.4 in connection with strained QW lasers.

8.6.2 Auger Recombination

The Auger process is the most detrimental factor in increasing the loss and thereby the threshold current density in InGaAsP–InP DH lasers. The same is true for InGaAs QW lasers.

Various band-to-band Auger processes in bulk semiconductors are shown in Figure 6.7. The same figures can be used to illustrate the Auger processes in a QW. The differences are that the transitions occur between different subbands and that the electrons and holes have 2-D motion.

We consider only the CCCH process to highlight the method of calculation of the Auger rate in QWs.

The expression for the Auger rate developed in Section 6.8.2 is modified for a QW in the following form:

$$R_a = \frac{2\pi}{\hbar}\left[\frac{1}{(2\pi)^2}\right]^4\frac{1}{d}\iiiint|M_{if}|^2 P(1,1',2,2')\delta(E_i - E_f)d\mathbf{k}_1 d\mathbf{k}_2 d\mathbf{k}_{1'}d\mathbf{k}_{2'} \tag{8.54}$$

where:

R_a is the Auger rate per unit volume

$|M_{if}|^2$ is the squared matrix element for transition from the initial (i) state to the final (f) state

P denotes the difference of probabilities of the Auger process and the inverse process of impact ionization

The well width d comes in the denominator to signify that the rate is per unit volume. The in-plane wave vectors are $\mathbf{k}_1$, $\mathbf{k}_2$, $\mathbf{k}_1'$, and $\mathbf{k}_2'$.

The squared matrix element is approximated as $|M_{if}|^2 \approx 4|\bar{M}_{if}|$, where

$$M_{if} = \iint \psi_{1'}^*(\mathbf{r}_1)\psi_{2'}^*(\mathbf{r}_2) \frac{e^2}{4\pi\varepsilon|\mathbf{r}_1 - \mathbf{r}_2|} \psi_1(\mathbf{r}_1)\psi_2(\mathbf{r}_2)d\mathbf{r}_1 d\mathbf{r}_2 \tag{8.55}$$

In Equation 8.55, $\psi(\mathbf{r})$'s are the QW wave functions, and in the Coulomb interaction term the effect of screening is not included. Using the Fourier transform of the Coulomb interaction

$$\frac{1}{|\mathbf{r}_1 - \mathbf{r}_2|} = \frac{1}{(2\pi)^3} \int \frac{4\pi}{q^2} \exp\left[j\mathbf{q}\cdot(\mathbf{r}_1 - \mathbf{r}_2) \right] d\mathbf{q} \tag{8.56}$$

one obtains

$$\bar{M}_{if} = \frac{e^2}{\varepsilon} \frac{1}{(2\pi)^3} \int \frac{1}{q^2} I_{1'1}(\mathbf{q}) I_{2'2}(\mathbf{q}) d\mathbf{q} \tag{8.57}$$

The integrals appearing in Equation 8.57 are expressed as

$$I_{mn}(\mathbf{q}) = \int \psi_m^*(\mathbf{r})\psi_n(\mathbf{r})\exp(j\mathbf{q}\cdot\mathbf{r})d\mathbf{r} \tag{8.58}$$

The wave functions in the QW are expressed as

$$\psi_m(\mathbf{r}) = U_m(\mathbf{r})\phi_m(z)\exp(j\mathbf{k}_m\cdot\mathbf{r}) \tag{8.59}$$

where:

U_m is the cell periodic part

ϕ_m is the envelope function, which will be approximated as sin function for infinite barrier

$\mathbf{k}_m$ and ρ are the 2-D wave vector and 2-D position vector, respectively

The integral given in Equation 8.58 now becomes

$$I_{mn}(\mathbf{q}) = (2\pi)^2 \delta\left(-\mathbf{k}_m + \mathbf{k}_n + \mathbf{Q}\right) F_{mn} G_{mn}\left(q_z\right) \tag{8.60}$$

where:

$$q^2 = Q^2 + q_z^2$$
$$F_{mn} = \int U_m^*(\mathbf{r}) U_n(\mathbf{r}) d\mathbf{r}$$
$$G_{mn} = \int \phi_m^*(z) \phi_n(z) \exp(jq_z z) dz$$

The in-plane component of q and its component along the z-direction are denoted, respectively, by Q and q_z. The integral F_{mn} is over a unit cell of the crystal. The matrix element may now be written as

$$\bar{M}_{if} = \frac{e^2}{2\varepsilon} V_0 \delta\left(\mathbf{k}_1 + \mathbf{k}_2 - \mathbf{k}_{1'} - \mathbf{k}_{2'}\right) \tag{8.61}$$

where

$$V_0 = 4\pi F_{11'} F_{22'} \int \frac{G_{11'} G_{22'}}{\left|\mathbf{k}_1 - \mathbf{k}_{1'}\right|^2 + q_z^2} dq_z .$$

In writing the denominator of Equation 8.61, we have used $q^2 = Q^2 + q_z^2 = \left|\mathbf{k}_1 - \mathbf{k}_{1'}\right|^2 + q_z^2$.

Using Equations 8.54 and 8.61, the Auger rate is expressed as

$$R_a = \frac{8\pi}{\hbar}\left(\frac{1}{4\pi^2}\right)^4 \frac{1}{d}\left(\frac{e^2}{2\varepsilon}\right)^2 I \tag{8.62}$$

where

$$I = \iiint \int P(1,1',2,2') d\mathbf{k}_1 d\mathbf{k}_{1'} d\mathbf{k}_2 d\mathbf{k}_{2'} \left|V_0(\mathbf{k}_1, \mathbf{k}_{1'})\right|^2 \delta(\mathbf{k}_1 + \mathbf{k}_2 - \mathbf{k}_{1'} - \mathbf{k}_{2'}) \delta(E_i - E_f) \tag{8.63}$$

Using the energy-conservation condition, we may write

$$\frac{\hbar^2}{2m_e} k_1^2 + \frac{\hbar^2}{2m_e} k_2^2 = \frac{\hbar^2}{2m_h} k_{1'}^2 - E_Q + \frac{\hbar^2}{2m_{et}} k_{2'}^2$$

where:

m_e and m_h are, respectively, the electron and hole effective masses at the band edges

m_{et} is the electron effective mass at energy E_2
E_Q is the effective gap in the QW

To evaluate the integral I, we introduce the following integration variables:

$$\mathbf{h} = \mathbf{k}_1 + \mathbf{k}_2, \; \mathbf{j} = \mathbf{k}_1 - \mathbf{k}_2$$

The integral now takes the form

$$I = \frac{m_e}{2\hbar^2} \iint d\mathbf{k}_{1'} d\mathbf{k}_{2'} \int_{-1}^{1} dt P(1,1',2,2') \left| V_o(\mathbf{k}_1, \mathbf{k}_{1'}) \right|^2 |j_0| \tag{8.64}$$

where:
$t = \cos\theta$
θ is the angle between $\mathbf{h}$ and $\mathbf{j}$

The integral is to be evaluated under the condition $j_0^2 \geq 0$, where

$$j_0^2 = 2\left(\beta_e k_{2'}^2 - \beta k_{1'}^2 - \frac{2m_e E_Q}{\hbar^2} \right) - \left| \mathbf{k}_{1'} + \mathbf{k}_{2'} \right|^2$$

$\beta_e = m_e/m_{et}$ and $\beta = m_e/m_h$. We further introduce two new variables:

$$z_1 = \mathbf{k}_{1'} + \frac{\mathbf{k}_{2'}}{1+2\beta}, \quad z_2 = \mathbf{k}_{2'} \tag{8.65}$$

and obtain

$$j_0^2 = \alpha_e z_2^2 - z_1^2(1+2\beta) - \frac{4m_e}{\hbar^2} E_Q \tag{8.66}$$

where $\alpha_e = \beta_e - [\beta/(1+2\beta)]$. In order that $j_0^2 \geq 0$, one must have $\alpha_e z_2^2 > 2m_e E_Q / \hbar^2$. Using the relation $E_{2'} = \hbar^2 z_2^2 / 2m_e$, we obtain

$$E_{2'} > E_T = \beta_e E_Q / \alpha_e \tag{8.67}$$

It appears therefore that the energy E_2' must reach a threshold energy E_T before the Auger process takes place.

When $E_Q \gg k_B T$, the inverse process of impact ionization is less probable. Ignoring this process and further using the nondegenerate distribution, the probability $P(1,1',2,2')$ may be expressed as

$$P(1,1',2,2') = \frac{n^2 p}{N_c^2 N_v} \exp\left(\frac{-E_{2'} + E_Q}{k_B T}\right)$$

(8.68)

where:

n and p are the electron and hole concentrations, respectively

N_c and N_v are the respective maximum DOS

The energy-conservation relation is

$$E_1 + E_2 = E_{2'} - E_Q - E_{1'}$$

(8.69)

It is to be noted that the probability P decreases rapidly as $E_{2'}$ increases. The $|V_0|^2$ term in Equation 8.64 is then replaced by its value at $E_{2'} = E_T$. Using Equations 8.65 and 8.68 in Equation 8.64, one finally obtains

$$I = \frac{m_e}{\hbar^2} \frac{n^2 p}{N_c^2 N_v} |V_T|^2 \exp\left(\frac{E_Q}{k_B T}\right) I_1$$

(8.70)

where

$$I_1 = \iint d\mathbf{z}_1 d\mathbf{z}_2 \exp\left(\frac{-\hbar^2 z_2^2}{2 m_{et} k_B T}\right)$$

$$= \frac{8\pi^2 \alpha_e}{1 + 2\beta}\left(\frac{2 m_{et} k_B T}{\hbar^2}\right)^2 \exp\left(-\frac{m_e}{m_{et} \alpha_e} \frac{E_Q}{k_B T}\right)$$

(8.71)

Here, $|V_T|^2$ is the value of $|V_0|^2$ evaluated at E_T. Using Equations 8.63 and 8.70, we obtain under the nondegenerate approximation

$$R_a \propto \frac{n^2 p}{N_c^2 N_v} \exp\left(-\frac{\Delta E_c}{k_B T}\right)$$

(8.72)

where

$$\Delta E_c = E_T - E_Q = \frac{m_{et} E_Q}{m_h + 2 m_e - m_{et}}$$

(8.73)

Equations 8.72 and 8.73 show that the Auger rate increases with decreasing bandgap and with increasing temperature. Further, if the hole mass is reduced, ΔE_c will increase and the Auger rate will be lowered.

8.7 MQW Laser

As seen in Equation 7.30, an increase in the mode confinement factor will lead to a reduction in the threshold current. This is due to gain saturation at a high carrier density. A better alternative is to distribute the carrier density over a number of QWs, N_W. By this, the gain per QW is reduced by less than N_W times, but the confinement factor is increased by N_W. The threshold-current expression is now modified by multiplying both the confinement factor and the volume by a factor N_W. One may thus write

$$I_{\text{thMQW}} \cong \frac{eN_W V_1 BN_{\text{th}}^2}{\eta_i} \exp\left\{2\left[\alpha_i + \alpha_m\right]/N_W \Gamma_1 g_0\right\} \tag{8.74}$$

In writing the expression, the structure considered is a SCH, so that optical mode pattern does not change as more QWs are added. The number of QWs is limited to the maximum number that can be accommodated within the maxima of the optical mode. The increased confinement factor also ensures higher power without moving too high in the gain curve.

The gain in one QW may be described to a good approximation by

$$g = \beta J_0 \ln\left(J / J_0\right) \tag{8.75}$$

where:
J_0 is the transparency current density
β is the gain constant

At threshold, the net gain G of a MQW laser having N_W number of wells overcomes the total loss, and thus

$$G = N_W \Gamma_s g_{\text{th}} = \alpha + \frac{1}{2L} \ln\left(\frac{1}{R_1 R_2}\right) \tag{8.76}$$

Using Equation 8.75, the threshold current density may be expressed as

$$J_{\text{th}} = \frac{N_W J_0}{\eta} \exp\left\{\frac{1}{N_W \Gamma_s J_0 \beta}\left(\alpha + \frac{1}{2L} \ln\left(\frac{1}{R_1 R_2}\right)\right)\right\} \tag{8.77}$$

The threshold current in a MQW is expressed, by noting that $I_{\text{th}} = J_{\text{th}} wL$, as [6]

$$I_{th} = \frac{wLN_W J_0}{\eta} \exp\left\{\frac{1}{N_W \Gamma_s J_0 \beta}\left[\alpha + \frac{1}{2L}\ln\left(\frac{1}{R_1 R_2}\right)\right]\right\}$$ (8.78)

From Equation 8.78, one notices that the threshold current increases linearly with the length L ($V = LWd$) for large L, when the mirror loss is negligible. However, for short L, the exponential term dominates and the gain decreases rapidly as L is increased slightly. Therefore, the threshold current must show a minimum for an optimum value of L. Using Equation 8.78 and putting $dI_{th}/dI = 0$, one may easily show that the minimum occurs for

$$L_{opt} = \frac{1}{2}\frac{1}{N_W \Gamma_s J_0 \beta}\ln\left(\frac{1}{R_1 R_2}\right)$$ (8.79)

and the expression for minimum threshold current becomes

$$I_{th,min} = \frac{1}{2\Gamma_s \beta \eta}\ln\left(\frac{1}{R_1 R_2}\right)\exp\left\{\left(1 + \frac{\alpha}{N_W \Gamma_s J_0 \beta}\right)\right\}$$ (8.80)

Example 8.5: The chosen values of parameters are $N_W = 8$, $\Gamma_s = 0.0152$, $J_0\beta = 516.6$ cm^{-1}, and $R_1 = R_2 = 0.3$. The optimum value of L for lowest threshold current is $L_{opt} = 0.19$ mm. The authors have given an empirical expression $L_{opt} = 1.63/N_W$ mm. From this, $L_{opt} = 0.2$ nm, in agreement with the value obtained from Equation 8.79.

Example 8.6: We estimate the values of threshold current for $N_W = 8$ using the values of parameters chosen in Example 8.5 and taking the width $w = 30$ μm. The calculated values for different lengths are tabulated below:

L in mm	5	2	1	0.2	0.1
I_{th} in A	0.86	0.36	0.20	0.087	0.113

It is clear from the table that I_{th} decreases with decreasing length, shows a minimum for $L \sim 0.2$ mm, and then increases.

Plots of threshold current versus length L for QWs with $N_W = 2$, 4, and 8 are given in Figure 8.10 for an InGaAs/InGaAsP QW, with $d = 0.2$ μm, $w = 30$ μm, and $R_1 = R_2 = 0.3$. Each curve shows a minimum value of threshold current. With increasing number of QWs, the value of L_{opt} shifts to

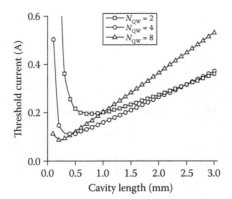

FIGURE 8.10
Threshold current–length plot for SQW and MQWs.

lower values. As expected, the threshold current is reduced with the increasing number of QWs.

8.8 Modulation Response of QW Lasers

As has been pointed out, due to the step-like DOS in QW-LDs, the differential optical gain $\partial g/\partial n$ is larger than that in bulk DH-LDs, leading to a larger relaxation oscillation frequency in QW-LDs, which in turn leads to high-speed modulations in QW-LDs.

Unfortunately, the argument presented above failed to explain why the modulation bandwidth for early QW-LDs was no better than that for DH-LDs. Several theoretical and experimental investigations have been performed to find the reasons for not obtaining the high modulation bandwidth of QW lasers, as predicted from early theory. There are three different arguments. The first reason is the neglect of any transport time for carriers to reach the active region. The transport time is negligible if the intrinsic region coincides with the active region, as in a DH laser. However, in SCH-QWs the time cannot be neglected. The second reason is the reduction of the differential gain when higher levels, particularly in the OCL, are occupied by the carriers. The third possibility is the presence of a hot carrier effect and spatial hole burning in the well–barrier interface.

In the following, all these arguments will be presented and discussed.

8.8.1 Transport in SCH Layers

The structure of an SCH-QW laser incorporating a single QW used for the study of transport effects is shown in Figure 8.11. The analysis may be extended to include MQWs [7,8].

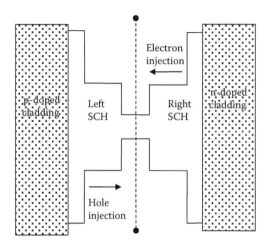

FIGURE 8.11
Structure of SCH-QW-LD.

As shown in the figure, electrons and holes are injected into the SCH layers by the two cladding layers at the two ends. The carriers are transported across the SCH regions by diffusion and drift, and at the same time undergo recombination. The transport may be studied by solving the continuity equation. The carriers are next captured by the QW. The process is governed by phonon scattering. The 3-D carriers at the bottom of the SCH layers are connected to the 2-D states just above the QW (virtual states), from where the 2-D electrons make a transition to the quantized energy levels in the QW. The capture process is a quantum mechanical process, and the capture time is calculated from the knowledge of the initial- and final-state wave functions, electron–phonon coupling, and phonon occupation numbers.

The opposite process to carrier capture is carrier escape from the QW to the SCH layers by thermionic emission. It reduces the degree of carrier capture in a single QW. However, in a MQW structure, this is the essential transport process between the QWs. The escape time depends on the barrier height and the temperature. In an MQW structure, tunneling between the wells causes transport of the carriers.

8.8.1.1 Continuity Equation in the SCH

Under a forward bias, the structure shown in Figure 8.9 essentially behaves as a p-i-n diode, in which the SCH region is the intrinsic (i) layer. Since electrons and holes enter into the i-region from opposite sides, the charge separation leads to a very high electric field. The laser structure can be analyzed like a forward-biased p-i-n diode under heavy injection conditions. The electron and hole current densities, including both drift and diffusion, are expressed as

$$J_n = eD_n \left(\frac{enF}{k_B T} + \frac{\partial n}{\partial x} \right) \tag{8.81a}$$

$$J_p = eD_p \left(\frac{epF}{k_B T} - \frac{\partial p}{\partial x} \right) \tag{8.81b}$$

where F is the electric field and the relation $D/\mu = k_B T/e$ has been employed. The continuity equations are

$$\frac{\partial n}{\partial t} = \frac{1}{e} \frac{\partial J_n}{\partial x} - R(n, p) \tag{8.82a}$$

$$\frac{\partial p}{\partial t} = -\frac{1}{e} \frac{\partial J_p}{\partial x} - R(n, p) \tag{8.82b}$$

where $R(n,p)$ is the net recombination rate. Assuming $n = p$, charge neutrality, that is, $\partial F/\partial x = 0$, and the steady-state condition $\partial n/\partial t = \partial p/\partial t = 0$, and combining Equations 8.81 and 8.82, the electric field term can be eliminated to give

$$\frac{d^2 p}{dx^2} - \frac{D_n + D_p}{2 D_n D_p} R(p) = 0 \tag{8.83}$$

We now introduce an ambipolar diffusion coefficient $D_a = 2 D_n D_p / (D_n + D_p)$, an ambipolar lifetime $\tau_a = p/R(p)$, and an ambipolar diffusion length $L_{a0} = \sqrt{D_a \tau_a}$, and obtain from Equation 8.83

$$\frac{d^2 p}{dx^2} - \frac{p}{L_{a0}^2} = 0 \tag{8.84}$$

8.8.1.2 Carrier Distribution in the SCH Layer

The general solution for the steady-state continuity equation is

$$p(x) = A_1 \exp(+x / L_{a0}) + A_2 \exp(-x / L_{a0}) \tag{8.85}$$

where the constants A_1 and A_2 are to be obtained from the relevant boundary conditions. Let the direct (dc) current entering the SCH region at $x = -L_s$ be I_s. The boundary condition is

$$J = \frac{I_s}{A} = -eD_a \frac{dp}{dx}\bigg|_{x=-L_s}$$

where A is the cross-sectional area of the laser diode. The second boundary condition is that at $x=0$, the carrier density should be equal to the density at the QW P_W. Using these conditions in the solution, we may express the coefficients as

$$A_1 = \frac{P_W \exp(+L_s / L_{a0}) - (I_s L_{a0} / eD_a A)}{\exp(+L_s / L_{a0}) + \exp(-L_s / L_{a0})} \tag{8.86a}$$

$$A_2 = \frac{P_W \exp(-L_s / L_{a0}) + (I_s L_{a0} / eD_a A)}{\exp(+L_s / L_{a0}) + \exp(-L_s / L_{a0})} \tag{8.86b}$$

The current flowing into the QW under steady state is given by

$$I_W = -eD_a A \frac{dp}{dx}\bigg|_{x=0} = I_s \operatorname{sech}(L_s / L_{a0})$$

$$- eAL_W (L_{a0} / L_W) \tanh(L_s / L_{a0})(P_W / \tau_a) \tag{8.87}$$

8.8.1.3 SCH Transport Factor

The differential SCH transport factor α_{SCH} is defined as

$$\alpha_{SCH} = \frac{\partial I_W}{\partial I_s} = \operatorname{sech}\left(\frac{L_s}{L_{a0}}\right) \tag{8.88}$$

This is analogous to the common base current gain of a bipolar junction transistor. When small-signal values of the transport factor are of interest, the dc value of diffusion length L_{a0} is replaced by the small-signal parameter, defined as $L_a = \sqrt{L_{a0}^2 / (1 + j\omega\tau_a)}$. The small-signal transport factor is then

$$\alpha_{SCH,ss} = \frac{1}{\cosh\left[(L_s^2 / D_a\tau_a)^{1/2}\sqrt{1 + j\omega\tau_a}\right]} \approx \frac{1}{1 + j\omega(L_s^2 / 2D_a)} \tag{8.89}$$

In arriving at Equation 8.89, it is assumed that $L_s \ll L_{a0}$.

8.8.1.4 Carrier Capture Time

The probability of scattering of an electron in the SCH region by emission of a LO phonon into a quantized state in the QW is the measure of capture time. The capture time has been found to be an oscillatory function of the QW width. With changing QW width, the quantized energy levels also change and thus may move in and out of the reach of any state in the SCH layer separated by the LO phonon energy. Calculations predict a capture time oscillating between 10 ps and 1 ns. For MQW structures, the number of final states in the QW satisfying the energy- and momentum-conservation conditions increases, and the calculated lifetime of 1 ps comes closer to the experimentally observed values.

8.8.1.5 Thermionic Emission

The expression for the thermionic emission time has been derived by assuming that the carriers in the barrier have 3-D character, and that the Boltzmann statistics is valid. The expression is

$$\tau_e = \left(\frac{2\pi m^* L_W^2}{k_B T} \right)^{1/2} \exp\left(\frac{E_B}{k_B T} \right) \tag{8.90}$$

where:
E_B is the effective barrier height
m^* is the carrier effective mass
L_W is the QW width

> **Example 8.7:** Let us consider electrons in a GaAs QW of width $L_W = 5$ nm. The effective mass is $m^* = m_e = 0.067\, m_0$ and the barrier height is 230 meV. The escape time is 0.384 ns.

8.8.2 Rate Equations

We denote the carrier density in the QW by P_W, that in the SCH layer by P_B, and the photon density in the cavity by S. The rate equations are written as

$$\frac{dP_B}{dt} = \frac{I}{eV_{SCH}} - \frac{P_B}{\tau_s} + \frac{P_W(V_W/V_{SCH})}{\tau_e} \tag{8.91}$$

$$\frac{dP_W}{dt} = \frac{P_B(V_{SCH}/V_W)}{\tau_s} - P_W\left[\frac{1}{\tau_n} + \frac{1}{\tau_e} \right] + \frac{v_g G(P)S}{1+\varepsilon S} \tag{8.92}$$

$$\frac{dS}{dt} = \frac{\Gamma v_g G(P)S}{1+\varepsilon S} - \frac{S}{\tau_{ph}}$$ (8.93)

where:

τ_n is the bimolecular recombination lifetime
τ_e is the escape time out of the QW
τ_{ph} is the photon lifetime
Γ is the optical confinement factor
$G(P)$ is the carrier-density-dependent optical gain
v_g is the mode velocity
ε is the gain compression factor
V_W and V are, respectively, the volume of the QW and the SCH region

The optical gain $G(P)$ is a function of both the electron and hole densities; however, under charge neutrality the electron and hole densities are equal.

Since we are interested in the small-signal solutions, we linearize all the quantities by writing $I=I_0+i\exp(j\omega t)$, $P_B=P_{B0}+p_B\exp(j\omega t)$, $P_W=P_{W0}+p_W\exp(j\omega t)$, $S=S_0+s\exp(j\omega t)$, and $G=G_0+g_0\exp(j\omega t)$.

Substituting the small-signal equations into the rate equations, and setting the steady-state quantities to zero, the small-signal equations may be written as

$$j\omega p_B = \frac{i}{eV_{SCH}} - \frac{p_B}{\tau_s} + \frac{p_W(V_W/V_{SCH})}{\tau_e}$$ (8.94)

$$j\omega p_W = \frac{p_B(V_{SCH}/V_W)}{\tau_s} - \frac{p_W}{\tau_n} - \frac{p_W}{\tau_e} - \frac{v_g g_0 S_0}{1+\varepsilon S_0}p_W - \frac{v_g G_0}{(1+\varepsilon S_0)^2}s$$ (8.95)

$$j\omega s = \frac{\Gamma v_g g_0 S_0}{1+\varepsilon S_0}p_W + \frac{s}{\tau_{ph}(1+\varepsilon S_0)} - \frac{s}{\tau_{ph}}$$ (8.96)

The steady-state solution to the photon density gives the following relation:

$$\frac{\Gamma v_g g_0}{1+\varepsilon S_0} = \frac{1}{\tau_{ph}}$$

which is the basic gain–loss relationship in the laser cavity. Eliminating p_B and p_W from the three small-signal rate equations (Equations 8.94 through 8.96), the following relation between the modulating current and the optical output is obtained:

$$\frac{s(\omega)}{i} = \left(\frac{1}{1+j\omega\tau_s}\right)$$

$$\frac{\dfrac{\Gamma v_g g_0 S_0}{eV_W}}{\left[j\omega\left\{1+\left(\dfrac{\tau_s}{1+j\omega\tau_s}\right)\dfrac{1}{\tau_e}\right\}+\dfrac{v_g g_0 S_0}{1+\varepsilon S_0}\dfrac{1}{\tau_n}\right]\left[j\omega(1+\varepsilon S_0)+\dfrac{\varepsilon S_0}{\tau_{ph}}\right]+\dfrac{v_g g_0 S_0}{\tau_{ph}(1+\varepsilon S_0)}}$$

$$(8.97)$$

If we write $M(\omega)=s(\omega)/i$, the modulation response is given by $|M(\omega)|$. We neglect the frequency dependence of τ_s and obtain the following expression:

$$M(\omega) = \left(\frac{1}{1+j\omega\tau_s}\right)\frac{A}{\omega_r^2 - \omega^2 + j\omega\gamma} \qquad (8.98)$$

where:

$$A = \frac{\Gamma(v_g g_0/\chi)S_0}{eV_W(1+\varepsilon S_0)} \qquad (8.99a)$$

$$\omega_r^2 = \frac{(v_g g_0/\chi)S_0}{\tau_{ph}(1+\varepsilon S_0)}\left(1+\frac{\varepsilon}{v_g g_0 \tau_n}\right) \qquad (8.99b)$$

$$\gamma = \frac{(v_g g_0/\chi)S_0}{(1+\varepsilon S_0)}+\frac{\varepsilon S_0/\tau_{ph}}{(1+\varepsilon S_0)}+\frac{1}{\chi\tau_s} \qquad (8.99c)$$

The quantity $\chi=1+\tau_s/\tau_e$ is called the transport factor, ω_r is the resonant frequency, and γ is the damping rate. The K-factor is defined as $\gamma = Kf_r^2 + \gamma_0$. Therefore, one obtains

$$K = 4\pi^2\left(\tau_{ph}+\frac{\varepsilon}{(v_g g_0/\chi)}\right) \qquad (8.100a)$$

$$\gamma_0 = \frac{1}{\chi\tau_n} \qquad (8.100b)$$

It is evident from Equatinos 8.100a and b that the K-factor may be seriously affected by the transport in the SCH layer. Several other conclusions may be drawn from the analytical expressions:

1. The low-frequency roll-off of the modulation response is only due to transport time in the SCH. Its effect cannot be distinguished from the effect due to parasitics.
2. The effective differential gain is reduced by the transport factor, that is, the differential gain is g_0/χ. The resonant frequency is also reduced by the transport factor.
3. The effective bimolecular recombination lifetime is increased to $\chi\tau_n$.
4. The gain compression factor remains unaltered.

The simple analytical expressions derived above are modified when the frequency dependence of τ_s is considered. The reader is referred to the work by Nagarajan et al. for the modified expressions [8,9]. The use of three rate equations has been examined in [10].

We are in a position now to study the effect of transport on the modulation response. The effect is illustrated by considering two GaAs–AlGaAs single QW lasers for which the SCH widths are (i) 75 nm (referred to as short SCH width), and (ii) 300 nm (referred to as long SCH width). The thickness of both the GaAs QW and the AlGaAs barrier is 8 nm. The SCH layers are $Al_{0.1}Ga_{0.9}As$. The values of diffusion constant, mobility, and so on of electrons and holes are obtained by using the expressions given by Nagarajan et al. [8,9]. The confinement factors for the short and long samples are, respectively, 2.9% and 1.7%.

The families of the calculated modulation response for the short SCH width are shown in Figure 8.12 for different values of output power. In Figure 8.12a, the response curves are plotted without considering the transport effect ($\chi = 1$), while the corresponding curves taking into account the effect of the transport factor are displayed in Figure 8.12b. It is easy to conclude from these two sets that the effect of the transport factor is not appreciable in this case. There is a slight decrease of the peak response in Figure 8.12b.

In contrast, the effect of transport looks more significant for the long SCH width. The response curves are shown in Figure 8.13a without the transport factor, and in Figure 8.13b including the transport effect. There is not only a shift of the resonance frequency but also a significant change in the response curves. The peaks now show negative values. The 3 dB bandwidths are significantly reduced also. As the SCH width is increased, the modulation response shows a detrimental effect of low-frequency roll-off due to carrier transport.

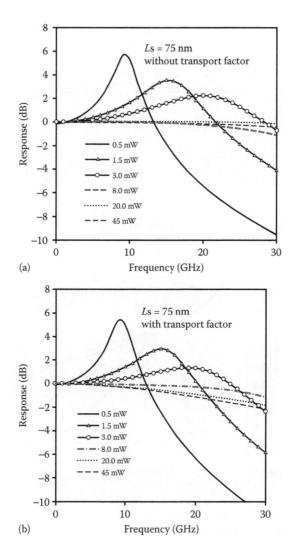

FIGURE 8.12
Modulation response of a QW laser with short SCH width: (a) without transport factor; (b) including the effect of transport factor.

Example 8.8: For typical SCH-SQW lasers at room temperature, $\tau_e \sim 100-500$ ps, $\tau_s \sim 20-100$ ps, and the transport factor is ~1.2.

8.8.3 Effect of State Filling on Modulation Response

The threshold current, modulation bandwidth, and other parameters depend on the gain coefficient or differential gain. The gain, in turn,

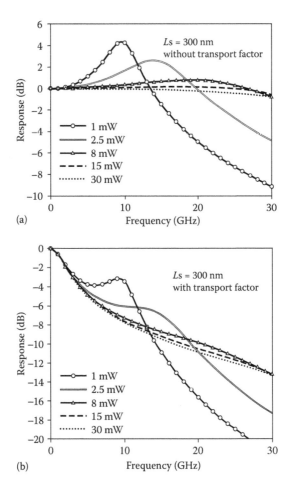

FIGURE 8.13
Modulation response of a QW laser with long SCH width: (a) without transport factor; (b) including the effect of transport factor.

depends on the DOS and number of subbands populated. The unexpected lower value of modulation bandwidth in comparison with DH lasers may partly be explained by carrier transport effects, as discussed in Section 8.8.2. The population in different bands, or, alternatively, the state-filling effect, also plays a role in limiting the bandwidth. Here we shall follow a simple treatment given by Zhao and Yariv (please see Reading List). In the simple two-level model, E_0 is the energy of the ground electron subband ($n = 1$), and E_1 is the upper energy states in the OCL, above the barrier of the QW. D is the effective number of these subbands. To simplify the analysis, $m_e = m_{hh}$, and the optical transition occurs between the ground subbands.

The maximum optical gain is expressed as

$$g_{max} = A_0 \rho_r \left(f_e - f_h \right) \tag{8.101}$$

where A_0 is the material dependent constant, $\rho_r = (1/2)(m_e/\pi\hbar^2)$ is the reduced DOS, and the Fermi function is given by

$$f_e = 1/1 + \exp(-F/k_B T) \tag{8.102}$$

F is the quasi-Fermi level from the first quantized level $E_0 = 0$. The 2-D carrier density is expressed by considering all the states as follows:

$$n = \rho_e k_B T \left\{ \ln\left[1 + \exp(F/k_B T)\right] + D\ln\left[1 + \exp(F - E_1/k_B T)\right] \right\} \tag{8.103}$$

where D is the degeneracy of the level with energy E_1. The expression for the differential gain may be obtained from Equations 8.101 through 8.103 and is given by

$$\frac{dg}{dn} = \frac{(A_0/k_B T)/\left[1 + \exp(F/k_B T)\right]}{1 + D\left[1 + \exp(F/k_B T)\right]/\left[\exp(E_1/k_B T) + \exp(F/k_B T)\right]} \tag{8.104}$$

For a given value of g_{max} (or F), Equations 8.103 and 8.104 indicate that the presence of upper subbands increases the injected carrier density and reduces the differential gain. The greater the difference $E_1 - E_0$, the larger is the value of differential gain or the smaller is the injected carrier density.

8.8.3.1 Hot Carrier Effect

Hot carrier effects have an impact on the high-speed modulation bandwidth of QW lasers [11–13]. As is seen from the structures of QW lasers, the carriers are injected from high energy levels, and, after losing their energy by phonon emission, the carriers come to the quantized energy levels in the active QW region. If the carriers cannot lose their energy fast enough in comparison with the stimulated emission lifetime and injected carrier modulation, a nonequilibrium carrier distribution results, which lead to an effective carrier temperature greater than the lattice temperature. This hot carrier effect has been introduced in Section 8.5. The hot carriers degrade the differential gain and introduce a large damping effect on the modulation response.

8.9 Strained QW Lasers

8.9.1 Gain Spectra

The gain or absorption in a strained QW may be calculated using Equations 8.25 or 8.29 by incorporating the strain- and quantum-confined-induced changes in the bandgap as well as strain-induced change in the in-plane effective mass. Neglecting coupling between HH and LH bands, the in-plane masses are $(\gamma_1 \pm 2\gamma_2)/m_0$ for both strained and unstrained QWs.

In the following, we shall discuss the special features of strained QW lasers (please see Reading List).

8.9.2 Bandgap Tunability

The bandgap of a semiconductor can be altered by tensile and compressive strain. It is usual practice to grow a suitably thick buffer layer on a substrate, so that the dislocations are confined near the interface of the substrate and the buffer. The lattice constant of the buffer is intermediate between those of the barrier B and the active layer A. The layers A and B are therefore in tension or compression. In this way, the bandgap can be tuned over a wide range.

8.9.3 Reduced DOS

In our discussion, we shall consider the QW geometry as shown in Figure 8.14, in which the layer is grown along the [001] axis, the z-direction, and the strain occurs along the same direction. The TE polarized wave has an electric field acting along the y-direction, while the field for TM acts along the z-direction.

Figure 3.15 shows the band diagrams of (a) a compressed layer, (b) an unstrained layer, and (c) a tensile-strained layer. As noted already, the strain removes the degeneracy of HH and LH bands. Under biaxial compression, the hydrostatic component of stress increases the mean bandgap, while the

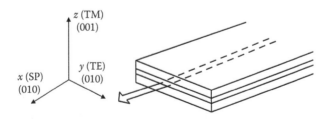

FIGURE 8.14
Strained QW: electric field for TE and TM polarized waves is along y- and z-directions, respectively.

axial component splits the degeneracy and introduces an anisotropic band structure. It may be proved (see Basu, Chuang in Reading List) that the highest band shown occurs for the highest m_z, while the in-plane mass is lower. According to convention, we shall call this the hh band. The mean bandgap decreases under biaxial compression, and the valence band splitting is reversed. The highest band is now light along z, but comparatively heavy in the plane. The two bands cross each other at some value of $k_\parallel$.

In DH lasers, the injected holes occupy both the hh and lh bands. Although in a QW laser the degeneracy is lifted, still the injected holes populate a number of hh and lh subbands. This leads to a higher value of transparency carrier density, as explained in Section 8.3. It is also clear that the carrier density is smaller in the lh band, in which the recombination lifetime is also shorter. Thus, the two bands make equal contributions to the threshold current density, while only one band is responsible for contributing to stimulated emission. In an unstrained DH laser, the isotropic nature of both polarization and gain increases the carrier and current densities needed for achieving threshold. However, when only one polarization component contributes significantly to spontaneous emission and gain, the threshold carrier and current are lower.

Thus, in an ideal semiconductor laser, one needs (1) only one valence band, (2) a low in-plane mass for lower threshold current density, and (3) anisotropic polarization and gain above transparency, that is, spontaneous emission and gain should be suppressed in all directions, except along the axis of the laser beam. All these requirements are satisfied by a strained QW laser.

It has been pointed out that axial compressive strain lifts the degeneracy of hh and lh bands, and that the highest valence subband has a lower in-plane effective mass. As discussed in Section 8.3.3, the lower in-plane effective mass reduces the transparency carrier density (see also plots of f_e, f_h in Figure 8.5 and related discussion). The loss mechanisms depend on the number of injected carriers, and as a result these losses are also reduced.

The situation is more complicated in tensile-strained QWs. Here, the lh subband becomes the highest-lying state. Quantum confinement, however, shifts the subbands in opposite directions, so that for some value of strain and well width, the two subbands belonging to the lh and hh bands may have identical energies. The effective mass in this situation is large. For higher values of tensile strain, however, the highest lh subband shifts significantly above the hh subbands, and again the DOS for lh states is reduced from the bulk value.

In an unstrained bulk material, there is equal contribution from $|X\rangle$-, $|Y\rangle$-, and $|Z\rangle$-like states, so that spontaneous emission comes from two bands and has equal components of polarization along the three principal axes. Hence with carrier injection, an equal contribution is made to the TM gain (polarized along z), TE (polarized along y), and spontaneous emission (along x).

Therefore, only one third of all the holes at the correct energy are in the right polarization state to contribute to the laser mode.

In the presence of compressive strain, the hh states move up, and as they have no $|Z\rangle$ character, but equal $|X\rangle$ and $|Y\rangle$ characters at $k\| = 0$, TE gain will be enhanced and TM gain suppressed. One in two injected carriers now contributes to the dominant polarization. The threshold current decreases, and the differential gain increases. With biaxial tension, the lh states move upward. As the states have $(2/3)|Z\rangle$ character, two out of three injected carriers contribute to the TM gain. There is some coupling between lh and hh states, but when the strain is large, the subbands separate out again.

8.9.4 Reduced Intervalence Band Absorption

As discussed in Chapter 3, a photon emitted due to a stimulated process may be reabsorbed by an electron in the SO band to land in a HH band. This process is known as IVBA, and constitutes an important loss process, particularly for lasers working in the long-wavelength region. The IVBA absorption depends on the energy of the electron, which is expressed as

$$E_{\text{IVBA}} = \frac{m_{\text{so}}\left(E_{\text{g}} - E_{\text{so}}\right)}{\left(m_{\text{hh}} - m_{\text{so}}\right)} \tag{8.105}$$

where the symbols have the usual meaning. As mentioned, m_{hh} is reduced along the plane under strain, and the energy E_{IVBA} becomes larger. The increased energy reduces the loss term, as the absorption coefficient depends on energy as $\exp(-E/k_{\text{B}}T)$. The threshold current is thereby reduced due to lower IVBA. The increase in E_{IVBA} in the strained layer is illustrated in Figure 8.15.

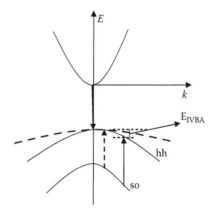

FIGURE 8.15
Reduction of IVBA with reduced in-plane mass.

8.9.5 Auger Recombination Rate

As discussed in Chapter 6, the Auger recombination rate depends on the densities of three types of carriers involved, and the current density J_{aug} in a device with undoped active region is written as

$$J_{aug} = C_{AR}(T)n_{th}^3 \qquad (8.106)$$

$C_{AR}(T)$ is the temperature-dependent Auger coefficient, and n_{th} is the threshold carrier density. As usual, it is assumed that the carrier density in the active layer is pinned at threshold. The current density needed to make up the loss due to Auger recombination changes in the presence of strain in the following two ways: (1) the reduced hole mass reduces n_{th} and thereby J_{aug}, and (2) the magnitude of $C_{AR}(T)$ changes due to strain. For a simple parabolic band using Boltzmann statistics, one may write

$$C_{AR}(T) = C_0 \exp\left(-E_a / k_B T\right) \qquad (8.107)$$

$$E_a(\text{CHCC}) = m_e E_g / \left(m_e + m_{hh}\right) \qquad (8.108)$$

$$E_a(\text{CHSH}) = m_{so}\left(E_g - E_{so}\right) / \left(2m_{hh} + m_e - m_{so}\right) \qquad (8.109)$$

These simple expressions suggest that a lower value of m_{hh} increases E_a in both cases and reduces the coefficient $C_{AR}(T)$.

In the expression for phonon-assisted Auger recombination rate, the appropriate coefficient is $C_p(T)$ instead of $C_0(T)$, where

$$C_p(T) = \frac{B}{e^x - 1}\left[\frac{1}{\left(E_1 + \hbar\omega_{LO}\right)^2} + \frac{e^x}{\left(E_1 - \hbar\omega_{LO}\right)^2}\right]; \quad x = \hbar\omega_{LO} / k_B T \qquad (8.110)$$

B is proportional to the transition matrix element, and E_1, the kinetic energy associated with the forbidden intermediate state, is expressed as

$$E_1 = \left(m_{so} / m_{hh}\right)\left(E_g - E_{so}\right), \quad \text{CHHS-ph} \qquad (8.111a)$$

$$E_1 = \left(m_e / m_{hh}\right)E_g, \quad \text{CHCC-ph} \qquad (8.111b)$$

If $\hbar\omega_{LO} \ll E_1$, then $C_p(T)$ is proportional to m_{hh}^2 and is reduced due to strain. However, this reduction is not as dramatic as in the direct process. The phonon-assisted Auger process may not be a significant loss mechanism in long-wavelength lasers.

Example 8.9: The CHHS process is the dominant Auger recombination process in an InGaAsP strained QW laser. Strain-induced reduction for hole mass is from $0.7m_0$ to $0.15m_0$. Using $m_e = 0.037m_0$, $m_{so} = 0.12m_0$, $E_g = 0.79$ eV, and $E_{so} = 0.39$ eV, E_1 increases from 0.0364 to 0.2212 eV. The coefficient $C_{AR}(T)$ then decreases from 2.5×10^{-1} to 2.02×10^{-4}: a decrease of three orders of magnitude [14].

8.9.6 Temperature Sensitivity

The temperature variation of the threshold current density in a laser is generally expressed as

$$J_{th}(T) = J_0 \exp(T / T_0) \tag{8.112}$$

where:
T_0 is the characteristic temperature
J_0 is an empirical constant

One may write from this

$$T_0 = \left[\frac{d}{dT} \ln J_{th}(T) \right]^{-1} = (T_2 - T_1) \left[\ln \frac{J_{th}(T_2)}{J_{th}(T_1)} \right]^{-1} \tag{8.113}$$

The symbols above are easy to define. For a better design of laser transmitter circuits, it is better to have a higher T_0 value. A typical value of T_0 is 75 K for long-wavelength lasers, while a somewhat higher value, ~300 K, has been reported for GaAs–GaAlAs systems. The temperature dependence of threshold current is governed by the Auger process, and therefore a reduction of its rate is desirable for increasing T_0.

The threshold current density is expressed as

$$J_{th} = (A_{nr} + SA)n_{th} + B(T)n_{th}^2 + C_{AR}(T)n_{th}^3 \tag{8.114}$$

The first term on the right-hand expresses trap- and surface-mediated nonradiative recombination processes, in which S denotes the surface recombination coefficient, A the surface area, and n_{th} the carrier density at threshold. $B(T)$ represents the temperature-dependent radiative coefficient. The last term is the nonradiative Auger recombination component, in which C_{AR} denotes the Auger coefficient. For an ideal QW structure, $B(T)$ decreases as n_{th} increases with T. We take the temperature dependence of n_{th} as $n_{th}(T) \propto T^{1+x}$ ($x > 0$), where x is the nonideality factor due to the occupancy of higher subbands, carrier spillover into the barrier or optical confining layer, and IVBA. The characteristic temperature may be written as (see section 16.4.5 of Basu in Reading List)

$$T_0 = \frac{T(1 + J_{AR}/J_R)}{1 + 2x + (3 + 3x + E_1/k_B T)(J_{AR}/J_R)} \tag{8.115}$$

In the ideal case, $J_{AR}/J_R = 0$, $E_1 = 0$, and $x = 0$. Thus, $T_0 = 300$ K, a value close to which has been observed in the GaAs/GaAlAs QW system. In the presence of Auger recombination, the value is considerably less.

Example 8.10: Let $J_{AR}/J = 3$, $x = 0$, and $E_1 = 0$. From Equation 8.115, $T_0 = 120$ K. On the other hand, when $J_{AR} \gg J_R$, the maximum value of T_0 is 100 K, even if $x = 0$ and $E_1 = 0$. Usually, both E_1 and x are nonzero, and therefore T_0 has a still lower value.

The reported value for InGaAsP DH lasers is quite low: ~40 K. The value increases somewhat for both compressive and tensile-strained layers, which may be attributed to the reduction of Auger recombination. However, T_0 values exceeding 100 K are not reported.

8.9.7 Linewidth, Chirp, and Modulation Bandwidth

A change in the carrier density n in the active region of a laser induces a change in the RI and hence a change in the wavelength of a longitudinal mode. Therefore, fluctuations in the photon density and carrier density lead to an increase in the linewidth of a laser. The modulation in the value of n leads to chirp also. These undesirable features are quantified in terms of the linewidth enhancement factor α, defined as

$$\alpha = -\frac{4\pi}{\lambda} \frac{d\eta/dn}{dg/dn} \tag{8.116}$$

Since the differential gain is larger in strained QW lasers, the α-factor is reduced, and the reduction of linewidth for a given power may be expressed as

$$\Delta v \approx \frac{1 + \alpha^2}{P} \tag{8.117}$$

The chirp width, which is proportional to $(1+\alpha^2)^{1/2}$, is also reduced.

The experimental data report a reduction of the linewidth enhancement factor by a factor of two when going from unstrained to strained systems.

The resonant frequency of a laser may be expressed as

$$f_r \propto \left(P \, dg / dn\right)^{1/2} \tag{8.118}$$

in particular in InGaAs QW lasers [15].

The enhanced differential gain in strained QWs therefore predicts a given relaxation frequency for a lower output power. This has been observed in different systems, in particular in InGaAs QW lasers. The 3 dB modulation bandwidth depends not only on f_r but also on damping rate, arising out of spectral hole-burning, carrier-heating, and carrier-transport effects. There is little change in K-value, and as such there is very little change in bandwidth of strained systems compared with unstrained systems.

The introduction of p-doping in addition to strain proves to be a route for increasing differential gain and decreasing K-factor. This combination of p-doping with strain seems to be the most important method for enhancing the modulation response of QW lasers (please see Reading List).

8.10 Type II Quantum Well Lasers

It has been mentioned in Chapter 3 that in Type II heterostructures and MQWs, electrons and holes are confined in different layers of the heterojunction and MQWs. The overlap between the electron and hole envelope functions is zero for the ideal case of infinite barrier heights. For finite barriers, however, the envelope functions penetrate through the barrier, and therefore the overlap function is finite but very small. As may be seen from Equation 8.25, the absorption or gain coefficient depends on the squared overlap function, and the value of the gain coefficient is quite low. Consequently, Type II QWs or superlattices do not appear to be attractive for laser applications.

There are some ranges in the EM spectrum in which Type II QWs may find some applications. In this section, we shall mention only one specific area. The elaborate theory will not be presented; we shall point out the area of application, the structure, and some design considerations [16–19].

There is a demand for lasers in the mid-infrared for use in medical sensing and surgery, biosensing, and contactless highly sensitive gas sensing. The Type I QWs normally used are based on quaternary-alloy GaInAsSb or interband cascade lasers, both grown on a GaSb material platform. However, the high cost of GaSb substrates and the complicated growth technology stand in the way of high-volume production.

The problem may be overcome by using InP as the substrate, for which the growth technology is already quite mature. The usual active material for long-wavelength devices is InGaAs. To extend the wavelength of operation to higher ranges, one needs to decrease the bandgap by increasing the In content in the ternary alloy. However, higher In content produces lattice mismatch with InP, and the maximum reported wavelength for this system is 2.33 µm, determined by the critical thickness for strain relaxation.

An effective way to reduce the bandgap and to work at higher wavelengths is to have active regions in the Type II configuration. In the case of InP substrate, the materials are InGaAs and GaAsSb. Example 3.6 has illustrated the Type II nature of the band alignment and bandgap.

The value quoted in the example for the energy difference between the CBs of the two materials is 0.36 eV. The value can be altered by changing the composition.

The following design rules have been suggested for optimization of the structure for useful laser operation.

1. Use of strained layers: The active regions are made of InGaAs and GaAsSb. Both these layers should be thin enough to allow quantum confinement so that the DOS becomes high. However, for unstrained layers, the DOS function for holes confined in the GaAsSb layer is higher than the electronic DOS in InGaAs. This makes the reduced DOS quite high, leading to larger threshold current density. A compressive strain in GaAsSb will reduce the hole effective mass and hence the DOS.

2. Increase of overlap function: Since electrons and holes are localized in different layers, the overlap between the two envelope functions is low. If very thin QWs are used, the lighter electrons can penetrate the barrier, and their envelope function exists at the hole-confining layer. This increases the overlap function, but increase in quantization energy for thinner wells reduces the wavelength of operation. A trade-off is therefore needed between the wave-function overlap and the emission wavelength. The problem is solved by applying compressive strain to the InGaAs layer also. This reduces the transition energy and at the same time keeps the overlap high.

3. Active region: The above design rules may be followed by using a superlattice consisting of InGaAs and GaAsSb layers with thickness in the range from 1.5 to 4.5 nm. Due to their higher effective mass, the holes are well confined in the GaAsSb layers, but the electrons penetrate into the hole-confining layers, and increased coupling between electrons leads to the formation of a miniband for the electrons. Though the overlap function is increased, the electron

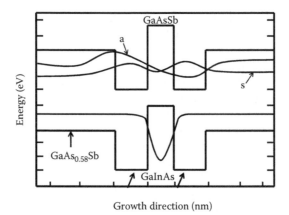

Growth direction (nm)

FIGURE 8.16
Band diagram for a W-shaped Type II QW structure. The wave functions for electrons (symmetric (s) and asymmetric (a)) and holes (symmetric) are shown.

miniband gives rise to spreading in the emission spectra, the width of which may be as high as 30 meV. The broad spontaneous emission spectra would increase the threshold current density of the laser.

To circumvent this problem, a W-shaped active region seems very useful. Figure 8.16 shows the band diagram. It consists of one layer of hole-confining GaAsSb layer surrounded by two compressively strained InGaAs layers, acting as QWs for electrons. The InGaAs layers are then surrounded by electron-confinement layers, such as tensile-strained GaAsSb, AlGaInAs, or AlGaAsSb, acting as strain-compensation layers at the same time [18].

The electrons in the coupled QWs possess two ground states: symmetric and asymmetric, as shown in Figure 8.16. The lowest hole state is a symmetric HH state. The overlap function is nonvanishing between the symmetric electron and hole functions, for symmetry reasons.

8.11 Tunnel-Injection QW Laser

The desirable features of a QW laser are a low threshold current density and a high modulation bandwidth. It is difficult to meet both these requirements simultaneously. In order to increase the optical confinement factor, a higher-bandgap semiconductor is needed as the cladding layer. Electrons and holes are injected into the QW from the cladding layers. The 3-D electrons injected into the OCL regions take some time (~1–2 ps) to thermalize, move by drift and diffusion, and then enter into the lower-lying states of the QW. The transport time is believed to shorten the modulation bandwidth. However, the measured value of diffusion time (<0.75 ps) does not totally account for the bandwidth, and it is felt that some other mechanism must play a role. A measurement of capture time of 3-D electrons in the barriers into the 2-D systems in the QW registers a high value (~3–15 ps). If the stimulated recombination time of EHPs is comparable to this time, the carrier distribution in the QW remains essentially "hot." As a result, gain compression, enhanced Auger rate, and other related hot carrier effects seriously limit the speed of the device.

An effective way to bypass the hot carrier injection into the QWs is to inject the carriers by resonant tunneling [20,21]. At low recombination rates, the carrier distribution is described by a quasi-Fermi distribution, but as the recombination rates get higher, the carrier distribution becomes monoenergetic. The distributions in a usual SCH-QW and in a QW laser, where carriers are injected by resonant tunneling, are shown in Figure 8.17.

The above idea has been tested experimentally by using lasers made with a variety of material systems, InGaAs–GaAs–AlGaAs, InGaAs–GaAs–InGaAsP–InGaP, and InGaAs–InGaAsP–InP, using both single-QW and

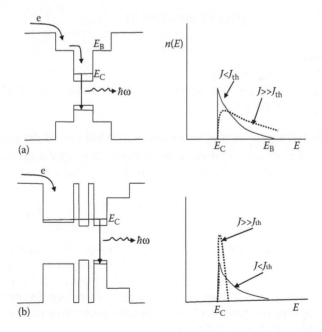

FIGURE 8.17
Structure and carrier distribution in (a) single SCH-QW laser and (b) tunnel-injection QW laser. Quasi-Fermi distribution in SCH-QW: (b) carrier distribution in tunnel injection case. (Sun, H. C., et al., Properties of tunneling injection quantum well laser: Recipe for a "cold" device with a larger modulation bandwidth, *IEEE Photonic. Tech. Lett.* © 1993 IEEE.)

MQW structures [21]. The structures show negligible gain compression, superior high-temperature performance, lower Auger recombination and wavelength chirp, and better modulation characteristics.

PROBLEMS

8.1. Calculate the range of photon energy for amplification in a GaAs QW at 0 K. Take $m_e = 0.067m_0$, $m_{hh\|} = 0.112m_0$, $m_{hhz} = 0.042m_0$, $d = 5$ nm, and $n = p = 10^{12}$ cm^{-2}.

8.2. Prove the equivalence of Equations 8.24 and 8.29.

8.3. Smaller in-plane mass leads to a higher quasi-Fermi level. In Example 8.2 this is proved at 0 K. Prove this statement for any finite temperature.

8.4. The 2-D electron density is 5×10^{11} cm^{-2}. Calculate the change in quasi-Fermi level if the in-plane mass is decreased from $0.1m_0$ to $0.05m_0$.

8.5. By using the approximate expressions for the Fermi functions, show that the gain in QW may be expressed as

$$g(\hbar\omega) = K(\Delta F - \hbar\omega)H(\hbar\omega - E_{mn})$$

where:
H is step function
E_{mn} is the gap corresponding to mth conduction and nth valence subbands

Sketch the gain curves for different injected carrier densities.

8.6. Calculate the absorption coefficient in a GaAs QW with $L = 10$ nm and infinite barrier height at an energy E_{g11}. Take $C_{11} = 1$, $A_{11} = 1.5$, $n = 3.6$, $f_h = 1$, and $f_e = 0$.

8.7. Prove the relation $R_{sp}(y) \propto \exp(-y/k_B T^*)H(y)$, given in Equation 8.50.

8.8. Obtain the expression for L_{opt} from Equation 8.79.

8.9. The expression for the optimum number of QWs that minimizes the threshold current is given in [22]. A tangent from the origin touches the g_{max}–J curve at g_0 and J_0. If all the wells in the MQW have same confinement factor Γ_w and current density, prove that the optimum number of wells to minimize current is given by $n_{opt} = (1/g_0\Gamma w)[\alpha + (1/L)\ln(1/R)]$.

8.10. Calculate the optimum number of QWs by using the expression given in Problem 8.9 and taking the following values for a 7.5 nm wide well: $g_0 = 1200$ cm^{-1}, $\Gamma_w = 0.022$, $\alpha = 10$ cm^{-1}, $L = 100$ μm, and $R = 0.3$.

8.11. The expression for gain is $g/g_0 = \ln(J/J_0) + 1$. Assuming that all QWs are identical, obtain the relation [23]

$$I_{th} = \eta_i J_0 n_w W L \exp\left\{\frac{\alpha_{fc}}{g_0 - 1} + \frac{\left[\alpha_{wg} + (1/2L)\ln(1/R_1 R_2)\right]}{n_w g_0 \Gamma_x \Gamma_y}\right\}$$

where:

W is the stripe width
L is the cavity length
α_{fc} is the loss in the active region
α_{wg} is the loss in the lateral waveguide (y-direction)
n_w is the number of wells
Γ_x and Γ_y are, respectively, the confinement factors along the x (transverse) and y (lateral) directions

8.12. An analytical formula for $\Gamma_y^2 = D^2\left[2(n_{eff2}/n_{eff1})^4 + D^2\right]$, where $D^2 = \left(4\pi^2 W^2/\lambda^2\right)\left(n_{eff2}^2 - n_{eff1}^2\right)$, has been given by [23].
Calculate and plot the values of threshold current as a function of width W (0.1 to 2 μm) by using the parameters $\lambda = 0.84$ μm, $n_{eff} = 3.3042$, $\Delta n_{eff} = 0.05$, $\Gamma_x = 0.033$, $R_1 = R_2 = 0.3$, $L = 500$ μm, $\alpha_{fc} = 100$ cm^{-1}, $\alpha_{wg} = 10$ cm^{-1}, and $\eta_i = 1$. A QW thickness of 7.5 nm is assumed, which implies that $g_0 = 1200$ cm^{-1} and $J_0 = 180$ A cm^{-2} [22].

8.13. The plot obtained in Problem 8.12 looks like the plots in Figure 8.10. By using the expressions for threshold current and Γ_y given in Problem 8.12, show that the width for which the minimum threshold current is obtained may be expressed as

$$W_{min}^2 = \left[\left(n_{eff2}/n_{eff1}\right)^4 \lambda^2/4\pi^2 n_w\left(n_{eff2}^2 - n_{eff1}^2\right)\right]$$
$$\times\left\{\rho + n_w + \left[\left(\rho + n_w\right)^2 + 4n_w\rho\right]^{1/2}\right\}$$

where $\rho = 2\alpha_{wg}/g_0\Gamma_x$.

8.14. Apply energy- and momentum-conservation conditions to arrive at IVBA expression.

8.15. Derive Equation 8.104 for dg/dn considering two subbands of energy E_0 and E_1.

8.16. The gain in a laser is expressed as $g = g_0(n - n_{tr})$, where n_{tr} is the transparency carrier density. Under tensile strain, $g_0(TM) > g_0(TE)$. Therefore, threshold occurs first for TM mode. However, for mirror reflectivity, $R(TE) > R(TM)$, holds, so that TE mode may be preferred to TM mode. Show, by equating the threshold carrier densities for TE and TM modes, that there exists a critical length of the cavity, L_c, given by

$$L_c = \frac{g_0(TE)\ln[1/R(TM)] - g_0(TM)\ln[1/R(TE)]}{g_0(TE)g_0(TM)\Gamma[n_{th}(TE) - n_{th}(TM)] + \alpha[g_0(TM) - g_0(TE)]}$$

For lengths below L_c, TE polarization mode will be supported. The two mirrors have the same reflectivity and the symbols have the usual meanings [24].

8.17. Estimate the factor by which the IVBA changes if the HH mass is reduced from $0.5m_0$ to $0.16m_0$. Use $m_{so} = 0.12m_0$, $E_{so} = 320$ meV, and $E_g = 0.8$ eV.

Reading List

Adams, A. R., E. P. OReilly and M. Silver, Chapter 2: Strained layer quantum well lasers. In *Semiconductor Laser I*, ed. Kapon, E., Academic, San Diego, pp. 123–176, 1999.

Agrawal, G. P. and N. K. Dutta, *Semiconductor Lasers*, 2nd edition. Van Nostrand Reinhold, New York, 1993.

Arakawa, Y. and A. Yariv, Theory of gain, modulation, response, and spectral line-width in AlGaAs quantum-well lasers, *IEEE J. Quantum Electron.* 21, 1666–1674, 1985.

Basu, P. K., *Theory of Optical Processes in Semiconductors*. Oxford University Press, Oxford, 1997.

Chuang, S. L., *Physics of Photonic Devices*. Wiley, New York, 2009.

Coldren, L. A., S. W. Corzine, and M. L. Mašanović, *Diode Lasers and Photonic Integrated Circuits*, 2nd edition. Wiley, New York, 2012.

Harrison, P., *Quantum Wells, Wires, and Dots, Theoretical and Computational Physics*. Wiley, New York, 2000.

Nagarajan, R. and J. E. Bowers, Chapter 3: High speed lasers. In *Semiconductor Laser I*, ed. Kapon, E. Academic, San Diego, pp. 177–290, 1999.

Numai, T., *Fundamentals of Semiconductor Lasers*. Springer, New York, 2004.

Singh, J., *Electronic and Optoelectronic Properties of Semiconductor Structures*. Cambridge University Press, Cambridge, 2003.

Zhao, B. and A. Yariv, Chapter 1: Quantum well semiconductor laser. In *Semiconductor Laser I*, ed. Kapon, E. Academic, San Diego, pp. 1–121, 1999.

Zory, P. S. (ed.), *Quantum Well Lasers*. Academic, San Diego, 1993.

References

1. Asryan, L. V., N. A. Gun'ko, A. S. Polkovnikov, G. G. Zegrya, R. A. Suris, P. K. Lau, and T. Makino, Threshold characteristics of InGaAsP/InP multiple quantum well lasers, *Semicond. Sci. Technol.* 15, 1131–1140, 2000.

2. Vahala, K. J. and C. E. Zah, Effect of doping on the optical gain and spontaneous noise enhancement factor in quantum well amplifiers by simple analytical expressions, *Appl. Phys. Lett.* 52, 1945–1947, 1988.

3. Makino, T., Analytical formulas for the optical gain of quantum wells, *IEEE J. Quantum Electron.* 32, 493–501, 1996.

4. Basu, P. K. and S. Kundu, Energy loss of two dimensional electron gas in GaAs–AlGaAs MQWs by screened electron-polar optic phonon interactions, *Appl. Phys. Lett.* 47, 264–266, 1985.

5. Asada, M., A. Kameyama, and Y. Suematsu, Gain and intervalence band absorption in quantum-well lasers, *IEEE J. Quantum Electron.* QE-20, 745–753, 1984.

6. Rozenweig, M., M. Mohrle, H. Duser, and H. Venghaus, Threshold current analysis of InGaAs–InGaAsP multiquantum well separate-confinement lasers, *IEEE J. Quantum Electron.* 27, 1804–1811, 1991.

7. Detemple, T. A. and C. M. Herzinger, On the semiconductor laser logarithmic gain current density relation, *IEEE J. Quantum Electron.* 29, 1246–1252, 1993.

8. Nagarajan, R., M. Ishikawa, T. Fukushima, R. S. Geels, and J. E. Bowers, High speed quantum well lasers and carrier transport effects, *IEEE J. Quantum Electron.* 28, 1990–2008, 1992.

9. Nagarajan, R. and J. E. Bowers, High speed lasers. In *Semiconductor Laser I*, ed. Kapon, E. Academic, San Diego, pp. 177–290, 1999.

10. McDonald, D. and R. F. O'Dowd, Comparison of two- and three-level rate equations in the modeling of quantum-well lasers, *IEEE J. Quantum Electron.* 31, 1927–1934, 1995.

11. Zhao, B., T. R. Chen, and A. Yariv, On the high speed modulation bandwidth of quantum well lasers, *Appl. Phys. Lett.* 60, 313–315, 1992.

12. Zhao, B., T. R. Chen, Y. Yamada, Y. H. Zhuang, N. Kuze, and A. Yariv, Evidence for state filling effect on high speed modulation dynamics of quantum well lasers, *Appl. Phys. Lett.* 61, 1907–1909, 1992.

13. Zhao, B., T. R. Chen, and A. Yariv, The extra differential gain enhancement in multiple quantum-well lasers, *IEEE Photonic. Tech. Lett.* 4, 124–126, 1992.

14. Thijs, P. J. A., L. F. Tiemeijer, J. J. M. Binsma, and T. Van Dongen, Progress in long-wavelength strained-layer InGaAs(P) quantum well semiconductor lasers and amplifiers, *IEEE J. Quantum Electron.* 30, 477–499, 1994.

15. Lau, K., S. Xin, and W. I. Wang, Enhancement of modulation bandwidth in InGaAs strained-layer single quantum well lasers, *Appl. Phys. Lett.* 55, 1173–1175, 1989.

16. Hu, J., X. G. Xu, J. A. H. Stotz, S. P. Watkins, A. E. Curzon, M. L. W. Thewalt, N. Matine, and C. R. Bolognesi, Type II photoluminescence and conduction band offsets of GaAsSb/InGaAs and GaAsSb/InP heterostructures grown by metal-organic vapor phase epitaxy, *Appl. Phys. Lett.* 73, 2799–2801, 1998.

17. Sprengel, S., A. Andrejew, K. Vizbaras, T. Gruendl, K. Geiger, G. Boehm, C. Grasse, and M.-C. Amann, Type-II InP-based lasers emitting at 2.55 μm, *Appl. Phys. Lett.* 100, 041109, 2012.

18. Pan, C.-H., C.-H. Chang, and C.-P. Lee, Room temperature optically pumped 2.56-μm lasers with "W" type InGaAs/GaAsSb quantum wells on InP substrates, *IEEE Photonic. Tech. Lett.* 24, 1145–1147, 2012.

19. Sprengel, S., C. Grasse, P. Wiecha, A. Andrejew, T. Gruendl, G. Boehm, R. Meyer, and M.-C. Amann, InP-based type-II quantum-well lasers and LEDs, *IEEE J. Sel. Top. Quantum Electron.* 19 (4), 1900909, 2013.

20. Sun, H. C., L. Davis, S. Sethgi, J. Singh, and P. Bhattacharya, Properties of tunneling injection quantum-well laser: Recipe for a "cold" device with a larger modulation bandwidth, *IEEE Photonic. Tech. Lett.* 5, 870–872, 1993.
21. Bhattacharya, P., J. Singh, H. Yoon, X. Zhang, A. G. Aitken, and Y. Lam, Tunneling injection lasers: A new class of lasers with reduced hot carrier effects, *IEEE J. Quantum Electron.* 32, 1620–1629, 1996.
22. McIlroy, P. W. A., A. Kurobe, and Y. Uematsu, Analysis application of theoretical gain curves to the design of multi-quantum-well lasers, *IEEE J. Quantum Electron.* QE-21, 1958, 1985.
23. Kapon, E., Threshold current of extremely narrow semiconductor quantum-well lasers, *Opt. Lett.* 15, 801–803, 1990.
24. Chong, T. C. and C. G. Fonstad, Theoretical gain of strained-layer semiconductor lasers in the large strain regime, *IEEE J. Quantum Electron.* 25, 171, 1989.

9

Quantum Dots

9.1 Introduction

The quantum well (QW) laser, first announced in 1975 with optical pumping [1], was soon improved by using the usual convenient electrical pumping [2]. Because of various advantages, the QW laser has been the standard diode laser and has almost totally captured the present commercial laser market.

The effect of reduced dimensionality on the performance of the lasers was studied soon after the announcement of QW lasers. The lasing characteristics of multidimensional QW lasers and, in particular, the temperature dependence of the threshold current in such lasers were first predicted by Arakawa and Sakaki in 1982 [3]. The authors introduced the term *multidimensional QW structures*, meaning that quantization takes place along two dimensions, as in a quantum wire (QWR) structure, or in all three dimensions, as in a quantum box (QB). It was predicted that the delta-function-like density of states (DOS) in QBs would lead to a reduced threshold current density and an infinite value for the characteristic temperature.

In spite of the predicted advantages, the progress in realizing practical QB or, more precisely, quantum dot (QD) lasers has been quite slow, and even today QD lasers have not achieved the status enjoyed by QW lasers. There are many reasons for this. First, the additional degrees of quantization are harder to obtain. Success of obtaining a QW structure is mainly due to the development of sophisticated epitaxial processes. Molecular beam epitaxy (MBE), metal organic chemical vapor deposition (MOCVD), atomic layer epitaxy (ALE), and chemical beam epitaxy (CBE) led to the growth of epitaxial layers with atomic-scale precision. On the other hand, growth of QWR or QD structures relied on patterning and etching in the plane of the structure; these processes did not achieve atomic-scale precision as quickly as epitaxy.

Despite the slow progress, many of the potential advantages of QD lasers have been demonstrated, at least in the laboratory. Many novel applications in which QD lasers play the key role have been proposed; some noteworthy application areas are short-haul optical networking, optical interconnects and quantum information processing, and quantum encryption with single photons.

In the present chapter, the growth of QDs will be very briefly described, mentioning the shape and size of the dots. The structures of QD lasers will also be introduced. A simple theoretical model for calculating the gain and threshold current of QB lasers will first be presented. Refinements of the simple model will then be introduced step by step. First, the effect of broadening on the gain spectra will be discussed. Thereafter, different models introduced in the literature to study QD lasers will be introduced. The modulation characteristics of QD lasers will be discussed by employing the rate equation model, and methods proposed and practically implemented to increase the bandwidth will be discussed. The temperature characteristics of QD lasers will then be presented, and the predicted ultrahigh characteristic temperature of tunnel-injection (TI)-QD lasers will be discussed. The chapter ends with a brief introduction to TI-QD lasers.

9.2 QD Growth Mechanisms and Structures

The laser structures incorporating QDs are more or less the same as those used in double heterostructure (DH) or QW lasers. There are contacts for applying bias, cladding layers for carrier and optical confinement, the active layer, having a stack of layers containing QD arrays, and the substrate. We first discuss briefly how the QD arrays are grown, the shape and size of the dots, and their influence on electronic structures and optical spectra. Next, we discuss the overall laser structure and, in particular, the methods to improve injection of carriers into the active layers.

9.2.1 Growth of QDs

As mentioned already, the fabrication of QD structures involves both epitaxy and patterning. In the case of epitaxy, the grown layer may be either lattice matched with the substrate or slightly lattice mismatched, giving rise to a strained layer. In the latter case, the thickness of the epilayer is kept below the critical layer thickness, in order to reduce the number of misfit dislocations at the heterointerface (see Reading List).

The 2-D layer-by-layer growth of lattice-matched or nearly lattice-matched layers on a substrate is known as Frank–van der Merwe growth. In this method, an entire atomic layer of the grown material is deposited first and is followed by the next atomic layer. The result is a uniform growth of the desired material, as shown in Figure 9.1a.

At the other extreme lies the growth of a highly mismatched (highly strained) layer on a substrate, known as Volmer–Weber growth. The growth

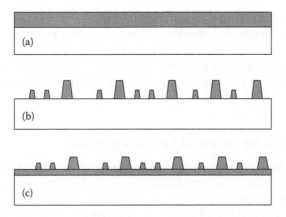

FIGURE 9.1
Illustration of (a) Frank–van der Merwe growth, (b) Volmer–Weber growth, and (c) Stranski–Krastanov growth.

of InAs on Si serves as an example of this process. In this case, the strain energy is too high to make a uniform growth of the deposited layer feasible. Instead, three-dimensional (3-D) clusters, or islands, of the material form on the substrate, as shown in Figure 9.1b. The islands form at random positions on the substrate and have random sizes.

In between the two extremes stated above, there exists an intermediate growth process, known as the Stranski–Krastanov (SK) process; different layers grown by this process are shown in Figure 9.1c. During the initial growth, the strain energy is not high enough, so that a 2-D layer is formed. After the growth of a few monolayers, the strain energy reaches a high enough level to force 3-D island growth. This mechanism is called the growth of *self-assembled* QDs; the thin 2-D layer grown initially is called the *wetting layer*.

In SK growth, the amount of material provided controls the growth process, measured in terms of equivalent monolayers (ML) of coverage. Unless a certain critical thickness is reached, there is no formation of dots. For growth of InAs QDs on GaAs, this thickness is about 1.3 ML. Provision of additional material after this critical thickness is obtained leads to the formation of islands with typical areal density $>10^{12}$ cm^{-2}. Further addition of material does not produce additional dots, but favors growth of larger islands containing dislocations.

As illustrated in Figure 9.1c, the dots are in the form of pyramidal or conical bumps. The bumps are typically about 3–4 nm high, and the extension of the base is about 25 nm; this is also the typical base diameter for conical-shaped dots. Each dot is, however, isolated, and, as the wetting layer provides a high barrier for the electrons in each dot, the dots are decoupled from their neighbors. The distributions of dot size and geometry are random. The irregular spacing between adjacent dots ensures an irregular probability of tunneling. Moreover, the size distribution of the dots gives rise to an appreciable distribution of electronic energy levels on account of the 3-D quantum confinement. This distribution affects the emission characteristics of the assembly of QDs and gives rise to a large, inhomogeneous broadening of the emission spectra. The role of this broadening will be discussed in Section 9.4.

9.2.2 Structures for Enhanced Carrier Collection

Basically, the laser structure remains the same as in single-QW or multiple-QW lasers. The only difference is that the active region is substituted by layers containing the QDs. However, the effective volume of the active region is drastically reduced for QD lasers in comparison with QW lasers. This is illustrated in Example 9.1.

> **Example 9.1:** Consider a single layer of QD array with areal density 5×10^{10} cm^{-2}. Each QD is assumed to be cylindrical with a base diameter of 25 nm and height 4 nm. The active region volume is thus $\pi(25/2)^2 \times 4 \times 10^{-27} \times 5 \times 10^{14}$ WL m$^3 \approx 10^{-9} WL$ m^3, where W and L are, respectively, the width and length of the active layer. With a QW of thickness 50 nm and the same W and L, the ratio is 1/50.

The obvious solution to increase the active volume is to increase the number of layers. The strain field from the underlying layers aids in the formation of dots in the subsequent layers. However, the size variation increases with the increase in the number of layers. With growth of multiple layers, strain builds up and more defects and recombination centers are introduced. Strain-compensation techniques need to be employed to suppress this.

Carriers injected into the active layer should be captured efficiently by the dots. Unfortunately, the capture process is not very efficient due to the small volume of the QDs and the spatial separation between them. Carrier capture by the dots is effective in the vicinity of the p and n layers. To increase the collection efficiency, a large density of carriers needs to be provided in the vicinity of the dots. Two different methods have been employed to provide a pool of carriers to be collected by the dots. The first method is to insert the dots in a QW; the well itself captures a high density of carriers and supplies the carriers to the dots. The corresponding structure is called a dot-in-a-well or DWELL structure, and its band diagram is shown in Figure 9.2a. In the other method, a structure is designed with a reservoir of carriers adjacent to the QDs, which are separated from the reservoir by a thin tunnel barrier. The band diagram is shown in Figure 9.2b. Here the well does not contain the dots, but is separated from the dots by the barrier. Carriers from the well tunnel into the dot energy states.

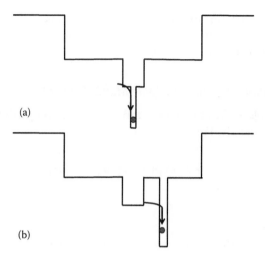

(a)

(b)

FIGURE 9.2
Schematic diagrams illustrating enhanced carrier collection by dots: (a) DWELL structure and (b) tunnel injection structure.

9.3 Introductory Model for QD Lasers

We shall first present a simple model for QD lasers, by assuming an assembly of identical cubic boxes. The expression for gain considering 3-D quantization, ideal DOS function, and polarization-dependent momentum matrix element will be derived first and will then be modified by considering homogeneous broadening. The expression for the threshold current density will then be obtained. It should be mentioned that the model is too simple to explain the experimental features. However, it serves as an introduction to QD laser theory, and the results point out some of the important features of the QD laser [4,5].

9.3.1 Absorption and Gain in QD Systems

The calculation of gain in a QD system is illustrated first by using a QB structure of dimension $d_x \times d_y \times d_z$, where the subscripts represent the width in the respective direction. The electron wave function in the conduction band is given by the following approximate form:

$$\Psi_{cnml} = U_c(\mathbf{r})\Phi_{cxn}(x)\Phi_{cym}(y)\Phi_{czl}(z) \tag{9.1}$$

where:
  c (or h) denotes the conduction band (or heavy-hole band)
  $n, m,$ and l denote the label of the quantized energy levels in the box

$U_c(\mathbf{r})$ is the periodic part of the bulk Bloch functions

Φ_{cxn}, Φ_{cym}, and Φ_{czl} are the envelope functions along the x-, y-, and z-directions, respectively

9.3.1.1 Polarization-Dependent Dipole Matrix Element

Each of the envelope functions is a standing wave composed of two waves propagating oppositely. For example, we may write

$$\Phi_{czl} = A\left[\exp(jk_z z) \pm \exp(-jk_z z)\right],\tag{9.2}$$

where:

$k_z = (2m_{cw}E_{czl})^{1/2}/\hbar$

m_{cw} denotes the effective electron mass in the well

The envelope function penetrates the finite potential barrier and decays exponentially there, as discussed in Chapter 3. The wave number k_z is quantized. One may write the other two envelope functions similarly as in Equation 9.2. Thus, the total wave function is divided into eight propagating wave functions, each of which is quantized due to quantization of E_{czl} and the other energy functions. Putting $\theta = 0$ at the y-axis and $\phi = 0$ at the z-axis, the eight directions of $\mathbf{k}$ are given by $(\theta, \pm\phi)$, $(\theta, \pi\pm\phi)$, $(\pi-\theta, \pm\phi)$, and $(\pi-\theta, \pi\pm\phi)$. The following relations are then obtained:

$$\cos^2\theta = \frac{k_y^2}{k^2} = \frac{E_{cym}}{E_{cnml}}\tag{9.3a}$$

$$\cos^2\phi = \frac{k_z^2}{\left(k^2 - k_y^2\right)} = \frac{E_{czl}}{(E_{cnml} - E_{cym})}\tag{9.3b}$$

$$E_{cnml} = E_{czl} + E_{cym} + E_{cxn}\tag{9.3c}$$

The dipole matrix element for transition between states in the conduction band and the heavy-hole band is given by the usual form:

$$R_{ch} = \left\langle\Psi_{cnml}\left|e\mathbf{r}\right|\Psi_{hn'm'l'}\right\rangle \approx \delta_{nn'}\delta_{mm'}\delta_{ll'}\sum_{\mathbf{k}}\left\langle U_c\left|e\mathbf{r}\right|U_h\right\rangle\tag{9.4}$$

In Equation 9.4, $\langle U_c|e\mathbf{r}|U_h\rangle$ represents the dipole matrix element for bulk at fixed wave vector $\mathbf{k}$ ($=\mathbf{k}_c=\mathbf{k}_h$), and the summation is made over all eight propagating vectors mentioned earlier; *lh* bands are not considered, since the energy levels are lower than the *hh* quantized levels.

The magnitude of the dipole matrix element of the bulk crystal as obtained by manipulating Equations 6.60 and 6.34 is given by

$$R^2 = \left(\frac{e\hbar}{2E_{ch}} \right)^2 \frac{E_g(E_g + \Delta_0)}{(E_g + 2\Delta_0 / 3)m_e} \tag{9.5}$$

The dipole moment rotates in a plane perpendicular to the wave vector **k**. The components of **R** are given by

$$R(\cos\theta \sin\phi + j\cos\phi), \quad \text{for } x\text{-direction} \tag{9.6a}$$

$$R\sin\theta, \quad \text{for } y\text{-direction} \tag{9.6b}$$

$$R(\cos\theta \cos\phi - j\sin\phi), \quad \text{for } z\text{-direction} \tag{9.6c}$$

The interaction between the dipole moment $\mathbf{R}_{ch}$ and the electric field vector **E** of the electromagnetic (EM) wave is given by $\mathbf{R}_{ch}\cdot\mathbf{E}$. For bulk semiconductors, both **k** and $\mathbf{R}_{ch}$ distribute over all directions, and therefore the effective dipole moment is obtained by averaging the component of the dipole moment parallel to **E** over all directions of **k**. For QB structures, the effective dipole moment is obtained by averaging dipole moments corresponding to eight directions of **k**. The magnitude of $\mathbf{R}_{ch}\cdot\mathbf{E}$ depends on the polarization direction of **E**. It has been found that the maximum of $\mathbf{R}_{ch}\cdot\mathbf{E}$ occurs when **E** is parallel to the longest dimension of the box. For cubic QB, this scalar product is isotropic. Assuming that the longest side of the box is along x and $\mathbf{E}\|x$, the square of the effective dipole moment is given by averaging Equation 9.4 over the eight fixed directions of **k** as

$$\langle R_{ch}^2 \rangle = R^2 \left(\cos^2\theta \sin^2\phi + \cos^2\phi \right) \delta_{ll'}\delta_{mm'}\delta_{nn'}$$

$$= R^2 \left[\left(k_y^2 + k_z^2 \right) / k^2 \right] \delta_{ll'}\delta_{mm'}\delta_{nn'} \tag{9.7}$$

It follows easily that $\langle R_{ch}^2 \rangle$ depends on the shape of the QB, and for cubic QB its value is $(2/3)\, R^2$: the value for bulk crystal.

9.3.1.2 Expression for Linear Gain

The expression for the linear gain may now be written as

$$g(\omega) = \frac{\omega}{\eta_r}\sqrt{\frac{\mu_0}{\varepsilon_0}}\sum_{lmn}\int_{E_g}^{\infty}\langle R_{ch}^2\rangle\frac{\rho_{ch}(f_e - f_h)(\hbar/\tau_{in})}{(E_{ch} - \hbar\omega)^2 + (\hbar/\tau_{in})^2}\,dE_{ch} \tag{9.8}$$

Most of the symbols are already defined. The intraband relaxation time is denoted by τ_{in}, and the DOS of the electron–hole pairs with the same quantized energy levels is given by

$$\rho_{ch}(E_{ch}) = \frac{2\delta(E_{ch} - E_{cnml} - E_{hnml} - E_g)}{d_x d_y d_z} \tag{9.9}$$

We assume charge neutrality, that is, $n=p$, and the carrier densities are related to the quasi-Fermi levels by the usual expressions:

$$n = \sum_{lmn}\frac{2}{\left[1+\exp\left(\dfrac{E_{cmnl} - F_e}{k_B T}\right)\right]d_x d_y d_z} \tag{9.10a}$$

$$p = \sum_{lmn}\frac{2}{\left[1+\exp\left(\dfrac{F_h - E_{hmnl}}{k_B T}\right)\right]d_x d_y d_z} \tag{9.10b}$$

9.3.2 Calculation of Subband Energies

The subband energies are calculated by a method similar to the equivalent refractive index (RI) method used in optical waveguide theory. The present method suits well when $d_x \gg d_y \gg d_z$ or when the well depth is larger than the resulting quantized energy levels (Figure 9.3).

In the calculation, the envelope function and energy levels are obtained for the z-direction by solving the following equation:

$$\left[-(\hbar^2/2m_e)\partial^2/\partial z^2 + V_{cz}\right]\Phi_{czl}(z) = E_{czl}\Phi_{czl}(z);$$

$$V_{cz}(z) = \begin{cases} 0, & (|z| \leq d_z/2 : \text{inside the well}) \\ \Delta E_c, & (|z| \leq d_z/2 : \text{outside the well}) \end{cases} \tag{9.11}$$

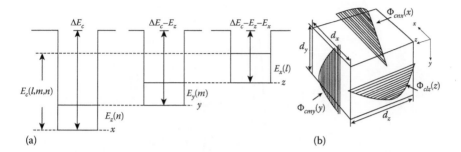

FIGURE 9.3
(a) The QB structure used in the model and (b) energy levels and barrier heights considered for calculation of subband energies for z-, y-, and x-directions.

The envelope function is given by

$$\Phi_{czl} = A \binom{\cos}{\sin} \left[\frac{(2m_{c1}E_{czl})^{1/2} z}{\hbar} \right], \binom{l:\text{even}}{l:\text{odd}}, \quad \text{for } |z| \le d_z/2; \quad (9.12a)$$

$$\Phi_{czl} = B \exp\left[-\left\{ 2m_{c2}\left(\Delta E_c - E_{czl} \right) \right\}^{1/2} |z| / \hbar \right], \quad \text{for } |z| \ge d_z/2 \quad (9.12b)$$

The energy level E_{czl} is obtained by solving the following equation:

$$\left[\left(\frac{m_{c1}}{m_{c2}} \right) \frac{\left(\Delta E_c - E_{czl} \right)}{E_{czl}} \right]^{1/2} = \binom{\tan}{-\cot} \left[\frac{d_z \left(2m_{c1}E_{czl} \right)^{1/2}}{\hbar} \right] \binom{l:\text{even}}{l:\text{odd}} \quad (9.13)$$

In the above equations, m_{c1} and m_{c2}, denote, respectively, the electron effective masses in the well and the barrier.

The calculation is repeated for the y-direction by using Φ_{cym} and E_{cym} in place of Φ_{czl} and E_{czl}, and using the effective barrier height $\Delta E_c - E_{czl}$. After obtaining the energy level, the process is repeated for the x-axis to calculate the envelope function Φ_{cxn} and energy E_{cxn} by considering the effective barrier height as $\Delta E_c - E_{czl} - E_{cym}$. The quantized energy level in the conduction band thus becomes

$$E_{cnml} = E_{czl} + E_{cym} + E_{cxn} \quad (9.14)$$

Quantized energy levels in the valence band may be calculated similarly.

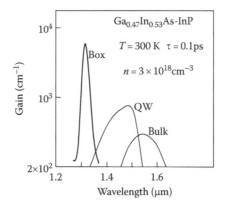

FIGURE 9.4
Calculated gain spectra for InGaAs cubic QB of 10 nm width in each direction. The gain spectra for 10 nm thick InGaAs QW and for InGaAs bulk are included for comparison.

9.3.3 Gain Spectra

The gain spectra calculated for a cubic $In_{0.53}Ga_{0.47}As/InP$ QB having width = 10 nm on each side are shown in Figure 9.4. The direction of the electric-field polarization has been chosen so that the gain becomes maximum. For comparison, gain spectra for 10 nm thick QW and bulk InGaAs are included in the figure. A value of $\tau_{in} = 1 \times 10^{-13}$ s was used for the intraband relaxation time in all cases. For each case, the electron density chosen is 3×10^{18} cm^{-3}.

As is found from Figure 9.4, the variation in gain is sharpest for the QB structure. This is expected, since the DOS function is almost delta function like, resembling the emission spectra of atoms. In fact, QBs are often called *superatoms*. The figure does not include the spectrum for a QWR, but it lies between those of QW and QB.

9.3.4 Calculation of Threshold Current

The cubic QBs form an array, as shown in Figure 9.5. The current flows along the y-direction and is related to the carrier density and other parameters as

$$J = \frac{e\eta N_{QB}d_y n}{\tau_s} \tag{9.15}$$

where:
  J is the average current density
  η is the ratio of surface area covered by QBs to the entire surface area

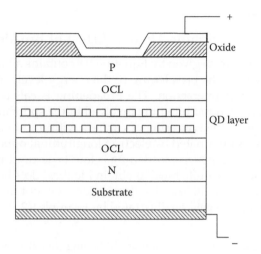

FIGURE 9.5
Schematic view of QB laser. Multiple layers containing QDs are sandwiched between two OCLs.

N_{QB} is the number of QB layers
τ_s is the carrier lifetime, consisting of both radiative and nonradiative components

It is assumed that the separation between QBs equals d_x and d_z along the respective directions, so that $\eta = 0.25$.
The following results may be obtained from the calculations:

1. The threshold current density (J_{th}) slowly decreases with decreasing thickness of the optical confinement layer (OCL) as the confinement factor increases slowly. However, for thickness below an optimum value, the confinement factor decreases rapidly with decreasing OCL thickness. The values of J_{th} increase rapidly with decreasing OCL thickness below this optimum value.

2. The variation of J_{th} with density of QBs is similar. For low values of density, optical confinement factor is very low and J_{th} is large. With increasing density, J_{th} gradually decreases, attains a minimum, and then rises. The rise of J_{th} for higher values of density is due to higher carrier consumption.

The nature of variation of J_{th} with the size of QBs, bandgap energy, and other variables is given in the paper by Miyamoto et al. [5].

9.4 Deviation from Simple Theory: Effect of Broadening

The expression for gain as given by Equation 9.8 contains a Lorentzian function that takes care of the broadening of the energy levels of the QD. This is due to light–matter interaction. The broadening is called homogeneous broadening. It is related to finite time involved in the spontaneous emission process, as well as to the scattering of carriers within individual dots. The scattering process is dominated by electron longitudinal optic phonon interaction. The linewidth for such broadening is low, typically 1 meV or less.

In actual practice, it is quite hard to obtain identical dots having the same base dimension and height. The size fluctuation from dot to dot results in a distribution of energy levels, as illustrated by Example 9.2.

> **Example 9.2:** Consider a rectangular QB having dimensions $d_x=5$ nm, $d_y=10$ nm, and $d_z=20$ nm. Using infinite barrier approximation, it is found that a change δd_x produces a change of subband energy by an amount $\delta E_{lmn}=-2(\hbar^2/2m_e)(\pi^2/d_x^3)\delta d_x$ obtained by differentiation. A change $\delta d_x/d_x=0.1$ changes the energy by 20%.

The variation of ground-state and excited-state electronic energies due to size fluctuation is shown in Figure 9.6. The result is a distribution of energy, which in many cases is approximated by a Gaussian distribution. There will be a similar distribution of hole energies, and, consequently, different dots will emit different wavelengths. Each individual dot has its own homogeneous broadening. The resulting emission spectra are shown in Figure 9.7.

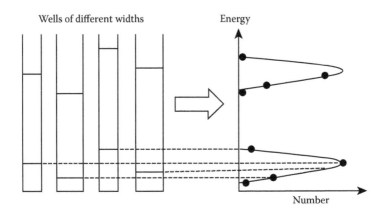

FIGURE 9.6
Illustration of distribution of energy levels for ground and excited states in the conduction band due to variation of QD size. The distribution of number of dots in a small energy interval is shown on the right.

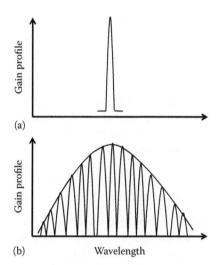

FIGURE 9.7

Gain profiles of (a) single dot showing homogeneous broadening and (b) ensemble of self-assembled QDs emitting at different wavelengths due to variation of size.

The overall emission from the ensemble is the independent sum of the individual spectra, which is quite narrow due to homogeneous broadening. The result is an inhomogeneously broadened spectrum, the linewidth of which may be as large as 20–30 meV.

The calculation of gain and threshold of QD assembly must therefore consider the large inhomogeneous broadening, in addition to the homogeneous broadening.

9.5 Subband Structures for Pyramidal QDs

In Section 9.3, a simple theory for gain and threshold current in a QB has been developed. The QB is assumed to have a cubic structure. In the calculation of gain spectra, homogeneous broadening has been included.

As discussed already, the shape of the dots is at best conical or pyramidal, and is in no way cubic. The simple theory for gain and threshold current presented above needs refinement due to several considerations. Calculation of the electron and hole levels in a QD requires knowledge of the size, shape, and chemical compositions of the materials. The electronic structure determines the optical transition energies and hence the emission wavelengths, oscillator strengths for transition, and the gain. Various levels of sophistication have been introduced for calculating the electronic states in QDs. Currently, the most advanced methods for handling arbitrary material distributions are the eight-band **k·p** theory [6,7] and the pseudopotential method [8]. These two methods give rather similar results for QDs in the 10 nm size range [6].

In the present text, we do not include these sophisticated methods and results obtained thereby. In the following, we shall rely on simple analytical or empirical results that are able to explain the experimental data. We introduce two methods for the calculation of subband structures, taking into consideration the shape and size of the dots.

9.5.1 Single-Band Constant-Confining Potential Model

In this model, proposed by Califano and Harrison [9], the envelope function $\Phi(x,y,z)$ is expanded in terms of a complete orthogonal set of solutions for a cubic dot having an infinite potential barrier. The method requires less computation time and less complex codes, and the results can reproduce the peak positions of experimental photoluminescence data. The numerical data of electron energy level, E_e, and of hole energy level, E_h, can be fitted with the following formula [10]:

$$E_i = A_i + B_i \exp(-L / L_i); \quad i = e, h \tag{9.16}$$

where:
A_i, B_i, and L_i are constants
L is the base length of the pyramidal QD

The numerical values for a GaAs QD are given in Example 9.3.

> **Example 9.3:** The energy levels for GaAs–InAs QDs are calculated using $A_e = -403$ meV, $B_e = 841.3$ meV, and $L_e = 9.35$ nm for electrons, and $A_h = -265$ meV, $B_h = 414.6$ meV, and $L_h = 5.4$ nm. The values for $L = 20$ nm are $E_e = -304$ meV and $E_h = -240$ meV. The energies are measured from the GaAs band edges. Using $\Delta E_c = 450$ meV, and $\Delta E_c = 265$ meV for InAs–GaAs systems, the transition energy is 0.886 eV and the emission wavelength is 1.4 μm.

9.5.2 Harmonic Oscillator Energy Levels

A careful analysis of the experimental data for QD lasers emitting at 1.3 μm reveals that the subband energy levels are equispaced. This has prompted workers [11–14] to argue that the confining potentials in pyramidal QDs are parabolic in nature. The subband energy levels for both electrons and holes are then expressed by Equation 9.17:

$$E(n_{x,a}, n_{y,a}, n_{z,a})$$

$$= \hbar\omega_{0xy,a}\left(n_{x,a}+1/2\right).\left(n_{y,a}+1/2\right)+\hbar\omega_{0z,a}\left(n_{z,a}+1/2\right); \quad a = e, h \quad (9.17)$$

The energy separation for parabolic potential in the x–y plane (base of the pyramid) is given by $\hbar\omega_{0xy}$ and that for the potential along the z-direction is given by $\hbar\omega_{0z}$.

9.6 Refined Theory for Gain and Threshold

The simple theory of gain and threshold current presented earlier for cubic QB structure must be modified to consider the shape and size of the dots, the energy levels, occupancy of multiple subbands, and recombination, capture, and emission mechanisms of carriers. We first discuss the points to be considered for developing a suitable model for QD lasers and then present a representative model that takes into consideration some of the important physical processes.

9.6.1 General Considerations

The simple model to calculate the gain, threshold current, and other parameters presented in Section 9.3, though instructive, is inadequate in many respects. The experimental results for gain and characteristic temperature are lower than the values predicted by this simple theory. The reasons for the disagreement are numerous: (1) neglect of excited states in the conduction and valence bands, (2) thermal broadening of hole energy states, (3) carrier occupation in the wetting layer, and (4) inhomogeneous broadening due to QD size distributions [11–25].

A very important consideration in developing an acceptable theory for QD lasers is the recombination mechanism. In the simple theory, the recombination mechanism considered is no different from the processes occurring in the bulk or QW systems and is due to electron–hole recombination. Though many authors have considered in their work the free electron–hole recombination mechanism to be the sole radiative process, a few authors, on the contrary, believe that the recombination in QDs is excitonic in nature. Since the confinement in QDs is 3-D, it is expected that excitonic effects will be significant, and experimental evidence suggests that excitons really do take part in the lasing mechanism. However, the relative importance of electron–hole recombination and excitonic recombination needs to be assessed.

Different models exist for the calculation of gain and threshold characteristics of QD lasers having a number of layers containing arrays of QDs, which are mostly pyramidal in shape and have a size distribution. We present below a representative model that attempts to include some of the important physical processes occurring in the QD system [26]. Here three different situations will be considered: (i) recombination by free carriers, (ii) excitonic recombination, and (iii) recombination of both free carriers and excitons. The model presented will be for illustration only and to help readers to follow more sophisticated models reported in the literature.

9.6.2 A Representative Model

As mentioned already, some authors have calculated gain by treating the electrons and holes as free carriers, while others assume that carriers form excitons. In this section, the role of carrier distribution in determining the gain and threshold characteristics of QD lasers will be discussed. The model considers free-carrier distribution, an excitonic model, and a combination of free-carrier and excitonic distribution [26].

In the model, the allowed energy levels are calculated by using a simple harmonic approximation. The threshold gain is determined from the internal loss and mirror loss. By choosing a particular carrier distribution, the quasi-Fermi level to achieve the threshold gain can be determined and can be used to calculate the threshold carrier density and the threshold current.

9.6.2.1 Expressions for Energy Levels and Gain

The structure considered and the band diagram are shown in Figure 9.8.

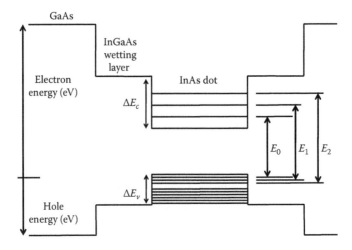

FIGURE 9.8
Band diagram of QD structure.

The energy levels for the pyramidal QD are calculated by using the harmonic oscillator Hamiltonian, and the corresponding energy levels are given by Equation 9.17.

Example 9.4: Consider growth of InAs dots (band gap$=0.847$ eV) with InGaAs wetting layer ($E_g=1.25$ eV). Using $\Delta E_c=0.8\Delta E_g$ and $\Delta E_v=0.2\Delta E_g$ and the level separation for electrons and holes, respectively, as 76 meV and 8 meV, the number of confined electron and hole levels is, respectively, 3 and 8.

The excited states have a higher degeneracy, leading to a higher DOS. For pyramidal QDs in which height is less than the base, the degeneracy of the first and second excited states is 2 and 3, respectively.

The carrier density in the QDs is expressed as

$$n = \int_{E_g}^{\infty} \sum_i 2N_{QD}G(E)f_n(E)dE \qquad (9.18)$$

where:

N_{QD} is the areal dot density
i is the energy level
$G(E)$ is the inhomogeneous broadening due to size fluctuations
$f_n(E)$ is the electronic Fermi distribution function involving the quasi-Fermi level F_n
the factor 2 is for spin degeneracy of each energy level

The inhomogeneous broadening is assumed to have the following Gaussian form:

$$G(E) = \frac{1}{\sqrt{2\pi}\sigma} \exp\left[\frac{-1}{2\sigma^2}(E - E_i)^2\right] \qquad (9.19)$$

where σ is related to the size fluctuations. Similar formulas can be written for the holes.

The gain at a particular injection (quasi-Fermi level) consists of two parts: the inversion factor, $[f_c+f_v-1]$, where fs are the Fermi functions, and the maximum gain. The expression for the maximum gain, considering contributions from all possible transitions, is

$$g_{max} = \sum_{i,j} \frac{\Gamma\pi e^2 \hbar |M_b|^2 N_{QD}N_L}{cm_0\varepsilon_0 n_r E_{i,j} t_{QD}} \langle \psi_i | \psi_j \rangle G(E_{i,j})S(E_{i,j}) \qquad (9.20)$$

where:

    Γ is the optical confinement factor
    n_r is the RI
    $\langle\psi_i|\psi_j\rangle$ is the wave function overlap for transition $i{\to}j$
    $|M_b|^2$ is the bulk matrix element
    N_L is the number of QD layers
    $S(E_{i,j})$ is the homogeneous broadening for a given transition

It is assumed that only transitions having the same quantum number are allowed, that is $n_{x,e}=n_{x,h}$, $n_{y,e}=n_{y,h}$ and $n_{z,e}=n_{z,h}$. The current density is calculated by considering the contributions from the dot layers as well as the wetting layer and is given by

$$J = \frac{eN_L}{\eta_i}\left(\frac{t_{QD}n_{QD}}{\tau_{QD}} + \frac{t_W n_W}{\tau_W}\right) \tag{9.21}$$

where:

    η_i is the injection efficiency
    t is the thickness of the respective layers
    n is the volume density of carriers
    τ is the recombination lifetime
    the subscripts QD and W refer, respectively, to the dot layer and the wetting layer

The recombination lifetime in the wetting layer is given by the usual expression $\tau^{-1}=A+Bn+Cn^3$.

> **Example 9.5:** We use Equation 9.20 to obtain the value of g_{max} in InAs–InGaAs QDs. The values chosen are $\Gamma=0.02$, $|M_b|^2=20\,eV$, $\langle\psi_i|\psi_j\rangle=0.2$, $N_{QD}=3.10^{14}\,m^{-2}$, $N_L=2$, $n_r=3.4$, $E_{i,j}=0.937$ eV, $t_{QD}=3$ nm, and $\sigma=40$ meV. The homogeneous broadening is not considered. The obtained value of g_{max} is 11 cm^{-1}.

9.6.2.2 Carrier Distribution, Gain, and Threshold

The values of gain and threshold current densities are now estimated by considering (i) free carrier distribution, (ii) excitonic distribution, and (iii) the presence of both free carriers and excitons.

9.6.2.2.1 Free-Carrier Model

In this case, electrons and holes are treated as free carriers. For a chosen separation of Fermi levels, the electron and hole concentrations are calculated assuming charge neutrality in the dots and using Equation 9.18. Since the hole levels are closely spaced and the degeneracy of the levels increases with the level index, the hole population in the higher states is larger. On the other hand, the excited electron states are not significantly occupied in spite of larger degeneracy, since the level separation is larger.

The gain spectrum is calculated from the expression

$$g_{free}(E) = g_{max} \cdot \left[f_c(E, F_n) + f_v(E, F_p) - 1 \right] \tag{9.22}$$

where:

  f represents Fermi function

  F_n and F_p denote, respectively, the quasi-Fermi levels for electrons and holes

The threshold gain, calculated by assuming $\alpha_i = 2$ cm^{-1}, mirror reflectivity of 0.32, cavity length $= 3.8$ mm, and optical confinement factor of 0.01/ per layer, turns out to be 5.15 cm^{-1}. It is to be noted that, due to occupation of excited states, it is difficult to increase the hole Fermi function and to make the inversion large. As observed by the authors, the gain in the ground state is below the threshold gain, and therefore lasing is not possible for transition between the pair of states labeled by $m = 0$, as well as by $m = 1$. The calculation shows that the gain exceeds the threshold only for the $m = 2$ state. The excited holes do not directly increase the threshold current, as there are few excited electrons to recombine with. The threshold current density of ~270 A cm^{-2} does not agree with the experimental values for lasers having dimensions and structures as assumed in the model.

9.6.2.2.2 Exciton Model

In the alternative approach, it is assumed that all the carriers in the dots form excitons, the distribution of which determines the gain. Instead of separate quasi-Fermi levels for electrons and holes, an excitonic Fermi level may be defined that may be used to calculate the excitonic distribution based on Fermi–Dirac (FD) statistics. In this case, the holes do not follow FD statistics, but follow the electron distribution. For the same injection level as in Section 9.6.2.2.1, the higher occupation of excited hole states is reduced. Since the holes follow the electrons, $f_c = f_v$, and the expression for the excitonic gain becomes

$$g_{exc}(E) = g_{max} \times \left[2 f_{ex} - 1 \right] \tag{9.23}$$

The dimensions of the InAs dot ~25 nm are smaller than the bulk exciton radius ~50 nm and the wave function overlap is ~1. The calculated ground-state gain is now substantially in excess of the threshold gain, and lasing occurs from this state. The calculated threshold current density is 20 A cm^{-2}. By repeating the calculation for different temperatures, a characteristic temperature $T_0 = 550$ K is obtained, which is too high compared with the experimental value.

9.6.2.2.3 Excitons and Free Carriers

In this case, it is assumed that both free carriers and excitons coexist. The expression for gain is now written as

$$g(E) = g_{exc}(E) \times \left(\frac{n_{ex}}{n} \right) + g_{free}(E) \tag{9.24}$$

where:

n_{ex} and n denote the exciton and total carrier density, respectively

g_{free} is the gain from free carriers

The ratio n_{ex}/n depends on the exciton binding energy. The ratio is calculated by using the following modified form of the Saha equation [27]:

$$\frac{n_{ex}}{n} = 1 - \sqrt{\frac{n_{ex}}{n} \frac{1}{\exp(E_B / k_B T) - (1 + E_B / k_B T)}} \tag{9.24a}$$

where E_B is the binding energy for a particular transition. For $E_B \gg k_B T$, the above equation reduces to the standard form of the Saha equation, noting that $n = n_{ex} + n_e$, where n_e is the electron density.

> **Example 9.6:** We assume that $E_B = 20$ meV. Solving the quadratic equation and putting $k_B T = 25.8$ meV at room temperature, the ratio $n_{ex}/n = 0.233$.

The peak gain considering free carriers and excitons as a function of average number of carrier pairs per dot is shown in Figure 9.9. The peak gain is reduced due to the sizable fraction of carriers in the ground state. However, this leads to a fair agreement with the experimentally observed value. The threshold current density is now 44 A cm^{-2}. The calculated value of $T_0 = 85$ K is in excellent agreement with the observed value of 83 K for undoped samples [28].

The model has been extended for p-doped QDs. Introduction of acceptors increases the free hole density and hence pushes down the quasi-Fermi level for holes. The inversion factor thereby increases, making the ground-state gain larger. The characteristic temperature also rises significantly, in agreement with the experimental observation.

9.7 Modulation Bandwidth: Rate Equation Analysis

The small-signal modulation characteristics and other time-dependent characteristics are obtained by solving the rate equations. In this connection also, different models have been developed by considering free carriers

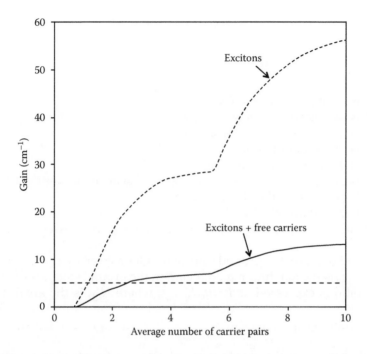

FIGURE 9.9
Peak gain using excitons only (*short dashed line*) and both excitons and free carriers (*solid line*). The horizontal dashed line represents the total loss of 5.15 cm⁻¹ to be overcome at threshold.

as well as excitons. In some cases, the number of rate equations is kept to a minimum by not considering the wetting layer and restricting the occupation to the ground state only. In more sophisticated models, multilevel occupancy is taken into account. It is not possible to discuss all the models; more sophisticated models need numerical solutions in many cases. In this work, we shall consider simple models that lead to analytical solutions to the problem.

9.7.1 Excitonic Model with Ground-Level Occupation

The model presented in this section leads to analytical form for various laser parameters. Let n_w, n_Q, and S denote, respectively, the carrier density in the wetting layer, the same in the ground state of the quantum dots, and the photon density. The rate equations for these densities may be written as [19]

$$\frac{dn_w}{dt} = \frac{\eta_{inj}I}{eV_a} - \frac{n_w}{\tau_s} - \frac{n_w}{\tau_c} + \frac{n_Q}{\tau_e} \tag{9.25a}$$

$$\frac{dn_Q}{dt} = \frac{n_w}{\tau_c} - \frac{n_Q}{\tau_e} - \frac{n_Q}{\tau_s} - v_g g_m S \tag{9.25b}$$

$$\frac{dS}{dt} = v_g g_m S - \frac{S}{\tau_p}$$

(9.25c)

where:

η_{inj} is the injection efficiency
I is the injection current
V_a is the volume of the active region
τ_s is the carrier lifetime
τ_c is the effective capture time
τ_e is the escape time from the QD ground state to the wetting layer
$\tau_s = [v_g(\alpha_i + \alpha_m)]^{-1}$ is the photon lifetime
v_g is the group velocity
α_i and α_m are, respectively, the intrinsic and the mirror losses

The modal gain of the ground state of the QD is given by $g_m = G_{max}(2P - 1)$, where G_{max} is the maximum modal gain in the ground state, and the occupancy ratio is expressed as $P = n_Q/2N_{QD}$, where N_{QD} is the areal density of the dots. In order to consider Pauli blocking (state filling), the effective capture time is expressed as $\tau_c = \tau_0/(1 - P)$, where τ_0 is an intrinsic capture time.

We first consider the steady-state solutions to Equations 9.25a through 9.25c by putting $d/dt = 0$ in (c), and noting that, for threshold $g_m = \alpha_i + \alpha_m$ and the relation between g_m and G_{max}, we may write [19]

$$P = \frac{1}{2}\left[1 + \frac{1}{v_g \tau_p G_{max}}\right] = \frac{1}{2}\left[1 + \frac{\alpha_i + \alpha_m}{G_{max}}\right]$$

(9.26)

at threshold $S = 0$. Putting this into Equation 9.25b, we may express n_{QD} in terms of n_W, and from (a) a relation between the injected current and n_w may be obtained. The expression for the threshold current density is obtained as [19]

$$J_{th} = \frac{2ePN_{QD}.N_L d}{\eta_{inj}\tau_s}\left[1 + \frac{A_2}{1 - P}\right];$$

(9.27)

where:

$$A_2 = \frac{\tau_0}{\tau_s} + \frac{\tau_0}{\tau_e}$$

$N_{QD}N_L d$ is the total areal density of the QDs
d is the thickness of the layer

The relationship between the current and the gain obtained at the nonlasing condition ($S=0$) is given by

$$g_m(J) = G_{max}\left\{J' + A_2 - \sqrt{(J'+1+A_2)^2 - 4J'}\right\}, \quad J' = \frac{\tau_s \eta_{inj} J}{2eN_{QD}N_L d} \quad (9.28)$$

Example 9.7: The threshold current density is calculated for InAs/GaAs QDs by using the values of parameters listed. $G_{max}=11$ cm^{-1}, $\alpha_i=5$ cm^{-1}, $\eta_{inj}=0.5$, $\tau_s=0.5$ ns, $\tau_0=3$ ps, $\tau_e=60$ ps, $L=500$ μm, $N_{QD}d=3.10^{10}$ cm^{-2}, $N_L=2$, and reflectivity $R=0.8$. The calculated mirror loss is 4.46 cm^{-1}, $P=0.93$, $A_2=56.10^{-2}$, and $J_{th}=128.6$ A cm^{-2}.

The threshold current density decreases monotonically with cavity length, since the mirror loss varies as L^{-1}. For a cavity of too short a length, the gain saturation effect makes the value of the threshold current density still higher.

The rate equations can be used in the usual manner to obtain the small signal response. The normalized modulation response, the resonance frequency, and the 3 dB bandwidth are expressed by the same formula as used for DH and QW lasers. Figure 9.10 illustrates the modulation response for an InGaAs–GaAs QD laser for three different values of the threshold current. The curves shift toward higher frequencies as the drive current is increased [29].

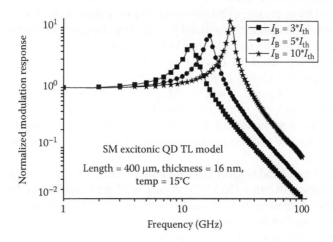

FIGURE 9.10
Normalized modulation response for InGaAs–GaAs QD laser.

9.7.2 Effect of p-Doping

QD systems possess higher differential gain than that exhibited by QW lasers. Based on this fact, it is predicted that the modulation bandwidth of QD lasers should be higher. Unfortunately, however, the observed bandwidths are not in conformity with this prediction. It has been proposed that slow carrier capture rate limits the device bandwidth. However, the measured values of capture and relaxation processes are quite low, so that this conjecture is not supported. It has been mentioned that the gain is greatly reduced due to the occupation of a few closely spaced hole levels. The role of p-doping in modulation bandwidth has been studied by a few authors [30,31]. The presence of excess built-in holes improves the gain characteristics significantly. The injected electrons prefer to stay in the ground state, and, in the case of p-doping, these electrons always find ground-state holes to recombine with. As a result, the room-temperature ground-state optical gain increases, and at the same time there is a drastic increase in the differential gain and hence the modulation bandwidth. This section outlines the theory on the effect of p-doping on the gain and modulation bandwidth of QD lasers [30,31].

> **Example 9.8:** The barrier region of a QD laser is 10 nm thick and is doped with 10^{18} acceptors cm^{-3}. Let the areal density of QDs be 5×10^{10} cm^{-2}. Each QD will then contain 20 holes.

The theory assumes that radiative transitions occur between electron and hole states having the same quantum number, m. Both the wetting layer and the number of QD layers are considered. A large number of quantum states in both the bands are considered. The rate equation for photon density is also considered in the usual way. The electron and hole quasi-Fermi levels are obtained by applying a charge neutrality condition to each QD layer. The abovementioned equations are used to calculate the modal gain for undoped samples and for doped samples having 50 holes per QD.

The parameters used for calculation are taken from experiments on QD lasers and are as follows: $\Gamma 0/d = 3.451 \times 10^6$ m^{-1}, $\lambda_0 = 1.29$ μm, $\hbar\Delta\omega = 30$ meV and 10 meV, $n_r = 3.4$, $s_0 = 2$, $s_1 = 8$, $n_{QD}/A_{WL} = 3.0 \times 10^{10}$ cm^{-2} per layer, and number of layers $N_L = 5$, $\tau_{sp,m} = \tau_{sp,WL} = 0.8$ ns for all ms satisfying $m_e = m_h$ and $= 0$ for $m_e \neq m_h$, $\Delta E_e = 80$ meV, and $\Delta E_h = 10$ meV. The waveguide loss is taken as $\alpha_{WG} = 3$ cm^{-1} for the p-doped active region but $\alpha_{WG} = 1$ cm^{-1} for the undoped sample.

The calculated modal gain first increases with number of injected electrons and then saturates [30,31]. The values of gain are larger for p-doped samples; the increase may be as much as 50 times. The importance of p-doping is therefore clearly demonstrated. The modal gain shows a substantial improvement when the linewidth is decreased from 30 to 10 meV, but the values are always lower when the temperature is increased from 300 to 373 K.

The calculated 3 dB modulation bandwidth is plotted against the cavity length for a drive current 10 times the threshold current. Three different cases are considered: undoped sample with 30 meV broadening, p-doped sample with 30 meV broadening, and p-doped sample with 10 meV broadening. For each case, two different temperatures, 300 and 373 K, are considered. The reflectivity values are $R_1 = R_2 = 0.3$. The plots show that there is an optimum cavity length for which the bandwidth is maximum. This is expected, as for long cavities the photon lifetime limits the bandwidth, whereas for shorter cavities gain saturation leads to a decrease in the differential gain and bandwidth.

The calculation may be extended to estimate the temperature dependence of the threshold current in the range 0–100°C. The optimized values of the cavity length are chosen in each case. The values of $T_0 = 303$, 351, and 343 K have been obtained for an undoped sample with 30 meV linewidth, p-doped with 30 meV linewidth, and p-doped with 10 meV linewidth.

9.8 Tunnel-Injection QD Lasers

The advantages of TI-QW lasers over normal single quantum well (SQW) or multiple quantum well (MQW) lasers have been pointed out in Chapter 8. In this section we shall briefly discuss a few investigations about TI-QD lasers [32].

9.8.1 Problems with Conventional QD Lasers

One of the important advantages of QD lasers predicted by Arakawa and Sakaki (1986) is the ultrahigh temperature stability, or, in other words, the infinite value for the characteristic temperature T_0. Unfortunately, however, the best results for the QD lasers come nowhere near this ideal value.

The dominant source of the temperature dependence of the threshold current in a QD laser is parasitic recombination outside the QDs, mainly in the OCLs. The component of current used for recombination in the OCL depends exponentially on temperature, making the total threshold current temperature dependent. Another source for temperature dependence of J_{th} in QD lasers is the inhomogeneous broadening due to QD size distribution. The physical mechanism that leads to temperature dependence of J_{th} is the same in both the OCL and inhomogeneous distribution. In both cases, there is a distribution of energy levels of the carriers. Only a small fraction of the carriers participate in recombination, corresponding to the lasing mode; the rest of the carriers are *wasted*. The fraction of J_{th} contributing to the recombination in nonlasing QDs depends on T, and therefore T_0 is no longer infinite.

The elimination of recombination in the OCL alone leads to a dramatic improvement of the value of T_0. This may be accomplished in a TI-QD laser, in which carriers tunnel into the QDs, where they recombine.

9.8.2 Tunnel-Injection QD Laser

We now consider a TI-QD laser proposed by Asryan and Luryi [32]. Consider the energy band diagram shown in Figure 9.11. Basically, it is a separate confinement DH laser, the n- and p-cladding layers of which inject electrons and holes, respectively. The QD layers (only one layer has been shown in Figure 9.11) have QWs on their two sides, separated by thin barriers. Injection of carriers into the QDs occurs by tunneling from the QWs. Electrons from the left can enter the right QW only through the confined states of the QDs. Similarly, holes entering from the right cannot directly approach the left QW.

The following conditions should be met in order to achieve the desired operation:

1. The lowest electron (or hole) confined state of the injecting QW should be aligned with the electron (or hole) level for the average-sized QD.
2. The barrier heights should be quite high to suppress thermionic emission of carriers from the QWs.

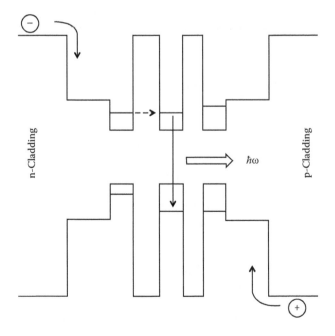

FIGURE 9.11
Energy band diagram of a TI-QD laser. The electron-injecting QW is wider than the hole-injecting QD.

3. Each QD layer should be separated from its neighboring QD layers by a material having a high band gap, so that all tunneling except that between QDs is suppressed.

4. The QDs should be sufficiently separated, so that no significant tunnel splitting occurs; otherwise the splitting may play the same role as inhomogeneous broadening. At the same time, the barrier between QW and QD should be thin to allow effective tunneling.

The hole subband in the right QW is aligned with the hole subband in the QD, but, as the electron subband is higher, electrons from QDs cannot escape through tunneling. The electron subband in the left QW is aligned with the electron subband in the QD. This does not prevent tunneling of QD holes into the electron-injecting QW. The barrier is made wider in order to suppress hole tunneling due to its higher mass, but allows electron tunneling to occur.

9.8.3 Performance

The advantages of TI-QDL, such as high modulation bandwidth, higher value of T_0, and reduced linewidth enhancement factor, have been demonstrated experimentally by several groups. Some representative figures may be found in Bhattacharya et al. [33,34].

PROBLEMS

9.1. Calculate the energy levels of an $In_{0.53}Ga_{0.47}As$ QB with InP as the barrier by using the method outlined in Section 9.3.2. The cubic box is 10 nm wide. Use a 40:60 ratio for the band offsets.

9.2. Calculate the energy levels for electrons and holes as a function of pyramidal base length L by using Equation 9.16 and the parameter values quoted in Example 9.4.

9.3. Show that the expressions for gain using dipole matrix element (Equation 9.8) and momentum matrix element (Equation 9.20) are equivalent when the inhomogeneous broadening parameter G is set to unity in Equation 9.20.

9.4. Obtain the expression for threshold current density (Equation 9.27) by solving the three rate equations (Equations 9.25a through Equation 9.25c).

9.5. Obtain the expression for steady-state solutions of the rate equations given in Eliseev [18].

9.6. Use the empirical expression for $g(J)$ given in Eliseev [18] to calculate the modulation bandwidth.

9.7. Use the expression for modulation bandwidth given in Fiore and Markus [25] and obtain the plot for modulation response for InGaAs QD lasers.

9.8. Prove that the absorption coefficient of a cubic box of side length *a* may be expressed as

$$\alpha = \left[\frac{\left(2\pi e^2 \langle |p_{cv}|^2 \rangle\right)}{\left(a^3 m_0^2 n_r c\omega\right)} \right] \sum_{n^2} g\left(n^2\right) \delta\left(\hbar\omega - E_g - \pi^2\hbar^2 n^2 / 2m_r a^2\right),$$

where $g(n^2)$ is the degeneracy of the energy level determined by n^2.

9.9. Assuming that fluctuation of side length in QDs is given by the following Gaussian distribution

$$P(a) = \frac{1}{D\sqrt{2\pi}} \exp\left[\frac{-(a - a_0)^2}{2D^2} \right],$$

where:

    a_0 is the average value
    D is the standard deviation

This shows that the absorption coefficient may be expressed as

$$\alpha = \left(\frac{\beta}{a_0} \right) \sum_{n^2} \left[\frac{g(n^2)}{\sigma n^2} \right] \exp\left\{ \frac{-(n / x - 1)^2}{2\sigma^2} \right\},$$

where:

$$x^2 = \left[\frac{(\hbar\omega - E_g)}{\left(\pi^2\hbar^2 / 2m_r a_0^2\right)} \right]$$

$$\sigma = \frac{D}{a_0}$$

$$\beta = \left(\frac{1}{\sqrt{2\pi}} \right)\left(\frac{Am_r}{\pi^2\hbar^2} \right)$$

$$A = \left[\frac{\left(2\pi e^2 \langle |p_{cv}|^2 \rangle\right)}{\left(a^3 m_0^2 n_r c\omega\right)} \right]$$

9.10. Establish the expression for the critical radius R_c for a spherical dot.

Reading List

Bhattacharya, P. and Z. Mi, Quantum dot optoelectronic devices, *Proc. IEEE*, 95, 9, 1723–1740, 2007.

Bhattacharya, P., S. Ghosh, and A. D. Stiff-Roberts, Quantum dot optoelectronic devices, *Annu. Rev. Mater. Res.* 34, 1–40, 2004 (review).

Bimberg, D. and U. W. Pohl, Quantum dots: Promises and accomplishments, *Mat. Today*, 14, 9, 388–397, 2011.

Bimberg, D., M. Grundmann, and N. N. Ledentsov, *Quantum Dot Heterostructures*. Wiley, Chichester, 1999.

Blood, P., Gain and recombination in quantum dot lasers, *IEEE J. Sel. Top. Quantum Electron.* 15, 3, 808–818, 2009.

Coleman, J. J., J. D. Young, and A. Garg, Semiconductor quantum dot lasers: A tutorial, *IEEE J. Lightwave Technol.*, 29, 4, 499–510, 2011.

Kapon, E., Chapter 4: Quantum wire and quantum dot lasers. In *Semiconductor Lasers*, ed. Kapon, E. Academic, San Diego, pp. 291–360, 1997.

Ledentsov, N. N., Quantum dot lasers, *Semicond. Sci. Technol.*, 26, 014001(1–8) 2011 (review).

Smyder, J. A. and T. D. Krauss, Coming attractions for semiconductor quantum dots, *Mat. Today*, 14, 9, 382–387, 2011.

Sugawara, M. (Ed.), *Self Assembled InGaAs/GaAs Quantum Dots, Semiconductors and Semimetals*, vol. 60. Academic, San Diego, CA, 1999.

References

1. Van der Ziel, J. P., R. Dingle, R. C. Miller, W. Wiegmann, and W. A. Nordland, Jr., Laser oscillation from quantum states in very thin GaAs–$Al_{0.2}Ga_{0.8}$As multilayer structures, *Appl. Phys. Lett.* 26, 463–465, 1975.

2. Dupuis, R. D., P. D. Dapkus, R. Chin, N. Holonyak, Jr., and S. W. Kirchoefer, Continuous 300 K laser operation of single-quantum-well $Al_xGa_{1-x}As$–GaAs heterostructure diodes grown by metalorganic chemical vapor deposition, *Appl. Phys. Lett.* 34, 265–267, 1979.

3. Arakawa, Y. and H. Sakaki, Multidimensional quantum well laser and temperature dependence of its threshold current, *Appl. Phys. Lett.* 40, 939–941, 1982.

4. Asada, M., Y. Miyamoto, and Y. Suematsu, Gain and threshold of three-dimensional quantum box lasers, *IEEE J. Quantum Electron.* QE-22, 9, 1915–1921, 1986.

5. Miyamoto, Y., Y. Miyake, M. Asada, and Y. Suematsu, Threshold current density of GaInAsP/InP quantum box lasers, *IEEE J. Quantum Electron.* 25, 9, 2001–2006, 1989.

6. Jiang, H. and J. Singh, Strain distribution and electronic spectra of InAs/GaAs self-assembled dots: An eight-band study, *Phys. Rev. B* 56, 4696–4701, 1997.

7. Stier, O., M. Grundmann, and D. Bimberg, Electronic and optical properties of strained quantum dots modeled by 8-band k.p theory, *Phys. Rev. B*, 59, 5688, 1999.

8. Zunger, A., Electronic-structure theory of semiconductor quantum dots, *MRS Bull.* 23, 35–42, 1998.

9. Califano, M. and P. Harrison, Quantum box energies as a route to the ground state levels of self-assembled InAs pyramidal dots, *J. Appl. Phys.* 88, 5870–5874, 2000.

10. Qasaimeh, O., Effect of inhomogeneous line broadening on gain and differential gain of quantum dot lasers, *IEEE Trans. Electron Dev.*, 50, 7, 1575–1581, 2003.

11. Park, G., O. B. Shchekin, D. L. Huffaker, and D. G. Deppe, Lasing from InGaAs/GaAs quantum dots with extended wavelength and well-defined harmonic-oscillator energy levels, *Appl. Phys. Lett.* 73, 23, 3351–3353, 1998.

12. Shchekin, O. B., G. Park, D. L. Huffaker, and D. G. Deppe, Discrete energy level separation and the threshold temperature dependence of quantum dot lasers, *Appl. Phys. Lett.*, 77, 4, 466–468, 2000.

13. Park, G., O. B. Shchekin, and D. G. Deppe, Temperature dependence of gain saturation in multilevel quantum dot lasers, *IEEE J. Quantum Electron.* 36, 9, 1065–1071, 2000.

14. Sugawara, M., Theory of spontaneous-emission lifetime of Wannier excitons in mesoscopic semiconductor quantum disks, *Phys. Rev. B*, 51, 16, 10743–10754, 1995.

15. Asryan, L. V., M. Grundmann, N. N. Ledentsov, O. Stier, R. A. Suris, and D. Bimberg, Effect of excited state transitions on the threshold characteristics of a quantum dot laser, *IEEE J. Quantum Electron.* 37, 418–425, 2001.

16. Deppe, D. G., D. L. Huffaker, S. Csutak, Z. Zou, G. Park, and O. B. Shchekin, Spontaneous emission and threshold characteristics of 1.3-mu m InGaAs–GaAs quantum-dot GaAs-based lasers, *IEEE J. Quantum Electron.* 35, 8, 1238–1246, 1999.

17. Park, G., O. B. Shchekin, and D. G. Deppe, Temperature dependence of gain saturation in multilevel quantum dot lasers, *IEEE J. Quantum Electron.* 36, 1065–1071, 2000.

18. Eliseev, P. G., H. Li, T. Liu, T. C. Newell, L. F. Lester, and K. J. Malloy, Ground-state emission and gain in ultralow-threshold InAs–InGaAs quantum-dot lasers, *IEEE J. Sel. Topics Quantum Electron.* 7, 2, 135–142, 2001.

19. Ishida, M., N. Hatori, T. Akiyama, K. Otsubo, Y. Nakata, H. Ebe, M. Sugawara, and Y. Arakawa, Photon lifetime dependence of modulation efficiency and K factor in 1.3 mm self-assembled InAs/GaAs quantum-dot lasers: Impact of capture time and maximum modal gain on modulation bandwidth, *Appl. Phys. Lett.* 85, 18, 4145–4147, 2004.

20. Lüdge, K., E. Schöll, E. Viktorov, and T. Erneux, Analytical approach to modulation properties of quantum dot lasers, *J. Appl. Phys.* 109, 103112 (1–8), 2011.

21. Kirstaedter, N., N. N. Ledentsov, M. Grundmann, D. Bimberg, V. M. Ustinov, S. S. Ruvimov, M. V. Maximov, et al., Low threshold, large T_0 injection laser emission from (InGa)As quantum dots, *Electron. Lett.* 30, 17, 1416–1417, 1994.

22. Bimberg, D., M. Grundmann, and N. N. Ledentsov, *Quantum Dot Heterostructures*. Wiley, Chichester, 1991.

23. Grundmann, M. and D. Bimberg, Theory of random population for quantum dots, *Phys. Rev. B*, 55, 15, 9740–9745, 1997.

24. Hatori, N., M. Sugawara, K. Mukai, Y. Nakata, and H. Ishikawa, Room temperature gain and differential gain characteristics of self-assembled InGaAs/GaAs quantum dots for 1.1–1.3 μm semiconductor lasers, *Appl. Phys. Lett.* 77, 6, 773–775, 2000.

25. Fiore, A. and A. Markus, Differential gain and gain compression in quantum-dot lasers, *IEEE J. Quantum Electron.* 43, 3, 287–294, 2007.
26. Dikshit, A. A. and J. M. Pikal, Carrier distribution, gain and lasing in 1.3 μm InAs–InGaAs quantum-dot lasers, *IEEE J. Quantum Electron.* 40, 2, 105–112, 2004.
27. Snoke, D. W. and J. D. Crawford, Hysteresis and MOTT transition between plasma and insulating gas, *Phys. Rev. E.* 52, 6, 5796–5799, 1995.
28. Shchekin, O. B. and D. G. Deppe, 1.3 μm InAs quantum dot laser with T = 161 K from 0 to 80 C, *Appl. Phys. Lett.* 80, 18, 3277–3279, 2002.
29. Basu, R., B. Mukhopadhyay, and P. K. Basu, Analytical theory of small signal modulation response of a transistor laser with dots-in-well in the base, *Semicond. Sci. Technol.*, 27, 015022 (1–7), 2012.
30. Shchekin, O. B. and D. G. Deppe, The role of p-type doping and the density of states on the modulation response of quantum dot lasers, *Appl. Phys. Lett.* 80, 15, 2758–2760, 2002.
31. Deppe, D. G., H. Huang, and O. B. Shchekin, Modulation characteristics of quantum-dot lasers: The influence of P-type doping and the electronic density of states on obtaining high speed, *IEEE J. Quantum Electron.* 38, 12, 1587–1593, 2002.
32. Asryan, L. V. and S. Luryi, Tunneling-injection quantum-dot laser: Ultrahigh temperature stability, *IEEE J. Quantum Electron,* 37, 7, 905–910, 2001.
33. Bhattacharya, P. and S. Ghosh, Tunnel injection $In_{0.4} Ga_{0.6} As/GaAs$ quantum dot lasers with 15 GHz modulation bandwidth at room temperature, *Appl. Phys. Lett.* 80, 3482–3484, 2002.
34. Bhattacharya, P., S. Ghosh, S. Pradhan, J. Singh, W. Zong-Kwei, J. Urayama, K. Kim, and T. B. Norris, Carrier dynamics and high-speed modulation properties of tunnel injection InGaAs–GaAs quantum-dot lasers, *IEEE J. Quantum Electron.* 39, 8, 952–962, 2003.

10

Quantum Cascade Lasers

10.1 Introduction

Semiconductor lasers, discussed in Chapters 7–9, involve recombination of electron–hole pairs, that is, transition from a conduction band (subband) state to a state in the valence band (subband). The processes are therefore interband processes.

In this chapter, we consider lasers based on transition between states in two subbands belonging to the same band, either conduction or valence bands. These lasers are called intersubband lasers, or unipolar lasers, as the charge particles have the same polarity, that is, either an electron or a hole makes stimulated transitions to give rise to laser action.

The present chapter begins with a brief introduction to the development of such lasers. Two main features of these lasers, unipolarity and quantum cascading, will then be introduced. After describing the basic structures and the principle of action, the theory of gain in quantum cascade lasers (QCLs), as they are now called, will be developed. The simplified rate equation model to describe the static and dynamic characteristics will then be presented. Some design and material issues will be discussed thereafter. The chapter ends with a brief discussion on QCLs using quantum dots (QDs).

10.2 A Brief History

In 1960, before the invention of diode lasers, Lax [1] put forward a proposal for semiconductor lasers based on transitions between quantized Landau levels in a semiconductor under a strong magnetic field. This was the first proposal for lasers based on transition between quantized levels in the same band and thus also for the unipolar laser. However, with the advent of diode lasers, interest shifted to these devices, which depend on electron–hole recombination. After the announcement of quantum wells (QWs), Kazarinov and Suris [2] suggested that optical gain could be achieved due to transitions between two-dimensional states in a superlattice biased by an external electric field.

There was no substantial progress in this area over the next 15 years or so following the first proposal. Researchers working on resonant tunneling rekindled the subject. In 1986, Capasso [3], working on resonant tunneling, put forward a proposal for a unipolar laser based on a superlattice structure. There had been several suggestions [4,5] and estimates for gain in intersubband lasers. It was Capasso who started working on the structure, and, with contributions from Faist and Sirtori, his group first announced unipolar laser action based on the quantum cascade structure [6].

Since its first announcement, several groups have reported QCL action using the GaAs/AlGaAs system [7], continuous wave operation [8], and room-temperature continuous-wave (CW) operation [9]. The operating wavelength range has been extended [10–14]. In the few years after the first announcement, dramatic improvements of the performance of QCLs were reported by several groups. The improvements came in the form of CW operation at room temperature and above [8,9], extending the wavelengths to as high as 11.5 µm, achieving QCL action in the terahertz range of frequencies [15–17], obtaining room-temperature high-power devices [8,9], and so on.

10.3 Basic Principle

As mentioned in the introduction, QCLs possess two unique features: unipolarity and cascading. In this section, these two features will be explained and illustrated. The conditions for population inversion between two subbands will then be established. After describing some of the structures used to achieve QCL action, the expressions for the intersubband gain and the threshold current density will be developed.

10.3.1 Unipolarity and Cascading

The unipolar or QC laser differs from the conventional diode lasers in many ways, the principal differences occurring due to two unique features: unipolarity (only electrons or only holes) and a cascading scheme (electron recycling). These two processes are shown in Figure 10.1 and are independent (please see Sirtori and Teissier in Reading List).

Figure 10.1 a shows the energy band diagrams of the conduction band and the valence band. In both the diagrams, a few subbands are shown. Also shown in this figure is the downward transition of an electron from an upper subband to the lower subband, giving rise to characteristic emission. A transition between a heavy-hole (HH) and a light-hole (LH) subband is shown in Figure 10.1a as an example, though transitions between two HH subbands or

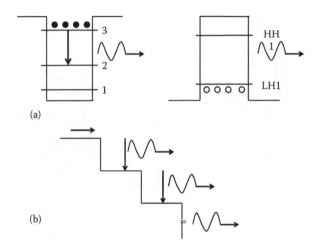

FIGURE 10.1
Schematic diagram showing (a) unipolarity and (b) cascading.

two LH subbands are also possible. The transitions illustrated in Figure 10.1a occur within the same band. The wavelength of emitted light is not dependent on the bandgap of the QW material, but is determined by the difference in subband energies. The energy separation between the two subbands is mostly dependent on the width of the QW, and, to some extent, the electron (or hole) effective masses in the well and the barrier, and the band discontinuities. If the E–$\mathbf{k}$ relation for the subbands is parabolic, then the emission spectra will be extremely sharp, in contrast to the case in interband lasers.

The second important feature of the QCL, cascadability, is illustrated in Figure 10.1b. The QCL structure usually consists of a multistage structure in which the electron responsible for transition in a QW is used in the next stage to induce a similar transition. Thus, if there are N stages, the same electron yields a single photon in each stage, thereby producing N photons, provided the above-threshold condition is maintained. This recycling of electrons increases the quantum efficiency and light power output, since both the quantities are proportional to the number of photons.

10.3.2 Basic Structure

Figure 10.2 shows the band diagram for electrons in one of the simplest QCL structures. In the figure, an active region is placed between two injector regions. The active region has two coupled QWs separated by a barrier layer. A typical material for the well is $In_{0.53}Ga_{0.47}As$, and for the barrier $In_{0.52}Al_{0.48}As$; these are both lattice matched to InP. QCLs with GaAs/AlGaAs QWs have also been demonstrated. The injector region is doped with impurities, and for InGaAs/InAlAs systems the Si-doping level is about 3–5×10^{16} cm^{-3}. The superlattice in the left part of the figure contains a miniband, which is aligned

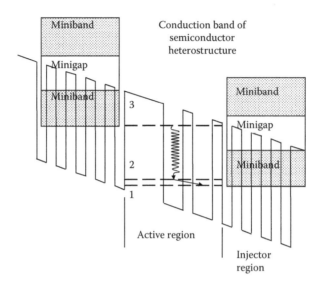

FIGURE 10.2
Basic QCL structure using conduction subbands.

with energy Level 3 in the coupled double QW. The electron is injected into
Level 3, from which it makes a radiative transition to Level 2 in the active
QW region. The electron then quickly transits to Level 1. With proper design
of the well and barrier widths, the separation between Levels 2 and 1 is made
equal to the longitudinal optic (LO) phonon energy, so that the electron in
Level 2 can very quickly scatter into Level 1. It then moves to the miniband in
the next superlattice by resonant tunneling. The injection region in the right-
hand side injects the electron into Level 3 of the next CDQW. Population
inversion is achieved through careful control of the lifetimes of the upper
and lower states of the CDQW. The process then repeats itself. The injectors
and the active region with QWs form one period of the laser. The pattern is
repeated so that the number of emitted photons is multiplied in the cascade
structure by utilizing the same carrier over many periods. The advantage is
that the same electron can be utilized to create another photon in the next
active QW. Although the process is illustrated using conduction subbands, it
may readily be applied to hole subbands also.

The number of active regions in a typical QCL structure is typically 20–30
and may be as high as 100, although laser action has been observed for a
single active region [11].

10.3.3 Condition for Population Inversion

A simplified band diagram for a typical period of the QCL is shown in
Figure 10.3. The injector injects an electron to Level 3 in the QW with

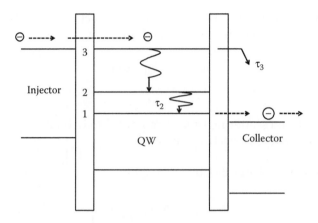

FIGURE 10.3
Active zone in a QCL.

energy E_3. The radiative transition occurs in the QW from Level 3 to Level 2 (energy E_2). Subsequently, the electron moves down to Level E_1 in the QW by scattering and is then injected into the collector, which acts as the injector for the next stage. The processes described above are then repeated in the next active zone. Denoting the population in Levels 3 and 2 by N_3 and N_2, respectively, we may write the rate equations in the QW as (please see Sirtori and Teissier in Reading List)

$$\frac{dN_3}{dt} = \frac{\eta_i J}{e} - \frac{N_3}{\tau_{31}} - \frac{N_3}{\tau_{32}} - G(N_3 - N_2) \tag{10.1}$$

$$\frac{dN_2}{dt} = \frac{N_3}{\tau_{32}} + G(N_3 - N_2)S - \frac{N_2}{\tau_{21}} \tag{10.2}$$

In Equations 10.1 and 10.2, the total current density in the injector is J and that injected into Level 3 is J_3, where $J_3 = \eta_i J$, and η_i is the injection efficiency. The time constant for 3→2 transition is τ_{32}, where other time constants may be similarly defined, S is the photon density, and G is the gain. If we neglect both S and G, the steady-state sheet density of electrons in Level 3 of the QW may be expressed as

$$N_3 = \left(\frac{\eta_i J}{e} \right) \tau_3 \tag{10.3}$$

where τ_3 is the total lifetime of the electron in Level 3, $\tau_3^{-1} = \tau_{32}^{-1} + \tau_{31}^{-1}$.

Assuming that Level 2 is populated only by direct scattering from Level 3, the population in Level 2 in steady state is given by

$$N_2 = N_3 \left(\tau_{21} / \tau_{32} \right) \tag{10.4}$$

where:

τ_{32} is the mean scattering time from Level 3 to Level 2
τ_{21} is the lifetime of electrons in Level 2

Using Equations 10.3 and 10.4, one may write

$$N_3 - N_2 = \frac{\eta_i J \tau_3}{e} \left(1 - \frac{\tau_{21}}{\tau_{32}} \right) \tag{10.5}$$

Equation 10.5 indicates that a high degree of population inversion may be achieved if the upper-state lifetimes τ_3 and τ_{32} are large and at the same time the lower-state lifetime τ_{21} is made very small. This requires that the electrons should stay in the upper state for a longer time, and the electrons in the lower state should be drained away as quickly as possible. The injection efficiency should also be made quite high.

If all electrons move from Level 3 to Level 2, making $\tau_3 = \tau_{32}$, the above equation reduces to

$$N_3 - N_2 = \frac{\eta_i J}{e} \left(\tau_3 - \tau_{21} \right) \tag{10.6}$$

Thus, the condition for population inversion is $\tau_3 > \tau_{21}$. If $E_2 - E_1 = \hbar\omega_{LO}$, where $\hbar\omega_{LO}$ is the energy of the LO phonon, τ_{21} may be made extremely short, and population inversion or optical gain is guaranteed from the first flow of electron. The light power then increases without any threshold, provided there are no material or mirror losses.

> **Example 10.1:** The widths of InGaAs QWs are 6.0 and 4.7 nm, and that of the InAlAs barrier is 1.6 nm. Calculation gives $E_{32} = 207$ meV, $E_{21} = 37$ meV: the LO phonon energy. Also, $\tau_{32} = 2.2$ ps, $\tau_{31} = 2.1$ ps, $\tau_3 = 1.1$ ps, and $\tau_{21} = \tau_2 = 0.3$ ps. This ensures that $\tau_{32} \gg \tau_2$, so that the condition for population inversion is satisfied (see Gmachl et al. in Reading List).

10.3.4 Intersubband Absorption and Gain

Let us first consider the transition from initial subband $|i\rangle$ to an upper final subband $|f\rangle$ by absorption of light of energy $\hbar\omega$. The transition rate W

for a given interaction potential V_p is expressed by using the Fermi golden rule as

$$W = \frac{2\pi}{\hbar} \sum_f \left| \langle \psi_f | V_p | \psi_i \rangle \right|^2 \delta(E_i - E_f - \hbar\omega) \tag{10.7}$$

The wave functions may be written as

$$\psi_i = S^{-1/2} U_c(\mathbf{r}) \exp(i\mathbf{k} \cdot \mathbf{\rho}) \phi_i(z)$$

$$\psi_f = S^{-1/2} U_c(\mathbf{r}) \exp(i\mathbf{k'} \cdot \mathbf{\rho}) \phi_f(z) \tag{10.8}$$

where:
   S is the surface area
   ϕs   are the envelope functions
   U_cs   are Bloch functions
   $\mathbf{k}$, $\mathbf{k'}$, and ρ refer to two-dimensional (2-D) vectors

One may note that the envelope functions ϕ are slowly varying in comparison to the cell periodic part Us. Therefore, the above matrix element may be expressed as [18]

$$\langle U_{c'}\phi_f | V_p | U_c\phi_i \rangle = \langle U_{c'} | V_p | U_c \rangle_{cell} \langle \phi_f | \phi_i \rangle + \langle U_{c'} | U_c \rangle_{cell} \langle \phi_f | V_p | \phi_i \rangle \tag{10.9}$$

The subscript *cell* indicates that the scalar product has been evaluated over a unit cell.

The first term in Equation 10.9 reduces to zero if a transition occurs within the same band, since ϕ_i and ϕ_f, being the eigenfunctions of the same effective-mass Hamiltonian, are orthogonal to each other. In the second term, the cell periodic parts of two different bands denoted by subscripts c and c' at the same point of the Brillouin zone are orthogonal, and therefore the inner product vanishes. The matrix element may then be written as

$$\langle \psi_f | V_p | \psi_i \rangle = \langle U_{c'} | U_c \rangle_{cell} \langle \phi_f | V_p | \phi_i \rangle \tag{10.10}$$

As may be noted, the cell periodic terms are considered for $\mathbf{k} \neq 0$. Using the expressions developed by **k.p** theory, one may write

$$U_{nk} = U_{n0} + \frac{\hbar}{m_0} \sum_{m0} \frac{\mathbf{k}.\langle m0 | p | n0 \rangle U_{m0}}{E_{n0} - E_{m0}} \tag{10.11}$$

Using the form for U_{ck} and $U_{ck'}$ to calculate the matrix element and noting that the effective mass is expressed as

$$m_e = \delta_{ij} + \frac{2}{m_0} \sum_{m0} \frac{\langle m0|\mathbf{p}|n0\rangle\langle n0|\mathbf{p}|m0\rangle}{E_{n0} - E_{m0}} \tag{10.12}$$

the matrix element in Equation 10.10 reduces to

$$\langle \psi_f|V_p|\psi_i\rangle = -\frac{eA_0}{m_e}\langle \phi_f|\boldsymbol{\varepsilon}\cdot\mathbf{p}|\phi_i\rangle \tag{10.13}$$

In writing Equation 10.13, the standard form of V_p is used, and the momentum of the photon is neglected. In Equation 10.13, the effective mass of the electron appears, whereas for interband transition the free electron mass is involved. The intersubband absorption is much like free-carrier absorption or high-frequency conduction, and therefore the presence of the effective mass is expected. In Equation 10.13, the matrix element is written in terms of the momentum operator $\mathbf{p}$. In some cases, this is written in terms of dipole matrix element:

$$\langle \psi_f|V_p|\psi_i\rangle = -\frac{eA_0(E_i - E_f)}{i\hbar}\boldsymbol{\varepsilon}\cdot\langle \phi_f|\mathbf{r}|\phi_i\rangle \tag{10.14}$$

where E_i and E_f denote, respectively, the energies of the initial and final states (subbands). Since the envelope functions are functions of z only and are orthogonal to each other, it follows easily that only the matrix element $\langle \phi_f|z|\phi_i\rangle$ is nonzero and that both $\langle \phi_f|x|\phi_i\rangle$ and $\langle \phi_f|y|\phi_i\rangle$ vanish. Absorption is thus forbidden when the light is polarized along the plane of the QW.

Let us assume that the incident radiation has an arbitrary direction of incidence and of polarization. Let the photon wave vector lie in the xy plane, and angles θ and β are defined such that $\cos^2\theta = A_{0x}^2/(A_{0x}^2 + A_{0z}^2)$ and $\cos^2\beta = A_{0z}^2/(A_{0x}^2 + A_{0z}^2)$, where A_{0x} and A_{0z} are the components of A_0. Thus, using the expression for A_0 and following the usual steps, the absorption coefficient may be expressed as [19,20]

$$\alpha_{if}(\hbar\omega) = \frac{\omega e^2 \cos^2\beta}{Ln_r\varepsilon_0 c} \sum_f (f_{ei} - f_{ef})\langle z\rangle^2 \delta(E_f - E_i - \hbar\omega) \tag{10.15}$$

where $\langle z\rangle = \langle \phi_f|z|\phi_i\rangle$.

In Equation 10.15, fs denote the Fermi occupational probabilities of electrons, and the summation over f is over all the final states, satisfying energy and momentum conservation. Converting the summation into an integration of 2-D wave vector $\mathbf{k}$ and making use of the energy-conserving δ-function, one may express

$$\alpha(\hbar\omega) = \sum_{i,f} \frac{\omega e^2 \cos^2\beta}{L n_r \varepsilon_0 c} \frac{m_e k_B T}{\hbar^2} \langle z \rangle^2$$

$$\times \ln \frac{1+\exp\left[(E_F - E_i)/k_B T\right]}{1+\exp\left[(E_F - E_f)/k_B T\right]} \frac{\Gamma/2}{(\hbar\omega - E_{fi})^2 + (\Gamma/2)^2} \qquad (10.16)$$

E_F in Equation 10.16 denotes the Fermi level for electrons. The logarithmic ratio is related to $(N_i - N_f)$: the difference in populations in the initial and final subbands. Using the relation and expressing the Lorentzian lineshape function by the symbol $L(\hbar\omega, E_{fi})$, the absorption coefficient is written in the following form:

$$\alpha(\hbar\omega) = \frac{\omega e^2 \cos^2\beta}{n_r \varepsilon_0 c L} \langle z \rangle^2 (N_i - N_f) L(\hbar\omega, E_{fi}) \qquad (10.17)$$

The expression for $\langle z_{mn} \rangle^2$ for transition between the mth and nth levels may be derived analytically for the simplest case of infinite barrier height. The envelope functions are written as

$$\phi_n(z) = \left(\frac{2}{L}\right)^{1/2} \cos\left(\frac{n\pi z}{L}\right): \quad \text{for } n = \text{odd integer} \qquad (10.18a)$$

$$\phi_n(z) = \left(\frac{2}{L}\right)^{1/2} \sin\left(\frac{n\pi z}{L}\right): \quad \text{for } n = \text{even integer} \qquad (10.18b)$$

The momentum matrix element is expressed as

$$\langle n|p_z|n'\rangle = \int_{-L/2}^{L/2} \left(\frac{2}{L}\right)^{1/2} \cos\left(\frac{n\pi z}{L}\right)\left(\frac{-i\hbar d}{dz}\right)\left(\frac{2}{L}\right)^{1/2} \sin\left(\frac{n'\pi z}{L}\right) dz$$

Performing the integration, we may write

$$\langle n|p_z|n'\rangle = -\frac{2i\hbar n'}{L} \left\{ \frac{\sin\left[(n'+n)\pi/2\right]}{n'+n} + \frac{\sin\left[(n'-n)\pi/2\right]}{n'-n} \right\} \qquad (10.19)$$

For $n=1$ and $n'=2$, the momentum matrix element is $-i8\hbar/3L$. For $2\rightarrow 1$ transition $\langle p_z \rangle = iE_{21} m_e \langle z \rangle / \hbar$ and $E_{21} = 3\hbar^2\pi^2/2m_e L^2$, we thus obtain $\langle z_{21} \rangle = 16L/9\pi^2$.

Example 10.2: The oscillator strength for transition from the first excited subband (2) to the ground subband (1) is expressed as $f_{osc} = (2m_0 E_{21}/\hbar^2)$ $\langle z_{21} \rangle^2$. The expression for $\langle z_{21} \rangle$ is given above; we obtain $f_{osc} = (m_0/m_e)$ $(256/27\pi^2) = 14.34$, using $m_e = 0.067 \, m_0$ for GaAs.

Example 10.3: The values of subband energies for a 6.5 nm wide GaAs QW are $E_1 = 54.7$ meV and $E_2 = 218.8$ meV. The same values are obtained for a thickness of 10.13 nm and infinite barrier height. The Fermi level is $E_F - E_1 = 6.49$ meV for an electron density of 1.6×10^{17} cm^{-3}. Using half-width $\Gamma/2 = 5$ meV and the above-quoted value of $\langle z_{21} \rangle$ in Equation 10.17, a value of 2.9×10^5 m^{-1} is obtained for the peak absorption coefficient (or gain).

10.3.5 Gain and Threshold Current

The expression for gain may be written from Equation 10.17 in terms of the dipole matrix element as

$$G(\hbar\omega) = \frac{2e^2\omega}{n_r \varepsilon_0 c 2\gamma L} \langle z_{32} \rangle^2 (N_3 - N_2)$$

(10.20)

where 2γ is the full width at half maximum (FWHM). The gain may be expressed in terms of the gain coefficient g and the drive current density J as

$$G = gJ$$

(10.21)

The gain coefficient may be defined as

$$g(\hbar\omega) = \frac{4\pi e \langle z_{32} \rangle^2}{n_r \varepsilon_0 \lambda L_p} \eta_i \tau_3 \left(1 - \frac{\tau_2}{\tau_{32}}\right) \frac{1}{2\gamma_{32}}$$

(10.22)

In this expression, L_p is the thickness of one period of active region and injector region, although the optical transition is taking place only in the region occupied by the coupled QWs. The gain coefficient is the gain per current density.

In order to relate the gain coefficient to the threshold current density, the losses encountered by the light wave in the waveguide are to be determined. The first loss component is the mirror loss α_m defined in terms of the reflectivities of the mirror and the length of the laser. The second important source of loss is the free-carrier absorption in the doped semiconductor region and the metallic contact layer. The influence of the latter may be minimized by careful design. However, the free-carrier absorption loss, which is proportional to λ^2, is important, especially at long wavelengths. For InGaAs/InAlAs systems, measurements indicate that the loss, termed the *waveguide loss*, is expressed as α_w (cm^{-1}) = 0.535λ (μm)^2. The third source of absorption losses arises due to the resonant intersubband transitions. The extrinsic electrons present in the doped injector layer may absorb the emitted laser light due to optical transitions resonant with the emitted laser wavelength. Proper design of the active and injector layers may avoid this loss.

The threshold current density may now be expressed as

$$J_{th} = \frac{\alpha_m + \alpha_w}{g\Gamma} \tag{10.23}$$

where $\Gamma = \Gamma_p N_p$ is the mode confinement factor, and we consider that the active region comprises N_p periods, each having the same confinement factor Γ_p.

Some representative structures and experimental data highlighting the important features of QCLs are given in [13–18] (and please see Reading List).

> **Example 10.4**: The value of the gain coefficient for the 3–2 transition will be calculated for a GaAs QW of infinite barrier height. The energy of the lowest subband is 56.1 meV. Therefore, the wavelength for the 3–2 transition is 4.41 μm. The value of $\langle z_{32} \rangle = (72L/25\pi^2)$. Let $\eta_i = 0.4$, $\tau_{32} = 2.1$ ps, $\tau_2 = 0.5$ ps, $\tau_3 = 3.4$ ps, $2\gamma = 10$ meV, and $L = 10$ nm. The calculated value is $g = 8.4 \times 10^{-3}$ m A^{-1}. This means that, in order to achieve a value of $G = 1.5 \times 10^3$ m^{-1}, with $\Gamma = 0.1$ the current density needed is $J = 1.79 \times 10^6$ A m^{-2}.

> **Example 10.5**: The emission wavelength of an InGaAs/InAlAs QCL is 6.0 μm. $n_r = 3.25$, $L_p = 47$ nm, $\tau_2 = 0.3$ ps, $\tau_{32} = 2.2$ ps, $\tau_3 = 1.1$ ps, $2\gamma_{32} = 20$ meV, and $z_{32} = 2.0$ nm. For $L = 2.5$ mm, $\alpha_m = 5.1$ cm^{-1}, $\alpha_w = 19$ cm^{-1}; let us take $\Gamma = 0.1$, and $\eta_i = 1$. Thus, $g = 30 \times 10^{-5}$ m A^{-1}, and the threshold current density is 8.2 kA cm^{-2}.

10.3.6 Temperature Dependence

It is safe to assume that the waveguide and mirror losses and the confinement factor are independent of temperature. The thermally generated free carriers and the electronic structure are neglected. Therefore, the temperature dependency of the LO phonon scattering time and the width of the gain spectrum should affect the threshold current density. In addition, since the electrons occupy higher energy states with a rise in temperature, the non-parabolicity effect may affect the scattering times and the broadening of the gain spectrum.

With a rise in temperature, extrinsic electrons from the injector can be back excited into the bottom laser level, thereby reducing the population inversion. The following expression for the threshold current density has been proposed by Faist et al. [10], which leads to good agreement with the experimental results:

$$J_{th} = \frac{1}{\tau_3(T)\left(1 - \dfrac{\tau_2(T)}{\tau_{32}(T)}\right)} \left[\frac{\varepsilon_0 n_r L_p \lambda \left(2\gamma_{32}(T)\right)}{4\pi e z_{32}{}^2} \cdot \frac{\alpha_m + \alpha_w}{\Gamma} + e n_g \exp\left(-\frac{\Delta}{k_B T}\right)\right]$$

(10.24a)

where:

$$\tau_i(T) = \frac{\tau_{i0}}{1 + 2/\left[\exp\left(E_{LO}/k_B T\right) - 1\right]}, \quad i = 1, 2, 3, \ldots$$

(10.24b)

τ_{i0} is the scattering time at low temperature
E_{LO} is the LO phonon energy
n_g is the sheet carrier density at the ground state of the injector
Δ is the energy separation between the injector ground state and Level 2 of the preceding active region

The temperature dependence of J_{th} may be expressed by the usual expression involving the characteristic temperature T_0 but slightly modified to fit the temperature dependence over the whole range of laser operation. Thus,

$$J_{th}(T) = J_0 \exp\left(\frac{T}{T_0}\right) + J_1$$

(10.25)

The temperature-independent part, J_1, proves to be useful for fitting the data for QCLs having a superlattice active region.

10.3.7 Carrier Transport

In the earlier subsections, it has been assumed that the injection of an electron from the injector into Level 3 of the active region and its removal from Level 1 to the next section occur very quickly. The injection from the injector ground state g to Level 3 occurs via resonant tunneling. The strong coupling between g and 3 induces a splitting ΔE, and the tunneling time is given by

$$\tau_{tun} = \frac{h}{2\Delta E}$$

(10.26a)

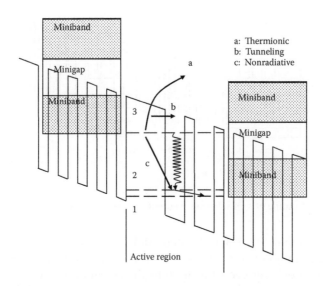

FIGURE 10.4
Current paths not contributing to the laser action.

The injected current density is given by

$$J = \frac{en_g}{2\tau_3} \tag{10.26b}$$

Typical tunnel time is a fraction of a picosecond, and the current density is determined by the doping level in the injector. The current density is chosen to be low enough to reduce waveguide losses but high enough to support the maximum current needed. The exit barrier is designed to achieve a tunneling time in the subpicosecond range. Equations 10.26a and 10.26b are applicable for fully resonant conditions. However, sufficient tunnel current may flow over a certain range of applied field around the current under fully resonant conditions.

We now include the current components, not included in earlier discussions, that do not contribute to the laser action. Figure 10.1 has been modified to include the detrimental current paths, and Figure 10.4 is the modified diagram. The additional current paths are as follows:

1. Scattering from the injector ground state directly into the bottom states 2 and 1 in the active region, bypassing the 3→2 laser transition. The effect is accounted for by including the injection efficiency η_i in the current density in Equation 10.1.

2. Electrons can scatter directly into Level 1. The upper-state lifetime τ_3 decreases, reducing the gain coefficient.

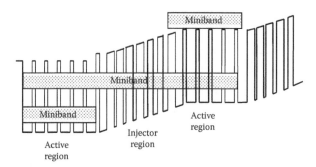

FIGURE 10.5
Conduction band profile of a QCL having a superlattice active region.

3. When Upper State 3 is located high up and close to the quasi-continuum above the barrier, electrons may reach it by thermionic emission or may even tunnel into the continuum. These electrons are accelerated by the electric field in the continuum and do not participate in laser action.

4. It has been assumed so far that electrons are directly injected following the $g \to 3$ path. Resonant tunnel injection may also occur along the path $g \to 3 + E_{LO}$.

10.4 Improved Design of Structures

The QCL structure discussed in the previous section comprises a double QW as the active layer. This structure was employed to demonstrate QCL action for the first time. Since then, many different structures and designs for the QCLs have been reported in the literature. In this section, a few of the structures characterized by improved design will be presented, and their important features will be discussed.

10.4.1 Three-Well Vertical Transition Active Region

In this structure, a thin well is inserted between the injector and the CDQW. The additional QW prevents electron scattering from the injector directly into the laser Ground State 2 as well as Level 1 in the active region. In the presence of the thin first injector QW, the overlap of the wave functions of the injector ground state g with the lower lasing state is not too large, and this maintains a high injection efficiency. This was found to be useful to obtain the first room-temperature operation of a QCL [10].

The structure uses InGaAs and InAlAs, respectively, as the well and barrier materials. The layer thicknesses in nanometers in a typical active region are **3.8**/2.1/**1.2**/6.5/**1.2**/5.3/**2.3**, where bold numbers correspond to the barriers. The n-doped injector superlattice has layer thicknesses 4.0/**1.1**/3.6/**1.2**/ 3.2/**1.2**/3.0/**1.6**/3.0, where the underlined layers are n-type doped. The total thickness of the active and injector layers is therefore 44.3 nm.

In the design, Level 1 of the preceding active region is aligned with Level 3 of the next active region. This allows resonant carrier transport between successive active regions with a very small transit time, ~0.5 ps. The carrier relaxation in the injector is insignificant. As the ground state of the injector is above Level 3 of the next active region, there is sufficient injection even when a larger electric field destroys the resonance between Levels 1 and 3 of successive active layers. This ensures a wide dynamic range of current and high optical power.

As expected, the threshold current density decreases with increasing number of periods, since the confinement factor increases. The decrease is linear at first. However, with increasing number of periods, the overlap factor saturates for higher N_p, and the decrease of J_{th} is less gradual. The characteristic temperature T_0 is typically ~100°C.

10.4.2 Superlattice Active Region

A schematic band diagram of a QCL, in which the active region is in the form of a superlattice (SL) rather than the CDQW or 3-QW active region as discussed before, is shown in Figure 10.6. Here, laser action takes place between minibands. The SL-QCLs possess high gain, large current-carrying capability, and a higher characteristic temperature, T_0. They are particularly useful for the longer-wavelength range, $\lambda \gg 7$ μm.

The active region consists of a few QWs having thin barriers, so that strong coupling between the wells gives rise to minibands, typically two as shown in Figure 10.5. Suitable doping in the SL allows extrinsic carriers to screen the externally applied electric field. Electrons are injected from the doped injector into the second upper miniband, where they occupy states near the bottom. The electrons then make an optical transition to the top of the lower miniband. In the momentum space, these transitions take place at the boundary of the mini-Brillouin zone.

The relaxation process is mediated by LO phonon scattering. It has been proved that the scattering time increases as the wave vector of the LO phonon increases. The injected electrons in the upper miniband quickly relax to the lower states by intraminiband scattering, since the emission of small-wave-vector phonons is involved. On the other hand, interminiband scattering, which is responsible for emission, involves larger wave vectors, and the scattering time is comparatively large, ~10 ps. This high ratio of relaxation times of inter- versus intraminiband processes guarantees population inversion across the minigap. The oscillator strength for

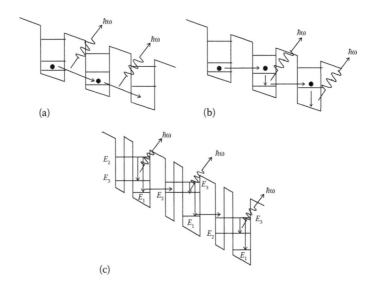

FIGURE 10.6
Schematic of three different QCL structures: (a) interwell, indicating diagonal transition in real space; (b) intrawell; (c) CDQW.

interminiband transition also increases with wave vector, and is maximal at the minizone boundary.

The energy levels and optical dipole matrix element in the presence of an electric field and variation of doping levels are calculated by solving the Schroedinger and Poisson equations self-consistently.

Equation 10.22 may be modified for the SL-QCL and is written as

$$g(\hbar\omega) = \frac{4\pi e \langle z_{21} \rangle^2}{n_r \varepsilon_0 \lambda L_p} \eta_i \tau_2 \left(1 - \frac{\tau_1}{\tau_{21}}\right) \frac{1}{2\gamma_{21}} \tag{10.27}$$

where τ_2 and τ_1 are, respectively, the lifetimes of the upper (2) and lower (1) states. In calculating τ_2, all scattering events from the upper miniband edge (Level 2) to all states in the lower miniband are considered, and the esti-mated value of τ_{21} is ~10 ps.

More improved designs for the SL-QCL structures have been reported in the literature. An important design is the *chirped-SL* QCL. In this design, the SL active region, which is undoped, consists of QWs with gradually decreasing well widths along the direction of electron flow. At zero bias, there is no coupling between wells, as the energies for different widths are out of tune with one another. At a particular electric field, the states are brought into resonance, and the strong coupling leads to the formation of minibands.

10.4.3 Diagonal Transition Active Region

In both two- and three-well active regions, and in the SL active region, the wave functions involved in the transition are located in the same region in real space, and the transition is accordingly termed the *vertical transition*. The applied electric field does not change the energy separation appreciably, and hence the wavelength is insensitive to variation of field as well as temperature. Since the overlap of wave functions is large, the optical dipole matrix element is large, but the electron scattering time is short.

In a different structure, the upper- and lower-state wave functions are located in different regions, as in a type II QW. The transition is diagonal, as shown in Figure 10.6a. The wavelength is now strongly dependent on the electric field due to voltage-induced Stark effect. The dipole matrix element is small, but the scattering time is large.

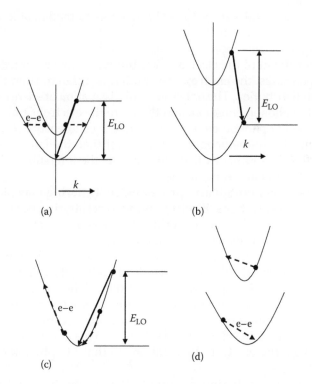

FIGURE 10.7

Inter- and intrasubband scattering mechanisms involving electron–LO phonon and electron–electron interactions: (a) Subband spacing < $\hbar\omega_{LO}$: ee scattering dominant; (b) subband spacing > $\hbar\omega_{LO}$: LO phonon scattering dominant; (c) intrasubband LO phonon scattering followed by ee scattering; (d) ee scattering involving two subbands.

10.5 Nonradiative Inter- and Intrasubband Transitions

As may be seen from Equation 10.6, the degree of population inversion depends on the different relaxation times, both radiative and nonradiative. These rates may be engineered by a proper design of the subband energies and wave functions. We now consider different intersubband scattering processes that determine the nonradiative relaxation rates.

10.5.1 Different Scattering Mechanisms

Two different situations are to be considered: (1) the subband separation $E_{fi} < \hbar\omega_{LO}$, where $\hbar\omega_{LO}$ is the LO phonon energy, and (2) $E_{fi} > \hbar\omega_{LO}$. The importance of the LO phonon in the second process has been firmly established, both theoretically and experimentally. The LO phonon scattering and electron–electron (ee) scattering processes are illustrated in Figure 10.7a and b.

When $E_{fi} < \hbar\omega_{LO}$, emission of the LO phonon is forbidden at low temperature. The nonradiative recombination rate is determined by a combination of ee scattering, electron–impurity scattering, and LO phonon emission from the high-energy tail of the electron distribution. It may be noted that, even at low temperatures, the electron distribution is different from the thermal equilibrium distribution. The only source of dissipation of energy of the electrons is by LO phonon emission, and therefore the electrons must be heated up so that the LO phonon emission rate increases to ensure a steady state. The resulting electron temperature T_e is larger than the lattice temperature T_L by 50–100 K. It is rather difficult to calculate the intersubband scattering rate for smaller subband separations.

As regards other nonradiative processes, ee scattering may play a role in high carrier concentrations. Electron–impurity scattering is, however, four times more effective than ee scattering. The reason is that in ee scattering an electron should interact with another electron of opposite spin; this restriction is absent in impurity scattering. Moreover, in impurity scattering, the electron effective mass, rather than half its value as in ee scattering, is involved. Though the QWs are modulation doped, so that the remoteness of electrons and impurities reduces the interaction and hence the linewidth of emission, calculations indicate a scattering time ~20 ps for transitions from Level 5 to Levels 4 and 3. This additional elastic scattering heats up the electron distribution, so that LO phonon emission is augmented.

Acoustic-phonon scattering is unimportant at low temperature; typical relaxation time is >100 ps. The effect of interface roughness scattering is difficult to estimate, but is >100 ps.

In addition to intersubband scattering, scattering by LO phonons and by ee interaction is also possible within a subband. These two intrasubband processes are illustrated in Figure 10.7c. The LO phonon scattering shown in Figure 10.7c is important in cooling the electron gas. Intrasubband ee scattering thermalizes the electron distribution in a particular subband. In Figure 10.7d, ee scattering involving two subbands is illustrated.

10.5.2 LO Phonon Scattering Rate

In this subsection, we outline the theory of calculating the scattering rate for electron–LO phonon interaction. It is assumed that the phonons are bulk-like. Multiple QWs (MQWs) also support confined phonons; however, it is established that LO phonon emission rates approach the rates for bulk GaAs for wide QWs and for bulk AlAs in the limiting case of narrow wells.

The scattering rate from an initial state of 2-D wave vector k_i to a final state of 2-D wave vector k_f is expressed by using Fermi's golden rule as

$$W_{if}(\mathbf{k_i}, \mathbf{k_f}) = \frac{2\pi}{\hbar} |\langle f, \mathbf{k_f} | H' | i, \mathbf{k_i} \rangle|^2 \delta[E_f(\mathbf{k_f}) - E_i(\mathbf{k_i}) \pm \hbar\omega_{LO}] \qquad (10.28)$$

where +(−) signs in the energy-conserving δ-function refer to the phonon emission (absorption) process. The interaction Hamiltonian is given by

$$H' = \sum_q \left[V(\mathbf{q})(e^{i\mathbf{q}\cdot\mathbf{r}}b_q + e^{-i\mathbf{q}\cdot\mathbf{r}}b_q^+) \right] \qquad (10.29)$$

where $b_q(b_q^+)$ is the phonon-creation (annihilation) operator. The electron–phonon interaction potential is expressed as

$$|V(\mathbf{q})|^2 = \frac{\hbar\omega_{LO}}{2} \frac{e^2}{q^2} \left(\frac{1}{\varepsilon_\infty} - \frac{1}{\varepsilon_0} \right) \qquad (10.30)$$

where ε_∞ and ε_0 are the high-frequency and static permittivity, respectively. The matrix element is written as

$$|\langle f, \mathbf{k_f} | H' | i, \mathbf{k_i} \rangle|^2 = \frac{e^2 \hbar\omega_{LO}}{2V} \left(\frac{1}{\varepsilon_\infty} - \frac{1}{\varepsilon_0} \right) \frac{1}{q_z^2 + Q^2} |A_{if}(q_z)|^2 \delta_{\mathbf{k_i}, \mathbf{k_f} \pm \mathbf{Q}} \left[n(\omega_{LO}) + \frac{1}{2} \mp \frac{1}{2} \right] \qquad (10.31)$$

In Equation 10.31, $\mathbf{Q}$ is the in-plane phonon wave vector, $q^2 = Q^2 + q_z^2$, the Kronecker δ denotes wave-vector conservation, $n(\omega_{LO})$ is the phonon number, and −(+) signs refer to phonon absorption (emission). The form factor is given by

$$A_{if}(q_z) = \int_{-\infty}^{\infty} dz \phi_f^*(z)\phi_i(z)e^{\pm q_z z} \tag{10.32}$$

where ϕs denote envelope functions. The scattering rate $W(k_i)$ for an initial wave vector k_i is obtained by integrating over phonon modes q and the final wave vector k_f. One may also write from energy conservation using parabolic bands

$$k_f^2 = k_i^2 + \frac{2m_e[E_f(0) - E_i(0) \mp \hbar\omega_{LO}]}{\hbar^2} \tag{10.33}$$

and from momentum conservation

$$Q^2 = |\mathbf{k}_i - \mathbf{k}_f|^2 = k_i^2 + k_f^2 - 2k_i k_f \cos\theta \tag{10.34}$$

where θ is the angle between $\mathbf{k}_i$ and $\mathbf{k}_f$. The transition rate may be expressed as

$$W_{if}(\mathbf{k}_i) = \frac{m_e e^2 \omega_{LO}}{8\pi\hbar^2}\left(\frac{1}{\varepsilon_\infty} - \frac{1}{\varepsilon_0}\right)\left[n(\omega_{LO}) + \frac{1}{2} \mp \frac{1}{2}\right]\int_0^{2\pi} d\theta B_{if}(Q) \tag{10.35}$$

The −(+) signs refer to the phonon absorption (emission) process, and B_{if} is expressed as

$$B_{if} = \int_{-\infty}^{\infty} dz \int dz' \phi_f^*(z)\phi_i(z)\phi_i^*(z')\phi_f(z')\frac{1}{Q}e^{-Q|z-z'|} \tag{10.36}$$

The scattering rate for intrasubband processes can be obtained by putting $i=f$ and $\phi_i=\phi_f$ in Equation 10.36.

The relaxation time is obtained by averaging the transition rate over the distribution function, and the expression is

$$\frac{1}{\tau_{if}} = \frac{\int_0^{\infty} dE_k f_i(E_k)W_{if}(E_k)}{\int_0^{\infty} dE_k f(E_k)} \tag{10.37}$$

When $E_{fi} < \hbar\omega_{LO}$, the relaxation time is simply written in terms of the number of states in the hot carrier distribution having energy in excess of the LO phonon energy. Thus,

$$\frac{1}{\tau_{if}} = W_{if}^{hot} \exp\left(\frac{E_{fi} - E_{LO}}{k_B T}\right) \tag{10.38}$$

10.5.3 Nonparabolicity Effect

Gelmont et al. [22] and Gorfinkel et al. [23] refined the simple model by considering nonparabolicity in the E–k relationship, phonon-induced broadening, and electron temperatures exceeding the lattice temperature.

The nonparabolicity in a QW is estimated by considering two degenerate conduction bands and six valence subbands. Assuming infinite barrier height for the QW, the dispersion relation for both the subbands may be expressed as

$$E_n(k) = \frac{E_g}{2}\left[1 + \frac{4E_n^0}{E_g} + \frac{2\hbar^2 k^2}{m_e E_g}\right]^{1/2} - \frac{E_g}{2} \tag{10.39}$$

where:
m_e is the effective mass at the conduction band edge
E_g is the bandgap of the semiconductor
$E_n^0 = \pi^2\hbar^2 n^2/2m_e a^2$
$n =$ 1,2, are the energies of the two subbands considered in this work

Expanding in k, the effective masses at the subband edges are obtained as

$$m_n = m_e\left(1 + \frac{2E_n(0)}{E_g}\right) \tag{10.40}$$

Example 10.6: Consider an InGaAs QW for which $E_g = 0.8$ eV, $m_e = 0.047m_0$, and $a = 10$ nm. The difference $E_2(0) - E_1(0) = 0.24$ eV. The effective mass ratio $m_2/m_1 = 1.5$ is rather large.

The expression for intersubband gain is obtained by Gelmont et al. by solving the density-matrix equations for the two-level system. However, the expression already obtained by using the conventional approach,

Equations 10.13 and 10.14, are identical. It is trivial to write the expression in the following form, as given by Gelmont et al.:

$$g = \frac{e^2 |z_{12}|^2 m_2 \omega}{\varepsilon_0 n_r c a \hbar \pi} \int_0^\infty \frac{d\varepsilon \left[f_2(\varepsilon) - f_1(\varepsilon) \right] \gamma(\varepsilon)}{\left[\hbar\omega - \hbar\omega_0 \right]^2 + \left[\gamma(\varepsilon) \right]^2} \tag{10.41}$$

Here, $\hbar\omega(\varepsilon) = \hbar\omega_0 + \varepsilon_2(k) - \varepsilon_1(k)$, $\varepsilon_n(k) = E_n(k) - E_n(0)$, $n = 1,2$ denote the kinetic energy in the nth subband and $\varepsilon = \varepsilon_2$. The value of ε_1 for a given k is determined from $\varepsilon = \varepsilon_2$.

To estimate the broadening parameter, it is assumed that only interaction with LO polar optic-phonon scattering is predominant, and therefore one may write

$$\gamma(\varepsilon) = \frac{e^2}{8\varepsilon_0 \hbar} \left(\frac{1}{\kappa_0} - \frac{1}{\kappa_\infty} \right) q_{ph} \times \begin{cases} N_{ph} \\ (N_{ph} + 1) \theta (\varepsilon - \hbar\omega_{ph}) \end{cases} \tag{10.42}$$

The top line corresponds to absorption and the bottom line to emission of an optical phonon of energy $\hbar\omega_{ph}$ and of wave vector $q_{ph} = \sqrt{2m_e \omega ph/\hbar}$. The phonon number is denoted by N_{ph}, and θ represents a step function.

> **Example 10.7:** An estimate of the broadening parameter for GaAs QW due to LO phonon scattering is made by using the following parameters: $\kappa_0 = 13.13$, $\kappa_\infty = 11.1$, $m_e = 0.067m_0$, and $\hbar\omega_{ph} = 36$ meV. The value $\gamma(\varepsilon) = 2 \times 10^{13} \, \text{s}^{-1}$ is obtained at room temperature for $\varepsilon > \hbar\omega_{ph}$.

10.6 Some Design Issues

We mention here some of the design issues related to QCL (please see Reading List).

10.6.1 Role of Conduction Band Discontinuity

Since the depth of the QW determines the maximum number of subbands, there is an upper limit to the photon energy that can be achieved. However, there is no fundamental limit to the long wavelength. Apart from these, the band offsets indirectly influence the design by modifying the threshold current and the injector barrier transparency.

10.6.1.1 Escape to the Continuum

As the subband energy gets closer to the top of the barrier, there is an enhanced probability of escape of carriers to the continuum by a thermal activation process.

The 2-D density of electrons localized in the QW and with energy greater than the band discontinuity ΔE_c is given by

$$n_{2D} = n_3 \exp\left[-\frac{(\Delta E_c - E_3)}{k_B T_e}\right] \tag{10.43}$$

where:
  n_{2D} is the total electron density in Level 3 per unit area
  T_e is the electron temperature

The leakage current density to the continuum is given by

$$J_{esc} = \frac{e n_{2D}}{\tau_{scatt}} \tag{10.44}$$

The applied electric field in the QCL is quite large, and therefore the probability of recapture of the electrons by the quantized level is small. The time τ_{scatt} is then the mean scattering time of the electrons from the confined state to the continuum states. It may be safely assumed to equal the scattering time by LO phonon scattering.

This escape current is to be compared with the cascade current J, which is governed by LO phonon scattering to the lower subbands. As noted already, the scattering rate depends on the phonon number N_{ph}. Therefore, the scattering time may be written as

$$\tau_3 = \frac{\tau_{3(0)}}{1 + 2/\left[\exp(\hbar\omega_{LO}/k_B T) - 1\right]} \tag{10.45}$$

10.6.2 Effective Mass

The effective mass of the active material plays a significant role in the transition energy E_{21}, the squared dipole matrix element $|z_{21}|^2$, and the lifetime of the excited state transition, τ_2. Under infinite barrier approximation, the quantity $1/m^* d^2$ is identical for any two materials, where m^* is the effective mass, and d is the QW width. Also, $|z_{21}|^2$ is proportional to L, and therefore $|z_{21}|^2 \propto 1/m^*$. The relaxation time $\tau_2 \propto 1/\sqrt{m^*}$, as it is due to electron–LO phonon interaction. Thus, from these considerations the gain $g \propto |z_{32}|^2 \tau_2 \propto (m^*)^{-3/2}$.

The simple relation stated above breaks down when the effect of non-parabolicity is considered. The nonparabolicity is more severe for lower-gap material. The dipole matrix element is now a complicated function of both the conduction and valence subband wave functions and their mixing. The higher the transition energy, the heavier is the effective mass. Consequently, gain as a function of energy can saturate, as may be the case with InAs, a lower-gap material.

10.6.3 Waveguides

A large optical confinement reduces the threshold current density. It is usual to have a multilayered active region as the core of the waveguide, which is separated on its two sides by the cladding or optical confinement layers (OCLs). Though an increase in the number of periods in the active region increases the confinement factor, it increases the applied voltage and hence power dissipation.

The major contribution to waveguide losses comes from the free carrier absorption (FCA) in the cladding layers, and, as the wavelength is large, FCA becomes severe ($\alpha_{FCA} \alpha \lambda^2$, see Equation 6.85), in particular in the terahertz frequency range. The optimum waveguide design depends on both the wavelength and the material system.

InP has a low refractive index and high thermal conductivity, almost twice that of GaAs and at least 10 times higher than other ternary and quaternary materials. It is therefore the preferred material as the OCL.

The refractive index of GaAs as the substrate material is higher than the index of the active region. Therefore alternate materials, like AlGaAs or GaInP alloys are used.

Plasmon-enhanced waveguides form the most attractive solution for InAs/AlSb QCLs. In THz QCLs, optical confinement is achieved through surface plasmons at the metal–semiconductor interface.

10.7 Frequency Response

In this section, the small-signal modulation response of QCLs will be studied by the standard rate equation model.

10.7.1 Rate Equation Model

The modulation response of the QCL is based on a three-level model [23]. A rigorous analysis of a real QCL, which consists of many periods, is quite difficult. The analysis is greatly simplified by assuming identical QWs and confinement factors. In addition, the removal rates for electrons are fast. Also, the coupling of the spontaneous emission to the lasing mode is ignored. This last assumption is valid above threshold when the stimulated emission rate exceeds the rate for spontaneous emission. The electron numbers are N_3, N_2, and N_1, respectively, in Levels 3, 2, and 1. The electron lifetime for transition between Levels 3 and 2 is expressed as

$$\frac{1}{\tau_n} = \frac{1}{\tau_{32}} + \frac{1}{\tau_{sp}} \tag{10.46}$$

In Equation 10.46, τ_{32} represents the nonradiative transition lifetime between Levels 3 and 2, and τ_{sp} is the spontaneous emission lifetime.

The rate equations are written as

$$\frac{dN_3}{dt} = \frac{I}{e} - \frac{N_3}{\tau_{31}} - \frac{N_3}{\tau_n} - G(N_3 - N_2)S \tag{10.47}$$

$$\frac{dN_2}{dt} = \frac{N_3}{\tau_n} + G(N_3 - N_2)S - \frac{N_2}{\tau_{21}} \tag{10.48}$$

$$\frac{dN_1}{dt} = \frac{N_2}{\tau_{21}} + \frac{N_3}{\tau_{31}} - \frac{N_1}{\tau_{out}} \tag{10.49}$$

$$\frac{dS}{dt} = G(N_3 - N_2)S - \frac{S}{\tau_p} \tag{10.50}$$

where:
- I is the injected current
- G is the gain coefficient
- S is the photon number
- N_1 is the electron number in Level 1
- τ_{31} and τ_{21} are, respectively, the electron lifetimes between 3 and 1 and between 2 and 1, both the lifetimes corresponding to nonradiative transitions
- τ_{out} is the electron removal lifetime
- τ_p is the photon lifetime

Equation 10.49 is needed to maintain current continuity. Putting

$$I_L = \frac{eN_3}{\tau_{31}} \tag{10.51}$$

in Equation 10.47, which does not participate in lasing action, the three-level model may be converted to an equivalent two-level model [24,25].

The rate equations, including that for photons and considering N identical QWs, are now written as

$$\frac{dN_3}{dt} = \frac{I}{e} - \frac{N_3}{\tau_{31}} - \frac{N_3}{\tau_n} - G(N_3 - N_2)S \tag{10.52}$$

$$\frac{dN_2}{dt} = \frac{N_3}{\tau_n} + G(N_3 - N_2)S - \frac{N_2}{\tau_{21}} \tag{10.53}$$

$$\frac{dS}{dt} = GN(N_3 - N_2)S - \frac{S}{\tau_p} \tag{10.55}$$

10.7.2 Steady-State Solutions

The steady-state quantities are denoted by subscript b. Setting all time derivatives to zero, we obtain from Equations 10.52–10.55

$$N_{3b} - N_{2b} = \frac{1}{GN\tau_p} \tag{10.56}$$

which signifies that the population difference saturates, although the individual-level population may increase above threshold.

Using Equations 10.52, 10.53, and 10.56, it may be shown that

$$S_b = K\left(\frac{I_b}{I_{th}} - 1\right) \tag{10.57}$$

where

$$I_{th} = \frac{e}{GN\tau_{21}\tau_p}\left(\frac{\xi + \xi'}{1 - \xi}\right) \tag{10.58}$$

The symbols in Equations 10.57 and 10.58 are defined as follows:

$$\xi = \frac{\tau_{21}}{\tau_n}, \xi' = \frac{\tau_{21}}{\tau_{31}}, \text{ and } K = \frac{\xi + \xi'}{(1 + \xi')G\tau_{21}} \tag{10.59}$$

The emitted power per facet, assuming equal reflectivities, is given by

$$P_e = \frac{1}{2}(v_g \alpha_m) \frac{hc}{\lambda} S_b \qquad (10.60)$$

where the symbols are already defined in Chapters 6–8.

10.7.3 Dynamic Response

10.7.3.1 Normalized Modulation Response

Assume now that a small deviation of current i superimposed on the direct-current (dc) value I_b introduces small deviations n_3, n_2, and p from the steady-state values N_{3b}, N_{2b}, and S_b, respectively. Taking the Laplace transform of Equations 10.52 through 10.54 and taking only first-order terms, one obtains the matrix equation

$$\begin{pmatrix} s + \left(\frac{\xi'}{\xi} + 1\right)\frac{1}{\tau_n} + GS_b & -GS_b & \frac{1}{N\tau_p} \\ -\left(\frac{1}{\tau_n} + GS_b\right) & s + \frac{1}{\xi\tau_n} + GS_b & -\frac{1}{N\tau_p} \\ -GNS_b & GNS_b & s \end{pmatrix} \begin{pmatrix} n_3 \\ n_2 \\ s_1 \end{pmatrix} = \begin{pmatrix} i/e \\ 0 \\ 0 \end{pmatrix} \qquad (10.61)$$

The normalized modulation response, obtained by solving Equation 10.61, is expressed as

$$\eta = \frac{s_1}{\left[i(1-\xi)/e(1+\xi')\right]N\tau_p} = \frac{(s/A)+1}{Bs^3 + Cs^2 + Ds + 1} \qquad (10.62)$$

where:

$$A = \left(\frac{1}{\xi} - 1\right)\frac{1}{\tau_n}, \ B = \frac{\tau_p \tau_n \xi}{GS_b(1+\xi')}, \ \text{and} \ C = B\left[2GS_b + \left(\frac{1+\xi'}{\xi} + 1\right)\frac{1}{\tau_n}\right]$$

$$D = B\left[GS_b\left(\frac{2}{\tau_p} + \frac{1+\xi'}{\xi\tau_n}\right) + \left(\frac{\xi'}{\xi} + 1\right)\frac{1}{\xi\tau_n^2}\right] \qquad (10.63)$$

10.7.4 Approximate Modulation Response

By considering the poles and zeros of the normalized modulation response given by Equation 10.62, the following approximate expression for the normalized modulation response has been obtained:

$$\eta = \frac{1}{Cs^2 + Ds + 1} \tag{10.64}$$

For low optical power, the expression takes a much more simplified form:

$$\eta = \frac{1}{Ds + 1} \tag{10.65}$$

Substituting $s = j\omega$ in Equation 10.64, the modulation response may be written in the standard form

$$\eta = \frac{1}{-(\omega/\omega_n)^2 + 2j\zeta(\omega/\omega_n) + 1} \tag{10.66}$$

where:
  $\omega_n = 1/\sqrt{C}$ is the natural undamped frequency
  $\zeta = D/(2\sqrt{C})$ is the damping ratio

The expression for 3 dB electrical bandwidth is given by

$$f_{3dB} = \frac{1}{2\pi} \left[\frac{\sqrt{(D^2 - 2C)^2 + 4C^2} - (D^2 - 2C)}{2C^2} \right]^{1/2} \tag{10.67}$$

The light power versus current curve shown in Figure 10.8 is obtained by using the following parameters: $\lambda = 5.0\ \mu m$, number of stages $N = 25$, waveguide loss coefficient $\alpha_w = 11\ cm^{-1}$, mirror loss coefficient $\alpha_m = 4.36\ cm^{-1}$, photon lifetime $\tau_p = 7.38\ ps$, $\tau_n\ (=\tau_{32}) = 2.1\ ps$, $\tau_{31} = 3.4\ ps$, $\tau_{21} = 0.5\ ps$, gain coefficient per stage including confinement factor, $G = 0.2 \times 10^4\ s^{-1}$, group index $= 3.4$.

The normalized modulation response calculated by using Equation 10.66 is shown in Figure 10.9 for three values of bias current. It is worth noting the absence of any peak in the response that is invariably present in double heterostructure (DH), QW, and QD lasers [24,25].

10.8 Terahertz QCL

The frequency range from 300 GHz to 10 THz, or the wavelength range from 30 to 1000 µm, is usually called the terahertz frequency range. This is still one of the least developed spectral regions, and suitable sources, detectors, and transmission technology for this spectral range are not available.

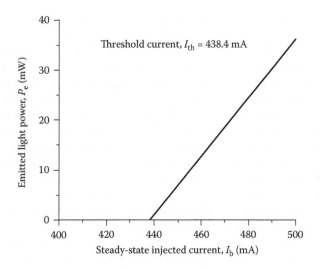

FIGURE 10.8
Light power output vs. injection current.

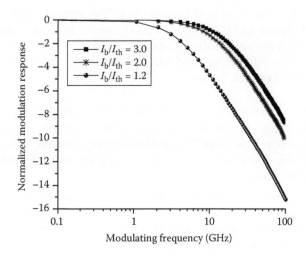

FIGURE 10.9
Normalized modulation response in terms of modulation frequency.

However, many new areas of potential applications have been identified, including astrophysics and atmospheric science, biological and medical sciences, security screening, toxic and illicit materials detection, ultrafast technology, and, not least, communication technology. Suitable solid-state devices are in demand for this range that should be compact, reliable, and coherent. In the first category, these sources may be of two different kinds.

They may be analogous to electronic devices such as high electron mobility transistors (HEMTs) or Gunn, IMPact ionization Avalanche Transit-Time (IMPATT), tunnel diode, or resonant tunneling diodes. Unfortunately, however, the power generated by these devices decreases as $1/f^2$ due to transit time or resistance–capacitance effects. The other approach may be the photonic approach, that is, a semiconductor laser operating at the required wavelength range. However, to realize the device, one needs a small-bandgap material. The longest-wavelength lead-salt lasers have a minimum limit of 15 THz and cannot go below this. The techniques adopted so far to generate waves above 1 THz are downconversion from the visible regime by using nonlinear or photoconductive effects or by upconversion or frequency multiplication of millimeter waves. Direct methods to generate terahertz frequencies are optical pumping of molecular gas lasers and free electron lasers. These sources are limited by size, cost and complexity, output power, inability for cw operation, and requirement for extensive cryogenic cooling.

As already mentioned in Chapter 1, the first QCL action in the THz range was demonstrated at 4.4 THz (67 μm) by Köhler et al. in 2002 [15]. Since then, many groups have achieved pulsed-mode operation at higher temperature, cw operation, and high power as well as covering a wide spectral range, for example from 0.84 to 5.0 THz.

In this section, we shall point out some of the features that characterize terahertz_QC lasers and distinguish them from the mid-infrared lasers discussed so far (please see Reading List).

The following major issues need to be addressed in order to realize a useful QCL, particularly for the long-wavelength range.

1. Since the separation between the levels involved in the transition is quite low, it is very difficult to achieve and maintain a high degree of population inversion. This problem is aggravated further as the broadening of the energy levels involved becomes comparable to the level separation.

2. Since the free-carrier absorption is roughly proportional to λ^2, the loss is extremely large. It is therefore necessary not to allow the electromagnetic modes to enter into the highly doped cladding layers. In other words, the waveguiding must be very tight.

Different designs of the active region and of the waveguides have been employed by different workers. The active layer design has taken place in three directions, described in Sections 10.8.1–10.8.3 [26].

10.8.1 Chirped SL

The intra- and interminiband transitions in the SL or chirped SL (CSL) active regions have been discussed already in Section 10.4.2. The radiative

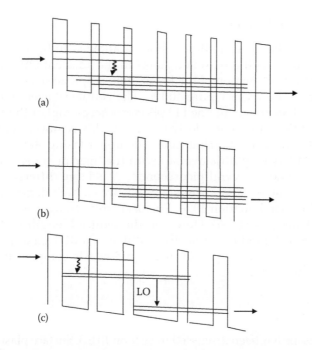

FIGURE 10.10
Schematic diagrams showing (a) transition in CSL, (b) bound-to-continuum transition, and (c) resonant-phonon-assisted transition.

transition takes place between the bottom of the upper miniband and the top of the lower miniband, as shown in Figure 10.10a. The energetic electrons cool via LO phonon scattering.

10.8.2 Bound-to-Continuum Transition

In this case, the lower radiative state and depopulation within the miniband remain the same, but the upper radiative state is designed to have an energy level in the minigap; in other words, the upper state is a *defect* state in the minigap. The effect is a radiative transition that is diagonal in real space. The schematic is shown in Figure 10.10b. Though the oscillator strength of the radiative transition in the bound-to-continuum (BCT) design is somewhat reduced from 2.5–3 in CSL to 1.5–2 in BCT, the upper-state lifetime increases as the rate of nonradiative scattering is similarly reduced. The injector states couple more strongly with the bound upper state than with the lower miniband. This selective injection process leads to better temperature sensitivity and power performance than can be obtained with the CSL design.

10.8.3 Resonant-Phonon

In this design, the lower radiative state is allowed to have a broad tunneling resonance with the excited states in the adjacent QWs, as a result of which the wave function of the lower state is spread over several QWs. The scheme is shown in Figure 10.10c. The coupled lower state and the states in the QWs are about one LO phonon energy higher than the injector state, and the lower-state electrons are quickly removed with a subpicosecond LO phonon scattering time. The upper lasing state is, however, localized and has very little overlap with the injector states; the lifetime in the upper state is several picoseconds. This large difference between lifetimes of the upper and lower states ensures a high degree of population inversion (see Equation 10.5). The oscillator strength is lower (0.5–1) than in BCT designs (1.5–2). However, the reduced length L of the resonant-phonon (RP) module, typically half that of a BCT module, produces higher gain ($g \propto L_p^{-1}$, see Equation 10.22) and nearly compensates the loss in oscillator strength.

10.8.4 Waveguide

The design issue has been discussed in Section 10.6.3. Surface plasmon waveguides seem to be the usual solution to achieve low loss.

10.9 QD QCL

As discussed in Section 10.5.1, the dominant nonradiative transition mechanism is the electron–LO phonon coupling. Since the electrons have in-plane motion, such transitions are always allowed, and the rate is even a few thousand times faster than the radiative process. This leads to a higher value of threshold current density.

The situation will be much improved if the QWs are replaced by coupled QDs. Here, both the excited and the ground states are discrete levels, and nonradiative transition by LO phonon emission is allowed only when the difference between the levels equals $\hbar\omega_{LO}$, the LO phonon energy. If the separation between the levels is larger than $\hbar\omega_{LO}$, the transition may occur via a multiphonon process, which has a low probability. The dominant decay mechanism is therefore the radiative transition.

The idea of replacing QWs by QD arrays was proposed by Wingreen and Stafford [27]. The authors estimated the size of the cubic dots and also the value of the threshold current density. They also gave a sketch of the structure.

There have been several attempts to realize such a QCL involving a stack of QDs. In particular, QDCLs may be an attractive solution for terahertz emitters. The reader is referred to some of the recent articles, news, and views appearing in the literature [28–30].

PROBLEMS

10.1. Obtain the expression for $\langle z_{32} \rangle^2$ for a QW of infinite barrier height.

10.2. Prove that when the barrier height is infinite, intersubband transition can occur between mth and nth subbands only when $m - n = $ odd.

10.3. Design a GaAs QW for which the transition from the 3rd to the 2nd subband gives rise to emission of 10 μm wavelength. Assume infinite barrier height.

10.4. An $In_{0.2}Ga_{0.8}As$–GaAs QW of width 20 nm is chosen as the active element of a QCL. Find out the emission wavelength for the 2→1 transition. Take into account the effect of strain and finite barrier height on the subband energies.

10.5. The well width in an $In_{0.53}Ga_{0.47}As$–$In_{0.52}Al_{0.48}As$ QW system is 5 nm. The band gap and electron mass values are 0.76 eV and $0.042m_0$ for InGaAs and 1.47 eV and $0.075m_0$ for InAlAs. The band offset is 550 meV. Calculate the oscillator strength for the 1–2 transition, the peak absorption coefficient, and the value of $\langle z \rangle$. Assume Gaussian line broadening with FWHM = 10 meV and electron density = 10^{18} cm^{-3}.

10.6. Prove that the recombination lifetime for the 2–1 transition is expressed as $\tau_r^{-1} = 2\pi n_r e^2 f_{21}/3m_e\varepsilon_0\lambda^2 c$. Obtain the value of the lifetime when $f_{21} = 15$, $n_r = 3.6$, and $\lambda = 10$ μm.

10.7. The measured V-J characteristic of a QCL is similar to the V-J characteristic of a p-n junction diode (Gmachl et al. 2001, see Reading List). The turn-on voltage is defined as the sum of energy drop/electron charge per stage and a small offset voltage. Give the expression for the turn-on voltage for different periods of the QCL.

10.8. The threshold voltage is the sum of the above turn-on voltage and a voltage drop over the dynamic resistance of the device. Using this, obtain the expression for the electrical power dissipation at threshold.

10.9. Show that the expression for population inversion is $n_3 - n_2 = J\tau_3/ed$ $[\eta\tau_3(1 - (\tau_2/\tau_{32})) - (1 - \eta)\tau_2]$.

10.10. The energy separation between the ground and the excited states in a cubic QB should be larger than the LO phonon energy. Calculate the maximum size of the GaAs QB to ensure this. Take $m_e = 0.067m_0$ and $\hbar\omega_{LO} = 36$ meV.

Reading List

Bhattacharya, P., X. Su, S. Chakrabarti, A. D. Stiff-Roberts, and C. H. Fischer, Chapter 8: Intersubband transitions in quantum dots. In *Intersubband Transitions in Quantum Structures*, ed. Paiella, R. McGraw Hill, New York, pp. 315–345, 2006.

Chakraborty, T. and V. M. Apalkov, Quantum cascade transitions in nanostructures, *Adv. Phys.* 52, 455–521, 2003.

Faist, J., T. Aellen, T. Gresch, M. Beck, and M. Giovannini, Progress in quantum cascade lasers. In *Mid-Infrared Coherent Sources and Applications*, ed. Ebrahim-Zadeh, M. and Sorokina, I. T. Springer, Berlin, pp. 178–192, 2008.

Gmachl, C., F. Capasso, D. L. Sivco, and A. Y. Cho, Recent progress in quantum cascade lasers and applications, *Rep. Prog. Phys.* 64, 1533–1601, 2001.

Sirtori, C., S. Barbieri, and R. Colombelli, Wave engineering with THz quantum cascade lasers, *Nature Photon.* 7, 691–701, 2013.

Sirtori, C. and R. Teissier, Chapter 1: Quantum cascade lasers: Overview of basic principles of operation and state of the art. In *Intersubband Transitions in Quantum Structures*, ed. Paiella, R. McGraw Hill, New York, pp. 1–44, 2006.

Tredicucci, A. and R. Kohler, Chapter 2: Terahertz quantum cascade lasers. In *Intersubband Transitions in Quantum Structures*, ed. Paiella, R. McGraw-Hill, New York, pp. 45–105, 2006.

Williams, B. S, Terahertz quantum-cascade lasers, *Nature Photonics* 1, 517–525, 2007.

References

1. Lax, B., Cyclotron resonance and impurity levels in semiconductors. Presented at the 1st International Conference on Quantum Electronics, 1959; published in *Quantum Electronics*, ed. C. H. Townes, Columbia University Press, New York, pp. 428–449, 1960.
2. Kazarinov, R. and R. A. Suris, Possibility of the amplification of electromagnetic waves in a semiconductor with a superlattice, *Fiz. Tekh. Poluprov.* 5, 797–800, 1971. Transl. in *Sov. Phys. Semicond.* 5, 707–709, 1971.
3. Capasso, F., K. Mohamed, and A. Y. Cho, Resonant tunneling through double barriers, perpendicular quantum transport phenomena in superlattices, and their device applications, *IEEE J. Quantum Electron.* 22, 1853–1869, 1986.
4. Liu, H. C., A novel superlattice infrared source, *J. Appl. Phys.* 63, 2856–2868, 1988.
5. Borenstain S. I. and J. Katz, Evaluation of the feasibility of a far infrared laser based on intersubband transitions in GaAs quantum wells, *Appl. Phys. Lett.* 55, 654–656, 1989.

6. Faist, J., F. Capasso, D. L. Sivco, C. Sirtori, A. L. Hutchinson, and A. Y. Cho, Quantum cascade laser, *Science* 264, 553–556, 1994.

7. Sirtori, C., J. Faist, F. Capasso, D. L. Sivco, A. L. Hutchinson, S. N. G. Chu, and A. Y. Cho, Continuous wave operation of mid infrared (7.4–8.6 µm) quantum cascade lasers up to 110 K temperature, *Appl. Phys. Lett.* 68, 1745–1747, 1996.

8. Evans, A., J. S. Yu, J. David, L. Doris, K. Mi, S. Slivken, and M. Razeghi, High-temperature, high power, continuous-wave operation of buried heterostructure quantum cascade lasers, *Appl. Phys. Lett.* 84, 314–316, 2004.

9. Beck, M., D. Hofsetter, T. Aellen, J. Faist, U. Oesterle, M. Ilegems, E. Gini, and H. Melchior, Continuous wave operation of a mid-infrared semiconductor laser at room temperature, *Science* 295, 301–305, 2002.

10. Faist, J., F. Capasso, C. Sirtori, D. L. Sivco, J. N. Baillargeon, and A. L. Hutchinson, High power mid-infrared ($\lambda \sim 5$ µm) quantum cascade lasers operating above room temperature, *Appl. Phys. Lett.* 68, 3680–3682, 1996.

11. Gmachl, C., F. Capasso, A. Tredicucci, D. L. Sivco, A. L. Hutchinson, S. G. Chu, and A. Y. Cho, Non-cascaded intersubband injection lasers at $\lambda = 7.7$ µm, *Appl. Phys. Lett.* 73, 3830–3832, 1998.

12. Scamarcio, G., F. Capasso, C. Sirtori, J. Faist, A. L. Hutchinson, and D. L. Sivco, High power infrared (8 µm wavelength) superlattice laser, *Science* 276, 773–776, 1997.

13. Scamarcio, G., C. Gmachl, F. Capasso, A. Tredicucci, A. L. Hutchinson, D. L. Sivco, and A. Y. Cho, Long-wavelength $\lambda = 11$ µm interminiband Fabry-Pérot and distributed feedback quantum cascade lasers, *Semicond. Sci. Tech.* 13, 1333–1339, 1998.

14. Faist, J., C. Sirtori, F. Capasso, D. L. Sivco, J. N. Baillargeon, A. L. Hutchinson, and A. Y. Cho, High-power long-wavelength (~11.5 µm) quantum cascade lasers operating above room temperature, *IEEE Photonic. Technol. Lett.* 10, 1100–1102, 1998.

15. Koehler, R., A. Tredicucci, F. Beltram, H. E. Beere, E. H. Linfeld, A. G. Davies, D. A. Ritchie, R. C. Iotti, and F. Rossi, Terahertz semiconductor heterostructures laser, *Nature* 417, 156–158, 2002.

16. Kumar, S., Q. Hu, and J. L. Reno, 186 K operation of terahertz quantum-cascade lasers based on a diagonal design, *Appl. Phys. Lett.* 94, 131105, 2009.

17. Brandstetter, M., C. Deutsch, M. Krall, H. Detz, D. C. MacFarland, T. Zederbauer, A. M. Andrews, W. Schrenk, G. Strasser, and K. Unterrainer, High power terahertz quantum cascade lasers with symmetric wafer bonded active regions, *Appl. Phys. Lett.* 103, 171113, 2013.

18. Coon, D. D. and R. P. G. Karunasiri, New mode of IR detection using quantum wells, *Appl. Phys. Lett.* 45, 649–651, 1984.

19. West, L. C. and S. J. Eglash, First observation of an extremely large dipole transition within the conduction band of a GaAs quantum well, *Appl. Phys. Lett.* 46, 1156–1158, 1985.

20. Ahn, D. and S. L. Chuang, Intersubband optical absorption in a quantum well with an applied electric field, *Phys. Rev. B.* 35, 4149–4151, 1987.

21. Smet, J. H., C. G. Fonstad, and Q. Hu, Intrawell and interwell intersubband transitions in multiple quantum wells for far-infrared sources, *J. Appl. Phys.* 79, 9305–9320, 1996.

22. Gelmont, B., V. B. Gorfinkel, and S. Luryi, Theory of the spectral lineshape and gain in quantum wells with intersubband transitions, *Appl. Phys. Lett.* 68, 2171–2173, 1996.

23. Gorfinkel, V. B., S. Luryi, and B. Gelmont, Theory of gain spectra for quantum cascade lasers and temperature dependence of their characteristics at low and moderate carrier concentrations, *IEEE J. Quantum Electron.* 32, 1995–2003, 1996.
24. Webb, J. F. and M. K. Haldar, Improved two level model of mid-infrared quantum cascade lasers for analysis of direct intensity modulation response, *J. Appl. Phys.* 111, 043110, 2012.
25. Biswas, A. and P. K. Basu, Equivalent circuit models of quantum cascade lasers for SPICE simulation of steady state and dynamic responses, *J. Optics A Pure Appl. Opt.* 9, 26–32, 2007.
26. Williams, B. S., Terahertz quantum-cascade lasers, *Nat Photon.* 1, 517–525, 2007.
27. Wingreen, N. S. and C. A. Stafford, Proposal for an ultralow-threshold semiconductor laser, *IEEE J. Quantum Electron.* 33, 7, 1170–1173, 1997.
28. Anders, S., L. Rebohle, F. F. Schrey, W. Schrenk, K. Unterrainer, and G. Strasser, Electroluminescence of a quantum dot cascade structure, *Appl. Phys. Lett.* 82, 3862–3864, 2003.
29. Dmitriev, I. A. and R. A. Suris, Quantum cascade lasers based on quantum dot superlattice, *Phys. Status Solidi A* 202, 987–991, 2005.
30. Tredicucci, A., Quantum dots: Long life in zero dimensions (News and Views), *Nat. Mater.* 8, 775–776, 2009.

11

Vertical-Cavity Surface-Emitting Laser

11.1 Introduction

Until now, we have considered semiconductor lasers in which the light beam comes out perpendicularly to the direction of current flow. These are all edge-emitting lasers. A different configuration, called the vertical-cavity surface-emitting laser (VCSEL), was proposed by Iga as early as 1977 [1]. The first such laser was realized by Soda et al. in 1979 [2]. In this structure, the cavity is formed by two surfaces of an epitaxial layer, and light output is taken vertically from one of the mirror surfaces. This structure is one member of a general class of surface-emitting lasers (SELs).

SELs offer many advantages:

1. The emission occurs perpendicularly to the semiconductor layers, and thus integration of the devices (1-D or 2-D array) becomes easy according to electrical packaging constraints.

2. The small size of the device and the perpendicular emission lead to planarization and a high level of integration.

3. Each member may be probe tested before separating individual devices into discrete chips.

4. The low volume of the active layer due to the presence of quantum wells (QWs) involves a submilliampere threshold current and low electrical power consumption.

5. The threshold current shows less thermal dependence near room temperature.

6. The serial fabrication reduces the cost and, as noted, allows on-wafer testing.

7. The light beam cross section is circular due to cylindrical geometry.

There are a number of applications of VCSELs today. They are important components in short-distance optical communications, such as local area and metro networks, avionics networks, and so on. VCSELs have been designed to fulfill the need for planar optoelectronic circuits. Today VCSEL emission covers a wide range, from blue-green to infrared. Apart from short-distance communication, VCSELs find wide use in consumer products such as laser printers, laser mice, and display devices.

In the present chapter, the structure, theory of operation, models for threshold and its temperature dependence, modulation bandwidth, and so on using QW structures will be discussed. Recent developments using quantum dots (QDs) as active materials, tunnel junctions, photonic crystal-embedded systems, and so on will be mentioned, indicating the theoretical models developed for the structures.

11.2 Structures and Basic Properties

The basic structure of a typical VCSEL is shown in Figure 11.1. The active layer is a multiple quantum well (MQW) 10 nm thick, which is sandwiched between two conductive layer stacks. These two multilayer stacks form the two mirrors of the laser cavity to provide feedback. The p-i-n-type doping configuration is similar to conventional edge-emitting lasers. The metal ring at the top layer serves as the ohmic contact and injects current into the p-i-n structure. Similarly, the bottom metal contact is for collection of current. As shown in the figure, current is confined within a predefined cross section in the active region. Various methods, such as mesa etching of the top mirror,

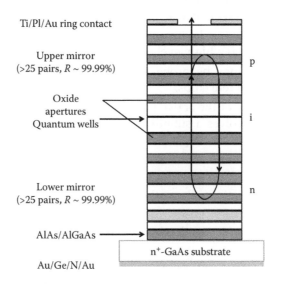

FIGURE 11.1
Typical structure of a VCSEL.

ion implantation or selective lateral oxidation, and so on, are employed to confine the current over a very small circular region. The methods will be described briefly in Section 11.4. Each method has its own advantages and disadvantages.

Reduction of the diameter of the active region to a few micrometers leads to a very low threshold current, less than 100 μA. For higher output power beyond 100 mW, the diameter exceeds 100 μm. When the active diameter is less than 4 μm, emission occurs in a single transverse mode. However, with larger diameter, higher-order radial and azimuthal modes appear. Typical light power output versus current curve is linear above threshold, but for higher drive current, the power shows a characteristic rollover due to heating of the device.

Depending on the wavelength and material composition, VCSELs may be top emitting through a ring contact or may be bottom emitting through a transparent substrate.

11.3 Elementary Theory of VCSEL

The expressions for the threshold current and quantum efficiency of a VCSEL may be developed from a simple model presented in this section (please see Reading List). The structure is shown in Figure 11.2, in which the active

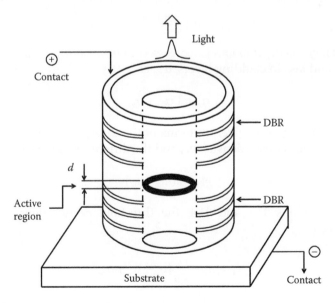

FIGURE 11.2
Schematic structure of a VCSEL for development of model.

region is in the form of a thin circular disc of diameter D. This active region may be buried in a material of smaller bandgap or may be surrounded by an insulating material. The injected carriers are confined in the circular active region. To reach the threshold, the modal gain should balance the optical loss at the resonant mode. Thus

$$\Gamma g_{th} = \alpha_a + \alpha_d + \alpha_m \tag{11.1}$$

where:

    α_a is the absorption loss
    α_d is the diffraction loss
    α_m is the mirror loss
    Γ is the mode confinement factor

The mode confinement factor or optical energy confinement factor Γ is the product of the longitudinal confinement factor Γ_1 and the transverse confinement factor Γ_t, so that $\Gamma = \Gamma_1.\Gamma t$.

The optical confinement factor is defined by the fraction by which the active or gain region occupies the total region covered by the mode or standing wave pattern. For a thick bulk semiconductor, the optical confinement factor is defined as $\Gamma_1 = d/L$, where d is the total active layer thickness and L is the effective cavity length. When a very thin layer (~10 nm) is placed at the maxima of the standing wave, the longitudinal confinement factor is expressed as

$$\Gamma_1 = 2d / L \tag{11.2}$$

The average absorption loss is expressed in terms of the loss in the active layer, α_{ac}, and loss in cladding layers, α_{ex}, as

$$\alpha_a = \Gamma\alpha_{ac} + \alpha_{ex} \tag{11.3}$$

The threshold gain is given in terms of the threshold carrier density, n_{th}, the transparency carrier density, n_{tr}, and the differential gain a ($= dg/dn$) as

$$g_{th} = a(n_{th} - n_{tr}) \tag{11.4}$$

Using Equations 11.1 and 11.4, the threshold carrier density may be expressed as

$$n_{th} = n_{tr} + \frac{1}{\Gamma a}(\alpha_a + \alpha_d + \alpha_m) \tag{11.5}$$

The threshold carrier density is related to the threshold current density by the relation

$$n_{th} = \frac{\eta_i \tau}{ed} J_{th} \tag{11.6}$$

where η_i is the injection efficiency. The carrier lifetime is given by

$$\frac{1}{\tau} = A + Bn + Cn^2 \tag{11.7}$$

where, as already pointed out in Chapter 2, the three terms on the right-hand side of Equation 11.7 stand, respectively, for nonradiative, spontaneous, and Auger recombination processes. If it is assumed that the radiative process is dominant, the lifetime is written approximately as

$$\frac{1}{\tau} = \frac{Bn}{\eta_{spon}} \tag{11.8}$$

where η_{spon} is the spontaneous emission efficiency. Using Equations 11.6 and 11.8, the threshold current density is given by

$$J_{th} = \frac{ed}{\eta_i \tau} n_{th} \cong \frac{edB}{\eta_i \eta_{spon}} n_{th}^2 \tag{11.9}$$

The threshold current of an SEL may now be expressed by using Equation 11.9 as

$$I_{th} = \pi \left(\frac{D}{2}\right)^2 J_{th} = \frac{eV}{\eta_i \tau} n_{th} \cong \frac{eVB}{\eta_i \eta_{spon}} n_{th}^2 \tag{11.10}$$

where V is the volume of the active region. It follows easily that the threshold current may be reduced if the volume of the active region is made small. To make a comparison, we show in Table 11.1 the typical dimensions and other parameters of a stripe-geometry Fabry–Perot (FP) laser and a VCSEL.

It is to be noted that the volume of the active region in a stripe laser may be as high as 90 μm^3, whereas that in a VCSEL may be 12 μm^3. The threshold

TABLE 11.1

Typical Dimensions in VCSELs and Stripe Geometry Lasers

Parameter	Symbol	Stripe Lasers	VCSELs
Active layer thickness	D	10 nm–0.1 μm	8 nm–0.5 μm
Reflectivity	R	0.3	0.99–0.999
Cavity length	L	300 μm	0.3 μm ($\lambda/2 - \lambda$)
Optical confinement factor	Γ	0.5%	3%×2
Area of active region	S	3×300 μm^2	5×5 μm^2

current in a VCSEL may be reduced by a factor of 7.5, considering the reduction of volume alone. The increase in confinement factor and mirror reflectivity will reduce the threshold current still further.

> **Example 11.1:** We calculate the threshold currents for a GaAs VCSEL and an FP stripe geometry laser. The values $N_{tr}=2.6\times10^{24}$ m^{-3}, differential gain $a=4.0\times10^{-20}$ m^2, and absorption loss $\alpha_a=10^3$ m^{-1} are assumed for both the lasers.
>
> For the VCSEL, assuming $R=0.99$ and $L=0.3$ μm, the mirror loss $\alpha_m=3.316\times10^4$ and the total loss $\alpha_a+\alpha_m=3.416\times10^4$. Taking $\Gamma=0.06$, and using Equation 11.5, the threshold carrier density is $n_{th}=1.683\times10^{25}$ m^{-3}. We now use Equation 11.10 to calculate the threshold current. We take $d=0.1$ μm, $D=2$ μm, and $\eta_i=\eta_{spon}=0.8$, and obtain $I_{th}=2.22$ mA.
>
> For the FP stripe geometry laser, we assume $R=0.3$ and $L=300$ μm and obtain $\alpha_m=4.01\times10^3$, total loss $=5.01\times10^3$. Also, $\Gamma=0.005$, $d=0.1$ μm, and the stripe width $w=3$ μm are used. Using the calculated value of $n_{th}=2.76\times10^{25}$ m^{-3} and $\eta_i=\eta_{spon}=0.8$, we obtain $I_{th}=1.714$ A.

This example suggests that mirror loss is higher in VCSEL, but the values of n_{th} are comparable. The reduction in I_{th} is due to drastic reduction in the volume of the active region in the VCSEL.

The threshold current may be reduced by reducing the diameter of the active region. However, with a decrease of the diameter, there is also a decrease in the optical confinement factor. In addition, for a smaller diameter, the lateral confinement factor is reduced, the loss due to scattering and diffraction increases, and nonradiative recombination starts to play a role. The threshold current first monotonically decreases with decreasing D, shows a minimum, and then increases rapidly with further decrease in diameter. This variation is similar to the behavior shown in Figure 8.10 for an MQW laser.

> **Example 11.2:** As an illustration of the effect of reduction of the diameter of the active region, we assume that $D=0.2$ μm: 10 times smaller than the value assumed in Example 11.1. The expected threshold current should be 100 times smaller, or $I_{th}=2.22\times10^{-5}$ A.
>
> We now include the effect of reduced transverse confinement factor and additional loss due to scattering and diffraction. The transverse confinement factor is assumed to be expressed by $\Gamma_t=\beta^2/(2+\beta^2)$, where
>
> $\beta=\left(2\pi D/\lambda_0\right)\sqrt{n_a^2-n_{cl}^2}$, λ_0 is the free-space wavelength, and n_a and n_{cl} are, respectively, the refractive indices (RIs) of the active and cladding layers.
>
> We assume $\lambda_0=0.85$ μm, $n_a=3.5$, $n_{cl}=1.5$, and $D=0.2$ μm, obtaining a value $\Gamma_t=0.915$. The mode confinement factor 0.06×0.915 is to be inserted in Equation 11.5 for calculation of n_{th}. We also assume $\alpha_d=10^3$ m^{-1}, so that total loss is now 3.516×10^4 m^{-1}. The value $n_{th}=1.861\times10^{25}$ is obtained, giving a value of 2.72×10^{-5} A, which is larger than 2.22×10^{-5} A. If we

take a value of $4.316 \times 10^4 \, \text{m}^{-1}$ for total loss, then $n_{th} = 2.225 \times 10^{25} \, \text{m}^{-3}$ and $I_{th} = 3.89 \times 10^{-5} \, \text{A}$.

The differential quantum efficiency of the SEL is expressed as

$$\eta_d = \eta_i \frac{(1/L)\ln\left(1/\sqrt{R_f}\right)}{\alpha(1/L)\ln\left(1/\sqrt{R_f R_r}\right)} \tag{11.11}$$

where R_f and R_r are, respectively, the reflectivity of the front and rear mirrors, and $\alpha = \alpha_a + \alpha_d$ is the total internal loss.

Example 11.3: Let $L = 0.3$ µm, $R_f = 0.99$, $R_r = 0.995$, $\alpha = 10^3 \, \text{m}^{-1}$, $\eta_i = 0.7$. Putting all these values into Equation 11.11, we obtain $\eta_d = 0.45$.

11.4 Requirements for Components

In order to improve the performance of a VCSEL, the following considerations must be applied for the selection of components and their properties:

1. The DBRs used as the mirrors should be highly reflective and transparent.
2. The optical losses must be minimized.
3. The overlap between the optical field and the gain profile must be maximal.
4. The electrodes should offer low resistance to achieve high-efficiency operation.
5. Proper heat sinking must be introduced for high-temperature and high-power operation.

11.4.1 Mirror Design

Two different material systems are employed to obtain highly reflective and transparent mirrors. The first option is to use dielectric multilayer mirrors. The advantages of using dielectrics are (1) relatively high index difference, (2) negligible optical loss, and (3) small effective penetration depth. An e-beam deposition technique is used that offers high controllability, and in situ optical path-length monitoring accurately controls the reflectivity and the center wavelength. Reflectivity values greater than 90% are achieved for TiO_2/SiO_2 and SiO_2/Si multilayered reflectors.

The other option is to use semiconductor multilayer structures. Several growth technologies, such as molecular beam epitaxy (MBE), metal organic chemical vapor deposition (MOCVD), and chemical beam technology (CBE), can provide the superlattice structures needed for the growth of distributed Bragg reflector (DBR) and distributed feedback (DFB)-type SELs. The MOCVD technique has been used to grow AlGaAs/AlAs reflectors that show reflectivities as high as 97% at 0.87 μm. The layers are appropriately doped to allow injection of carriers through them into the active region. Semiconductor reflectors using GaInAsP and InP pairs have been grown by CBE, and reflectivities in the range 95%–99% have been attained.

11.4.2 Gain Medium

The gain medium is usually a QW or, better, an MQW structure. As is well known, QWs provide high gain, low threshold current, high relaxation oscillation frequency, and high characteristic temperature. Recently, VCSELs employing multiple stacks of QDs have been reported and studied. Also, tunnel-injection (TI) structures have been employed. The special features of QD- and TI-VCSELs will be mentioned in Sections 11.8 and 11.9.

11.4.3 Current and Optical Confinement

Methods for carrier confinement have been introduced in connection with stripe-geometry double heterostructure (DH) structures in Chapter 1. Similar methods are also employed here for effectively confining the carriers over a narrow width. The ring electrode at the top is the simplest one and is easily fabricated. However, the current diffuses out into the outer regions. The width of the carrier-confining region may be restricted by proton bombardment, which causes the regions surrounding the active region to become insulating. The buried heterostructures (BHs) provide ideal confinement, but the reproducible growth of 2-D BHs poses technical difficulties. The air-post structure confines the current into a small post, and it can be fabricated by etching the sides of a circular or rectangular mesa.

The oxide confinement structure provides both current and optical confinements and is widely used to make very-low-threshold devices. The configuration is shown in Figure 11.3.

GaAs-based VCSELs operating in the range 0.65–1.3 μm employ oxide confinement. An extra AlAs layer is grown just above the top cladding layer grown over the active region. The AlAs layer is selectively oxidized to form an Al_xO_y insulating layer. The oxide aperture thus defines the diameter of the current-confinement region. The oxide has a lower RI, and it thus provides transverse confinement by forming a lateral waveguide consisting of the active region as the core and the oxide as the cladding. Note that the DBRs are made of semiconductors.

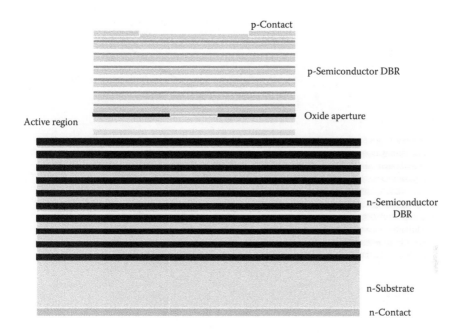

FIGURE 11.3
Cross-sectional view of an oxide-confined VCSEL.

A buried tunnel junction (BTJ) is the commonly employed confinement structure for InP-based VCSELs (covering 1.3–2.3 μm) and GaSb-based VCSELs (covering 2–3 μm). The structure is illustrated in Figure 11.4. Here, the reflectors are made of dielectrics.

11.5 Characteristics of VCSELs

11.5.1 Optical Confinement

A schematic diagram of the active region of the VCSEL enclosed between two Bragg mirrors is shown in Figure 11.5.

The electric field amplitude has maxima at both ends of the inner cavity of length L. The resonance condition in the cavity is expressed by the usual expression

$$\langle n \rangle L = \frac{m\lambda}{2} \qquad (11.12)$$

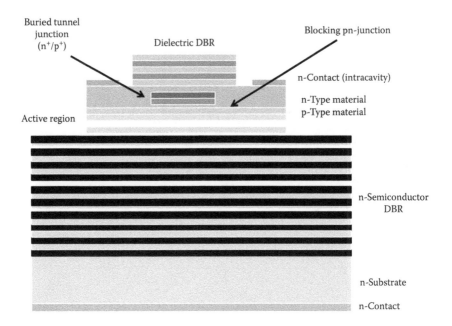

FIGURE 11.4
Schematic diagram of a BTJ VCSEL.

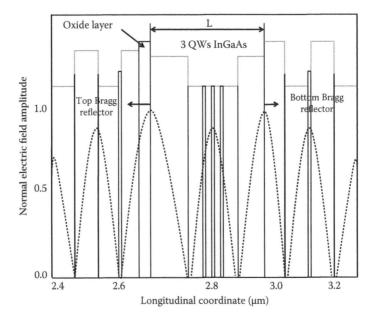

FIGURE 11.5
Electric-field amplitude in the active layer and Bragg mirrors.

where:
- $\langle n \rangle$ is the spatially averaged RI
- m is an integer
- λ is the wavelength

The above condition indicates that an integer number of half wavelengths must fit within the cavity boundaries. In order to provide good coupling with the electromagnetic (EM) field, the active layer is to be placed in an antinode of the standing wave pattern. The shortest symmetric cavity length is just one wavelength thick, meaning that $m = 2$.

The active gain region is limited by the heteroboundaries between the larger gap-surrounding materials. Therefore, for an arbitrary position and total thickness d_a of the active layer, the mode confinement factor may be expressed as

$$\Gamma_r = \frac{L}{d_a} \frac{\int\limits_{d_a} |E(z)|^2 \, dz}{\int\limits_{L_a} |E(z)|^2 \, dz} \tag{11.13}$$

The electric field profile in the central $\lambda/(2\langle n \rangle)$ part of the cavity may be well approximated as

$$E(z) = E_o \cos\left(\frac{2\pi \langle n \rangle z}{\lambda}\right) \tag{11.14}$$

Let the center of the inner cavity be located at $z = 0$. The expression for the mode confinement factor may then easily be written as

$$\Gamma_r = 1 + \frac{\sin\left(2\pi \langle n \rangle d_a / \lambda\right)}{2\pi \langle n \rangle d_a / \lambda} \tag{11.15}$$

For a thin QW ($d_a \ll \lambda$), $\Gamma_r = 2$. If we have M_a number of QWs (MQWs) in the active region, each having the same gain and located at positions $z_{il} \le z \le z_{ih}$ with $i = 1, \dots M_a$, then

$$\Gamma_r = 1 + \frac{\lambda}{4\pi \langle n \rangle} \frac{\sum\limits_{i=1}^{M_a} \sin\left(4\pi \langle n \rangle z_{ih} / \lambda\right) - \sin\left(4\pi \langle n \rangle z_{il} / \lambda\right)}{\sum\limits_{i=1}^{M_a} z_{ih} - z_{il}} \tag{11.16}$$

Example 11.4: Consider three QWs, each 8 nm thick, separated by 10 nm barriers. The above gives a value of 1.8 for the mode confinement factor. This example shows that the gain can be almost doubled.

11.5.2 Bragg Reflectors

As mentioned already, the mirrors used in VCSELs are Bragg reflectors, which consist of alternate layers of high- and low-RI materials with quarter-wavelength thickness. More than 20 of each such pair are needed to yield high reflectivity.

The first layer of the top and bottom mirror, as seen from the inner cavity, must have a lower RI than that of the neighboring carrier-confinement layer. This demands an integral number of Bragg pairs for the top mirror, which has air above it. For an AlGaAs-based VCSEL, a single low-index quarter-wave plate is to be added to the bottom mirror just above the high-index GaAs substrate.

We denote the RIs of the lower-RI and higher-RI materials by n_1 and n_2, respectively, while those of substrate and cladding layers are denoted by n_s and n_c, respectively. For the top (t) mirror, the index sequence as reckoned from the cladding is $n_c | (n_1 | n_2)^{M_{Bt}} | n_s$, and for the bottom mirror it is $n_c | (n_1 | n_2)^{M_{Bb}} | n_1 | n_s$. The number of layer pairs at the top (bottom) Bragg reflector is M_{Bt} (M_{Bb}). The peak reflectivities at the Bragg wavelength λ_B are written as [3]

$$R_{t,b} = \left(\frac{1 - b_{t,b}}{1 + b_{t,b}} \right)^2 \tag{11.17}$$

The expressions for b are given by

$$b_t = \frac{n_s}{n_c} \left(\frac{n_1}{n_2} \right)^{2M_{Bt}} \quad \text{and} \quad b_b = \frac{n_1^2}{n_c n_s} \left(\frac{n_1}{n_2} \right)^{2M_{Bb}} \tag{11.18}$$

The layer thicknesses are chosen as $d_{1,2} = \lambda_B / (4 n_{1,2})$.

Waves are incident from the AlGaAs cladding layer and are transmitted to air, and out of GaAs substrate, in the case of the top and bottom mirror, respectively. For the bottom mirror, an additional AlAs quarter-wave layer is inserted on top of the GaAs substrate.

The peak reflectivity of GaAs–AlAs Bragg reflectors versus the number of mirror pairs, assuming that the mirrors are lossless, is plotted in Figure 11.6.

11.5.3 Threshold

The threshold condition is achieved when the gain in the cavity equals the total loss. Let the losses in the active and passive regions be denoted by α_a

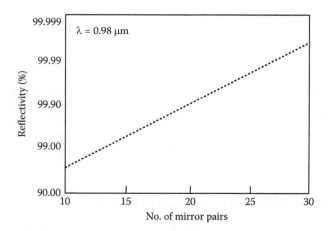

FIGURE 11.6
Peak reflectivity of GaAs–AlAs Bragg reflector.

and α_i, respectively. The active region thickness is d_a, and the mode confinement factor is Γ. A point to note in this connection is that the optical wave penetrates into the Bragg mirror, so that the effective cavity length is written as (please see Reading List)

$$L_{\text{eff}} = L + l_{\text{eff,t}} + l_{\text{eff,b}} \qquad (11.19)$$

where:

L is the separation between two mirrors

$l_{\text{eff,t}}$ and $l_{\text{eff,b}}$ denote, respectively, the effective penetration depths in the top and bottom mirrors

The threshold condition may therefore be written as

$$g_{\text{th}} = \alpha_a + \frac{1}{\Gamma_r d_a} \left[\alpha_i \left(L_{\text{eff}} - d_a \right) + \ln \frac{1}{\sqrt{R_t R_b}} \right] \qquad (11.20)$$

The intensity reflection coefficients are valid for lossless mirrors. If one can neglect the losses in the mirrors, then the expression for the threshold gain reduces to

$$g_{\text{th}} = \alpha_a + \frac{1}{\Gamma_r d_a} \left[\alpha_i \left(L - d_a \right) + \ln \frac{1}{\sqrt{R_t R_b}} \right] \qquad (11.21)$$

The photon lifetime is expressed as

$$\frac{1}{\tau_p} = \frac{d_a}{L_{eff}} \langle v_g \rangle \Gamma_r g_{th} \approx \langle v_g \rangle \left[\alpha_i + \frac{1}{L_{eff}} \ln \frac{1}{\sqrt{R_t R_b}} \right] \qquad (11.22)$$

Example 11.5: Let the average group index $n_r = 3.6$. This gives the average group velocity $\langle v_g \rangle = c/n_r$. With $L_{eff} = 1$ μm, $R_t = R_b = 99.5\%$, $\alpha_a = \alpha_i = 10$ cm^{-1}, $\Gamma_r = 1.8$, and $d_a = 20$ nm, one obtains $g_{th} = 1668$ cm^{-1} and $\tau_p = 2.0$ ps. Note that for edge-emitting lasers a typical value of L is 200 μm, and therefore a large value of gain is achieved for smaller values of reflectivity.

11.5.4 Output Power and Conversion Efficiency

The power coming out of the top (bottom) mirror $P_{t(b)}$ above threshold is expressed in terms of drive current and threshold current as

$$P_{t(b)} = \eta_{ex,t(b)} \frac{\hbar\omega}{e} (I - I_{th}) \qquad (11.23)$$

The external quantum efficiency is the product of differential quantum efficiency η_d and injection efficiency η_i, which accounts for leakage current and carrier overflow over the confining barrier. Thus,

$$\eta_{ex,t(b)} = \eta_{dt(b)} \eta_i \qquad (11.24)$$

The conversion or wallplug efficiency is the ratio of light power output from a mirror and the electrical input power. Therefore,

$$\eta_{c,t(b)} = \frac{P_{t(b)}}{IV} \qquad (11.25)$$

The applied voltage across the VCSEL ideally is given by

$$V \approx V_k + R_s I \qquad (11.26)$$

where V_k is the kink voltage related to the separation of the quasi-Fermi levels, but may be written as $V_k \approx \hbar\omega/e$, and $R_s = dV/dI$ is the differential series resistance. The conversion efficiency may now be expressed as

$$\eta_{c,t(b)} = \eta_{d,t(b)} \eta_i \frac{\hbar\omega}{e} \frac{I - I_{th}}{IV_k + I^2 R_s} \qquad (11.27)$$

It is evident that when $I \gg I_{th}$, the conversion efficiency decreases due to the series resistance. It is easy to prove that the conversion efficiency is maximal for the optimum current given as

$$I_{opt} = I_{th}.\left(1 + \sqrt{1+p}\right) \text{ with } p = V_k / I_{th} R_s \tag{11.27a}$$

The expression for peak conversion efficiency is

$$\eta_{c,t(b)} = \eta_{d,t(b)} \eta_i \frac{\hbar\omega}{eV_k} \frac{p}{\left(1 + \sqrt{1+p}\right)^2} \tag{11.28}$$

Example 11.6: Let us take $\eta_{db} = 0.9$, $\eta_i = 0.9$, $I_{th} = 3$ mA, $\hbar\omega = 1.43$ eV, and $R_s = 40\ \Omega$. Thus, $V_k = \hbar\omega/e = 1.43$ V and $p = 11.9$. The calculated conversion efficiency is 0.456 or 46%.

11.6 Modulation Bandwidth

The modulation response of the VCSEL is obtained by considering the structure shown schematically in Figure 11.7 (please see Michalzik and Ebeling in Reading List). The carrier density in the barrier of volume V_b is n_b, while that in the well is n_w; the total volume of the well is V_w, in which the active layer is composed of a few QWs of total thickness d_a. The resonator of length L_{eff} contains a cavity of one-wavelength thickness and the penetration depths of the two mirrors. The photon density is S within the photon volume V_p. The confinement factor is defined as

$$\Gamma = \frac{V_w}{V_p} = \Gamma_z.\Gamma_t = \frac{d_a}{L_{eff}}.\Gamma_z \tag{11.29}$$

The transverse confinement factor Γ_t is limited to $0 < \Gamma_t \leq 1$, and often $\Gamma_t \approx 1$ is used.

11.6.1 Rate Equation

The three rate equations are

$$\frac{dn_b}{dt} = \frac{\eta_i I}{eV_b} + \frac{V_w}{V_b}\frac{n_w}{\tau_e} - \frac{n_b}{\tau_s} - \frac{n_b}{\tau_{sp,b}} \tag{11.30a}$$

$$\frac{dn_w}{dt} = \frac{V_b}{V_w}\frac{n_b}{\tau_s} - \frac{n_w}{\tau_e} - \frac{n_w}{\tau_{sp,w}} - \Gamma_r v_g \sum_m g_m(n_w, S)S_m \tag{11.30b}$$

$$\frac{dS_m}{dt} = \beta_{sp,m}\Gamma_m \frac{n_w}{\tau_{sp,w}} - \frac{S_m}{\tau_{p,m}} + \Gamma_m \Gamma_r g_m(n_w, S)v_g S_m \tag{11.30c}$$

In the above, we assume that different transverse modes of mode number m exist. The time constant τ_s is the time for supply of carriers to the well by the barriers, which is the diffusion-dominated transport time in undoped potential wells. Carriers escape from the wells into the barriers with time constant τ_e and are removed from the well and barrier regions by spontaneous emission with respective time constants $\tau_{sp,w}$ and $\tau_{sp,b}$. The current entering into the barrier is I, while η_i is the injection efficiency (Figure 11.7).

The analysis becomes greatly simplified if emission at a single transverse mode is considered. The modal material gain is now approximated as

$$g(n_w, S) = \frac{g_0 \ln(n_w / n_{wtr})}{1 + \varepsilon S} \tag{11.31}$$

where:

   g_0 denotes the gain constant
   n_{wtr} is the transparency carrier density in the well
   ε is the gain compression factor

11.6.2 Small-Signal Modulation Response

The standard method for small-signal analysis is now applied, in which the carrier and photon densities and the injected current are composed of a direct current (dc) part and a small-signal part that has an $\exp(j\omega t)$-like variation. The carrier-dependent term of the gain coefficient is linearized in the following form:

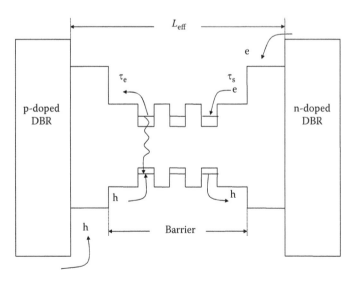

FIGURE 11.7
Model structure to calculate the modulation bandwidth.

$$g(n_{w0}) = g_0 \ln\left(\frac{n_{w0}}{n_{wtr}}\right) \approx a(n_{w0} - n_{tr}) \tag{11.32}$$

The differential gain a and the transparency carrier density n_{tr} are defined as

$$a = \frac{g_0}{n_{w0}} \text{ and } n_{tr} = n_{w0}\left[1 - \ln\left(\frac{n_{w0}}{n_{wtr}}\right)\right] \tag{11.33}$$

The modulation transfer function, neglecting the term containing β_{sp}, is obtained by standard analysis and is written as

$$M(f) = \frac{\Delta S(\omega)}{\Delta I(\omega)/e} = \frac{1}{1 + j\omega\tau_s} \frac{A}{\left(\omega_r^2 - \omega^2\right) + j\omega\gamma} \tag{11.34}$$

The amplitude factor is given by

$$A = \frac{\eta_i v_g \Gamma_r a S_0}{V_p \chi (1 + \varepsilon S_0)} \tag{11.35}$$

The damping coefficient is expressed in terms of the transport factor $\chi = (1 + \tau_s/\tau_e)$ and photon lifetime τ_p as

$$\gamma = \frac{1}{\chi \tau_{sp,w}} + A V_p + \frac{\varepsilon S_0}{\tau_p (1 + \varepsilon S_0)} \tag{11.36}$$

The expression for the resonant frequency is

$$\omega_r = \sqrt{A \frac{V_p}{\tau_p}\left(1 + \frac{\varepsilon}{\tau_{sp,w} v_g \Gamma_r a}\right)} \approx \sqrt{\frac{A V_p}{\tau_p}} \tag{11.37}$$

Using the expression for K-factor, given as

$$K = 4\pi^2\left(\tau_p + \frac{\varepsilon}{v_g \Gamma_r a/\chi}\right) \tag{11.38}$$

the damping constant may be rewritten as

$$\gamma = K\left(\frac{\omega_r}{2\pi}\right)^2 + \frac{1}{\chi \tau_{sp,w}} \tag{11.39}$$

The maximum 3 dB modulation corner frequency of the modulation transfer function $|M(\omega)|^2$ is related to K as

$$f_{max} = \sqrt{2}\,\frac{2\pi}{K} \tag{11.40}$$

11.6.3 Optimization of f_{max}

The photon lifetime is expressed as $\tau_p = (\Gamma\Gamma_g g_{th} v_g)^{-1}$ and $a = g_0/n_{th}$. Defining the normalized threshold gain as $g_n = g_{th}/g_0$, and writing $n_{th} = n_{w0}$, we may write $n_{th} = n_{wtr}\exp(g_n)$. Using these in Equation 11.37, we obtain the following expression for the K-factor:

$$K = 4\pi^2\left(\frac{1}{\Gamma\Gamma_r g_{th} v_g} + \frac{\chi\varepsilon n_{wtr}\exp(g_{th}/g_0)}{v_g\Gamma_r g_0}\right) \tag{11.41}$$

The first term within parentheses decreases with increasing g_{th}, while the second term increases. Therefore, there should be a minimum value for K_{min}. Putting $dK/dn_{th} = 0$, we obtain the characteristic equation as

$$\Gamma\varepsilon\chi n_{wtr}g_n^2 = \exp(-g_n) \tag{11.42}$$

Putting this in Equation 11.41, we may write

$$K_{min} = \frac{4\pi^2}{v_g\Gamma\Gamma_r g_0}\frac{g_n+1}{g_n^2} = 4\pi^2\tau_p\left(1+\frac{g_0}{g_{th}}\right) \tag{11.43}$$

Example 11.7: As an example, we chose the following values for an InGaAs QW-VCSEL: $d_a = 24$ nm, $L_{eff} = 1.25$ µm, $\Gamma_r = 1.8$, $\varepsilon = 10^{-23}$ m^{-3}, $\chi = 1$, $n_r = 3.6$, $n_{wtr} = 1.8\times10^{24}$ m^{-3}. The value of $\Gamma = \Gamma_z = d_a/L_{eff} = 19.2\times10^{-3}$. The characteristic equation becomes $0.3456g_n^2 = \exp(-g_n)$. Solving, we get $g_n = 1.02$, so that $g_{th} = g_0 g_n = 2.142\times10^5$ m^{-1}. Using the chosen values of parameters and the calculated value of g_n, we get $K_{min} = 0.13$ ns and $f_{max} = 70$ GHz. By changing ε, the corresponding values for K_{min} and f_{max} may be obtained.

The dependence of resonant frequency on the current above threshold is given by

$$f_r = \frac{1}{2\pi}\sqrt{\frac{\eta_i\,\Gamma v_g}{eV_a}\frac{\partial g/\partial n}{\chi}}\sqrt{I - I_{th}} = D\sqrt{I - I_{th}} \tag{11.44}$$

where the symbols are already defined in this section.

A higher intrinsic bandwidth is expected because of a small photon lifetime, small gain compression, and large differential gain, which results in a small *K*-factor and therefore less damping at a given resonance frequency. However, these fundamental laser parameters cannot be independently optimized. A smaller photon lifetime leads to higher cavity loss, which results in a higher threshold carrier density and consequently a lower differential gain. This calls for tradeoffs in the design.

Since VCSELs have a relatively high series resistance, current-induced self-heating has a more pronounced effect on speed than for edge-emitting lasers. With a rise in temperature gain, differential gain and internal quantum efficiency decrease, and the threshold current increases. This eventually causes the photon density and the output power, and therefore the resonance frequency, to saturate at a certain current, referred to as the thermal roll-over current. Therefore, a large *D*-factor may increase the thermally limited bandwidth, which enables the VCSEL to reach a high resonance frequency before saturation. This again emphasizes the importance of large differential gain. In addition, smaller-aperture VCSELs can reach a higher resonance frequency, since the *D*-factor is inversely proportional to the square root of the active region volume. However, too small an aperture leads to very high resistance and therefore excessive self-heating. The optimum aperture size is determined by how resistance and thermal impedance scale with aperture size. For GaAs-based oxide-confined VCSELs, the highest resonance frequency is typically reached for an aperture size in the range 3–7 μm.

With the strong effects of self-heating on the dynamics of VCSELs (the resonance frequency in particular), it is clear that thermal effects impose severe limitations on the achievable modulation bandwidth. High-speed VCSELs should, therefore, be designed for low heat generation (low resistance, low nonradiative recombination rates, low internal optical loss, and low thermal impedance. This calls for the use of DBRs with high thermal conductivity, integrated metallic heat spreaders, mounting for efficient heat sinking, and so on, to minimize the current-induced temperature rise.

The modulation bandwidth is also limited by electrical parasitics (resistances and capacitances) associated with the VCSEL chip and the bond pad.

11.7 Temperature Dependence

The emission wavelength in an edge-emitting FP laser is determined by the wavelength at which the peak gain occurs. On the other hand, since the optical resonator is of short dimensions in VCSELs, it is the cavity resonance that determines the emission wavelength. The wavelength shift due to a change in temperature in VCSELs is mainly governed by changes in the average RI in the resonator and to a lesser extent (~10%) by the thermal expansion of the semiconductor layers. Thus, the wavelength shift depends on the material composition of the Bragg resonators and the inner cavity. In the wavelength range between 800 and 1000 nm, the mode shift is typically given by $\partial \lambda / \partial T \approx 0.07$ nm K^{-1}. On the other hand, the peak material gain wavelength in QWs, λ_p, changes as $\partial \lambda_p / \partial T \approx 0.32$ nm K^{-1}, the change being basically due to bandgap shrinkage. Since the mode shift and gain-peak shifts are different, increasing current, and consequent temperature rise, changes the lasing mode and gain spectrum.

The gain bandwidth in 980 nm VCSELs with $In_{0.18}Ga_{0.82}As$ quantum wells spectrally shifts at 0.33 nm C^{-1}, whereas for 850 nm VCSELs with GaAs quantum wells the shift is 0.32 nm C^{-1}. However, the cavity resonances for both VCSELs shift at an approximate rate of 0.07 nm C^{-1}. The misalignment between the peak gain and cavity resonance with changing temperature causes the threshold current to increase. The change in the threshold current is captured by Equation 11.45

$$I_{th}(T) = \alpha + \beta (T - T_{min})^2 \tag{11.45}$$

where:

α, β are parameters

T_{min} denotes the temperature at which the threshold current reaches the minimum value

Typical values for a 980 nm laser are $\alpha = 1.103$ mA, $\beta = 0.354$ µA C^{-2}, and $T_{min} = 51.55°C$.

With increasing drive current above threshold and consequent temperature rise, the output power is no longer a linear function of the drive current. The deviation from linearity is due to higher carrier densities to maintain the threshold gain, carrier and current leakage, and increased nonradiative recombination. The output power characteristics with changing drive current for different heat-sink temperatures are shown in Figure 11.8. As may be noted, the linearity is severely restricted at higher temperatures. Also, all the curves show decreasing optical power after the peak power is attained.

11.8 Tunnel Junction

The typical structure of a BTJ VCSEL is shown in Figure 11.4. The short-wavelength lasers employ GaAs substrate, corresponding lattice-matched alloys, and compounds such as AlGaAs and AlAs. Higher wavelengths can be achieved by using InP substrate and InGaAs or InAlAs. The real

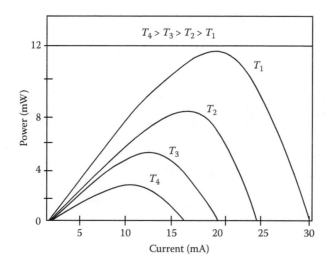

FIGURE 11.8

Typical variation of continuous-wave optical power with drive current for various heat-sink temperatures.

challenge lies in achieving long-wavelength VCSELs. Though short-wavelength VCSELs were commercialized in the 1990s, it was not until considerably later that the first long-wavelength (>1.3 μm) VCSELs with satisfactory performance were presented.

LW-VCSELs offer a low-cost alternative to their in-plane counterparts in the access and metro-area network component market, but performance of these devices has been historically limited by the high resistance and excessive optical loss associated with high acceptor-doping levels in the p-type DBR and current-spreading layers. At longer wavelengths, poor thermal conductivity of the materials, increased thickness of DBRs, and stronger temperature dependence of the threshold current lead to laser heating and make the realization of such VCSELs quite difficult.

More recently, several groups have incorporated TJ structures into long-wavelength (LW)-VCSELs in order to realize reduced optical loss as well as current and optical confinement [4–7]. In the last several years, groups incorporating TJ structures in LW-VCSELs have demonstrated multimode (MM) output power in excess of 9 mW and single-mode (SM) output power greater than 2.5 mW at 20°C and 1.5 mW at 70°C [4–6,8,9]. Several groups have also reported 3.125 Gb s^{-1} SM transmission up to 70°C [5,7]. Several other groups have also developed LW-VCSEL structures incorporating TJ layers that are lattice matched to InP [6].

In this section, we shall briefly discuss an approach that overcomes these problems. The structure employs a high-reflective and thermally well-conducting hybrid metallic/dielectric mirror together with a BTJ to enable low-resistance lateral current confinement and waveguiding.

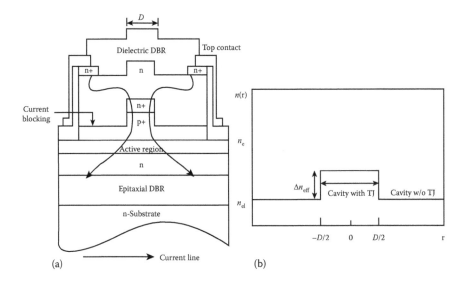

FIGURE 11.9

Schematic diagram of a BTJ VCSEL (a) and the RI variation in its lateral direction (b). (S. Arafin, A. Bachmann, and M.-C. Amann, Transverse-mode characteristics of GaSb-based VSELs with buried-tunnel junctions, *IEEE J. Sel. Top. Quantum Electron.* 9, 1576–1583. © 2011 IEEE.)

   The typical structure of a GaSb-based LW-VCSEL using BTJ is shown in Figure 11.9. Here, heat is drained out from the active region through the thin but highly reflective dielectric mirror into the integrated gold heat sink. The BTJ accomplishes both current confinement to the active region and waveguiding by the RI distribution to achieve low threshold currents. Furthermore, it allows substitution of p-doped device parts by more suitable n-doped material.

   As a consequence, extremely low electrical series resistance between 20 and 50 Ω is obtained, and only ultralow Joule heating occurs. Additionally, the BTJ causes a transverse waveguiding, as shown in Figure 11.9b, which yields both a stable built-in transverse mode profile and small threshold currents.

11.9 QD-VCSEL

Ideally, QD lasers are characterized by lower threshold current, higher modulation bandwidth, and very high (ideally infinite) characteristic temperature T_0. However, in practice the ideal performance is yet to be achieved, the main reason being size fluctuation.

In spite of this, there is growing interest in developing QD-VCSELs for short-distance data communication with bit rates beyond 10 Gb s^{-1}. In order to make the systems cost-effective, the lasers should be directly modulated. Since VCSELs are the most attractive source for this purpose, incorporation of QDs into the VCSEL structure has received attention from workers.

Keeping in mind that commercial local area networks (LANs) achieve a fourfold increase in speed every five years, a fourfold increase in speed requires a 16-fold increase in current density and a proportionate increase in power (please see Reading List).

The modulation speed of Stranski–Krastanov (SK)-grown QD lasers is limited by size fluctuation; a relatively small area reduces the differential gain and limits the speed. Furthermore, electron scattering time from wetting layer to QDs leads to gain saturation. Improvements are sought by using p-doping, using TI structures, and reducing defects. The area is still in the developmental stage. In addition to the article by Ledentsov et al., we include another reference as a representative of the publications in this area [10].

11.10 Microcavity Effects and Nanolasers

The cavity dimensions in VCSELs are of the order of a few micrometers. Some interesting physical phenomena occur in such cavities. With cavity dimensions reduced to the range of nanometers, such phenomena can be studied in more detail. In this section, we shall introduce two such phenomena: the Purcell effect [11,12] and photon recycling [13,14]. The effects can be exploited to produce microcavity or nanocavity light-emitting diodes (LEDs) and lasers, the basic structures of which are not much different from the VCSEL structures. Some features of nano-LEDs and nanolasers will also be mentioned in this section.

11.10.1 Control of Spontaneous Emission

In the earlier discussions on basic light–matter interactions, it was mentioned that emission takes place in two different ways: spontaneous and stimulated. However, the spontaneous process may be thought of as a special case of stimulated emission: the stimulus comes from the vacuum field fluctuations that give rise to zero-point energy of harmonic oscillators $(1/2)\hbar\omega$. The mode and lifetime of the spontaneous emission can be modified by controlling the vacuum field within an optical cavity. This is referred to as cavity quantum electrodynamics.

11.10.2 Fermi Golden Rule

Consider a two-level system having ground (g) and an excited (x) states, as shown in Figure 11.10, and the energy difference is $\hbar\omega = E_x - E_g$. The

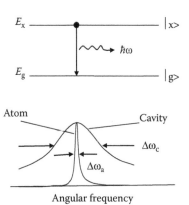

FIGURE 11.10
Two-level atomic system: Emission spectra.

resonance spectral widths of the optical cavity and of the emission spectra are, respectively, $\Delta\omega_c$ and $\Delta\omega_a$, and, as shown, $\Delta\omega_a \ll \Delta\omega_c$ or $\tau_c \ll \tau_a$. Since the lifetime for atomic transition is quite low, the emitted light readily comes out of the optical cavity and is not reabsorbed.

From the second-order perturbation theory, the transition rate of the atoms can be expressed as (see Numai in Reading List)

$$W = \frac{2\pi}{\hbar}\sum_i \mu_i^2 \left(\frac{\hbar\omega_i}{2V\varepsilon}\right)\sin^2(\mathbf{k_i}\cdot\mathbf{r})(n_i+1)L\left(E_x - E_g - \hbar\omega_i\right) \qquad (11.46)$$

where

  μ_i is the electric dipole moment for the *i*th cavity mode
  ω_i is the angular frequency of the *i*th cavity mode having wave vector $\mathbf{k_i}$
  V is the volume of the optical cavity
  ε is the permittivity of the material of the cavity
  n_i is the photon density
  $L(x)$ is a lineshape function

It appears from Equation 11.45 that even when there is no photon ($n_i=0$), there is a finite probability of transition rate. The rate for spontaneous transition is then expressed as

$$W_{sp} = \frac{2\pi}{\hbar}\sum_i \mu_i^2 \left(\frac{\hbar\omega}{2V\varepsilon}\right)\sin^2(\mathbf{k_i}\cdot\mathbf{r})(n_i+1)L\left(E_x - E_g - \hbar\omega_i\right) \qquad (11.47)$$

A free space is regarded as an optical cavity with quite large dimensions compared with the wavelength of light and with modes so close that the dispersion is quasi-continuous (see discussions in Chapter 5). The sum over *i* in Equation 11.46 is then replaced by integration with respect to k. The spontaneous emission rate in free space is therefore written as

$$W_{sp,free} = \frac{\omega_0^3 n_r^3}{3\pi c^3 \hbar \varepsilon} \langle \mu \rangle^2 \tag{11.48}$$

where:

$\langle \mu \rangle^2$ is the squared electric dipole moment averaged over all directions of polarization

$\sin^2(\mathbf{k}_i . \mathbf{r})$ is replaced by its average value ½

The transition rate for each mode is extremely small, proportional to $(1/V)$, where V is the volume of the cavity. Since the number of modes is proportional to V, the spontaneous emission rate in free space is independent of the volume of the cavity. Furthermore, the spontaneous emission rate does not depend on the position of the atom and is therefore governed by the properties of the atoms themselves.

The size of a microcavity is of the order of the wavelength. Hence, the resonant modes are discrete. From Equation 11.46, putting the spontaneous emission coupling factor $\beta_{sp} = 1$, the spontaneous emission rate in the microcavity is expressed as

$$W_{sp,mc} = \frac{\pi}{\hbar} \mu_i^2 \left(\frac{\hbar \omega_i}{V \varepsilon} \right) \left(\frac{1}{\hbar \Delta \omega_c} \right) \sin^2 (\mathbf{k}_i . \mathbf{r}) \tag{11.49}$$

where $\Delta \omega_c$ is the spectral linewidth of the resonant mode of the optical cavity.

Using Equations 11.47 and 11.48, when $\omega_0 = \omega_i$, that is, the light frequency equals the frequency of the ith mode in the cavity, one obtains the ratio

$$\frac{W_{sp,mc}}{W_{sp,free}} = \frac{3}{8\pi} \frac{\lambda_i^3}{n_r^3} \frac{1}{V} \left(\frac{\omega_i}{\Delta \omega_c} \right) \sin^2 (\mathbf{k}_i . \mathbf{r}) \tag{11.50}$$

Equation 11.49 indicates that under certain conditions the spontaneous emission rate in a microcavity may exceed the corresponding rate in free space. The conditions are (1) the size of the cavity should be comparable to the light wavelength, that is, $V \approx \lambda_i^3 / n_r^3$; (2) the Q-value of the cavity should be extremely large, that is, $\omega_i / \Delta \omega_c \gg 1$; and (3) the excited atoms should be placed at the antinode of the standing waves in the optical cavity, so that $\sin^2(\mathbf{k}_i . \mathbf{r}) = 1$.

In contrast, spontaneous emission will be inhibited if the light frequency and the frequency of the cavity mode are different ($\omega_0 \neq \omega_i$) and the excited atoms are placed at the nodes of the standing wave pattern ($\sin^2(\mathbf{k}_i . \mathbf{r}) = 0$).

11.10.3 Microcavity Lasers

It was shown in Chapter 7 that a low mode density and a large spontaneous emission rate lead to a larger stimulated emission rate and hence a lower threshold current density. The spontaneous emission rate increases with larger Q-value of the cavity. Therefore, VCSELs are suitable for microcavity laser diodes.

The rate equations for a laser diode are reproduced from Chapter 7 and are given below:

$$\frac{dn}{dt} = \frac{J}{ed} - G(n)S - \frac{n}{\tau_n} \tag{11.51}$$

$$\frac{dS}{dt} = G(n)S - \frac{S}{\tau_{ph}} + \beta_{sp}\frac{n}{\tau_n} \tag{11.52}$$

where the symbols are already defined. Here, $G(n)$, the amplification rate due to the stimulated emission, includes the enhancement or suppression of the spontaneous emission in the microcavity, and τ_n is the radiative recombination lifetime.

For a continuous-mode distribution, the spontaneous-emission coupling coefficient has been expressed earlier as

$$\beta_{sp} = \frac{\Gamma}{4\pi^2 n_r^3 V}\frac{\lambda_0^4}{\Delta\lambda} \tag{11.53}$$

It follows that for small mode volume V and a narrow full width at half maximum (FWHM) $\Delta\lambda$, the value of β_{sp} becomes large. Figure 11.11 shows the carrier concentration n and the photon density S in the microcavity as a

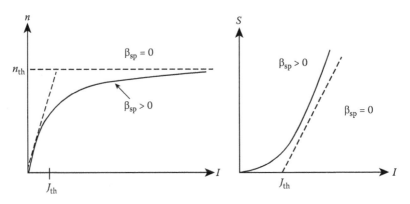

FIGURE 11.11
Variation of carrier density (a) and photon density (b) with current density.

function of the injected current density J. It is found that when β_{sp} is large, the decrease in the threshold current density is not as distinct as in the case of negligible β_{sp}.

The enhancement of the spontaneous emission rate in a microcavity decreases the carrier lifetime τ_n, thereby increasing the stimulated emission rate. This in turn increases the differential gain dG/dn. The resonance frequency of the laser diode is related to the differential gain by the following relation:

$$f_r = \frac{1}{2\pi}\sqrt{\frac{1}{\tau_n \tau_{ph}}\frac{J - J_{th}}{J_{th} - J_0'}} = \frac{1}{2\pi}\sqrt{\frac{\partial G}{\partial n}\frac{S_0}{\tau_{ph}}} \tag{11.54}$$

The large differential gain in a microcavity laser therefore increases the resonance frequency or the modulation bandwidth. However, for high-Q cavities, the photon lifetime increases to reduce the resonance frequency.

Microcavity lasers have their limitations also. Since the size of the microcavity is of the order of the light wavelength, the smaller size gives rise to higher electrical resistance. This in turn increases the injection current and the associated Joule heating. This may sometimes make room-temperature operation difficult or at least cause high power consumption. Also, the small emission area leads to low light output.

11.10.4 Photon Recycling

The current applied to a semiconductor laser injects excess carriers into its active region, in which the carrier pairs recombine spontaneously. The spontaneous emission occurs in all the modes supported by the cavity, and the emitted light propagates in all directions. Only a tiny fraction of the spontaneously emitted photons are coupled to the lasing mode, and this fraction acts as the seed to trigger stimulated emission. The emitted light is repeatedly amplified in the optical cavity, and when the gain equals the loss, self-sustained laser oscillation starts.

The remaining portion of the spontaneously emitted light, which is not coupled to the lasing mode, comes out of the cavity. The injected carriers consumed in such spontaneous emission do not contribute to the laser oscillation. However, there are means to absorb the wasted spontaneous emission to generate carrier pairs. The process is known as photon recycling, and the carrier concentration in the cavity in the presence of photon recycling is higher than the concentration in its absence. As a consequence, the threshold current is lowered by this process.

Photon recycling can be done by depositing highly reflecting layers on all but the output facet of the laser. Almost all the spontaneously emitted photons are now returned to and reabsorbed in the active region and ultimately exit the laser medium in the desired mode.

To model the photon-recycling effect in a VCSEL, we note that the gain spectrum in a QW active layer is smeared due to broadening mechanisms and looks like the spectrum for a DH laser. The model for a DH laser may be used by applying a scaling factor to the gain spectrum of a QW laser.

In the schematic diagram of a DH laser shown in Figure 11.12, L, W, and D denote, respectively, the length, width, and height of the DH structure, and d is the thickness of the active layer. Spontaneously emitted photons, moving in random directions, may come out of the structure from the sides and end facets over an area $2DL + 2DW$ with transmittivity T_d. They may be absorbed in the contacts of area $2LW$, characterized by absorption coefficient A_d; they may be absorbed in the bulk, characterized by absorption coefficient α_{av} averaged over the volume of the active and cladding regions; and lastly, the photons may be reabsorbed in the active layer, in which the absorption coefficient is $\alpha_i(E)$ at photon energy E.

The fraction of the spontaneously emitted photons of energy E that is reabsorbed in the active layer is

$$\theta = \frac{\alpha_i(E)dLW}{\alpha_i(E)dLW + \alpha_{av}DLW + (T_d/4)(2DL+2DW) + (A_d/4)(2LW)}$$

$$\approx \frac{\alpha_i(E)d}{\alpha_i(E)d + \alpha_{av}D + (1/2)A_d} \tag{11.55}$$

The last approximation holds when $L \gg W \gg D$. The factor ¼ arises from the flux across a surface with randomly directed velocities.

The schematic variation of $\alpha_i(E)$ and the spontaneous emission rate are shown in Figure 11.13. The plots may be obtained for a material with a specified injection-current density. It is observed that there is a wide range of

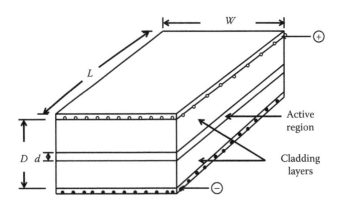

FIGURE 11.12
Schematic diagram of laser structure.

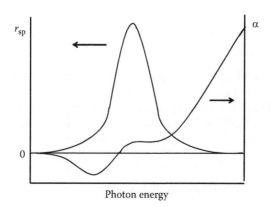

FIGURE 11.13

Typical variation of interband absorption coefficient and spontaneous emission rate r_{sp} vs. photon energy.

spontaneous spectrum for which $\alpha_i(E)$ is positive. In this range, photons may be reabsorbed to generate additional electron–hole pairs in the active layer.

We may express the fraction of all the spontaneously emitted photons that is reabsorbed in the active layer as

$$\theta_{av} = \frac{\int \theta(E) r_{sp}(E)dE}{\int r_{sp}(E)dE} \tag{11.56}$$

The effective current density, J_{eff}, needed to reach a given excitation level is the sum of externally supplied current density, J_{ext}, and a current density $\eta_{sp}\theta_{av}J_{eff}$, where η_{sp} is the quantum efficiency. Thus,

$$J_{eff} = J_{ext} + \eta_{sp}\theta_{av}J_{eff}, \quad J_{eff} \le J_{th} \tag{11.57}$$

At the lasing threshold, $J_{eff} = J_{th}$. This gives

$$J_{ext,th} = J_{th}\left(1 - \eta_{sp}\theta_{av}\right) \tag{11.58}$$

indicating a reduction in the external current density to reach threshold by a fraction $\eta_{sp}\theta_{av}$.

> **Example 11.8:** The reduction in the threshold current density due to the recycling effect in a GaAs/AlGaAs DH laser has been estimated by Stern and Woodall [13]. The values $d=0.4\ \mu m$, $D=4.4\ \mu m$, $W=20\ \mu m$, $L=500\ \mu m$, $\alpha_{av}=20\ cm^{-1}$, and $A_d=0.05$ are chosen. The value of $\alpha_i(1.5\ eV)=4700\ cm^{-1}$. This gives a value of $\theta=0.069$. From the calculated absorption and spontaneous

emission spectra, a value of $\theta_{av}=0.41$ was obtained. Using $\eta_{sp}=0.5$, the externally supplied threshold current density is reduced by 20%.

The usual method to confine the spontaneously emitted photons within the cavity is to coat the sidewalls as well as the top DBR with metals such as gold or silver. There is almost no transmission through the upper DBR. The light comes out of the lower DBR. Photonic crystals are also used to increase the confinement and at the same time introduce Purcell enhancement.

11.10.5 Modulation Bandwidth of Nanolasers

There are many well-known advantages of electronic integrated circuits. There have been many efforts to achieve photonic integrated circuits in the emerging field of photonic signal processing. For this purpose, small and efficient light emitters having low threshold current and high modulation bandwidths are in great demand. Because of the small volume, low threshold current, and planar nature of processing, VCSELs are the most suitable candidates for such emitters. Today's fastest laser diodes exhibit modulation speed in the order of a few tens of gigahertz. Nano-LEDs and nanolasers are expected to surpass this limit on bandwidth. With rapid progress in the process technology in the past two decades, it is now possible to fabricate active cavities with high quality factors and low mode volumes. In these structures, the Purcell enhancement of spontaneous emission can play a significant role in the dynamics of the device [3,8,9,15]. In this subsection, we shall briefly point out some theoretical predictions about the modulation bandwidth of nanolasers that take account of the Purcell enhancement factor.

Altug et al. made an analysis of the laser rate equations with a phenomenological Purcell enhancement of the spontaneous emission rate and predicted that the modulation speeds could exceed 100 GHz [16]. In a more recent paper, Lau et al. investigated the same system numerically and pointed out the importance of gain suppression [17]. They indicated that the ultrahigh modulation speeds reported by Altug et al. were the result of the measurement method, and that in reality these speeds are only achievable in nonlasing devices with ultralow mode volume.

A detailed analysis of the dynamics of the device has been presented by Suhr et al., including a rigorous treatment of spontaneous emission based on fundamental principles [18]. The Purcell enhancement factor was treated using a linear model as well as a model dependent on the optical density-of-states function. An important finding of their analysis is that the Purcell enhancement saturates when the cavity quality factor is increased, which limits the maximum achievable spontaneous recombination rate. The calculated modulation bandwidth of QW nano-LED and nanolaser devices is found to be limited to a few tens of gigahertz (~35 GHz) for realistic devices.

PROBLEMS

11.1. Use the simple formula for the transverse confinement factor given in Example 11.2. Prove that the threshold current shows a minimum for an optimum value of diameter D.

11.2. Obtain a plot for I_{th} vs. D using the parameter values given in Example 11.2.

11.3. Calculate the value of the reflectivity for a GaAs–AlAs DBR

11.4. Calculate and plot the values of the mode confinement factor against the number of QWs. Each QW is 8 nm thick and the barrier width is 10 nm.

11.5. Calculate the approximate value of the bandwidth of a GaAs–AlAs Bragg mirror at $\lambda_B = 0.98$ μm. The RI difference is 0.56, and the spatially averaged group index is 3.6.

11.6. Plot $f(p) = p / (1 + \sqrt{1+p})^2$ against p, where p is defined in Equation 11.27a, and obtain the maximum value of $f(p)$, the maximum value of peak conversion efficiency given by Equation 11.28, using ideal values of conversion efficiencies and of V_k.

11.7. The threshold current in VCSELs is larger than in edge-emitting lasers due to large sheet resistance offered by multilayer stack and also because of smaller gain due to thin QW. Prove now that the conversion efficiency is lower in VCSELs than in edge-emitting lasers.

11.8. Obtain plots of K_{min} given by Equation 11.42 as a function of gain compression factor. Show also in the plot the values of f_{max}.

11.9. Obtain a plot of I_{th} vs. T for a VCSEL emitting at 980 nm. Use the parameter values given after Equation 11.43 and the minimum value of $I_{th}(min) = 1$ μA.

11.10. The spontaneous emission rate needs to be 100 times larger than the rate in free space. The emission wavelength is 1.55 μm, the volume of the cavity is 3 μm in height and 3 μm in diameter, and the RI = 3.5. Determine the Q-value needed.

Reading List

Dallesasse, J. M. and D. G. Deppe, III-V oxidation: Discoveries and applications in vertical-cavity surface-emitting lasers, *Proc. IEEE* 101, 2234–2242, 2013.

Iga, K., Surface emitting laser-its birth and generation of new optoelectronics field, *IEEE J. Sel. Top. Quantum Electron.* 6, 1201–1215, 2000.

Iga, K., Vertical-cavity surface-emitting laser (VCSEL), *Proc. IEEE* 101, 2229–2233, 2013.

Iga, K., and F. Koyama, Chapter 5: Surface-emitting lasers. In E. Kapon (ed.), *Semiconductor Lasers II*, pp. 323–372. Academic Press, San Diego, 1999.

Jewell, J. L., J. P. Harbison, A. Scherer, Y. H. Lee, and L. T. Florez, Vertical-cavity surface-emitting lasers: Design, growth, fabrication, characterization, *IEEE J. Quantum Electron.* 27, 1332–1346, 1991.

Larsson, A., Advances in VCSELs for communication and sensing, *IEEE J. Sel. Top. Quantum Electron.* 17, 1552–156, 2011.

Ledentsov, N. N., F. Hopfer, and D. Bimberg, High-speed quantum-dot vertical-cavity surface-emitting lasers, *Proc. IEEE* 95, 1741–1756, 2007.

Michalzik, R. (ed.), *VCSELs: Fundamentals, Technology and Applications of Vertical-Cavity-Surface Emitting Lasers*. Springer, New York, 2013.

Michalzik, R., and K. J. Ebeling, Operating principles of VCSEL. In H. Li and K. Iga (eds), *Vertical Cavity Surface Emitting Devices*, Springer Series in Photonics, vol. 6, Springer, Berlin, pp. 53–98, 2003.

Numai, T., *Fundamentals of Semiconductor Lasers*. Springer, New York, 2004.

Yu, S. F., *Analysis and Design of Vertical Cavity Surface Emitting Lasers*. Wiley, New York, 2003.

References

1. Iga, K., *Laboratory Notebook of P&I Laboratory*, Tokyo Institute of Technology, Tokyo, 1977.

2. Soda, H., K. Iga, C. Kitahara, and Y. Suematsu, GaInAsP/InP surface emitting injection lasers, *Jpn. J. Appl. Phys.* 18, 2329–2330, 1979.

3. Corzine, S. W., R. H. Yan, and L. A. Corzine, A tanh substitution technique for the analysis of abrupt and graded interface multilayer dielectric stacks, *IEEE J. Quantum Electron.* 27, 2086–2090, 1991.

4. W. Hofmann, High speed buried tunnel junction vertical cavity surface emitting lasers, *IEEE Photon. J.* 2, 802–815, 2010.

5. Tiwari, M., D. Feezell, D. A. Buell, A. W. Jackson, L. A. Coldren, and J. E. Bowers, Electrical design optimization of single-mode tunnel-junction-based long-wavelength VCSELs, *IEEE J. Quantum Electron.* 42, 675–682, 2006.

6. Amann, M. C. and W. Hofmann, InP-based long-wavelength VCSELs and VCSEL arrays, *IEEE J. Sel. Topics Quantum Electron.* 15, 861–868, 2009.

7. Arafin, S., A. Bachmann, and M.-C. Amann, Transverse-mode characteristics of GaSb-based VCSELs with buried-tunnel junctions, *IEEE J. Sel. Top. Quantum Electron.* 9, 1576–1583, 2011.

8. Hepburn, C. J., R. Sceats, D. Ramoo, A. Boland-Thoms, N. Balkan, M. J. Adams, A. J. Dann, et al., Temperature dependent operation of 1.5 μm GaInAsP/InP VCSELs, *Superlatt. Microstr.* 32, 103–116, 2002.

9. Nishiyama, N., C. Caneau, M. Sauer, A. Kobyakov, and C. E. Zah, InP based long wavelength VCSELs: Their characteristics and applications, *Proc. SPIE* 6782, 67820M-1–67820M-9, 2007.

10. Xu, D., C. Tong, S. F. Yoon, W. Fan, D. H. Zhang, M. Wasiak, Ł. Piskorski, et al., Room-temperature continuous-wave operation of the In(Ga)As/GaAs quantum-dot VCSELs for the 1.3 μm optical-fibre communication, *Semicond. Sci. Technol.* 24, 055003, 2009.

11. Purcell, E. M., Spontaneous emission probabilities at radio frequencies, *Phys. Rev.* 69, 681, 1946.

12. Yamamoto, Y., S. Machida, and G. Björk, Microcavity semiconductor laser with enhanced spontaneous emission, *Phys. Rev. A* 44, 657–668, 1991.

13. Stern, F. and J. M. Woodall, Photon recycling in semiconductor lasers, *J. Appl. Phys.* 45, 3904–3906, 1974.

14. Chan, S.-H. G and R. A. Linke, Photon recycling for threshold reduction in semiconductor lasers. In *Proceedings of 9th Annual IEEE Princeton Section Sarnoff Symposium*, pp. 1–5, Princeton, 1993.

15. Chen, P., P. O. Leisher, A. A. Allerman, K. M. Geib, and K. D. Choquette, Temperature analysis of threshold current in infrared vertical-cavity surface-emitting lasers, *IEEE J. Quantum Electron.* 42, 1078–1083, 2006.

16. Altug, H., D. Englund, and J. Vučković, Ultrafast photonic crystal nanocavity laser, *Nat. Phys.* 2, 484–488, 2006.

17. Lau, E. K., A. Lakhani, R. S. Tucker, and M. C. Wu, Enhanced modulation bandwidth of nanocavity light emitting devices, *Opt. Express* 17, 7790–7799, 2009.

18. Suhr, T., N. Gregersen, K. Yvind, and J. Mørk, Modulation response of nanoLEDs and nanolasers exploiting Purcell enhanced spontaneous emission, *Opt. Express* 18, 11230–11241, 2010.

12

Single-Mode and Tunable Lasers

12.1 Introduction

It has been mentioned in Chapter 1 that the Fabry–Perot (FP) laser supports multiple longitudinal modes. Single-mode operation of lasers is essential for optical fiber telecommunication. In the elementary discussion presented in Chapter 1, the distributed feedback (DFB) laser was introduced, and a qualitative discussion of how the structure supports a single longitudinal mode was presented.

In the present chapter, we shall discuss in more detail the operation of lasers supporting a single mode. First, the necessity of using a single-mode laser of very narrow linewidth is pointed out. The limitations of the FP resonator in achieving the required characteristics will then be discussed. The two forms of single-mode laser, the distributed Bragg reflector (DBR) and the DFB laser, will then be introduced, and for each type, the structure, the principle of operation, and the method of analysis will be described.

There is a need to tune the emission wavelength of lasers in communication systems, an example being dense wavelength division multiplexing (DWDM) systems. The present chapter will also introduce tunable lasers. The principle of changing the emission wavelength, the structures used, and the characteristics, such as tuning range and output power, will also be discussed.

12.2 Need for Single-Mode Laser

In optical fiber telecommunication, the signal to be transmitted is in the digital format. The digitized signal modulates the output power of the laser. In the simplest modulation scheme, a bit stream of binary 0 and 1 is superimposed on the direct current (dc) bias current. As a result, one obtains a variation of optical power replicating the impressed digital electrical signal. The conversion has been illustrated in Figure 7.10. The optical bit stream then propagates through a section of a fiber and suffers attenuation and dispersion. As a result, the well-defined input optical pulses smear out due to dispersion, and after traversing a length L of the fiber, the dispersion may be so high that a binary 1 encroaches into the time slot allotted to the neighboring pulses. This gives rise to intersymbol interference, and a bit error occurs. The bit rate–length product is approximately expressed as

$$BL \leq \frac{0.5}{D\Delta\lambda} \tag{12.1}$$

where:
 B is the bit rate
 L is the length of the fiber
 D is the dispersion parameter
 $\Delta\lambda$ is the linewidth of the source

It follows at once that the BL product will increase if the linewidth is kept very small.

> **Example 12.1:** Assume that $B=10$ Gb s^{-1}, $D=17$ ps (nm km)$^{-1}$ at 1.55 μm, and $\Delta\lambda=1$ nm. The maximum allowable length of the fiber for error-free operation is $L=2.94$ km. If the linewidth is 0.1 nm, then the maximum length will increase tenfold.

The example indicates the importance of reducing the linewidth.

12.3 Limitation of FP Laser

An FP resonator having mirror separation, L, supports a large number of longitudinal modes. The resonance condition is that an integral number of half wavelengths should fit in the cavity. This gives the expression for the wavelength of the kth mode as

$$\lambda_k = \frac{2n(\lambda_k)L}{k} \tag{12.2}$$

where $n(\lambda_k)$ is the real part of the complex (wavelength-dependent) refractive index (RI) of the material in the cavity. The longitudinal mode spacing is given by

$$\Delta\lambda_m = \frac{\lambda_0^2}{2n_g L} \tag{12.3}$$

where:

λ_0 is the center wavelength

n_g is the group index

Only those modes that fall under the gain spectra are supported by the gain medium.

Example 12.2: Suppose that the width of the gain spectra is 10 nm. Let $L=300$ μm, $n_g=4$, and $\lambda_0=1.55$ μm, so that $\Delta\lambda_m=1.0$ nm. Then, the number of longitudinal modes supported by the gain spectra is $10/1\sim10$.

The spectral purity of the laser is determined from the side-mode suppression ratio (SSR), which is defined as

$$SSR = \frac{S_0}{S_1} \tag{12.4}$$

where S denotes the photon density, and the subscripts 0 and 1 correspond to the central mode 0 and the first side mode 1. The photon densities are obtained by solving rate equations. The expressions are

$$S_1 = -\frac{n_{sp}\alpha_{tot}}{\Delta g(\lambda_1)} \text{ and } S_0 = \frac{2P}{\hbar\omega v_g n_{sp}\alpha_{tot}\alpha_m}. \tag{12.5}$$

Here, n_{sp} is the spontaneous emission factor (~2), $\alpha_{tot}=\alpha_i+\alpha_m$, α_m is the mirror loss and α_i is the other loss, P is the output power, and $\Delta g(\lambda_1)$ is given by

$$\Delta g(\lambda_1) = \Delta g(\lambda_0) - \delta g \approx \delta g \tag{12.6}$$

$\Delta g(\lambda_0)$ is the difference between actual gain and total loss, and hence for lasing it equals zero. The SSR is now expressed as

$$SSR = \frac{2P\delta g}{\hbar\omega v_g n_{sp}\alpha_{tot}\alpha_m} \tag{12.7}$$

Using the relation between the modal and material gains and assuming that the material gain varies parabolically around the center wavelength, we may write

$$g(\lambda) = \Gamma g_m(\lambda) = \Gamma\left[g_m(\lambda_0) - a(\lambda - \lambda_0)^2\right] \tag{12.8}$$

so that the gain difference may be expressed as

$$\delta g = \Gamma a \Delta \lambda_m^2 \tag{12.9}$$

Putting Equation 12.9 into Equation 12.7, the final expression for SSR reads as

$$\text{SSR} = \frac{2P\Gamma a \Delta \lambda_m^2}{\hbar \omega v_g n_{sp} \alpha_{tot} \alpha_m} \tag{12.10}$$

Example 12.3: We consider an InGaAsP/InP FP laser operating at $\lambda_0 = 1.55$ μm and giving 5 mW of optical power per mirror. The other parameters are $L = 300$ μm, $n_g = 4$, $\Gamma = 0.2$, $n_{sp} = 2$, $R_1 = R_2 = 0.36$, $a = 0.15$ cm^{-1} nm^{-2}, and $\alpha_i = 30$ cm^{-1}. The mirror loss is $\alpha_m = 34$ cm^{-1}. The mode spacing is $\Delta \lambda_m = 1.0$ nm. The gain difference is $\delta g = 0.03$ cm^{-1}. The obtained value of SSR = 71.8 or 18.56 dB.

The value of SSR for optical fiber communication must be at least 30 dB; however, the value obtained in Example 12.3 is only 18 dB. It seems therefore that FP lasers are not suited for applications where a high degree of spectral purity is needed.

12.4 Distributed Feedback

The discussion in Section 12.3 leads to the conclusion that the side modes should be suppressed with respect to the dominant mode by a larger amount than is provided by the FP resonator. In other words, the difference in gain between the dominant and the side modes should be higher. This higher gain difference can be achieved by making the loss wavelength dependent. If the mirror reflectivity is made wavelength dependent, then the amount of feedback can be reduced for side modes, thereby suppressing their power output. It has been mentioned in Chapter 1 that a distributed feedback structure provides the required wavelength-dependent feedback. When the Bragg reflection condition is satisfied, only the dominant mode gets the proper feedback for self-sustained oscillation.

The Bragg grating filter can be placed in the structure in two different ways. If it is placed outside the active region at the ends of the laser cavity, as shown in Figure 12.1a, the structure is called the DBR laser. In the other configuration, shown in Figure 12.1b, the Bragg grating is collocated with the active region along the entire laser length. This is known as the DFB laser structure. Thus, in both the DFB and DBR structures, feedback is provided in a distributed way. In contrast, in an FP laser, feedback is localized at the two end mirrors.

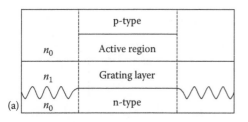 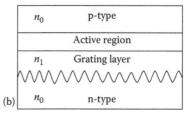

FIGURE 12.1
Schematic diagram of (a) a DBR laser and (b) a DFB laser structure. The periodic RI variation is introduced by thickness variation of the grating layer with RI $n_1 \neq n_0$ of the lower layer.

In both the structures, a longitudinal index variation is accomplished by growing a grating layer of RI n_1 with periodic thickness variation on top of a semiconductor layer of RI n_0.

12.5 DBR Laser

The DBR laser is effectively an FP laser in which the two end mirrors are replaced by Bragg reflectors, which show wavelength-dependent reflectivity.

12.5.1 Coupled-Mode Theory for DBR Laser

The gratings in both the DBR and DFB structures introduce an RI perturbation that induces a coupling between the forward- and backward-propagating waves of a particular laser mode. This implies that the optical feedback is not localized at the two end mirrors, but is distributed throughout the entire laser cavity (DFB) or in a part of it (DBR).

The analysis is based on coupled-mode theory, proposed by Kogelnik and Shank [1]. The model developed is based on the schematic diagram given in Figure 12.2, in which the DBR extends from $z=0$ to $z=L_B$. The effective RI n_{eff} varies sinusoidally along z with a spatial period Λ and is expressed as (see Amann in Reading List)

$$n_{eff}(z) = \bar{n}_{eff} + \frac{\Delta n_{eff}}{2} \sin(k_g z) \qquad (12.11)$$

where:
$\bar{n}_{eff}$ and Δn_{eff} denote, respectively, the average effective RI and the effective RI difference (peak-to-peak)
k_g denotes the grating wave vector, defined as $k_g = 2\pi/\Lambda$

The square of the z-dependent wave vector is written as

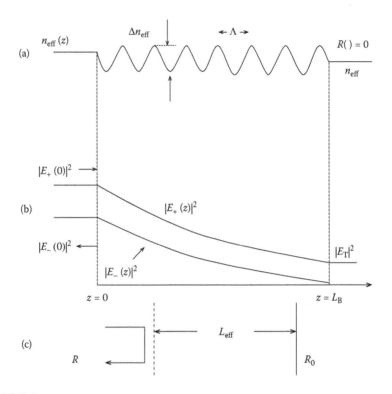

FIGURE 12.2
(a–c) Diagram showing the RI variation along the length of the DBR. The variations of the intensities for forward- and backward-propagating waves are shown.

$$k(z)^2 = k_0^2[n(z)]^2 \approx k_0^2\left(\bar{n}_{\text{eff}}^2 + \bar{n}_{\text{eff}}\Delta n_{\text{eff}}\sin k_g z\right) \qquad (12.12)$$

where the contribution of Δn_{eff}^2 has been neglected; k_0 is the wave vector in free space.

We may now write the one-dimensional wave equation as

$$\frac{d^2E}{dz^2} + k_0^2\left(\bar{n}_{\text{eff}}^2 + \bar{n}_{\text{eff}}\Delta n_{\text{eff}}\sin k_g z\right)E = 0 \qquad (12.13)$$

The z-dependent electric field is decomposed into a forward wave (E_+) propagating along the +z-direction and a backward wave (E_-) propagating along the −z-direction. Thus,

$$E(z) = E_+(z) + E_-(z) \qquad (12.14)$$

We define the Bragg wavelength as $\lambda_B = 2\text{Re}(\bar{n}_{\text{eff}})\Lambda$, and the Bragg wave vector as $k_B = k_g/2 = 2\pi\text{Re}(\bar{n}_{\text{eff}})/\lambda_B$, where Re denotes the real part. Both the forward and the backward waves may be written in terms of the Bragg wave vector as

$$E_+(z) = A(z)\exp(-jk_B z) \tag{12.15}$$

$$E_-(z) = B(z)\exp(+jk_B z) \tag{12.16}$$

In Equations 12.15 and 12.16, $A(z)$ and $B(z)$ represent functions that vary slowly with z. We write

$$\frac{d^2 E}{dz^2} = \frac{d^2 E_+}{dz^2} + \frac{d^2 E_-}{dz^2} \tag{12.17}$$

In obtaining the second derivative of $E_\pm(z)$ with respect to z, the second derivatives of A and B are neglected, so that

$$\left| \frac{d^2 A}{dz^2} \right| \ll k_B \left| \frac{dA}{dz} \right| \quad \text{and} \quad \left| \frac{d^2 B}{dz^2} \right| \ll k_B \left| \frac{dB}{dz} \right|$$

Therefore, from Equations 12.15 and 12.16, one obtains

$$\frac{d^2 E_+}{dz^2} \approx \left\{ -2jk_B \frac{dA}{dz} - k_B^2 A \right\} \exp(-jk_B z) \tag{12.18}$$

$$\frac{d^2 E_-}{dz^2} \approx \left\{ 2jk_B \frac{dB}{dz} - k_B^2 B \right\} \exp(jk_B z) \tag{12.19}$$

We may also write

$$k_0^2 \left(\bar{n}_{\text{eff}}^2 + \bar{n}_{\text{eff}} \Delta n_{\text{eff}} \sin k_g z \right) E$$

$$= k_0^2 \left\{ \bar{n}_{\text{eff}}^2 + \frac{\bar{n}_{\text{eff}} \Delta n_{\text{eff}}}{2j} \left[\exp(j2k_B z) - \exp(-j2k_B z) \right] \right\}$$

$$\times \left[A\exp(-jk_B z) + B\exp(jk_B z) \right] \tag{12.20}$$

The right-hand side of the above equation may be written, after neglecting the rapidly varying terms such as $\exp(\pm j3k_B z)$ obtained by multiplication, as follows

$$\left\{ k_0^2\bar{n}_{\text{eff}}^2 A + jk_0^2\left(\frac{\Delta n_{\text{eff}}\bar{n}_{\text{eff}}}{2}\right)B \right\}\exp(-jk_\text{B}z) + \left\{ k_0^2\bar{n}_{\text{eff}}^2 B - jk_0^2\left(\frac{\Delta n_{\text{eff}}\bar{n}_{\text{eff}}}{2}\right)A \right\}\exp(+jk_\text{B}z)$$

$$(12.21)$$

where the first (second) term corresponds to forward- (backward-)propagating waves. Substituting Equations 12.18, 12.19, and 12.21 into Equation 12.13, one obtains

$$\left\{ \left(k_0^2\bar{n}_{\text{eff}}^2 - k_\text{B}^2\right)A - j2k_\text{B}\frac{\text{d}A}{\text{d}z} + jk_0^2\frac{\Delta n_{\text{eff}}\bar{n}_{\text{eff}}}{2}B \right\}\exp\left(-jk_\text{B}z\right)$$

$$+ \left\{ \left(k_0^2\bar{n}_{\text{eff}}^2 - k_\text{B}^2\right)B + j2k_\text{B}\frac{\text{d}B}{\text{d}z} + jk_0^2\frac{\Delta n_{\text{eff}}\bar{n}_{\text{eff}}}{2}A \right\}\exp\left(jk_\text{B}z\right) = 0$$

$$(12.22)$$

This equation is satisfied only if the expressions for forward- and backward-propagating waves are individually zero. Let us put

$$k_0\bar{n}_{\text{eff}} = k_\text{B} + \Delta\beta \qquad (12.23)$$

and, using Equation 12.23, express $\Delta\beta$ in terms of the wavelength deviation from the Bragg wavelength $\delta\lambda = \lambda - \lambda_\text{B}$ as

$$\Delta\beta = 2\pi\left\{ \frac{\bar{n}_{\text{eff}}(\lambda_\text{B} + \delta\lambda)}{\lambda_\text{B} + \delta\lambda} - \frac{\text{Re}[\bar{n}_{\text{eff}}(\lambda_\text{B})]}{\lambda_\text{B}} \right\} \approx jk_0(\lambda_\text{B})\text{Im}(\bar{n}_{\text{eff}}) - k_0(\lambda_\text{B})\bar{n}_{\text{eff,g}}\frac{\delta\lambda}{\lambda_\text{B}} \quad (12.24)$$

Im denotes the imaginary part. The group effective RI at Bragg wavelength is

$$\bar{n}_{\text{eff,g}} = \text{Re}\left[\bar{n}_{\text{eff}}(\lambda_\text{B})\right] - \lambda_\text{B}\frac{\text{d}\,\text{Re}(\bar{n}_{\text{eff}})}{\text{d}\lambda}\bigg|_{\lambda=\lambda_\text{B}} \qquad (12.25)$$

Since, in the neighborhood of λ_B, $\Delta\beta \ll k_\text{B}$ and hence $k_0^2\bar{n}_{\text{eff}}^2 - k_\text{B}^2 \approx 2k_\text{B}\Delta\beta$, Equation 12.22 may be rewritten in terms of the following two equations

$$\Delta\beta A - j\frac{\text{d}A}{\text{d}z} + j\kappa B = 0 \qquad (12.26)$$

$$\Delta\beta B + j\frac{\text{d}B}{\text{d}z} - j\kappa A = 0 \qquad (12.27)$$

The coupling coefficient κ is defined as

$$\kappa = \frac{k_0 \Delta n_{\text{eff}}}{4} \tag{12.28}$$

where the approximation $k_0 \bar{n}_{\text{eff}} \approx k_B$ has been introduced in Equation 12.22. The electric fields are written in terms of vector

$$\mathbf{E} = \begin{vmatrix} A \\ B \end{vmatrix} \tag{12.29}$$

Equations 12.26 and 12.27 may be combined in the following form

$$\frac{d}{dt}\begin{vmatrix} A \\ B \end{vmatrix} = \begin{bmatrix} -j\Delta\beta & \kappa \\ \kappa & j\Delta\beta \end{bmatrix}\begin{vmatrix} A \\ B \end{vmatrix} \tag{12.30}$$

The eigenvalues of the above 2×2 matrix are

$$s_{1,2} = \pm s, \quad \text{where } s = \sqrt{\kappa^2 - \Delta\beta^2}, \tag{12.31}$$

and the corresponding eigenvectors read

$$E_1 = E_0 \begin{vmatrix} 1 \\ (j\Delta\beta + s)/\kappa \end{vmatrix} \quad \text{and} \quad E_2 = E_0 \begin{vmatrix} 1 \\ (j\Delta\beta - s)/\kappa \end{vmatrix} \tag{12.32}$$

where E_0 is a normalization constant having the dimension of the electric field (V/length). The general solution of Equation 12.29 is now expressed as

$$E = C_2 E_1 \exp(sz) + C_2 E_2 \exp(-sz) \tag{12.33}$$

where the constants C_1 and C_2 are chosen to meet the boundary conditions. The constants are, however, eliminated by expressing the fields at $z = 0$ by the fields at $z = L_B$. We thus obtain

$$E(0) = [T] \cdot E(L_B) \tag{12.34}$$

The transfer matrix $\mathbf{T}$ of the DBR is obtained from Equations 12.29 and 12.31 as

$$\mathbf{T} = \begin{bmatrix} \cosh(sL_B) + (j\Delta\beta/\kappa)\sinh(sL_B) & -(\kappa/s)\sinh(sL_B) \\ -(\kappa/s)\sinh(sL_B) & \cosh(sL_B) - (j\Delta\beta/\kappa)\sinh(sL_B) \end{bmatrix} \tag{12.35}$$

Assuming now that $R(L_B)=0$, and correspondingly $B(L_B)=0$, that is, there is no reflection at the right end of the DBR, and further assuming that there is no optical loss or gain, so that both $n_{eff}(z)$ and $\bar{n}_{eff}(z)$ are real, we may rewrite Equation 12.31 as

$$\begin{vmatrix} A(0) \\ B(0) \end{vmatrix} = \mathbf{T} \cdot \begin{vmatrix} A(L_B) \\ 0 \end{vmatrix} \tag{12.36}$$

The amplitude reflectivity at $z=0$, $r=B(0)/A(0)=T_{21}/T_{11}$, and it is easily obtained from Equation 12.32 as

$$r = -\frac{\kappa \sinh(sL_B)}{j\Delta\beta \sinh(sL_B) + s \cosh(sL_B)} \tag{12.37}$$

At the Bragg wavelength, $\Delta\beta=0$ and then $s=\kappa$, the expression for the power reflection coefficient becomes

$$R(\lambda_B) = |r|^2 = \tanh^2(\kappa L_B) \tag{12.38}$$

The intensity distributions of the forward- and backward-propagating waves within the DBR are shown in Figure 12.2b, in which the amplitude of a wave transmitted through the DBR is denoted by E_T.

12.5.2 Wavelength Selectivity

Bragg gratings show strong wavelength selectivity at the Bragg wavelength. The following example illustrates this behavior.

> **Example 12.4:** The power reflectivity of a DBR is calculated using Equation 12.38 by taking $L=400$ µm, $\bar{n}_{eff,g} = 3.4$, $\lambda=1.55$ µm, and $\kappa L_B=2$. The reflectivity at the Bragg wavelength is 0.93. Let the frequency deviation be $\delta\lambda=0.5$ nm. The calculated values are: $\kappa=5000$ m^{-1}, $\Delta\beta=4.445\times10^3$ m^{-1}, and $s=2.289\times10^3$ m^{-1}, and $R=0.84$. For $\delta\lambda=0.55$ nm, $\Delta\beta=4.8895\times10^3$ m^{-1}, $s=1.044\times10^3$ m^{-1}, and $R=0.81$.

The calculated curves for power reflection coefficient of a DBR laser are shown in Figure 12.3 for three different values of κL_B: 2, 1, and 0.5. The parameters are $L=400$ µm, $\bar{n}_{eff,g} = 3.5$, and $\lambda=1.55$ µm. The reflectivity is plotted as a function of deviation from Bragg wavelength for different values of κL_B. The curves show that the power reflectivity is maximum at the Bragg wavelength, and that a sharp fall of the reflectivity occurs as the wavelength deviates from λ_B, indicating that the bandwidth is ~1 nm.

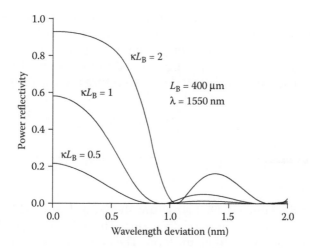

FIGURE 12.3
Power reflectivity of a DBR vs. deviation of wavelength from the Bragg wavelength.

12.5.3 Structure of DBR Lasers

As noted already in Figure 12.1, in a DBR laser both the mirrors in an FP laser are replaced by Bragg reflectors. In some cases, only one mirror is replaced by a Bragg grating. The DBRs provide losses to all other modes, and only the mode satisfying the Bragg condition is sustained.

The schematic cross-sectional diagram of an InGaAsP/InP DBR is shown in Figure 12.4. In this structure, the mirror on the right hand is replaced by a DBR. The Bragg mirror is formed between p-InP and the waveguiding InGaAsP layer. The waveguiding InGaAsP has a bandgap wavelength

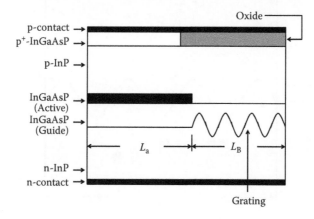

FIGURE 12.4
Schematic diagram of an InGaAsP/InP DBR laser.

$\lambda_{WG} < \lambda_0$, the laser wavelength, and an RI larger than that of InP. This layer acts as a passive waveguide layer surrounding InP. The active InGaAsP layer with emission wavelength of 1.55 μm is placed at one end of the cavity. An oxide isolation technique is employed to confine the drive current to the active region only. The lateral device structure is buried heterostructure type [2].

12.6 DFB Laser

We consider the simplest DFB structure shown in Figure 12.5, in which a Bragg grating is present in the gain region. The gain along the length L of the medium is assumed to be uniform. The end mirrors are nonreflecting ($R_1 = R_2 = 0$).

12.6.1 Solutions to Coupled-Mode Equations

The general solution to the coupled-mode equations may be written in the following form

$$A(z) = A_+(z)\exp(jqz) + A_-(z)\exp(-jqz) \tag{12.39a}$$

$$B(z) = B_+(z)\exp(jqz) + B_-(z)\exp(-jqz) \tag{12.39b}$$

where

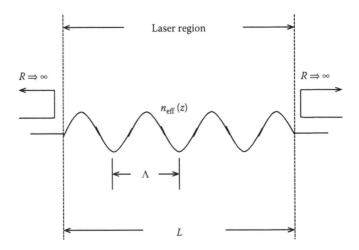

FIGURE 12.5
Schematic of an idealized DFB laser; the end facets are nonreflecting.

$$\frac{B_+}{A_+} = \frac{\Delta\beta - q}{-\kappa} = \frac{\kappa}{\Delta\beta + q} \quad \text{for the exp}(jqz) \text{ eigenmode} \qquad (12.40a)$$

and

$$\frac{B_-}{A_-} = \frac{\Delta\beta + q}{-\kappa} = \frac{\kappa}{\Delta\beta - q} \quad \text{for the exp}(jqz) \text{ eigenmode} \qquad (12.40b)$$

The DFB reflection coefficient for the + eigenmode is defined as

$$r_p(q) = \frac{B_+}{A_+} = \frac{\kappa}{\Delta\beta + q} \qquad (12.41a)$$

and for the − eigenmode the reflection coefficient is

$$r_m(q) = \frac{A_-}{B_-} = \frac{-\kappa}{\Delta\beta + q} \qquad (12.41b)$$

Using these definitions, Equations 12.36a and 12.36b may be rewritten as

$$A(z) = A_+(z)\exp(jqz) + r_m(q)B_-(z)\exp(-jqz) \qquad (12.42a)$$

$$B(z) = r_p(q)A_+(z)\exp(jqz) + B_-(z)\exp(-jqz) \qquad (12.42b)$$

The boundary conditions at the two ends of the DFB structure are

1. At $z = 0$, $A(0) = r_1 B(0)$. This gives

$$\left(1 - r_1 r_p\right)A_+ + \left(r_m - r_1\right)B_- = 0 \qquad (12.43a)$$

2. At $z = L$, $B(L) = r_2 A(L)$, and thus

$$\left(r_2 - r_p\right)e^{j2qL}A_+ - \left(1 - r_2 r_m\right)B_- = 0 \qquad (12.43b)$$

It is to be noted that both r_1 and r_2 are complex in general. In order to obtain nontrivial solutions for A_+ and B_-, the determinants of the coefficients must be zero. Therefore, one obtains

$$\left(1 - r_1 r_p\right)\left(1 - r_2 r_m\right) + \left(r_2 - r_p\right)\left(r_m - r_1\right)e^{j2qL} = 0 \qquad (12.44)$$

Rearranging terms, Equation 12.44 may be rewritten as

$$\left[\frac{r_1 - r_m(q)}{1 - r_1 r_p(q)}\right]\left[\frac{r_2 - r_p(q)}{1 - r_2 r_m(q)}\right]e^{j2qL} = 1 \qquad (12.45)$$

Since the medium possesses gain, the gain coefficient should be included in the propagation constant, and therefore

$$\Delta\beta = \beta - \beta_0 \rightarrow \left(k_0 n - j\frac{G}{2}\right) - \beta_0 = \delta - j\frac{G}{2} \qquad (12.46)$$

where:
  δ is the detuning parameter
  $G = \Gamma g$ is the modal gain

Since $q = js$, we write using Equation 12.31

$$q = \sqrt{(\delta - jG/2)^2 - |\kappa|^2} \qquad (12.47)$$

12.6.2 Laser Oscillation Condition

If the DFB structure is absent, $\kappa = 0$, $r_p = r_m = 0$, and by putting these in Equation 12.45 one recovers the eigenequation for the FP laser

$$r_1 r_2 \exp(j2qL) = 1 \qquad (12.48)$$

Again, if there are no sharp boundaries at $z = 0$, and $z + L$, $r_1 = r_2 = 0$ and one obtains from Equation 12.45

$$r_p(q) r_m(q) \exp(j2qL) = 1 \qquad (12.49)$$

This implies that $r_p(q)$ and $r_m(q)$ act, respectively, as the reflection coefficients for the forward- and backward-propagating waves. Using Equations 12.41a and 12.41b, we may rewrite Equation 12.49 as

$$\frac{|\kappa|^2}{(q + \Delta\beta)^2} \exp(j2qL) = 1 \qquad (12.50)$$

Noting that $q^2 = (\Delta\beta)^2 - |\kappa|^2$, we obtain $q = j\Delta\beta \tan qL$, and multiplying this by L, we obtain for the eigenvalue equation

$$\Delta\beta L = -jqL \cot(qL)$$

and using Equation 12.46 we obtain the following

$$\delta L - j\frac{GL}{2} = -j\sqrt{\left(\delta L - j\frac{GL}{2}\right)^2 - (\kappa L)^2} \ \cot\sqrt{\left(\delta L - j\frac{GL}{2}\right)^2 - (\kappa L)^2} \quad (12.51)$$

Equation 12.51 is to be solved to obtain δL, and GL for a given value of κL. In general, the solution must be obtained numerically, and the task is complicated due to complex quantities. An approximate solution may be obtained analytically for the high gain condition $GL \gg \kappa L$, which leads to $q \approx \Delta\beta$. From Equation 12.50, one may write

$$\kappa^2 e^{j2\Delta\beta L} = (2\Delta\beta)^2 \quad (12.52)$$

Comparing the magnitude and phase of the above equation, we obtain

$$\kappa^2 e^{GL} = G^2 + 4\delta^2 \quad (12.53a)$$

and

$$\delta L = \left(m - \frac{1}{2}\right)\pi + \tan^{-1}\left(\frac{2\delta}{G}\right), \quad m = \pm1, \pm2,\ldots \quad (12.53b)$$

For small values of δ, we approach the Bragg condition

$$\delta = \frac{1}{L}\left(m - \frac{1}{2}\right)\pi \quad (12.54)$$

The frequency of the modes is given by

$$f = f_B \pm \frac{c}{2n_{r0}L}\left(m - \frac{1}{2}\right) \quad (12.55)$$

Note that there is no allowed wave at the Bragg frequency. The modes are situated symmetrically around the Bragg frequency, and the mode spacing is equal to the spacing in an FP laser, according to the approximation introduced.

Example 12.5: We calculate the threshold gains for three different modes: $m = 0$, 1, and 2. The length $L = 1000$ μm, $\lambda_0 = 1.55$ μm, and two values of $\Delta n_{r1} = 0.001$ and 0.002 are considered. The value of gain is obtained by solving Equation 12.53. The following values are obtained.

	Value of g ($\Delta n_{r1}=0.001$) (cm^{-1})	Value of g ($\Delta n_{r1}=0.002$) (cm^{-1})
$m=0$	28.7	9.67
$m=1$	46.8	31.81
$m=2$	56	41.58

12.6.3 Results for Threshold Gain

Figure 12.6 shows the calculated values of the threshold gain versus normalized frequency deviation $\delta\lambda/\Delta\lambda_m$ for a few lowest-order longitudinal modes for two different values of κL product.

The following conclusions may be derived by examining the figures:

1. There is no lasing mode at the Bragg wavelength ($\delta\lambda=0$).
2. The smallest modal threshold gain is obtained for the two modes nearest to the Bragg wavelength.
3. The two modes having smallest threshold gain are separated by a stop band $\Delta\lambda_s$, which increases with increasing κL product.
4. Far away from the Bragg wavelength, the mode separation approaches $\Delta\lambda_m$: the mode spacing in an FP laser.

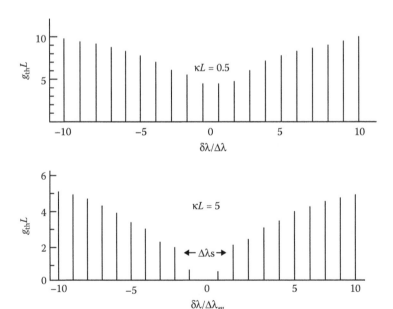

FIGURE 12.6
Threshold gain–length product vs. normalized wavelength deviation in a DFB laser for different values of κL.

5. The lowest threshold gain is much lower than the threshold gain for the pair having the next higher longitudinal mode number. The difference increases with increasing κL product.

Example 12.6: A calculation similar to those shown in Figure 12.6, for $\kappa L = 2$, indicates that the lowest value of $g_{th}L$ is ~2, and the value for the next higher mode is ~3.5. Taking $L = 300$ μm, the values of threshold gain are 67 and 117 cm^{-1}. The gain difference is 50 cm^{-1}: more than three orders of magnitude higher than in the FP laser considered in Example 12.3 ($\delta g = 0.03$ cm^{-1}).

As noted from Figure 12.6, the pure DFB laser oscillates with two wavelengths symmetrically located around the Bragg wavelength. It therefore cannot be used as a single-moded laser. Some modification of the DFB grating is to be introduced in order to achieve emission at a single mode. The structure is discussed in the following subsection.

12.6.4 Quarter-Wave-Shifted DBR Laser

The grating structure used to break the threshold degeneracy is shown in Figure 12.7. Near the cavity center, a quarter-wavelength shift, or in other words a shift of $\Lambda/2$, is introduced in the grating. The end facets are non-reflecting. The gain spectrum obtained by solving the coupled-mode equation is shown in Figure 12.7b [3]. The lowest modal threshold gain occurs exactly at the Bragg wavelength, but at both sides the modal threshold gain is degenerate.

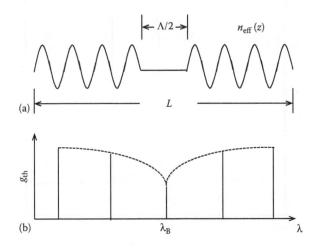

FIGURE 12.7
(a) DFB structure with λ/4 phase-shifted Bragg grating and nonreflecting end facets. (b) The gain spectrum at threshold.

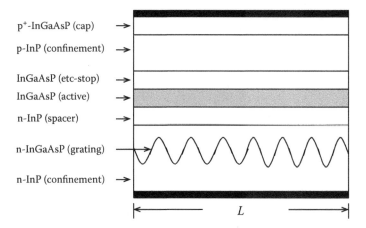

p⁺-InGaAsP (cap)
p-InP (confinement)
InGaAsP (etc-stop)
InGaAsP (active)
n-InP (spacer)
n-InGaAsP (grating)
n-InP (confinement)

L

FIGURE 12.8
Schematic longitudinal cross section of an InGaAsP/InP DFB laser.

12.6.5 Representative DFB Laser Structure

A typical DFB laser structure is shown in Figure 12.8. In addition to the usual layers, an n-type InP spacer layer has been inserted between the active and the grating layers. It ensures a small optical confinement in the active layer and a small coupling coefficient. A typical longitudinal mode spectrum is shown in Figure 12.9, from which single-moded behavior is evident. Though the

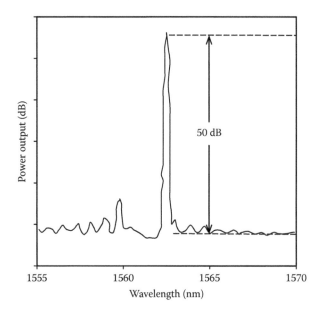

FIGURE 12.9
Typical longitudinal mode spectrum of a DFB laser.

difference of power between the dominant mode and the feeble side modes depends on the drive current, a 30 dB difference is almost always present.

12.7 Tunable Lasers

Many applications demand single-mode laser diodes, the wavelengths of which can be tuned electronically. There are many applications. One important area is coherent optical communication, in which the wavelength of the local oscillator should be continuously tunable. Tunable lasers are also needed in DWDM communication systems. Other application areas include wavelength conversion, optical sensing techniques such as range sensing by frequency-modulated continuous-wave radar methods, fiber measurements, spectroscopy, reflectometry, and short-pulse generation.

For all these applications, a high spectral purity, a linewidth as narrow as 25 MHz, and an SSR in excess of 30 dB are required.

12.7.1 Physical Mechanism for Tuning

A change in the wavelength of emission is introduced by changing the RI of the gain medium. Three different methods are employed to induce the RI change: injection of carriers, applying an electric field, and heating the material. In this section, the methods will be discussed, pointing out their advantages and disadvantages.

12.7.1.1 Carrier-Induced Change

The change in RI, which in turn introduces the required change in the wavelength, may be achieved by injection of carriers (the plasma effect) and by exploiting the quantum-confined Stark effect (QCSE).

So far, the largest RI change and tuning range have been achieved by using the plasma effect. The free-carrier-induced change in RI is expressed as (see Equation 6.86)

$$\Delta n = -\frac{e^2 \lambda_0^2}{8\pi^2 c^2 n_r \varepsilon_0} \left(\frac{\Delta N}{m_e} + \frac{\Delta P}{m_h} \right) \tag{12.56}$$

where ΔN (ΔP) and m_e (m_h) denote the density and effective mass of injected electrons (holes).

> **Example 12.7:** The change in RI for $In_{0.58}Ga_{0.42}As_{0.9}P_{0.1}$ lattice matched to InP with a bandgap of 0.8 eV ($\lambda_0 = 1.55$ μm) is calculated by using $n_r = 3.5$,

$m_e = 0.045 m_0$, $m_{hh} = 0.438 m_0$, and $\Delta N = \Delta P = 3 \times 10^{18}$ cm^{-3}. The calculated value is $\Delta n = -0.022$.

In addition to the plasma effect, the bandgap shrinkage and band-filling effect caused by injected carriers introduce changes in RI. The injected carriers also contribute to free-carrier absorption and intervalence band absorption (IVBA). The IVBA for InP and InGaAsP at 1.55 μm has been found to be about 20–40 cm^{-1}. The linewidth enhancement factor (LEF) is one of the advantages of this method. A typical value for InGaAsP with a bandgap wavelength of 1.3 μm in the 193 THz range is −20.

The injected carrier density, N, is determined from the rate equation given by

$$\frac{dN}{dt} = \frac{I}{eV} - (AN + BN^2 + CN^3) \tag{12.57}$$

where:
  I is the current
  V is the volume

the terms in brackets represent the recombination rate

A sustained carrier is needed to maintain the carrier density. The injection–recombination process has a time constant in the nanosecond range, which limits the tuning speed. The nonzero series resistance of the tuning diode and Joule heating of the waveguide material limit the power output from the laser and reduce the differential quantum efficiency.

12.7.1.2 Electric-Field-Induced Change

An electric field applied to a bulk III–V semiconductor can change its RI by (i) the linear electro-optic (EO) or Pockel effect and (ii) the quadratic EO or Franz–Keldysh effect. Typical changes of RI for InGaAsP at photon energies below the bandgap are

$$\Delta n_{Pockels} \approx -3.10^{-11} \, (\text{m V}^{-1}) \, F \, (\text{V m}^{-1}) \tag{12.58a}$$

$$\Delta n_{FK} \approx +1.10^{-18} \, (\text{m V}^{-1})^2 \, F^2 \, (\text{V m}^{-1})^2 \tag{12.58b}$$

Hence, these two effects counter each other and even for a large electric field $F = 10^7$ V m^{-1}, the RI change is only ~10^{-4}.

The electric-field-induced change is stronger in a quantum well (QW) due to QCSE. The effective bandgap is lowered with the rise in field and the change in absorption spectra. This in turn introduces a change in RI via the Kramers–Kronig relationship. The closer the wavelength is to the bandgap, the larger the change in RI. Hence, a careful match between the laser frequency and the QW bandgap is needed. The change in RI may be in the order of 10^{-3} to 10^{-2}. The LEF is only 10, but by using an asymmetric QW structure its value can be increased, giving a lower loss.

The effective RI change is, however, much reduced because of the low value of the optical confinement factor. The value of Γ may be increased by using a multiple quantum well (MQW) structure, in which its value is multiplied by the number of QWs. Even then, it is hard to achieve the value for bulk waveguides exploiting the carrier-induced change of RI. The typical value of the index change is 10^{-3}.

The QCSE-induced RI change is achieved in a p-i-n structure in which the MQWs comprise the intrinsic layer. A reverse bias is applied to the p-i-n structure, so that there is negligible current flow and hence negligible heating. There is no buildup of carriers. The time constant is determined by parasitic inductance and capacitance. The speed of the device is accordingly quite high.

12.7.1.3 Thermally Induced Change

A change in temperature induces a change in the RI of III–V semiconductors and their alloys. The emission wavelength of an InGaAsP/InP laser working at 1.55 μm increases with temperature at an approximate rate of 0.1 nm K^{-1}. In other words, the temperature coefficient of RI $\partial n/\partial T \approx 2.10^{-4} K^{-1}$. With thermal tuning, the confinement factor is unity, as both the active and cladding layers are heated. However, heating the entire laser increases the threshold current and reduces the differential quantum efficiency. If the active region is thermally insulated from the tuning section, and heating is applied only to the tuning section, then a larger temperature rise is permissible. The heating is applied by using a reverse-biased diode, by integrating a resistor in the top of the InP cladding layer, or by depositing a thin-film resistive heater on top of the waveguide. The slow response speed, ranging between microseconds and milliseconds, is the major drawback of thermal tuning.

Table 12.1 gives a summary of the characteristics of the three methods for changing RI.

TABLE 12.1

Comparison between Three Methods for Changing RI

Parameter	Carriers	Electric Field	Temperature
Change in RI	−0.05	−0.01	0.01
Confinement factor	0.5	0.2	1
Effective change in RI	−0.025	−0.002	0.01
LEF	−20	−10	Large
3 dB bandwidth	100 MHz	>10 GHz	<1 MHz
Power consumption	Large	Negligible	Very large

12.7.2 Principle of Wavelength Tuning

The lasing wavelength λ is simply related to the mode number, m, the effective index, n_{eff}, and the effective cavity length, L, by the relation

$$\frac{m\lambda}{2} = n_{\text{eff}}L \qquad (12.59)$$

The wavelength may be changed by changing m, n, or L. The relative change in wavelength is given by

$$\frac{\Delta\lambda}{\lambda} = \frac{\Delta n_{\text{eff}}}{n_{\text{eff}}} + \frac{\Delta L}{L} - \frac{\Delta m}{m} \qquad (12.60)$$

The mode number is changed by mode-selection filter (via index change or grating angle). The physical length is changed electronically or by using an external mirror, and the effective index is changed by carriers, electric field, or temperature. In the following sections, some of the methods employed will be described.

12.8 Characteristics of Tunable Lasers

12.8.1 Tuning Range

Normally, three different types of tuning are achieved: continuous tuning, discontinuous tuning, and quasi-continuous tuning. Figure 12.10 shows how the wavelength changes with change in control current or voltage for the three schemes.

Continuous wavelength tuning is preferred in almost all applications. In this scheme, the laser emits in the same longitudinal mode throughout the whole tuning range, and no mode changes or jumps occur. This means that the tuning range cannot exceed the longitudinal mode separation. Assuming that a DFB laser operates exactly at the Bragg wavelength ($\lambda_0 = \lambda_B$), the tuning range for continuous tuning may be given by

$$\frac{\Delta\lambda}{\lambda_0} = \frac{\text{Re}(\delta n_{\text{eff}})}{n_{\text{eff,g}}} \qquad (12.61)$$

where:
  Re stands for real part
  $n_{\text{eff,g}}$ is the effective group index

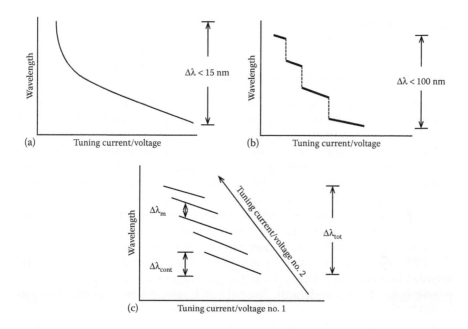

FIGURE 12.10
Variation of wavelength with control current/voltage: (a) continuous, (b) discontinuous, and (c) quasi-continuous.

Using typical values for RI change by carrier effect ~10^{-3}, the tuning range can, at most, be about 15 nm at 1.55 μm.

If a larger tuning range is needed, for example in DWDM systems, a change in mode number is unavoidable. The tuning in this scheme is discontinuous, as shown in Figure 12.10b, and there are a number of mode jumps. The maximum tuning range is limited by the gain bandwidth of the active region, which is of the order of 100 nm at 1.55 μm wavelength.

The third technique, quasi-continuous tuning, has the characteristics shown in Figure 12.10c. Here, two control currents (or voltages) are employed: one allows continuous change over a particular mode, and the other provides discontinuous tuning over the large wavelength range λ_{tot} via changes in the longitudinal modes. When the continuous tuning range exceeds the mode spacing, overlapping continuous tuning ranges are obtained, as shown in Figure 12.10c. In this way, a complete coverage of the tuning range λ_{tot} is obtained.

12.9 Methods and Structures for Continuous and Discontinuous Tuning

Different methods have been employed for tuning the wavelength of DFB or DBR lasers. In the following, we shall present a few structures, from earlier simple structures to the more sophisticated ones, mentioning the special features and the range of tunability of the structures.

12.9.1 Two- and Three-Section DFB Structures

The most effective way to bring about a change in the RI is current injection. Typically, this is achieved by employing a multiple-electrode DFB or DBR structure, as shown in Figure 12.11 [4,5].

A three-section DFB laser is shown in Figure 12.11a, in which a large current is applied to one electrode and a small current to another electrode. The smaller current, which is below the threshold, overcomes the absorption loss and simply acts as the Bragg mirror, changing the RI and wavelength. The gain is provided by the other section, which is pumped above the threshold. The maximum continuous tuning range is, however, limited to 3–4 nm.

A three-section DBR laser is shown in Figure 12.11b. The rightmost region, the DBR region, is decoupled from the left region, the gain region. The forward-bias current I_g injected into the gain region only changes the output power, but not the wavelength. The current I_b injected into the Bragg region controls the wavelength independently of the output power.

As in a conventional DBR laser, the laser has multiple closely spaced cavity modes. The lasing mode corresponds to the wavelength peak of the Bragg grating. With the variation of I_b, the laser hops from one cavity mode to another.

In order to obtain continuous tuning over the entire wavelength range, the third section shown in the middle of Figure 12.11b is introduced. The current I_p introduced in this section, called the phase section, allows control of the cavity-mode spacing. A change of effective cavity length by half a wavelength is sufficient to obtain tuning across the entire free spectral range (FSR). This is achievable by current injection into the phase section.

Two- and three-section DBRs capable of tuning over 32 channels in 50 GHz increments are commercially available.

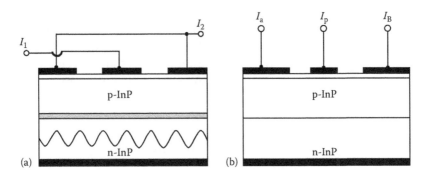

FIGURE 12.11
Tunable laser diode structures.

12.9.2 Methods for Obtaining Larger Tuning Range

The change in RI by current injection is only 0.5%–2%, and this change leads to a tuning range of about 10–15 nm. To achieve a larger tuning range, several new tricks need to be introduced. In one such method, the laser wavelength is made to depend on the difference between the RIs of two different regions. The vertical grating-assisted coupler filter (VGF) makes use of this, and the overall variation is much larger than the variation in individual regions [6,7]. The second method makes use of the Vernier effect, in which there are two combs of wavelength, each with slightly different comb spacing. The combination of the two combs gives rise to another periodic comb with a much higher wavelength spacing between its peaks. Even if each comb has a small tuning range, the combination yields a much higher tuning range. The sampled grating (SG) [8,9] and superstructure grating (SSG) [10] DBRs use the Vernier effect.

In the following subsections, these structures and their working principles will be introduced.

12.9.3 VGF Lasers

The VGF laser structure, the schematic diagram of which is shown in Figure 12.12, consists of two waveguides with a coupling region in between. The wavelength λ is coupled from waveguide 1 of RI n_1 to waveguide 2 of RI n_2, provided

$$\lambda = \Lambda_B(n_1 - n_2) \tag{12.62}$$

where Λ_B is the period of the Bragg grating. If the RI of waveguide 1 is changed by Δn_1, there is a change in wavelength given by

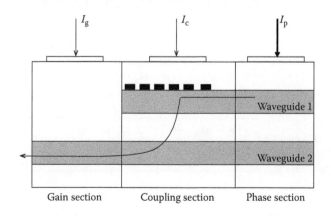

FIGURE 12.12
Tunable VGF laser structure.

$$\frac{\Delta\lambda}{\lambda} \approx \frac{\Delta n_1}{(n_1 - n_2)} \tag{12.63}$$

The tuning range is significantly larger than the range $\Delta\lambda/\lambda \approx \Delta n/n$ achievable in the two- and three-section DBRs considered earlier.

As shown in Figure 12.12, the index of the waveguide is controlled by the current I_v, and the current I_g provides the current to the gain region of the other waveguide. The third current I_p controls the cavity-mode spacing. Lasers with a tuning range of more than 70 nm have been reported using this approach.

In order to have good coupling between the waveguides, the cavity length needs to be quite long, typically 800–1000 µm. The cavity modes are therefore closely spaced. The laser may hop from one cavity mode to another very easily even though all control currents are steady. This results in poor SSR, and these lasers are not suitable for high-bit-rate long-haul communication.

12.9.4 Vernier Effect

The Vernier effect is illustrated in Figure 12.13. There are two gratings with different reflectivities R_a and R_b at the two ends of the active region, as shown in Figure 12.13a. Since the two gratings are different, the mode spacings for the two will be different, that is, $\Delta\lambda_a \neq \Delta\lambda b$, as shown in Figure 12.13b and c. In this structure, lasing is possible only at those wavelengths at which both

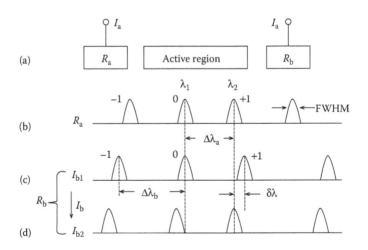

FIGURE 12.13
(a) Schematic structure of a widely tunable laser with two different Bragg gratings at the two ends; (b) reflection spectra of left grating; (c) and (d) reflection spectra of the right grating for two different values of the bias current.

the spectra show a reflection peak simultaneously, since a close feedback loop exists only at these resonance wavelengths.

The presence of two reflectors having different mode spacings ensure that small shifts of one of the comb-reflection spectra may introduce a large change in the resonance wavelength. This is much like the Vernier effect in slide calipers. However, lasing should be avoided at multiple wavelengths. Therefore, the spacing $\Delta\lambda_r$ between the multiple resonance wavelengths must be kept larger than the gain bandwidth $\Delta\lambda_{gain}$ of the active region. Thus (see Problem 12.7),

$$\Delta\lambda_r = \frac{\Delta\lambda_a \cdot \Delta\lambda_b}{|\Delta\lambda_a - \Delta\lambda_b|} \gg \Delta\lambda_{gain} \qquad (12.64)$$

Another condition to be fulfilled is to achieve sufficiently small overlap between the reflection peaks of R_a and R_b. Therefore, the differences of the periods $\Delta\lambda_a$ must be larger than the full width at half maximum (FWHM) of the reflection peaks, that is,

$$|\Delta\lambda_a - \Delta\lambda_b| \gg \text{FWHM} \qquad (12.65)$$

Consider Figure 12.13b and c. The two gratings are resonant at a wavelength λ_1, and each of the reflection peaks is labeled by 0. The adjacent reflection peaks (denoted by ±1) are mismatched by $|\Delta\lambda_a - \Delta\lambda b|$, and their overlap should be very small to prevent laser operation at these wavelengths. The FWHM can be estimated from Equation 12.65, and the reflection spectra are plotted in Figure 12.13. The zeros of R correspond to $\Delta\beta = \pm\kappa$. Therefore,

$$\text{FWHM} < \frac{\kappa\lambda_B^2}{\pi\bar{n}_{eff,g}} \qquad (12.66)$$

The coupling coefficient κ is assumed to be equal for both the gratings. Equation 12.65 may then be rewritten as

$$|\Delta\lambda_a - \Delta\lambda_b| \gg \frac{\kappa\lambda_B^2}{\pi\bar{n}_{eff,g}} \qquad (12.67)$$

If the reflection spectrum of R_a and R_b shifts by $\delta\lambda = |\Delta\lambda_a - \Delta\lambda_b|$, the resonance wavelengths shift by $\Delta\lambda_b$ or $\Delta\lambda_a$. If $\Delta\lambda_a \approx \Delta\lambda_b \gg |\Delta\lambda_a - \Delta\lambda_b|$, a relatively small change in the reflection spectra introduces a large change in the lasing wavelength. The ratio of the shift of the laser wavelength to the shift of the reflection spectrum is called the tuning enhancement factor, and for $\Delta\lambda_a \approx \Delta\lambda_b$ it is expressed as

$$F = \frac{\Delta\lambda_a}{|\Delta\lambda_a - \Delta\lambda_b|} \qquad (12.68)$$

A typical value of F is about 15.

The reflection spectra can be shifted along the wavelength axis by carrier injection into the Bragg reflector sections via one or both of the two bias currents I_a and I_b. The spacing of the comb modes remains almost unaffected, however. Suppose the bias currents are I_a and I_{b1}, as shown in Figure 12.13b and c. The device then lases at wavelength λ_1 (denoted by 0) in the figure, as the two spectra coincide only at this wavelength. If the current I_b is slightly increased to I_{b2}, as shown in Figure 12.13d, the effective index in R_b decreases, shifting the reflection peaks by $\delta\lambda = |\Delta\lambda_a - \Delta\lambda_b|$ toward lower wavelengths. As a result, the lasing wavelength shifts to a higher wavelength λ_2 corresponding to mode +1. The change in laser wavelength $\Delta\lambda = \Delta\lambda_a$ is much larger than the wavelength shift $\delta\lambda$ induced in R_b. It may be noticed that the wavelength tuning is discontinuous, because the longitudinal mode spacing of the entire laser cavity is usually much smaller than the bandwidth of the reflector peaks, and consequently, several longitudinal modes may have equal and similar reflections at the same time.

12.9.5 SG-DBR and SSG-DBR

The spatial period of a Bragg grating may be modulated in two different ways. The devices are termed SG when amplitude modulation is introduced. In the case of frequency modulation, the device is called the SSG-DBR. In the following we describe briefly the operation of a spatially amplitude-modulated Bragg grating. Detailed discussions may be found in papers by Jayaraman et al. [8,9].

Figure 12.14 illustrates the operating principle of the SG-DBR. The Bragg grating period is Λ, as indicated in Figure 12.14a. It is modulated by the sampling function shown in Figure 12.14b with sampling period $\Lambda_s = \Lambda_1 + \Lambda_2$. The modulated SG is shown in Figure 12.14c. By a spatial Fourier transform, the SG may be shown to be composed of homogeneous gratings with different amplitudes and Bragg wavelengths. If the constituent Bragg wavelengths are far apart from each other, each Fourier component will lead to independent Bragg reflections.

Consider Figure 12.14b. The variation of the effective RI along z may be written as

$$n_{\text{eff}}(z) = \bar{n}_{\text{eff}} + \left(\frac{\Delta n_{\text{eff}}}{2}\right)\sin(k_g z) \times \begin{bmatrix} 0 & \text{for } n\Lambda_s + \Lambda_1 \leq z \leq (n+1)\Lambda_s \\ 1 & \text{for } n\Lambda_s \leq z \leq n\Lambda_s + \Lambda_1 \end{bmatrix} \qquad (12.69)$$

where n is an integer. Writing $n_{\text{eff}}(z)$ as a Fourier sum,

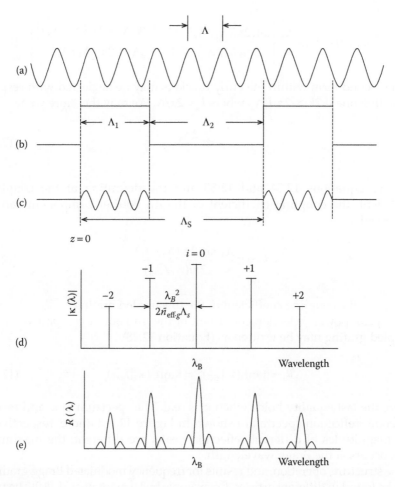

FIGURE 12.14
Sampled BG. (a) BG; (b) sampling function; (c) sampled grating; (d) mode pattern; (e) reflectivity pattern.

$$n_{eff}(z) = \bar{n}_{eff} + \sum_{i=0,\pm1,\pm2} \left[\frac{\Delta n_{eff}(i)}{2} \right] \sin \left[k_g + \left(\frac{2\pi i}{\Lambda_s} \right) \right] z \qquad (12.70)$$

The Fourier components of the RI change are obtained from

$$\Delta n_{eff}(i) = \Delta n_{eff} \frac{2}{\Lambda_s} \int_0^{\Lambda_1} \sin k_g z \, \sin \left(k_g + \frac{2\pi i}{\Lambda_s} \right) z \qquad (12.71)$$

Performing the integration, one obtains

$$\Delta n_{eff}(i) = \Delta n_{eff} \frac{\Lambda_1}{\Lambda_s} \left\{ \frac{\sin 2\pi i(\Lambda_1/\Lambda_s)}{2\pi i(\Lambda_1/\Lambda_s)} - \frac{\sin\left[2k_g\Lambda_1 + 2\pi i\left(\Lambda_1/\Lambda_s\right)\right]}{2k_g\Lambda_1 + 2\pi i\left(\Lambda_1/\Lambda_s\right)} \right\} \quad (12.72)$$

The second term within the curly brackets may be neglected with respect to the first one if $2k_g \gg 2\pi|i|\Lambda_s$, where $k_g = 2\pi/\Lambda$. We may therefore write

$$|i| \ll \frac{2\Lambda_s}{\Lambda} \quad (12.73)$$

Using Equations 12.54 and 12.55 and the definition of the coupling coefficient, the coupling coefficient of the *i*th Fourier component can be expressed as

$$k_i = \kappa \frac{\Lambda_1}{\Lambda_s} \frac{\sin 2\pi i(\Lambda_1/\Lambda_s)}{2\pi i(\Lambda_1/\Lambda_s)} \quad (12.74)$$

where κ is the coupling coefficient of the unsampled grating.

The peak power reflection coefficient of the *i*th Fourier component of the sampled grating may be written as (Equation 12.38)

$$R_i = |\tanh(\kappa_i L_{SG})|^2 = \tanh^2(|\kappa_i| L_{SG}) \quad (12.75)$$

where the last equality holds when κ_i is real. The spectrum of κ_i and resulting comb-reflection spectra are shown in Figure 12.14d and e, respectively. The magnitudes of different reflection peaks are different; the maximum peak occurs at the Bragg wavelength.

The structure, operation, and results for frequency modulated Bragg gratings may be found in different articles, for example by Jayaraman et al. [8,9].The tuning range, SSRs, and other design issues and results may also be found there.

12.10 Tunable Vertical-Cavity Surface-Emitting Laser

The main challenges to realizing vertical-cavity surface-emitting lasers (VCSELs) at 1.55 µm are obtaining sufficient cavity gain, obtaining high-reflectivity mirrors, draining away the heat generated, and operating in a single longitudinal mode. A VCSEL structure that attempts to solve these problems and at the same time provides tunability has been developed [11]. Tuning is provided by using a movable microelectromechanical (MEM) membrane as the upper mirror. The cavity spacing can be tuned by applying a voltage between the upper and lower mirrors. Thus, the wavelength is changed by a slight change ΔL of the cavity length, as indicated by Equation 12.60.

To conduct the heat away from the bottom mirror, a hole is etched in the InP substrate. The 980 nm pump creates electron–hole pairs in the active region. The pump spot size is matched with the size of the fundamental lasing mode, thereby making the laser single moded and suppressing the higher-order FP modes. The laser, described in [11], gives continuous-wave power at 0 dBm over a tuning range of 50 nm centered on 1.55 μm. For details, the reader is referred to the review article on tunable VCSELs by Chang-Hasnain [12].

A report on 50 nm continuously tunable MEMS VCSEL devices with surface micromachining operating at 1.95 μm emission wavelength has appeared recently [13]. The substrate is InP, and the DBR at the top side is substituted by a dielectric SiO_x/SiN_y membrane. The deflection of the membrane tunes the cavity and causes a continuous change of the emission wavelength. The membrane actuation may be achieved either by electrostatic attraction or by electrothermal expansion. The maximum tuning range is 50 nm, and SSR exceeding 50 dB has been achieved over the whole tuning range.

PROBLEMS

12.1. For error-free transmission in fiber-optic communication, there should be no overlap of adjacent bits that spread due to chromatic dispersion. This is expressed by the inequality $|D|LB\Delta\lambda \leq 0.491$, where the symbols are defined after Equation 12.1. Assuming that the spectral width of the modulated source is 2.5 times the bit rate B, calculate the maximum length L of the fiber for error-free transmission at 1.55 μm, when $B = 10$ Gb s^{-1} and $D = 17$ ps (nm km)$^{-1}$.

12.2. Obtain the mode number that will give a 3 dB lower gain than the maximum. Use the parameters given in Example 12.3.

12.3. Examine the effect of changing the values of different parameters in Equation 12.10 to obtain a higher value for SSR. Assume unity mode confinement factor, $a = 0.5$ cm^{-1}nm^{-2}, $R_1 = 0.36$ and $R_2 = 0.99$, $\alpha_i = 5$ cm^{-1}, and three values for L: 100, 300, and 700 μm.

12.4. Calculate the SSR for a gain difference of 50 cm^{-1}, obtained in Example 12.6, by assuming the same parameter values as given in Example 12.3. Express the difference of SSR values obtained in the two examples in decibels.

12.5. The change in RI in a tunable VGF laser is 10^{-3}. The RIs in the two waveguides are 3.5 and 3.48. Calculate the tuning range, and compare the value with the range obtained for two- and three-section DBR lasers with $n = 3.5$ and the same change of RI.

12.6. Derive Equation 12.64.

12.7. Show that if the two filters have periods 500 and 600 GHz, their cascade will be periodic with a period of 3000 GHz.

Reading List

Amann, M. C., Chapter 3: Single-mode and tunable laser diodes. In *Semiconductor Lasers II: Materials and Structures*, ed. Kapon, E. Academic, San Diego, pp. 157–258, 1999.

Buus, J., M.-C. Amann, and D. J. Blumenthal, *Tunable Laser Diodes and Related Optical Sources*, 2nd edition. Wiley-IEEE, Piscataway, NJ, 2005.

Coldren, L. A., G. A. Fish, Y. Akulova, J. S. Barton, L. Johansson, and C. W. Coldren, Tunable semiconductor lasers: A tutorial, *J. Lightwave Technol.*, 22, 193–202, 2004.

Sarlet, G., J. Buus, and P.-J. Rigole, Chapter 4: Tunable laser diodes. In *WDM Technologies: Active Optical Components*, ed. Dutta, A. K., Dutta, N. K. and Fujiwara, M. Academic (Elsevier), San Diego, 2002.

References

1. Kogelnik, H. and C. V. Shank, Coupled wave theory of distributed feedback lasers, *J. Appl. Phys.*, 43, 2327, 1972.
2. Tohmori, Y. and M. Oishi, 1.55 μm butt-jointed distributed Bragg reflector lasers grown entirely by low-pressure MOVPE, *Jpn. J. Appl. Phys.*, 27, L693, 1988.
3. Utaka, K., S. Akiba, K. Sakai, and Y. Matshuhita, λ/4 shifted InGaAs/InP DFB lasers, *IEEE J. Quantum Electron.*, QE-22, 1042–1051, 1986.
4. Koch, T. L. and U. Koren, Semiconductor lasers for coherent optical fiber communication, *J. Lightwave Technol.*, 8, 274–293, 1990.
5. Kaminow, I. P., C. R. Doerr, C. Dragone, T. L. Koch, U. Koren, A. A. M. Saleh, A. J. Kirby, C. M. Ozveren, et al., A wideband all-optical WDM network, *IEEE JSAC/JLT Special Issue on Optical Networks*, 14, 780–789, 1996.
6. Alferness, R. C., U. Koren, L. L. Buhl, B. I. Miller, M. G. Young, T. L. Koch, G. Raybon, and C. A. Burrus, Broadly tunable InGaAs/InP laser based on a vertical coupled filter with 57 nm tuning range, *Appl. Phys. Lett.*, 60, 3209–3211, 1992.
7. Amann, M. C. and S. Illek, Tunable laser diodes utilizing transverse tuning, *J. Lightwave Technol.*, 11, 1168–1182, 1993.
8. Jayaraman, V., L. A. Coldren, S. Denbaars, A. Mathur, and P. D. Dapkus, *Wide Tunability and Large Mode-Suppression in a Multi-Section Semiconductor Laser Using Sampled Gratings*, Integrated Photonics Research, OSA, Washington, DC, pp. 306–307, 1992.
9. Jayaraman, V., Z.-M. Chuang, and L. A. Coldren, Theory, design, and performance of extended tuning range semiconductor lasers with sampled gratings, *IEEE J. Quantum Electron.*, 29, 1824–1834, 1993.

10. Tohmori, Y., Y. Yoshikuni, H. Ishii, F. Kano, T. Tamamura, Y. Kondo, and M. Yamamoto, Broad range wavelength-tunable superstructure grating (SSG) DBR lasers, *IEEE J. Quantum Electron.*, 29, 1817–1823, 1993.
11. Vakhshoori, D., P. Tayebati, C.-C. Lu, M. Azimi, P. Wang, J.-H. Zhou, and E. Canoglu, 2 mW CW single mode operation of a tunable 1550 nm vertical cavity surface emitting laser, *Electron. Lett.* 35, 900–901, 1999.
12. Chang-Hasnain, C. J., Tunable VCSEL, *IEEE J. Sel. Top. Quantum Electron.*, 6, 978–987, 2000.
13. Gruendl, T., K. Zogal, P. Debernardi, C. Gierl, C. Grasse, K. Geiger, R. Meyer, et al., 50 nm continuously tunable MEMS VCSEL devices with surface micromachining operating at 1.95 μm emission wavelength, *Sem. Sci. Technol.*, 28, 012001, 2012.

13

Nitride Lasers

13.1 Introduction

Group III–based nitrides were identified as a potential material for light emission by Maruska and Tietjen in 1969 [1], by Pankove et al. in 1971 [2], and in later years by Akasaki et al. [3]. The workers investigated the properties of light-emitting diodes (LEDs) made of GaN. As the bandgap of GaN is ~3.4 eV, it is useful for emission in the ultraviolet (UV) region of the EM spectrum. The growth of single crystals of GaN and observation of the first optically pumped stimulated emission were reported by Dingle et al. in 1971 [4]. After this, the progress in realizing electrically injected devices such as LEDs and laser diodes (LDs) became quite slow for various reasons, as will be discussed in this chapter. The real progress started in the 1990s [5,6], and since then there has been phenomenal growth in the area of nitride-based optoelectronic and electronic devices. Many devices have been commercialized and employed in day-to-day life. At present, the market for nitride-based devices and products is at the level of several billion dollars.

Since GaN, AlN, InN, and their alloys cover the UV, blue, and green regions of the EM spectrum in particular, and the entire visible range in general, devices made out of these materials find a number of applications, a few of which are listed below:

1. High-density data storage (digital versatile discs (DVD) and Blu-ray discs): The data storage density in optical discs is inversely proportional to λ^2. Discs having data written by blue rays can therefore store four times more data than discs written by red rays.

2. Lasers can act as light sources for high-definition TVs with large projection areas. Nitrides are suitable candidates for emission in the blue and green regions: two primary colors.

3. Lasers are also needed for portable projectors and, more specifically, picoprojectors. Green light emitters are essential component for picoprojectors. At present, frequency doubling of AlGaAs-based red

light is used to produce green and blue rays. GaN-based LDs serve as a viable replacement for these emitters.

4. Continuous research is ongoing to realize faster read/write times and 3-D storage of data. This motivates further research in Blu-ray technology based on nitride lasers.

5. In addition to LDs, there is a need to develop environmentally friendly white LEDs. Blue and green LEDs using nitrides act as the primary source, and use of suitable phosphors extends the color to longer wavelengths. The ultimate goal is to integrate LDs emitting primary colors. Nitride-based LDs form the essential components for these white LEDs.

6. Nitride-based materials are also useful in producing solar cells of higher efficiency.

As may be noted from the above list, nitride-based LDs are of tremendous commercial value and form an important area of current research.

As mentioned earlier in this section, the journey of nitride-based emitters from early development to their current importance has not been smooth. Even now, progress toward achieving longer wavelengths is hindered by several factors.

The basic theory for nitride-based lasers is no different from the theory of DH, QW, and QD lasers developed in earlier chapters. However, the crystal structure of nitrides (wurtzite, rather than zinc blende (ZB) as in other III–V and II–VI compounds and their alloys) gives rise to a few physical phenomena not present in the usual ZB structures. In this chapter, we shall point out these processes: in particular, the spontaneous and piezoelectric polarization and quantum-confined Stark effect (QCSE) will be introduced and explained. The special structures for nitride lasers, particularly the type of substrate, special considerations for dopants used, growth conditions, and orientation of the emitter surface, among others, will be mentioned. Some problems related to achieving longer wavelengths will be pointed out.

It must be mentioned that the subject is growing at a fast pace. The present chapter attempts to address the basic issues only; it does not present detailed theory and practice. The aim is to introduce readers to the vast amount of work already accumulated and to encourage them to study the devices in more detail.

13.2 Polar Materials and Polarization Charge

It is well known that in polar semiconductors, there is a transfer of charge from one atom to another in the basis atoms. This transfer leads to the formation of a negatively charged cation and a positively charged anion. In unstrained ZB structures, the anion and cation sublattices are arranged in

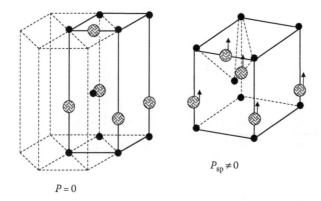

$P = 0$

$P_{sp} \neq 0$

FIGURE 13.1
Spontaneous polarization in the presence of shift in sublattices.

such a way that there is no net polarization in the material. In wurtzite (WZ) crystals such as AlN, GaN, and InN, on the other hand, a net spontaneous polarization exists. This is illustrated in Figure 13.1. When there is no shift in the cation–anion sublattice, as shown in Figure 13.1a, there is no polarization; however, when there is a relative shift in the two sublattices, as shown in Figure 13.1b, a net spontaneous polarization develops.

The presence of strain can cause a relative shift of the two sublattices, thereby creating a net polarization in the material. We discuss below how the shifts lead to the development of charges at a heterointerface.

13.2.1 Polar Charge at a Heterointerface

As discussed in Section 13.2 and illustrated in the right part of Figure 13.1, a polarization field develops when there is a net movement of one sublattice from another. The field gives rise to a positive and a negative polar charge. The polar charges at the free surface are neutralized by the charges in the atmosphere. In the case of a heterojunction, the two materials usually have different polarization, and there is a net polar charge and polarization at the heterointerface. This is illustrated in Figure 13.2.

If the polarization in the two materials is P_A and P_B, the net interface charge density is $P = P_A - P_B$ and the built-in electric field is

$$F = \frac{P}{\varepsilon} \tag{13.1}$$

13.2.2 Piezoelectric Effect

Application of strain introduces polarization in the medium: a phenomenon known as the piezoelectric effect. The value of polar charges induced

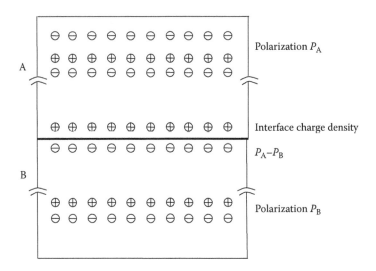

FIGURE 13.2
Development of interface charge density at a heterointerface of two polar semiconductors.

by strain depends on the strain tensor. In strained nitride structures grown along the *c*-axis, the strain tensor components are given as

$$\varepsilon_{xx} = \varepsilon_{yy} = \frac{a_f}{a_s} - 1 \tag{13.2a}$$

$$\varepsilon_{zz} = -2\frac{C_{13}}{C_{33}}\varepsilon_{xx} \tag{13.2b}$$

where:
ε_{xx} and ε_{yy} are the in-plane strain
a_f and a_s, are, respectively, the lattice constants of the grown film and the
   substrate

C_{13} and C_{33} are the elastic constants for the nitride materials grown in the hexagonal close-packed (hcp) structure
The piezoelectric polarization is related to the strain tensor by

$$\mathbf{P}_{pz} = e_{33}\varepsilon_{zz} + e_{31}(\varepsilon_{xx} + \varepsilon_{yy}) \tag{13.3}$$

where *e*s denote piezoelectric constants
In nitride-based heterostructures, polarization charges appear at the interface due to both the strain-related piezoelectric effect and spontaneous polarization.

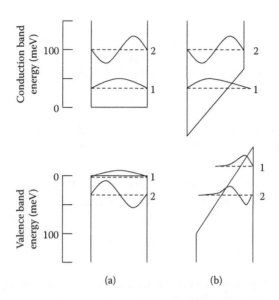

FIGURE 13.3

Subband energies and envelope functions in a QW: (a) without any field; (b) with an electric field applied perpendicular to the QW layer plane.

In face-centered cubic (fcc)-based semiconductors, there is no spontaneous polarization, but strain effect can induce a net polarization. The polarization difference gives rise to a built-in field that produces a band bending, as shown in Figure 13.3. The fixed charge developed plays the role of a dopant.

The general expression for the polarization developed due to strain is given by

$$P_i = \sum_{k,l} e_{ikl} \varepsilon_{kl} \tag{13.4}$$

For ZB structures, there is only one piezoelectric constant $e_{14} = e_{25} = e_{36}$; the notation $xx \Rightarrow 1$, $yy \Rightarrow 2$, $zz \Rightarrow 3$, $yz \Rightarrow 4$, $zx \Rightarrow 5$, $xy \Rightarrow 6$ is used and $e_{ikl} \Rightarrow e_{im}$, $m = 1 - 6$. Thus, piezoelectric polarization occurs only in the presence of shear strain. For growth along the (100) direction, the strain tensor is diagonal, and thus there is no piezoelectric polarization for the ZB structure. However, when these materials are grown along, say, the (111) direction, the shear component does produce piezoelectric polarization. For (111) growth of a strained layer of ZB structure, there is a strong dipole moment and band bending, as shown in Figure 13.3b. The built-in field is given by

$$F = \sqrt{3} \frac{e_{14} \varepsilon_{xy}}{\varepsilon_s} \tag{13.5}$$

where:

e_{14} is the piezoelectric coefficient
ε_{xy} is the off-diagonal strain component
ε_s is the relative permittivity of the semiconductor

> **Example 13.1:** For GaAs, $e_{14}=-0.16$ C m^{-2} and $\varepsilon_s=13.1$. Let the strain $\varepsilon_{xy}=0.01$. The built-in electric field is $F=2.39\times10^7$ V m^{-1}.

The growth of WZ structures usually occurs along the c-axis, that is, the (0001) or (000 $\bar{1}$) direction. In this case, the strain tensor is

$$\varepsilon_{xx} = \varepsilon_{yy} = \frac{a_s}{a_0} - 1 \tag{13.6}$$

where a_s and a_0 are, respectively, the lattice constants of the substrate and the material to grow on the substrate.

For the growth direction, the strain is expressed as

$$\varepsilon_{zz} = -\frac{2C_{13}}{C_{33}}\left(\frac{a_s}{a_0} - 1\right) \tag{13.7}$$

The piezoelectric polarization in a WZ system is given by

$$\mathbf{P}_{pz} = e_{33}\varepsilon_{zz} + e_{31}(\varepsilon_{xx} + \varepsilon_{yy}) \tag{13.8}$$

13.3 Quantum-Confined Stark Effect

An electric field applied perpendicular to the layer plane of a QW changes the subband energies and the value of the absorption coefficient and slightly alters the excitonic binding energy. These are the manifestations of the QCSE, as detailed later in this section. We have noted in Section 13.2 that the spontaneous and piezoelectric polarizations in the heterostructures and QWs give rise to an electric field, which should then introduce QCSE.

To calculate the change in the subband energies analytically, we assume that the external electric field, applied perpendicular to the QW layer plane, is directed along the z-direction and that the barrier height is infinite. The origin of the coordinate system and of electrostatic potential is taken at the center of a well of width d. The Hamiltonian in the presence of an electric field F is then

$$H = -\frac{\hbar^2}{2m_e}\frac{\partial^2}{\partial z^2} + eFz = H_0 + eFz \tag{13.9}$$

where H_0 is the Hamiltonian without field. If the field is weak enough, so that the condition $eFd \ll (\hbar^2/2m_e)(\pi/d)^2$, then the change in energy may be calculated by using second-order perturbation theory. The change for the nth subband is written as

$$\Delta E_n = \sum_m{}' \frac{|\langle m0|eFz|n0\rangle|^2}{E_{m0} - E_{n0}} = \frac{2e^2F^2m_ed^2}{\hbar^2\pi^2} \sum_m{}' \frac{|\langle m0|z|n0\rangle|^2}{n^2 - m^2} \tag{13.10}$$

where:

$$E_{n0} = \frac{\hbar^2}{2m_e}\left(\frac{n\pi}{d}\right)^2$$

is the unperturbed eigenvalue, and the subscript 0 refers to the zeroth-order eigenfunctions and eigenvalues. Since $|n0\rangle$ and $|m0\rangle$ are sine functions, the matrix element in Equation 13.10 may easily be calculated to give

$$\langle m0|z|n0\rangle = \frac{8d}{\pi^2}\frac{nm}{n^2 - m^2} \tag{13.11}$$

Putting $n=1$ for the lowest subband and summing over all permissible values of m starting from 2, the change in energy may be expressed as

$$\Delta E_1 = -\frac{1}{24\pi^2}\left(\frac{15}{\pi^2} - 1\right)\frac{e^2F^2m_ed^4}{\hbar^2} \tag{13.12}$$

It appears therefore that the energy of the lowest subband decreases quadratically with the applied electric field. Since for holes the energy is opposite, the principal effect of the field is to decrease the effective bandgap.

Example 13.2: Let a field strength of 10^5 V cm^{-1} be applied to a GaAs QW of width $d=10$ nm. The change in subband energy is 1.93 meV, taking $m_e = 0.067m_0$.

It is now clear that the effect of a perpendicular electric field is to shift the energies of the electron and hole subbands. Figure 13.4a shows the subband structure and envelope functions without an electric field, while Figure 13.4b

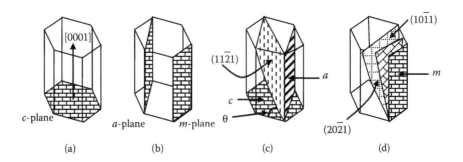

FIGURE 13.4
Different crystallographic planes in WZ structure.

shows the same in the presence of an electric field. Figure 13.4b indicates that the peaks of the electron and hole envelope functions are shifted in opposite directions in the presence of a field.

The shift in exciton peak is due to the Stark shift induced by the field. The exciton peaks should occur at an energy given by

$$E = E_g + E_{1e} + E_{1,hh} - E_B, \tag{13.13}$$

where:

E_g is the bandgap in the bulk
$E_{1,e}$ and $E_{1,hh}$ are, respectively, the field-dependent subband energies for the lowest electron and lowest heavy-hole subbands
E_B is the binding energy of the exciton, which is also dependent on the field

The reason for the change in exciton binding energy may be understood from Figure 13.4b. Since the envelope functions for electrons and holes move in opposite directions, there is a reduction of Coulombic attraction between the two in the presence of the field and hence a decrease in the binding energy. The change from the value without a field is, however, small.

> **Example 13.3:** The changes in subband energies E_{1e} and $E_{1,hh}$ for a field of 10^5 V cm^{-1} are approximately 2 and 13 meV from Equation 14.12. The binding energy $E_B \approx 10$ meV without a field and is slightly lowered with a field; hence the shift is mainly due to changes in subband energies.

The effect of a perpendicular electric field on the change in peak exciton absorption position is called the QCSE because of the prominence of the Stark shift. The reason why exciton peaks survive in a very high field is that the walls of the wells prevent the electrons and holes from being torn apart, although the electric field tends to separate them from each other. When the field is applied parallel to the QW plane, there are no such walls to confine the electron and the hole, and as a result, field ionization becomes easy. As is evident from the nature of wave functions shown in Figure 13.4, the overlap between the electron and the hole decreases with increasing field, thereby decreasing the absorption oscillator strength. The peak absorption value decreases with a field. Typical absorption spectra for GaAs–AlGaAs systems are given in [7–9] as well as in several books (see Reading Lists in Chapters 3 and 6).

Different methods have been used by several authors to model the absorption in both the presence and the absence of the electric field [7–9].

The presence of carriers screens the Coulomb interaction and quenches the excitonic absorption.

13.4 Early Work and Challenges

GaN is a wide-bandgap material and is suitable as a light-emitting material in the UV region. Once its importance was realized, the study of electronic and optical properties of nitrides was undertaken as early as 1969. However, the development of nitrides as a suitable laser material faced a number of challenges, which have been overcome by a number of breakthroughs. In this section, we shall outline the path of developments, mentioning the difficulties and the methods used to eliminate them.

13.4.1 Crystal Growth

It is difficult to grow a large bulk single crystal of GaN. Therefore, the chosen method from the early days until a few years ago was heteroepitaxy. The chosen substrate was sapphire, on which a thin GaN film was grown by Maruska and Tietjen in 1969 by using hydride vapor phase epitaxy (HVPE) [1]. However, the large lattice mismatch between the film and the substrate made it difficult to grow flat surfaces free of cracks. The density of threading dislocations was enormously large: $\sim 10^{10}\ cm^{-2}$. Furthermore, the residual donor concentrations in GaN higher than $10^{19}\ cm^{-3}$ did not allow control of n-type conductivity and made it impossible to achieve p-type conductivity.

With the advent of modern growth techniques such as molecular beam epitaxy (MBE) and metal organic chemical vapor deposition (MOCVD), the possibility of growing good-quality thin films of GaN on sapphire opened up. However, the metalorganic vapor phase epitaxy (MOVPE) method proved to be most successful, with a growth rate between MBE (slow) and HVPE (fast). MOVPE had the added advantage that alloy composition and impurity doping could be readily controlled by varying the flow rates. Later, the low-temperature (LT)-deposited buffer layer technology was developed by

Amano et al. to grow an AlN layer on sapphire [10], and then by Nakamura to grow a GaN buffer layer [6].

13.4.2 p-Type Layer

As mentioned already, the high concentration of residual donors made conductivity control and p-type conductivity difficult to achieve. However, it was realized that improvement of crystal quality could alter the situation. To achieve p-type material, Zn was tried. Mg was soon found to be a more suitable dopant to achieve p-type conductivity [11]. The ionization energy of Mg is quite high, which leads to other problems.

13.4.3 Control of n-Type Conductivity

The huge amount of residual donors in early GaN crystals did not allow proper control of conductivity. With successful growth of LT-AlN buffer layers on GaN, it became possible to control n-type conductivity by using SiH_4 as a Si dopant.

13.5 Some Useful Properties of Nitrides

13.5.1 Elastic Constants and Bandgap

Table 13.1 gives values of the elastic constants and piezoelectric constants of the three important nitride materials. The values are compared with similar values for III–V compounds. The three nitrides possess WZ structure, as indicated in the table. GaN can, however, also be grown in cubic structures. In this section, we shall give a brief introduction of the structural and some electronic properties of the nitrides. The corresponding parameters for the alloys are sometimes obtained by using Vegard's law, and for more accurate evaluation, empirical laws fitting experimental data are used.

The bandgap of GaN was measured, and the data can be fitted with the expression

$$E_g(T) = 3.503 + \frac{5.08 \times 10^{-4} T^2}{(T - 996)} \tag{13.14}$$

Equation 13.15 represents the bandgap of $In_xGa_{1-x}N$ [12]:

$$E_g(x) = 3.493 - 2.843x - 2.5x(1 - x) \tag{13.15}$$

TABLE 13.1

Piezoelectric and Elastic Constants and Spontaneous Polarization

Material	AlAs	GaAs	GaSb	GaP	InAs	InP	AlN	GaN	InN
Structure	ZB	ZB	ZB	ZB	ZB	ZB	WZ	WZ	WZ
C_{11} (N m^{-2})		1.2×10^{11}							
C_{12}									
C_{41}									
C_{13}							12.7	9.4	10.0
C_{33}							38.2	39.0	39.2
e_{14} (C m^{-2})	−0.23	−0.16	−0.13	−0.10	−0.05	−0.04			
e_{31}							−0.6	−0.49	−0.57
e_{33}							1.46	0.73	0.97
Psp (C m^{-2})							−0.081	−0.029	−0.032

This equation gives 0.65 eV as the bandgap of InN. In an earlier work, however, the bandgap of InGaN is expressed as [13]

$$E_g(x) = xE_g(\text{InN}) + (1-x)E_g(\text{GaN}) - bx(1-x) \tag{13.16}$$

with $b = 1.0$ eV. These two expressions give different values for the bandgap of InN: 0.65 and 2.03 eV.

13.5.2 Structures and Planes

GaN and other nitrides possess WZ structure, and the most thermodynamically stable structure of GaN is grown on the basal c-plane, as shown in Figure 13.4a. The plane is called a polar plane due to its noncentrosymmetric nature. It lacks inversion symmetry along the c-direction and gives rise to high built-in spontaneous polarization. The c-plane, however, has rotational symmetry, and thus the optical properties are isotropic in the plane.

Most of the LEDs and LDs studied earlier and until recently have been grown on the c-plane along the perpendicular c-direction.

There are a few nonpolar planes perpendicular to the c-plane. As shown in Figure 13.4b, these are the a-plane (11 $\overline{2}$ 0) and the m-plane (10 $\overline{1}$ 0). When an m-plane is rotated by 30°, it coincides with an a-plane. Each of these planes replicates itself when rotated by 60° around the c-axis. The crystal properties do not change when a nonpolar plane rotates by 180° about the c-axis. This means that there is symmetry in the direction perpendicular to the nonpolar planes. There is no spontaneous polarization for crystals grown on these planes.

13.6 First Laser Diode

The first electrically injected device was achieved and improved during the early part of the 1990s [14–16]. The device structure used by the authors resulted through a straightforward development of their earlier QW LED structures. The active region consisted of 26 periods of an $In_{0.2}Ga_{0.8}N$ QW (25 Å) with $In_{0.5}Ga_{0.94}N$ barriers (50 Å). This was surrounded by 0.1 μm GaN guiding layers and 0.4 μm $Al_{0.15}Ga_{0.85}N$ cladding layers. The substrate was c-plane sapphire, on top of which a 0.1 μm $In_{0.1}Ga_{0.9}N$ buffer layer was grown to prevent cracking of the relatively thick (3 μm) GaN layer grown on top of it and used as the base for the whole structure. A 200 Å thick $Al_{0.2}Ga_{0.8}N$ layer was grown immediately following the multiple quantum well (MQW) structure, to prevent dissociation of the InGaN layers during the growth of the p-type layers. The structure was mesa etched to form a stripe laser 1500 μm × 30 μm in area. Al/Ti and Ni/Au contacts were made to the n- and p-sides of the device. Because c-plane sapphire does not easily cleave, mirror facets were formed by reactive ion etching and their reflectivities increased to 60%/70% by suitable coatings.

The laser operated at room temperature, in pulsed mode, with an emission wavelength of 417 nm (2.97 eV), threshold current density of 4 kA cm^{-2}, and threshold voltage of 34 V. The slope efficiency was 13% per facet, and a pulsed power of 215 mW per facet was obtained at a drive current of 2.3 A.

13.7 Violet c-Plane Laser

At present, violet lasers have a potential market due to their use in high-density optical data storage. All the commercially available LDs are grown on GaN substrates by MOCVD on the c-plane of the materials. A typical structure of violet LD is shown in Figure 13.5 [15]. Both the p and n cladding layers are made up of AlGaN/GaN short-period superlattices to provide low refractive index (RI). The p and n waveguiding layers comprise GaN or $In_xGa_{1-x}N$ ($x = 0.03–0.07$) layers. Sandwiched between these two layers is the active region, typically having two or three QWs, though a single QW is not uncommon. The p waveguiding layer uses a very low concentration of Mg to reduce the Mg-related optical loss. Just above the bulk GaN substrate, a strain-compensating InN layer having an In concentration of about 0.05 and thickness of about 100 nm is grown.

The optical loss is reduced by decreasing the optical mode overlap with Mg-doped regions. This can be done by reducing Mg doping, moving the Mg doping farther from the active region, or shifting the optical mode toward the n side of the device by suitably designing the waveguide structure.

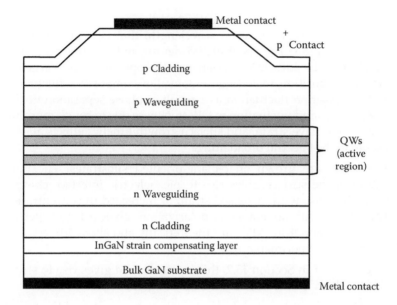

FIGURE 13.5
Cross-sectional diagram of a typical violet laser on the *c*-plane of the GaN substrate.

The state-of-the art continuous wave output power is 500 mW for single-mode devices [17].

13.8 Blue and Green Lasers

Blue and green are the two important primary colors for red-green-blue (RGB) light sources needed for portable projectors or picoprojectors. Before 2009, the maximum achievable wavelength was below 500 nm. Any attempt to increase the emission wavelength was associated with a dramatic increase in the threshold current (see Reading List).

There are several challenges in producing green LDs:

1. Emission at higher wavelength requires that the bandgap should be lowered. This means that the active region should contain more In. The lattice mismatch between the in-plane lattice constants of InN and GaN leads to 1.8% strain in $In_{0.18}Ga_{0.82}N$ (blue) and 3.0% strain in $In_{0.3}Ga_{0.7}N$ (green) QWs, which is, indeed, quite high. The high degree of strain degrades the material quality. The higher In concentration requires a lower growth temperature for growth of InGaN on GaN.

2. The spontaneous emission spectra and the gain spectra of InGaN QWs broaden with emission wavelength, that is, with increasing In concentration. The spectral broadening reduces the peak gain, which must be sufficient to overcome the optical losses. Variations in QW composition, including random alloy composition fluctuation and composition fluctuation as a result of phase separation during QW growth, and fluctuation in QW thickness due to rough morphology, compressive strain, and surface growth kinetics effects all contribute to the spectral broadening.

3. The layers are grown on *c*-plane-oriented substrates. The polar nature of the surface gives rise to piezoelectric interface charges, which increase with increased strain or increased In concentration. This, along with spontaneous polarization, gives a large interface field that is about $16x$ MV cm^{-1} theoretically and about $10x$ MV cm^{-1} as observed experimentally.

   As explained in Section 13.2, this interface field gives rise to strong QCSE, which in turn reduces the optical gain.

4. The QCSE has a beneficial effect in the case of green LDs operating at low current density. As the emission wavelength is redshifted (higher wavelength), a smaller In concentration is needed. Results show that, for a given wavelength, the In concentration needed may be reduced by 4% on account of QCSE. However, for higher current density, increased injection leads to screening of the polarization field and consequent blueshift of the emission wavelength. Therefore, this beneficial effect is present only for low-threshold lasers.

5. Higher In concentration leads to higher band offsets between barriers and wells. In the case of MQWs, it is necessary to ensure uniform carrier injection into all the active carriers. In green-light emitters, carrier escape is not as fast as in shorter-wavelength LDs in which the well depth is lower. This causes problems for carrier transport in MQWs. The issue is more serious for holes due to their low values of mobility. The interface charges of the *c*-plane LDs introduce spikes in the QW barriers, affecting carrier transport.

6. The RI of both InGaN and AlGaN decreases with wavelength at a given alloy composition. This lowers the RI contrast between the core and cladding regions of the waveguide and hence reduces the value of the optical confinement factor.

7. The low conductivity of the p-type layer seriously affects the wall-plug efficiency and causes heat-dissipation problems. The efficiency and dissipation are related to operation voltage. As the ionization energy of Mg, the usual dopant for achieving p-type, is quite high ($\sim$170 meV), a hole concentration no higher than 10^{17} cm^{-3} can be

achieved, even with a Mg concentration of 10^{19} cm^{-3}. This high concentration of acceptors leads to low hole mobility. This low conductivity leads to a high operation voltage. For higher wavelength, layers are grown at lower temperature. This gives rise to N vacancies that compensate the acceptors, thus lowering the conductivity further.

Careful design methodologies have been employed to circumvent these problems. It has been possible to fabricate good-quality *c*-plane LDs at an operating wavelength of 529 nm. The characteristics are summarized in table 1 of Sizov et al. (see Reading List).

13.9 Nonpolar and Semipolar Growth Planes

Figure 13.4 illustrates the different polar, semipolar, and nonpolar planes in WZ crystals. The *c*-plane is a polar (P) plane, while the *m*- and *a*-planes, orthogonal to the *c*-plane, are nonpolar (NP). Any plane inclined between the P and NP planes is called semipolar (SP). The inclination angle is indicated by θ, and it equals 90° for nonpolar *m*- and *a*-planes.

It was predicted theoretically that polarization-related fields would be totally absent for films grown along nonpolar planes (Figure 13.6). On the other hand, such fields show a maximum for the *c*-plane ($\theta = 0$), decrease to reach zero for $\theta \sim 50°$, change sign to attain another maximum, and reach

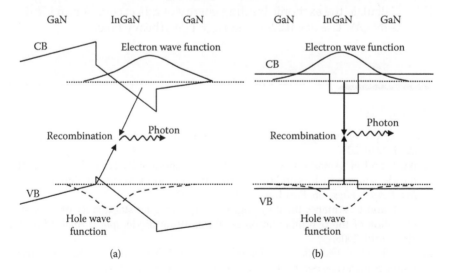

FIGURE 13.6

(a) Band diagram for growth along the polar plane; the overlap between the electron and hole wave functions is less due to QCSE; (b) absence of QCSE for growth along the NP plane.

zero for $\theta = 90°$. The maximum value increases with increasing x in $In_xGa_{1-x}N$ on GaN and in $Al_xGa_{1-x}N$ on GaN.

Since the polarization-related field is absent in nonpolar planes, there is no QCSE. This is illustrated in Figure 13.6, in which the reduced overlap between electron and hole wave functions for polar planes is clearly indicated. There have been attempts to grow NP-plane-oriented GaN films on sapphire, but the high stacking fault densities ($\sim 10^3$ to 10^5 cm^{-1}) seriously limited the device performance. The situation improved with the successful growth of freestanding GaN films, and the first successful violet m-plane laser was demonstrated in 2007 [18].

Green lasers grown on both polar and semipolar planes have been studied. Both the structures have advantages and disadvantages. A useful comparison has been presented in table 1 of the paper by Sizov et al. (see Reading List), and the interested reader is referred to this work.

The gain values for polar- and nonpolar-oriented GaN lasers are compared in [19].

PROBLEMS

13.1. Calculate the excitonic linewidth of GaN at 300 K. The contribution for fluctuations is 5 meV, that for acoustic phonons is 30 μeV K^{-1}, and the prefactor for LO phonons that is to be divided by the Bose factor is 400 meV. LO phonon energy is 91.8 meV.

13.2. Calculate the composition of an $Al_xIn_{1-x}N$ lattice matched to GaN. a(GaN) = 3.112, a(InN) = 3.54, and a(AlN) = 3.112 A.

13.3. Calculate the excitonic binding energy of a 1s exciton for an ideal GaN QW. Use $m_e = 0.22m_0$ and $m_{hh} = 1.0m_0$ (heavy hole).

Reading List

Akasaki, I., GaN-based p–n junction blue-light-emitting devices, *Proc. IEEE*, 101, 2200–2210, 2013.

Akasaki, I. and H. Amano, Crystal growth and conductivity control of group III-nitride semiconductors and their application to short wavelength light emitters, *Jpn. J. Appl. Phys.*, 36, Part I, A, 5394–5408, 1997.

Akasaki, I. and H. Amano, Breakthroughs in improving crystal quality of GaN and invention of the p–n junction blue-light-emitting diode, *Jpn. J. Appl. Phys.* 45, 9001–9010, 2006.

Hardy, M. T., D. F. Feezella, S. P. DenBaarsa, and S. Nakamura, Group III-nitride lasers: A materials perspective, *Mater. Today* 14, 408–415, 2011.

Nakamura, S., InGaN-based blue laser diodes, *IEEE J. Sel. Topics Quantum Electron.* 3, 712–718, 1997.

Nakamura, S., InGaN-based violet laser diodes, *Semicond. Sci. Technol.* 14, R27–R40, 1999.

Nakamura, S. and G. Fasol, *The Blue Laser Diode*, Springer, Berlin, pp. 216–219, 1997.

Nakamura, T. and K. Motoki, GaN substrate technologies for optical devices, *Proc. IEEE* 101, 2221–2228, 2013.

Orton, J. W. and C. T. Foxon, Group III nitride semiconductors for short wavelength light-emitting devices, *Rep. Prog. Phys.* 61, 1–75, 1998.

Piprek, J. (ed.) *Nitride Semiconductor Devices: Principles and Simulation*, Wiley, New York, 2007.

Sizov, D., R. Bhat, and C.-E. Zah, Gallium indium nitride-based green lasers, *J. Lightwave Technol.* 30, 679–699, 2012.

References

1. Maruska, H. P. and J. J. Tietjen, The preparation and properties of vapor-deposited single-crystalline GaN, *Appl. Phys. Lett.* 15, 327–329, 1969.
2. Pankove, J. I., E. A. Miller, and J. E. Berkeyheiser, GaN electroluminescent diodes, *RCA Rev.* 32, 383–392, 1971.
3. Akasaki, I., H. Amano, Y. Koide, K. Hiramatsu, and N. Sawaki, Effects of AlN buffer layer on crystallographic structure and on electrical and optical properties of GaN and $Ga_{1-x}Al_xN$ $(0 < x \leq 0.4)$ films grown on sapphire substrate by MOVPE, *J. Crystal Growth* 89, 209–219, 1989.
4. Dingle, R., K. L. Shaklee, R. F. Leheny, and R. B. Zetterstrom, Stimulated emission and laser action in gallium nitride, *Appl. Phys. Lett.* 19, 5–7, 1971.
5. Amano, H., M. Kito, K. Hiramatsu, and I. Akasaki, p-Type conduction in Mg-doped GaN treated with low-energy electron beam irradiation (LEEBI), *Jpn. J. Appl. Phys.* 28, L2112–L2114, 1989.
6. Nakamura, S., GaN growth using GaN buffer layer, *Jpn. J. Appl. Phys.* 30, L1705–L1707, 1991.
7. Schmitt-Rink, S., D. S. Chemla, and D. A. B. Miller, Linear and nonlinear optical properties of semiconductor quantum wells, *Adv. Phys.* 38, 89–188, 1989.
8. Miller, D. A. B., D. S. Chemla, T. C. Damen, A. C. Gossard, W. Wiegmann, T. H. Wood, and C. A. Burrus, Electric field dependence of optical absorption near the band gap of quantum-well structures, *Phys. Rev. B.* 32, 1043–1060, 1985.
9. Bandyopadhyay, A. and P. K. Basu, Modeling of electrorefraction in InGaAsP multiple quantum wells, *IEEE J. Quantum Electron.* 29, 2724–2730, 1993.
10. Amano, H., N. Sawaki, I. Akasaki, and T. Toyoda, Metalorganic vapor phase epitaxial growth of a high quality GaN film using an AlN buffer layer, *Appl. Phys. Lett.* 48, 353–355, 1986.
11. Nakamura, S., N. Iwasa, M. Senoh, and T. Mukai, Hole compensation mechanism of P-type GaN films, *Jpn. J. Appl. Phys.* 31, 1258–1266, 1992.
12. Davydov, V. Y., A. A. Klochikhin, V. V. Emtsev, D. A. Kurdyukov, S. V. Ivanov, V. A. Vekshin, F. Bechstedt, et al., Band gap of hexagonal InN and InGaN alloys, *Phys. Status Solidi B Basic Solid State Phys.* 234, 787–795, 2002.
13. Nakamura, S., T. Mukai, and M. Senoh, Si-doped InGaN films grown on GaN films, *Jpn. J. Appl. Phys. Part 2*, 32, L16–L19, 1993.

14. Akasaki, I., H. Amano, M. Kito, and K. Hiramatsu, Photoluminescence of Mg-doped p-type GaN and of GaN p-n junction LED, *J. Lumin.* 48–49, 666–670, 1991.

15. Nakamura, S., T. Mukai, and M. Senoh, High-power GaN p-n junction blue-light-emitting diodes, *Jpn. J. Appl. Phys.* 30, L1998–L2001, 1991.

16. Nakamura, S., M. Senoh, S. Nagahama, N. Iwasa, T. Yamada, T. Matsushita, H. Kiyoku, and Y. Sugimoto, InGaN-based multi-quantumwell-structure laser diodes, *Jpn. J. Appl. Phys.* 35, L74–L76, 1996.

17. Kozaki, T., H. Matsumura, Y. Sugimoto, S.-I. Nagahama, and T. Mukai, High-power and wide wavelength range GaN-based laser diodes, *Proc. SPIE* 6133, 613306-1–613306-12, 2006.

18. Schmidt, M. C., K. C. Kim, R. M. Farrell, H. Sato, N. Fellows, H. Masui, S. Nakamura, S. P. DenBaars, and J. S. Speck, Demonstration of nonpolar *m*-plane InGaN/GaN laser diodes, *Jpn. J. Appl. Phys.* 46, 9, L190–L191, 2007.

19. Melo, T., Y.-L. Hu, C. Weisbuch, M. C. Schmidt, A. David, B. Ellis, C. Poblenz, Y.-D. Lin, M. R. Krames, and J. W. Raring, Gain comparison in polar and non-polar/semipolar gallium-nitride-based laser diodes, *Semicond. Sci. Technol.* 27, 024015, 2012.

14

Group IV Lasers

14.1 Introduction

The theory of semiconductor lasers, both double heterostructures (DHs) and quantum structures, which was developed in earlier chapters, considered direct-bandgap semiconductors. It is well known that although Si is the most used material for electronics, it is not suitable for photonic device applications, particularly in the form of emitters, due to the indirect nature of its bandgap. The same is true for Ge, the other Group IV semiconductor that is used in electronic devices.

The last two decades have witnessed the widespread use of optoelectronic devices in many areas, such as the telecommunication, computer, and entertainment industries. In order to achieve high-speed data transfer in electronic equipment, the idea of integrating optical and electronic devices together on a silicon platform was introduced. Such an optoelectronic integrated circuit (OEIC) is likely to offer all the benefits of electronic ICs, that is, low cost, batch fabrication, high packing density, reliability, and so on.

In the past one or two decades, extensive theoretical and experimental works have been performed by many workers in academics and industries throughout the globe in the area of Group IV photonics. The main challenge in this work is to realize a suitable, electrically pumped laser on the Si platform. If the goal is achieved, the laser along with all other photonic components, such as modulators, detectors, and waveguides, may form a link for optical fiber communication as well as intrachip, chip-to-chip, or board-to-board interconnects in very-large-scale ICs.

In the present chapter, we will discuss in further detail the need for Si photonics. Then, the various attempts that have been made so far to realize light emitters using Si or related materials will be described. Currently, more emphasis is given to the use of Ge or alloys of Ge with Si and Sn grown on Si substrate as the active material for achieving laser action. The built-in strain in Ge or GeSn causes an indirect-to-direct crossover of the conduction band, making it feasible to observe stimulated emission. The band structure modification due to strain will be discussed. Finally, the condition for laser action and challenges in achieving a practical laser by using the method will be described.

14.2 Need for Si (Group IV) Lasers

Silicon is one of the most widely studied materials in the history of civilization. Since the first demonstration of planar ICs around the 1960s, industry has never given a second thought to using materials other than silicon. The present-day information age would not have come into existence in the absence of the electronics revolution, which was brought about by the maturity of silicon-based microelectronics. The famous Moore's law, which has been followed by the industry over the last four or five decades, states that the number of transistors in an IC chip doubles every 12 months (since revised to every 18 months). However, during the last few years there is an indication of the decline of Moore's law. There are doubts about whether, in future, silicon-based ICs will deliver the same advantages and functionalities as shown today.

The weakest point of silicon is that efficient light emitters and high-speed modulators cannot be realized using silicon due to its indirect bandgap nature. On the other hand, there is a steady increase in the use of silicon in the area of photonics, in the form of optical communication and networking, optical information processing, and consumer electronics based on light. Present-day photonic devices such as lasers, modulators, and photodetectors or even passive waveguide devices are made with compound semiconductors and their alloys. Although discrete devices using these materials show very good performance, when it comes to the integration of these devices, preferably on the same substrate, the level of integration and performance is far below what has already been achieved in electronic integration. It is natural to expect that monolithic OEICs will provide the same advantages, that is, low cost due to batch fabrication, high functionality, scaling for denser integration, and so on, as provided by silicon ICs.

If it is possible to grow OEICs on silicon and to integrate them with electronic ICs by using the same microelectronic production facilities, Si-based systems in their integrated form will then be used in all fields of electronics, computers, and communications.

The other complexity of present-day ICs has reached such a high level that the interconnects within it are formed on a number of levels. Currently, the number is six, but it is expected to increase twofold within a few years. The metallic interconnects, mainly Cu, provide delay due to resistor-capacitor (RC) time constants, which far exceeds the transit time delay associated with individual transistors. If the increase in speed is to be maintained at the same rate for the next generation of ICs, the interconnect bottleneck must be properly removed. Optics is believed to be the right solution to the problem, since light beams move faster and cross each other.

The development of Si-based active photonic devices has remained a challenge over the last few decades. The principal aim is naturally to realize efficient light emitters, modulators, and photodetectors using Si and its alloys on Si substrate and using the existing microfabrication facilities.

Therefore, Si photonics holds promise for next-generation telecommunications. To be useful for telecommunications or as intrachip interconnections, it is desirable for a silicon-based laser to satisfy the following conditions: (1) it can be monolithically grown on silicon substrates, which cost much less than any other semiconductor material; (2) it is compatible with complementary metal–oxide semiconductor (CMOS) processing, which allows for the integration of the laser with other electronic or optoelectronic components; (3) it emits light around the telecommunication wavelengths of 1300 or 1550 nm; and (4) it is desirable that it is electrically pumped, so it requires no additional pump light source. In all such attempts, heterostructures formed by Si and its alloy with Ge, $Si_{1-x}Ge_x$, are the most studied material systems.

14.3 Problems Related to Group IV Semiconductors: Indirect Gap

Group IV semiconductors such as silicon (Si) and germanium (Ge) are well known for their electronic applications. As already mentioned, the main difficulty in applying them to active devices in the optoelectronic area is their indirect bandgaps. In these cases, the conduction bands are characterized by several minima. Under normal conditions, electrons occupy the lowest minimum. The minima located away from the center of the zone are usually on the axes of symmetry and are multinumbered. Si has a principal bandgap of 1.12 eV at 300 K, located in the Γ–X ($\langle 100 \rangle$) equivalent **k**-space directions, and thus there are six equivalent principal conduction bands for Si. Ge, on the other hand, has a principal bandgap of 0.66 eV at 300 K, located in the Γ–L ($\langle 111 \rangle$) equivalent **k**-space directions, and thus there are four degenerate minimum energy levels in Ge.

In these materials, the probability for radiative recombination is low, which means that the electron-hole radiative lifetime is long, of the order of some milliseconds. But typical nonradiative recombination lifetimes are several nanoseconds. So the problem is the fact that while waiting for radiative recombination to take place, both the electrons and the holes move around. If they encounter a defect or a trapping center, they can recombine nonradiatively. As a result, the internal quantum efficiency is low. That is why they are a poor luminescent material: nonradiative recombination is more efficient than radiative. In these cases, a photon cannot directly lift an electron to the conduction band. In order to conserve crystal momentum or wave vector, it is necessary that some momentum-conserving agency should participate in the absorption process. The required momentum may come through scattering with an impurity or alloy disorder; however, in most cases of pure materials, phonons give the necessary momentum. The photon brings an electron in the valence band to an intermediate state, by the process of which there is negligible change of momentum. The electron then has its momentum changed by an amount $\mathbf{q}=\mathbf{k}_0$, by colliding with a phonon and comes to one of the six degenerate conduction band valleys. The wave vector conservation then takes the form $\mathbf{k}_c=\mathbf{k}_v\pm\mathbf{q}$, where $\mathbf{q}$ is the phonon wave vector and v is the respective valley. The energy conservation is now expressed as

$$\hbar\omega = \mathbf{E}_c(\mathbf{k}_C) - \mathbf{E}_v(\mathbf{k}_v) \pm \hbar\omega_q \tag{14.1}$$

where $\hbar\omega_q$ is the energy of the phonon. In these cases, + and − refer to phonon absorption and emission processes, respectively. This type of transition is referred to as a second-order process and the transition probability is calculated by second-order perturbation theory. The probability is extremely low in comparison with that in direct-bandgap semiconductors. The absorption coefficient being directly related to the probability of transition is also very low. The low values of the absorption in Si make the absorption length longer, thereby increasing the size of the optoelectronics device.

The variation of the absorption coefficient can be understood using the expression obtained by second-order perturbation theory. Instead, we can follow the empirical relation deduced by Bucher et al. [1] for the absorption coefficient as

$$\alpha(T) = \sum_{\substack{i=1,2 \\ j=1,2}} A_j C_i \left\{ \frac{\left[\hbar\omega - E_{gj}(T) + E_{phi}\right]^2}{\exp\left(E_{phi}/k_B T\right) - 1} + \frac{\left[\hbar\omega - E_{gj}(T) - E_{phi}\right]^2}{1 - \exp\left(-E_{phi}/k_B T\right)} \right\}$$

$$+ \frac{A_d \left(\hbar\omega - E_{gd}\right)^{3/2}}{\hbar\omega} \tag{14.2}$$

where i and j denote, respectively, the types of phonons and bandgaps. E_{g1} is the difference between the lowest Γ point and the conduction band minima in the Δ point $(\Delta_1 - \Gamma_{25})$, whereas E_{g2} is that for the conduction band minima in the L point $(L_1 - \Gamma_{25})$; E_{gd} is the direct bandgap, and E_{ph} is the energy of the phonon. The first term in Equation 14.2 describes the absorption of phonons and the second the emission of phonons. The third term gives the absorption for a direct (forbidden) transition; $E_g(T)$ is the bandgap at temperature

T related to the bandgap at $T=0$. The phonons of energy $E_{ph}/k=212$ K (TA), 670 K (TO), 1030 K, and 1420 K take part in the transitions. The values of the constants are $A_1=253$ cm^{-1} eV^{-2}, $A_2=3312$ cm^{-1} eV^{-2}, $A_d=2.3\times10^7$ cm^{-1} eV^{-2}, $E_{g1}=1.16$ eV, $E_{g2}=2.25$ eV, and $E_d=3.2$ eV.

As the absorption coefficient has been obtained, it is straightforward to relate the gain coefficient with the absorption coefficient, as we know that the condition of population inversion can create an amplification of electromagnetic radiation. So, it is obvious that the low values of the absorption coefficients (gain coefficients) are the main problem in using Si and Ge as a laser source.

> **Example 14.1:** Considering the abovementioned values of different parameters, the value of α at 300 K for photon energy $= 1.2$ eV is 40.42 cm^{-1} for TA phonons.

14.4 Recent Challenges

The realization of Si-compatible, Group IV–based photonic devices such as lasers, detectors, and modulators has remained a challenge so far. On the other hand, if properly developed, Group IV photonic devices using the IC fabrication process may provide a cheap solution to short-haul chip-to-chip and board-to-board communications as well as long-haul optical fiber communication links. This has motivated a significant research effort in recent years.

The most recent progress in this field has occurred in the development of low-threshold Si Raman lasers with racetrack ring resonator cavities [2]; Ge-on-Si lasers operating at room temperature; rare earth element, for example, Er-doped Si/SiO$_2$ light-emitting devices [3]; and hybrid Si lasers [4]. The first continuous-wave Raman laser, which was announced by Intel, used effective pump powers of ~182 mW to provide Raman gain for the Stokes signal at 1542 nm. The silicon-on-insulator (SOI) waveguide has a lateral p-i-n structure, an applied reverse bias to which the free carriers sweep, generated by two-photon absorption. Recently, the threshold pump power has been reduced by using a high Q-factor racetrack ring resonator cavity and an optimized p-i-n diode structure.

Transitions between different levels in Er ions give rise to amplification at 1554 nm. Recently, Er-containing SiO$_2$ layers, in which Si nanoclusters act as sensitizers, have shown laser action under electrical pumping.

In hybrid Si lasers, III–V compounds are wafer bonded to SOI structures. The compound semiconductor acts as laser material, whereas the SOI structure forms the waveguide. Recently, a compact hybrid Si microring laser has shown better performance.

14.4.1 Alloying Si–Ge–Sn

An interesting alternative would be a direct-gap alloy based on Group IV materials Si, Ge, and Sn, which are generally compatible with silicon technology, and this has been widely investigated. However, early studies of epitaxial SiGeSn alloys have revealed the difficulties of their growth because of a large lattice mismatch between α-Sn (6.489 Å) and Ge (5.646 Å) or Si (5.431 Å), approximately 15% and 17%, respectively. Furthermore, because of the lower surface free energy of alpha-tin and germanium, there can be segregation on the surface. These difficulties have been overcome by low-temperature molecular beam epitaxy (MBE), which has enabled epitaxial growth of, for example, strained $Ge_{1-x}Sn_x$ superlattices and random $Ge_{1-x}Sn_x$ alloys on a Ge substrate. However, these are not expected to show an indirect–direct transition, because of their compressive strain: it is presently believed that a direct gap can only appear in tensile-strained or relaxed $Ge_{1-x}Sn_x$. Further advances were made by ultrahigh vacuum chemical vapor deposition (UHV-CVD) and uniform, homogeneous, relaxed $Ge_{1-x}Sn_x$ alloys with $x < 0.2$ being grown on silicon [5,6]. Experimental investigations revealed significant changes in optical constants and redshifts in the interband transition energy as x varied [7], indicating the wide tunability of the bandgap of these alloys.

Soref and Perry [8] used the linear interpolation scheme to calculate the electronic band structure and optical properties of $Ge_{1-x-y}Si_xSn_y$ alloys and concluded that these will be tunable direct-bandgap semiconductors. Furthermore, both the direct and indirect bandgaps in Ge decrease with tensile strain, but the former (initially 140 meV above) does so faster, eventually delivering a direct-gap material. Therefore, one can use strained Ge, grown on $Ge_{1-x-y}Si_xSn_y$ ternary alloys [9].

14.5 Use of Heterostructure for Direct Bandgap Type I Structure

Menéndez and Kouvetakis [9] theoretically predicted that the direct gap nature of the bandgap may be observed in the tensile-strained Ge layer. The lowest direct gap in Ge (L point) is only 140 meV above the direct gap (Γ point). With the application of tensile strain, the energy of both the edges decreases, but the Γ conduction band edge decreases more rapidly than the L valley to effect indirect-to-direct crossover. This idea cannot be applied to the Ge-$Si_{1-x}Ge_x$ system, since the smaller lattice constant of Si makes the Ge layer compressively strained. The lattice constant of Sn, on the other hand, is larger and hence the growth of the Ge layer on $Ge_{1-y}Sn_y$ is expected to solve the problem. However, the direct gap in GeSn decreases very rapidly and therefore the direct gap in Ge may not be the lowest gap. The incorporation of Si in GeSn raises the bandgap, but at the same time reduces the tensile strain. Menéndez and Kouvekatis predicted that, under suitable conditions, the direct bandgap in Ge is also the lowest bandgap in Ge/Ge$_{1-x-y}$Si$_x$Sn$_y$ multilayers and further that the band lineup is Type I.

In the following, we have essentially reproduced the theory of band structure and band lineup presented by Menéndez and Kouvetakis.

The method is a generalization of van de Walle's approach [10]. The average of the three valence bands is taken as the reference level. In the work, the strain-dependent spin-orbit Hamiltonian has also been included. Considering the growth of the (001)-oriented Ge layer, lattice matched to a relaxed Ge$_{1-x-y}$Si$_x$Sn$_y$ layer, the following expressions for the band edges of Ge are derived:

$$E_{v\Gamma}(Ge) = -\frac{\Delta_0(x,y)}{3} - \Delta E_{v,av}(x,y) - \frac{\Delta_0(Ge)}{6} + \delta E_h^0 + \frac{1}{4}\delta E_{001}$$

$$+\frac{1}{2}\sqrt{\left(\Delta_0(Ge) + \frac{1}{2}\delta E_{001}\right)^2 + 2(\delta E'_{001})^2} \tag{14.3}$$

$$E_{c\Gamma}(Ge) = E_{v\Gamma}(Ge) + E_0(Ge) + \delta E_h^{c\Gamma} \tag{14.4}$$

$$E_{cL}(Ge) = E_{v\Gamma}(Ge) + E_{ind}(Ge) + \delta E_h^{cL} \tag{14.5}$$

All the energies are measured using the valence band edge of the alloy SiGeSn as the reference. Here, $E_{v\Gamma}$, $E_{c\Gamma}$, and E_{cL} have the usual meanings; $E_0(Ge) = E_{c\Gamma} - E_{v\Gamma} = 0.805$ eV is the direct gap in Ge and $E_{ind}(Ge) = E_{cL} - E_{v\Gamma} = 0.664$ eV is the indirect gap in Ge. Using linear interpolation, the spin-orbit splitting in the alloy is expressed as $\Delta_0(x,y) = \Delta_0(Ge) - 0.26x + 0.47y$, with $\Delta_0(Ge) = 0.30$ eV. Also $\Delta E_{v,av}(x, y) = 0.69y - 0.48x$. The strain shifts are given by

$$\left.\begin{aligned}
\delta E_{001} &= -2(b_1 + 2b_2)\left[\frac{2C_{12}}{C_{11}} + 1\right]e_{11} \\[2mm]
\delta E'_{001} &= -2(b_1 - b_2)\left[\frac{2C_{12}}{C_{11}} + 1\right]e_{11}
\end{aligned}\right\} \tag{14.6}$$

$$\left.\begin{aligned}
\delta E_h^v &= 2a_v\left[1 - \frac{C_{12}}{C_{11}}\right]e_{11} \\[2mm]
\delta E_c^{c\Gamma} &= 2a_c\left[1 - \frac{C_{12}}{C_{11}}\right]e_{11} \\[2mm]
\delta E_c^{cL} &= 2\left[\Xi_d + \frac{1}{3}\Xi_u\right]\left[1 - \frac{C_{12}}{C_{11}}\right]e_{11}
\end{aligned}\right\} \tag{14.7}$$

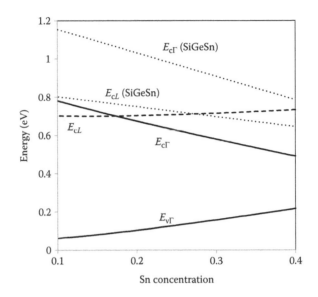

FIGURE 14.1
Band edge variation of Γ and L points in tensile-strained Ge/SiGeSn layers.

The in-plane strain is calculated from the relation:

$$e_{\parallel} = \left[a(\mathrm{Ge}_{1-x-y}\mathrm{Si}_x\mathrm{Sn}_y) - a(\mathrm{Ge})\right]/a(\mathrm{Ge}) \tag{14.8}$$

where a is the lattice constant.

The band edges in the ternary alloy are calculated using linear interpolation between the constituents (Figure 14.1). The expressions are

$$\left.\begin{array}{l} E_{c\Gamma}(x,y) = E_c(\mathrm{Ge}) + 3.14x - 1.225y \\ E_{cL}(x,y) = E_{cL}(\mathrm{Ge}) + 1.27x - 0.524y \end{array}\right\} \tag{14.9}$$

Example 14.2: The values of the band-edge energies will be calculated for $x = 0.15$ and $y = 0.3$. Using the parameters given in Table 2.2 (Appendix II), we obtain the following values: $E_{v\Gamma}(\mathrm{Ge}) = 0.158$ eV, $E_{c\Gamma}(\mathrm{Ge}) = 0.58$ eV, and $E_{cL}(\mathrm{Ge}) = 0.715$ eV. The corresponding values of $E_{c\Gamma}(x,y) = 0.909$ eV and $E_{c\Gamma}(x, y) = 0.697$ eV. This example indicates that the band structure in strained Ge is indeed direct in nature.

Example 14.3: The offsets in the valence band and the conduction band can easily be calculated from Figure 14.1 for a particular value of Sn mole fraction with $x = 0.15$. For example, with $y = 0.24$, $\Delta E_v = E_{v\Gamma} = 0.125$ eV and $\Delta E_c = E_{c\Gamma}(\mathrm{SiGeSn}) - E_{c\Gamma} = 0.346$ eV.

It has been found that Ge becomes a direct-gap semiconductor for a wide range of concentrations and also that the band alignment is Type I, with both the valence and conduction band edges localized in the Ge layer.

14.6 Ge Laser at 1550 nm

Among Group IV semiconductors, germanium is a promising material for an efficient light emitter. Germanium is a quasi-direct-bandgap material, that is, the direct conduction band edge lies only 134.5 meV above the lowest indirect conduction band edge at room temperature. Therefore, it is possible to convert the quasi-direct bandgap of Ge into a direct bandgap through proper engineering of the bandgap, allowing it to serve as a gain medium. Another particular characteristic of Ge is that its direct bandgap is about 0.8 eV, corresponding to an emission wavelength of 1550 nm, the most popular wavelength in telecommunications. Those advantages make Ge a potential candidate for an efficient silicon-based laser for telecommunications. But, theoretically, Ge becomes a direct-gap material at 2% tensile strain according to the deformation potential theory as described in Section 14.4 and the bandgap shrinks to ~0.5 eV in that case, corresponding to a wavelength of 2500 nm instead of 1550 nm.

Liu et al. [11] first predicted the laser action in heavily doped Ge at 1550 nm using a lesser amount of strain and adequate carrier injection. Adequate strain and n-type doping engineering can effectively provide population inversion in the direct bandgap of Ge. Liu et al. achieved 0.25% tensile strain in epitaxial Ge layers on Si using the thermal expansion mismatch between Ge and Si. With this 0.25% tensile strain, the difference between the Γ and L valleys of Ge can be decreased to 115 meV, as shown in Figure 14.2. An additional benefit is that the top of the valence band is determined by the light-hole band with a very small effective mass under tensile strain. The optical gain increases faster with the injected carrier density due to the low density of states associated with the light-hole band. To achieve efficient light emission while still keeping the emitted wavelength around 1550 nm, Liu et al. proposed to compensate for the rest of the difference between the Γ and L valleys of Ge by n-type doping instead of further increasing the tensile strain, as is schematically shown in Figure 14.2. For 0.25% tensile-strained Ge, they compensated for the 115 meV energy difference between the Γ valley and the L valley by filling the L valleys with 7.6×10^{19} cm^{-3} electrons (Figure 14.2).

Example 14.4: For a strained layer, the conduction band edge is shifted by $\delta E_c = 2a_d(1 - (C_{12}/C_{11}))\varepsilon$, where a_d is the deformation potential for the respective band, C_{11} and C_{12} are the elastic constants, and ε is the

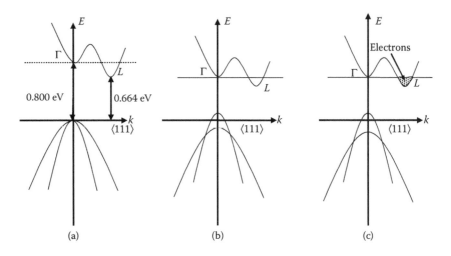

FIGURE 14.2
Band structure of Ge: (a) bulk Ge, (b) tensile-strained intrinsic Ge, and (c) tensile-strained, heavily doped n-type Ge.

strain. If we consider for Ge, $C_{11}=128.53$ GPa and $C_{12}=48.26$ GPa, the deformation potentials for the Γ and L band are, respectively, −8.24 and −1.54 eV, and the calculated difference between L and Γ for 0.12% strain is 144.56 meV.

14.7 Mid-Infrared Laser Based on GeSn

Silicon-germanium-tin technology is an excellent new Group IV technology for the monolithic integration on silicon or SOI. In this respect, GeSn is the best-known Sn-containing alloy and for this material the infrared region is most natural for radiative band-to-band emission because the direct bandgap of this crystalline alloy is narrower than that of elemental Ge. Sun et al. [12] presented the conception, modeling, and simulation of an Si-based Group IV semiconductor injection laser diode in which the active region of the GeSn alloy has a direct bandgap wavelength in the 1.8–3 μm midwave infrared for 6%–12% α-Sn. They proposed a p-i-n structure that is grown lattice matched on a relaxed GeSn buffer on SOI, believed to be manufacturable in CMOS technology. The compositions of GeSn and SiGeSn are chosen to provide Type I band alignment at the Γ point as well as lattice matching. The schematic diagram of their proposed structure is given in Figure 14.3.

To determine the compositions of $Ge_{1-z}Sn_z$ and $Ge_{1-x-y}Si_xSn_y$, the following equation based on the theory of Jaros [13], Weber [14], and D'Costa et al.

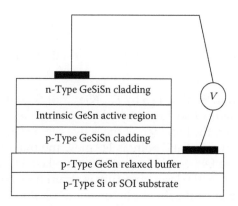

FIGURE 14.3
p-i-n structure of GeSn-based mid-infrared laser.

[15,16] can be used. The minima of the conduction band at points L and Γ are given by

$$E_c(Ge_{1-x-y}Si_xSn_y) = E_{Ge}(1-x-y)$$
$$+ E_{Si}x + E_{Sn}y - b_{GeSi}(1-x-y)x - b_{GeSn}(1-x-y)y - b_{SiSn}xy$$
$$(14.10)$$

The values of the different parameters are given in Table 14.1.

> **Example 14.5:** The conduction band edge for $Ge_{1-x-y}Si_xSn_y$ for $x=0.25$ and $y=0.03$ is 0.982 and 1.396 eV at L and Γ points, respectively. But for $x=0.15$ and $y=0.1$, the edges lie at 0.818 and 0.796 eV at L and Γ points, respectively. It is clear from this example that $Ge_{1-x-y}Si_xSn_y$ behaves as a direct-bandgap semiconductor for the later composition.

The top of the valence for $Ge_{1-x-y}Si_xSn_y$ is determined by the relation $E_{v,av}$

$$E_v\left(Ge_{1-x-y}Si_xSn_y\right) = \frac{\Delta_{so}\left(Ge_{1-x-y}Si_xSn_y\right)}{3} + E_{v,av}\left(Ge_{1-x-y}Si_xSn_y\right) \quad (14.11)$$

TABLE 14.1

Energy Bandgaps of Si, Ge, and Sn and the Bowing Parameters of Their Alloys in Different Valleys

Valley	E_{Ge} (eV)	E_{Si} (eV)	E_{Sn} (eV)	b_{GeSi} (eV)	b_{GeSn} (eV)	b_{SiSn} (eV)
L	0.66	2.0	0.14	0.0	−0.11	0.0
Γ	0.795	4.06	−0.413	0.21	1.94	13.2

where

$$\Delta_{so}(Ge_{1-x-y}Si_xSn_y) = 0.295(1-x-y)+0.043x+0.800y \qquad (14.12)$$

and $E_{v,av}$ is as given in Section 14.5 by $E_{v,av}(Ge_{1-x-y}Si_xSn_y)=0.69y-0.48x$.

Finally, a laser made of lattice-matched $Ge_{0.94}Sn_{0.06}/Ge_{0.75}Si_{0.15}Sn_{0.1}$ DHs has been proposed where the bandgap of the active region $Ge_{0.94}Sn_{0.06}$ is 0.619 eV, the conduction band offset at Γ point is 149 meV, and that of the valence band is 54 meV. The entire DH would be grown on a relaxed $Ge_{0.94}Sn_{0.06}$ buffer on either Si or SOI. Hence, the laser structure is strain free. For a given injected carrier density, the optical gain at photon energy E due to e-hh recombination can be calculated by the relation:

$$g = \frac{e^2\,\hbar^2\,|M_b|^2}{\varepsilon_0 m_0^2 c \bar{n} E}\left(\frac{2m_r}{\hbar^2}\right)^{3/2}(E-E_g)^{1/2}\left[f_c(E_e)+f_v(E_{hh})-1\right] \qquad (14.13)$$

where:

e   is the electron charge
$\hbar$ is the Planck constant
$|M_b|^2$ is the average matrix element for the Bloch states
m_r is the reduced mass
ε_0 is the permittivity of vacuum
m_0 is the free electron mass
c   is the speed of light in vacuum
$\bar{n}$ is the refraction index
E_g is the bandgap
$f_c(E_e)$ and $f_v(E_{hh})$ are the occupation probabilities of an electron and a hole at states that are separated by photon energy E with the same k in the reciprocal space

The gain is shown in Figure 14.4 for different carrier concentrations.

When different losses are considered, the threshold current density needed to compensate for the loss of 20 cm^{-1} at room temperature is too high (>10 kA cm^{-2}). Sun et al. substituted the DH with a p-i-n diode $Ge_{0.94}Sn_{0.06}/Ge_{0.75}Si_{0.15}Sn_{0.1}$ multiple quantum well (MQW) active region where the direct band $Ge_{0.90}Sn_{0.10}$ quantum wells (QWs) are confined by $Ge_{0.75}Si_{0.15}Sn_{0.1}$ barrier layers. In this case, the MQW laser with an active region consisting of 20 nm $Ge_{0.90}Sn_{0.10}$ wells and 20 nm $Ge_{0.75}Si_{0.15}Sn_{0.1}$ barriers is considered and the barriers are optically confined by a thick $Ge_{0.75}Si_{0.15}Sn_{0.1}$ cladding layer [17]. The same calculations were done to obtain the direct bandgap Type I alignment and, due to the quantum confinement, an energy separation between the ground-state electron and HH subbands was obtained as

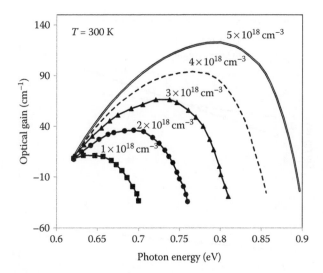

FIGURE 14.4
Optical gain spectra of a GeSn-based p-i-n DH laser for different carrier concentrations.

0.541 eV yielding to a lasing wavelength of 2.3 μm. The gain is calculated by the relation:

$$g = \frac{e^2 \left| M_b \right|^2 m_r}{\varepsilon_0 m_0^2 c \hbar \bar{n} d_w} \left[f_c(E_e) + f_v(E_{hh}) - 1 \right] \tag{14.14}$$

For this structure, the gain can exceed 100 cm⁻¹ for a pumping current density of 3 kA cm⁻², which is sufficient to compensate for losses in mid-IR semiconductor lasers. This type of electrically injected Group IV near/mid-IR laser can be operated at room temperature.

> **Example 14.6:** For a GeSn-based p-i-n DH, the gain is 61.27 cm⁻¹ for a carrier concentration of 3×10^{18} cm⁻³ at a photon energy of 0.7 eV.

14.8 Incorporation of C

One problem related to the above discussions is that the maximum direct bandgap in a Ge layer is ~0.6 eV; therefore, the layers are not suitable for emission at 1550 nm, which is the wavelength for present-day telecommunications. It occurred to us that a slight amount of C in the active Ge layer may increase the gap to 0.8 eV, corresponding to 1550 nm [18]. It was found

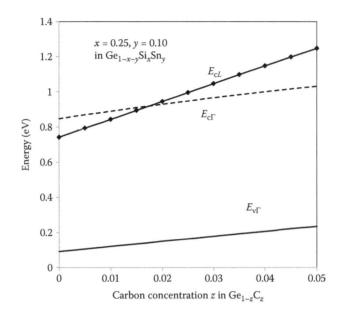

FIGURE 14.5
Variation of band-edge energies in the GeC layer as a function of C composition z. The barrier is $Ge_{0.65}Si_{0.25}Sn_{0.10}$.

that the direct gap may indeed be 0.8 eV corresponding to 1550 nm and the amount of C needed is only 0.015, which does not exceed the present practical limit (~0.03 in SiGeC). It should be pointed out that Kolodzey et al. [19] theoretically predicted indirect to direct crossover in GeC alloys, but the C concentration that is needed is extremely high. Our results are obtained when all parameters for alloys are obtained by linear interpolation.

The values of different band edges calculated by the abovementioned procedure are shown in Figure 14.5 for fixed values of the Si concentration x (=0.25) and Sn concentration y (=0.10). It appears that the direct-to-indirect crossover occurs at the C concentration $z \approx 0.017$.

> **Example 14.7:** The conduction band edges at the L and Γ valleys for $z = 0.03$ are 1.02 and 0.95 eV, respectively.

PROBLEMS

14.1. Using Equation 14.2, plot the absorption coefficient for Si with photon energy ranging from 1 to 1.4 eV considering TA and TO phonons at $T = 300$ K.

14.2. Calculate the in-plane strain as a function of (x,y) of $Ge_{1-x-y}Si_xSn_y$ grown on a Ge substrate. Use the values of the parameters given in Appendix II.

14.3. Calculate the values of valence band splitting for all three bands as a function of z for strained $Ge_{1-z}Sn_z$ on relaxed $Ge_{1-x-y}Si_xSn_y$ ($x=0.15$, $y=0.25$). Compare the results with that given in Figure 14.1.

14.4. Determine the minimum value of z for the unstrained $Ge_{1-z}Sn_z$ alloy for which it becomes a direct-bandgap material.

14.5. The lattice constants of Si, Ge, and Sn are, respectively, 5.431, 5.658, and 6.489 Å. Determine the composition of $Ge_{1-x-y}Si_xSn_y$ for perfect lattice matching with Ge. Use Vegard's law.

14.6. Determine the values of (x,y) of $Ge_{1-x-y}Si_xSn_y$ on which strained $Ge_{0.97}C_{0.03}$ becomes a direct-bandgap material. Also, determine the corresponding values of the direct bandgaps for each pair of (x,y).

14.7. Calculate the position of the conduction band minima at L and Γ points as a function of (x,y) of unstrained $Ge_{1-x-y}Si_xSn_y$. Use the values of the parameters given in Table 14.1. Check whether it can behave as a direct bandgap or not.

14.8. Derive the expression of the optical gain coefficient as given in Equation 14.13 for a DH laser at photon energy E. Plot the gain coefficient with different carrier concentrations for a fixed value of photon energy.

14.9. Plot the band-edge energies of strained Si_zGe_{1-z} on relaxed $Ge_{1-x-y}Si_xSn_y$ ($x=0.15$, $y=0.25$). Determine the value of z for which it becomes a direct-bandgap material.

14.10. Plot the band-edge energies of strained $Ge_{1-x}Sn_x$ on relaxed $Ge_{1-y}Sn_y$. Determine the value of x and y for which $Ge_{1-x}Sn_x$ becomes a direct-bandgap material.

14.11. Repeat Problem 14.7 for unstrained $Si_{1-x}Sn_x$.

14.12. Explore the design space for a heterostructure based on lattice-matched pairs of GeSiSn so that one of these pairs would act as a QW and the other as the barrier material.

Suppose a strain-balanced $Ge_{1-z}Sn_{0.z}/Si_{0.09}Ge_{0.8}Sn_{0.11}$ MQW structure is grown on a fully strain-relaxed $Ge_{0.88}Sn_{0.12}$ buffer layer in such a way that compressive strain introduced in the $Ge_{1-z}Sn_{0.z}$ well is compensated for by the tensile stain introduced in the $Si_{0.09}Ge_{0.8}Sn_{0.11}$ barrier. The strain-balanced condition is given by $\Sigma_{i=1}^n A_iL_i\varepsilon^i/a_i = 0$, with $A_i = C_{11}^{(i)}+C_{12}^i-(2C_{12}^{(i)2}/C_{11}^i)$, where L_i is the thickness of the ith layer, ε_i is the strain, and a_i is the lattice constant of the ith layer.

If the well thickness is 100 Å and the barrier thickness is 90 Å, find the composition z for the well. Use the linear interpolation technique with the values of the lattice constants and elastic constants of Si, Ge, and Sn given in Appendix II.

Reading List

Basu, P. K., *Theory of Optical Processes in Semiconductors*. Oxford University Press, Oxford, 1997.

Deen, M. J. and P. K. Basu, *Silicon Photonics: Fundamentals and Devices*. Wiley, Chichester, 2012.

Kasper, E., M. Kittler, M. Oehme, and T. Arguirov, Germanium tin: Silicon photonics toward the mid-infrared, *Photon. Res.* 1, 69–76, 2013.

Michel, J., J. Liu, and L. C. Kimerling, High-performance Ge-on-Si photodetectors, *Nat. Photonics* 4, 527–534, 2010.

Pavesi, L. and D. J. Lockwood, *Silicon Photonics*. Springer, New York, 2004.

Reed, G. T. and A. P. Knights, *Silicon Photonics: An Introduction*. Wiley, Chichester, 2004.

Reed, G. T., G. Mashanovich, F. Y. Gardes, and D. J. Thomson, Silicon optical modulators, *Nat. Photonics* 4, 518–526, 2010.

Saito, S., F. Y. Gardes, A. Z. Al-Attili, K. Tani, K. Oda, Y. Suwa, T. Ido, Y. Ishikawa, S. Kako, S. Iwamoto, and Y. Arakawa, Group IV light sources to enable the convergence of photonics and electronics, *Frontiers Mat.* 1, Article 15, 1–15, 2014.

Shimizu, T., N. Hatori, M. Okano, M. Ishizaka, Y. Urino, T. Yamamoto, M. Mori, T. Nakamura, and Y. Arakawa, Multichannel and high-density hybrid integrated light source with a laser diode array on a silicon optical waveguide platform for interchip optical interconnection, *Photon. Res.* 2, A14–A19, 2014.

Soref, R., The achievements and challenges of silicon photonics, *Adv. Opt. Technol.* 2008, 472305, 2008.

Soref, R., Mid-infrared photonics in silicon and germanium, *Nat. Photonics*, 4, 495–497, 2010.

Sukhdeo, D. S., D. Nam, J. H. Kang, M. L. Brongersma, and K. C. Saraswat, Direct bandgap germanium-on-silicon inferred from 5.7% ⟨100⟩ uniaxial tensile strain, *Photon. Res.* 2, A8–A13, 2014.

Urino, Y., T. Usuki, J. Fujikata, M. Ishizaka, K. Yamada, T. Horikawa, T. Nakamura, and Y. Arakawa, High-density and wide-bandwidth optical interconnects with silicon optical interposers, *Photon. Res.* 2, A1–A7, 2014.

Ye, H. and Yu, J., Germanium epitaxy on silicon, *Sci. Technol. Adv. Mater.* 15, 024601–024601-9, 2014.

References

1. Bucher, K., J. Bruns, and H. G. Wagemann, Absorption coefficient of silicon: An assessment of measurements and the simulation of temperature variation, *J. Appl. Phys.* 75, 1127–1132, 1994.

2. Rong, H., S. Xu, Y. H. Kuo, V. Sih, O. Cohen, O. Raday, and M. Paniccia, Low threshold continuous wave Raman silicon lasers, *Nat. Photonics* 1, 232–237, 2007.

3. Jambois, O., F. Gourbilleau, A. J. Kenyon, J. Montserrat, R. Rizk, and B. Garrido, Towards population inversion of electrically pumped Er ion sensitized by Si nanoclusters, *Opt. Express* 18, 2230–2235, 2010.

4. Liang, D. and J. E. Bowers, Recent progress in lasers on silicon, *Nat. Photonics* 4, 511–517, 2010.
5. Kouvetakis, J., J. Menéndez, and A. G. V. Chizmeshya, Tin-based Group IV semi-conductors: New platforms for opto- and microelectronics on silicon, *Annu. Rev. Mater. Res.* 36, 497–554, 2006.
6. Tolle, J., R. Roucka, A. G. V. Chizmeshya, J. Kouvetakis, V. R. D'Costa, and J. Menéndez, Compliant tin-based buffers for the growth of defect-free strained heterostructures on silicon, *Appl. Phys. Lett.* 88, 252112–252112-3, 2006.
7. Cook, C. S., S. Zollner, M. R. Bauer, P. Aella, J. Menéndez, and J. Kouvetakis, Optical constants and interband transitions of $Ge_{1-x}Sn_x$ alloys ($x < 0.2$) grown on Si by UHV-CVD, *Thin Solid Films* 455–456, 217–221, 2004.
8. Soref, R. A. and C. H. Perry, Predicted bandgap of the new semiconductor SiGeSn, *J. Appl. Phys.* 69, 539–547, 1991.
9. Menéndez, J. and J. Kouvetakis, Type-I $Ge/Ge_{1-x-y}Si_xSn_y$ strained-layer hetero-structures with a direct Ge bandgap, *Appl. Phys. Lett.*, 85, 1175–1177, 2004.
10. Van de Walle, C. G., Band lineups and deformation potentials in the model-solid theory, *Phys. Rev. B* 39, 1871–1883, 1989.
11. Liu, J., X. Sun, D. Pan, X. Wang, L. C. Kimerling, T. L. Koch, and J. Michel, Tensile-strained, n-type Ge as a gain medium for monolithic laser integration on Si, *Opt. Express* 15, 11272–11277, 2007.
12. Sun, G., R. A. Soref, and H. H. Cheng, Design of an electrically pumped SiGeSn/GeSn/SiGeSn double heterostructure midinfrared laser, *J. Appl. Phys.* 108, 033107–033107-6, 2010.
13. Jaros, M., Simple analytic model for heterojunction band offsets, *Phys. Rev. B* 37, 7112–7114, 1988.
14. Weber, J. and M. I. Alonso, Near-band-gap photoluminescence of Si-Ge alloys, *Phys. Rev. B* 40, 5683–5693, 1989.
15. D'Costa, V. R., C. S. Cook, A. G. Birdwell, C. L. Littler, M. Canonico, S. Zollner, J. Kouvetakis, and J. Menendez, Optical critical points of thin-film $Ge_{1-y}Sn_y$ alloys: A comparative $Ge_{1-y}Sn_y/Ge_{1-x}Si_x$ study, *Phys. Rev. B* 73, 125207–125207-6, 2006.
16. D'Costa, V. R., C. S. Cook, J. Menéndez, J. Tolle, J. Kouvetakis, and S. Zollner, Transferability of optical bowing parameters between binary and ternary Group-IV alloys, *Solid State Commun.* 138, 309–313, 2006.
17. Sun, G., R. A. Soref, and H. H. Cheng, Design of a Si-based lattice-matched room temperature GeSn/GeSiSn multi-quantum-well mid-infrared laser diode, *Opt. Express* 18, 19957–19965, 2010.
18. Ghosh, S. and P. K. Basu, The calculated composition of $Ge_{1-z}C_z/Ge_{1-x-y}Si_xSn_y$ heterostructure grown on Si for direct gap emission from $Ge_{1-z}C_z$ at 1:55 µm. *Solid State Commun.* 150, 844–847, 2010.
19. Kolodzey, J., P. R. Berger, B. Orner, D. Hits, F. Chen, A. Khan, X. Shao, M. M. Waite, S. Ismat Shah, C. P. Swann, and K. M. Unruh, Optical and electronic properties of SiGeC alloys grown on Si substrates, *J. Cryst. Growth* 157, 386–391, 1995.

15

Transistor Lasers

15.1 Introduction

The transistor laser (TL) is a new device that has the unique feature of working as both an electronic and a photonic device. After the announcement of the light-emitting transistor [1], the group led by Feng and Holonyak at the University of Illinois at Urbana–Champaign (UIUC) announced the TL in 2004 [2]. Basically, it is a heterojunction bipolar transistor (HBT) in which the emitter and collector terminals are used for electrical signal input and output, respectively, while an optical signal, either incoherent or coherent (self-sustained or amplified), comes out of the base region.

Since the device is quite new but has potential applications in several fields of electronics, both analog and digital systems, as well as in photonics, including fiber-optic communication, it is felt that an introduction of the working principle of the device, presentation of simple models to predict its performance, and mention of some possible applications of the device will be useful for uninitiated readers. The present chapter attempts to provide this.

15.2 Structure and Basic Working Principle

Several workers have developed TLs using different combinations of materials [2–12]. We consider here a typical layer structure of a TL, grown on GaAs substrate, as shown in Figure 15.1. The basic structure is an HBT, in which the higher-bandgap material, n-type InGaP in the illustration, forms the emitter layer, while the lower-gap material, p-type GaAs, acts as the base material. The n-type GaAs layer beneath the p-GaAs base acts as the collector. Embedded within the p-GaAs base is an InGaAs quantum well (QW), which acts as the active layer for light emission.

As in a normal transistor, the carriers injected into the base of a TL undergo recombination, as a result of which the characteristic spontaneous emission occurs. However, when the current injected into the base is

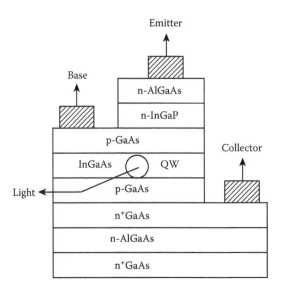

FIGURE 15.1
Basic structure of TL.

sufficient to create population inversion between conduction and valence subbands, stimulated emission becomes dominant, giving rise to amplification of light. If the medium is enclosed between a pair of Fabry–Perot (FP) mirrors, coherent light comes out in the direction perpendicular to the flow of electric current. The FP resonator provides the usual feedback for the light waves generated by stimulated recombination of electron–hole pairs in the QW.

15.3 Principle of Operation: Model Description

The experimentally obtained results for TLs have been used in a few analyses to establish novel features related to TLs. Initially, by using the charge control model [4], the UIUC group was able to show that the charge distribution in the base is not a simple triangle as in a normal HBT; rather, the distribution consists of two straight lines with a change in slope at the position of the QW. Also, with the onset of stimulated emission, the collector current (I_c) versus the emitter–base voltage (V_{CE}) characteristics become closer with the same increase in the base current. Thus, there is a drop of the current gain of the transistor: a phenomenon known as current-gain (beta) compression.

Analytical and numerical models for TLs have been developed by a few groups [13–17]. In the following, we shall present our own model [18–22], which has been successful in explaining both the abovementioned features of the TL.

We shall consider an HBT consisting of N-InGaP/p-GaAs/n-GaAs layers, in which an $In_{0.2}Ga_{0.8}As$ QW is inserted in the base. The transistor operates in its normal, active mode, that is, the base–emitter junction is forward biased and the base–collector junction is reverse biased.

In the one-dimensional model, the QW is assumed to be situated at $x=0$, the emitter is at the left of the base $(x=-W_B/2)$, and the collector is at the right side $(x=+W_B/2)$. Figure 15.2 shows the conduction energy band of the base and the direct current (dc) excess minority carrier distribution $(\delta N(x))$ in the base region. Under forward bias, the carriers injected from the emitter diffuse across the base and reach the QW. These unbounded, three-dimensional (3-D) carriers may either undergo quantum capture to the bound states in the QW with a lifetime of τ_{cap} or diffuse to the collector–base junction, where they are swept out by the reverse bias. The carriers captured by the QW may escape from it with a lifetime τ_{esc}. The 3-D carriers at $x=0$ are located at the virtual bound states.

These states are localized at the QW but occupy energies higher than the conduction band edge of the barrier material and aid in the conversion of carriers from the 3-D states (n_{vs}) above the well to the 2-D states within the QW (n_{QW}) and vice versa [13,14].

The QW adds another recombination channel in the base. The recombination rate is limited by spontaneous emission below the laser threshold and

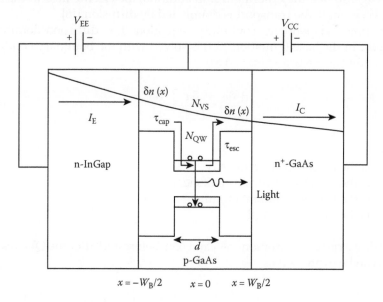

FIGURE 15.2
Schematic of carrier diffusion and quantum capture in the QW and the conduction band energy diagram of the base region.

mainly by stimulated emission above the laser threshold. The base current, as a consequence, consists of two components: (1) the regular base region recombination current, which is characterized by a base region recombination lifetime (τ_B) and a diffusion coefficient (D_n), and (2) the current needed to drive the laser.

Lengthwise, the base of the transistor also forms a resonant cavity in which two cleaved sides of the device act as mirrors. Because of electron recombination due to spontaneous and stimulated emission, the QW plays an important role in determining the base current. The electronic characteristics of the HBT deviate from normal behavior. This is evidenced by the reduction of the electrical gain, $\beta = \Delta I_C / \Delta I_B$, due to enhanced carrier recombination in the base region when stimulated emission occurs [4,7].

15.3.1 Terminal Currents

The minority carrier transport in the base may be described by the drift–diffusion equation, provided the base width $W_B \gg D/v_T$, where D is the diffusion coefficient and v_T is the thermal velocity. Taking electron mass (m_e) = 0.063 m_0 and $D = 26$ cm^2 s^{-1}, the length D/v_T is 5.6 nm, which is very small indeed compared with the base width ~100 nm in the present analysis. Further, under low-level injection from the emitter into the base by thermionic emission and tunneling, the carriers occupy energy levels very close to the conduction band edge, so that they thermalize. In addition, the electric field in the base being very small, the transport is dominated by diffusion [18].

The time-independent continuity equation for diffusion-dominated transport in the base (Equation 15.1) is solved with the boundary condition $\delta n(z_Q^-) = \delta n(z_Q^+) = N_{VS}$ and $\delta n(W_B) = 0$.

$$\frac{d^2 \delta n}{dz^2} = \frac{\delta n}{D_n \tau_B} = \frac{\delta n}{L_D^2} \tag{15.1}$$

where:
  δn is the excess electron density
  D_n is the electron diffusion coefficient
  τ_B is the base recombination time
  L_D is the diffusion length

The dc carrier concentrations δN_1 and δN_2, before and after the QW, respectively, can be expressed as

$$\delta N_1 = \left[\frac{N_{VS} - (L_D/qD_n)J_E.e^{-z_Q/L_D}}{2\cosh(z_Q/2L_D)} \right] e^{z/L_D} + \left[\frac{N_{VS} + (L_D/qD_n)J_E.e^{-z_Q/L_D}}{2\cosh(z_Q/2L_D)} \right] e^{-z/L_D}$$

$$\tag{15.2}$$

$$\delta N_2 = \frac{N_{VS}e^{-W_B/L_D}}{2\cosh(z_Q - W_B/L_D)}e^{z/L_D} + \frac{N_{VS}e^{W_B/L_D}}{2\cosh(W_B - z_Q/2L_D)}e^{-z/L_D} \quad (15.3)$$

The virtual state (VS) current density is

$$J_{VS} = qD_n\left[\frac{d}{dz}(\delta N_1)_{z=z_Q} - \frac{d}{dz}(\delta N_2)_{z=z_Q}\right] \quad (15.4)$$

From the above two equations, the terminal currents may be expressed as

$$I_E = BN_{VS}\left[\sinh(z_1) - \coth(z_2)\cosh(z_1)\right] - AJ_{VS}\cosh(z_1) \quad (15.5)$$

$$I_C = BN_{VS}\left[\tanh(z_1) - \cosh(z_1)\coth(z_2)\right] - AJ_{VS}\cosh(z_1) \quad (15.6)$$

$$I_B = BN_{VS}\left[\sinh(z_1) + \coth(z_1) - 2\cosh(z_1)\coth(z_2)\right] - 2AJ_{VS}\cosh(z_1) \quad (15.7)$$

where:
$B \quad = A_q D_n/L_D$
$z_1 \quad = z_Q/L_D$
$z_2 \quad = z_Q - W_B/L_D$
$A \quad$ is the area of the base
$q \quad$ is the electronic charge

The connections between the carrier densities and current densities related to the QW and the VS are made through the relations [14]

$$\frac{J_{VS}}{qd} = \frac{J_{QW}}{qd} - \frac{N_{VS}}{\tau_S} \quad \text{and} \quad \frac{J_{QW}}{qd} = \frac{N_{VS}}{\tau_{cap}} - \frac{N_{QW}}{\tau_{esc}}$$

where τ_s, τ_{cap}, and τ_{esc} denote, respectively, the spontaneous recombination lifetime, the capture time of electrons into the QW, and the escape time of electrons from the QW to the VS. The first and second terms in Equation 15.7 for the base current represent, respectively, the usual recombination current and the laser current. The recombination current is dominant below the laser threshold ($J_B < J_{B,th}$) because there is no stimulated emission in the QW. Above the laser threshold, the laser current is much larger than the recombination current.

15.3.2 Optical Characteristics

15.3.2.1 Strain Effect on the Band-Edge Energies

The insertion of the $In_{1-x}Ga_xAs$ QW in the undoped GaAs base leads to strain in the QW. The in-plane component of the built-in strain is given by

$$\varepsilon = \varepsilon_{xx} = \varepsilon_{yy} = \frac{a_{GaAs} - a(x)}{a_{GaAs}} \tag{15.8}$$

where a_{GaAs} and $a(x)$ are the lattice constants of the GaAs substrate and $In_{1-x}Ga_xAs$ QW, respectively. The strain in the perpendicular direction is

$$\varepsilon_{\perp} = \varepsilon_{zz} = -2\frac{C_{12}}{C_{11}}\varepsilon \tag{15.9}$$

where C_{11} and C_{12} are the elastic constants.

Figure 15.3 shows the schematic band diagram for the strained QW. The conduction band edge, the valence band edge, and the bandgap of the unstrained alloy are denoted respectively by E_C, E_V, and $E_g(x)$, while δE_C and δE_{HH} are, respectively, the strain-induced shifts of the conduction (C) and heavy-hole (HH) band edges. The energies are measured by considering $E_V = 0$. The symbols $E_{cqw,n}$ and $E_{hqw,m}$ stand for the energies of the nth conduction subband and the mth hole subband, respectively, and U is the band offset in the conduction band.

For a strained $In_{1-x}Ga_xAs$ layer, the conduction band edge is shifted by

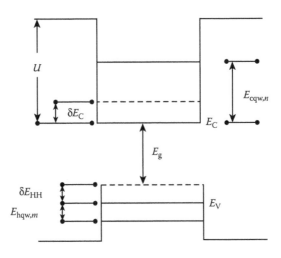

FIGURE 15.3
Band structure of the model.

$$\delta E_C = a_c \left(\varepsilon_{xx} + \varepsilon_{yy} + \varepsilon_{zz} \right) = 2a_c \left(1 - \frac{C_{12}}{C_{11}} \right) \varepsilon \tag{15.10}$$

and the valence subbands are shifted by

$$\delta E_{HH} = -P - Q, \ \delta E_{LH} = -P + Q \tag{15.11}$$

where:

$$P = -a_v \left(\varepsilon_{xx} + \varepsilon_{yy} + \varepsilon_{zz} \right) = -2a_v \left(1 - \frac{C_{12}}{C_{11}} \right) \varepsilon \tag{15.12}$$

$$Q = -\frac{b}{2} \left(\varepsilon_{xx} + \varepsilon_{yy} - 2\varepsilon_{zz} \right) = -b \left(1 + \frac{2C_{12}}{C_{11}} \right) \varepsilon \tag{15.13}$$

where:
a_c and a_v are the hydrostatic deformation potential for the conduction band and the valence band, respectively
b is the shear deformation potential

The new band edges are therefore

$$E_C = E_g \left(x \right) + \delta E_C \tag{15.14}$$

$$E_{HH} = \delta E_{HH} \text{ and } E_{LH} = \delta E_{LH} \tag{15.15}$$

For a compressively strained layer, the heavy-hole band is the topmost valence band. So the effective bandgap will be

$$E_{C-HH} = E_g \left(x \right) + \delta E_C - \delta E_{HH} \tag{15.16}$$

The quantum levels for the conduction and valence bands are given by

$$E_{c,n} = E_{cqw,n} + \delta E_C \tag{15.17a}$$

and

$$E_{h,m} = E_{hqw,m} + \delta E_{bH}, \ b = L \text{ or } H \tag{15.17b}$$

In calculating $E_{cqw,n}$ and $E_{hqw,m}$, the finite heights of the barriers (U in Figure 15.3) are duly considered. The values of δE_C and δE_{bH} are calculated following Equations 15.10 and 15.11, considering the different types of strain.

In this case, the effective bandgap should be modified as

$$E_{h,m}^{c,n} = E_g(x) + E_{c,n} - E_{h,m} \tag{15.18}$$

15.3.2.3 Gain Spectrum

The optical gain analysis in the QW is performed by using the technique described in [23]. It is assumed that the quantization occurs along the x-direction, and the wave function for electrons is expressed as $|\mathbf{k}_t, \mathbf{r}_t\rangle = \exp(\mathbf{k}_t, \mathbf{r}_t)$ $\phi_n(z)$, where $\mathbf{k}_t$ and $\mathbf{r}_t$ are, respectively, the 2-D wave vector and the position vector (in the transverse direction) and ϕ_n is the envelope function for the z-direction for the nth subband. A similar equation is written for 2-D holes. The momentum is conserved for motion along the 2-D plane (k-conservation, which gives rise to the appearance of reduced mass m_r in the energy-conservation process; see Chapter 8). It is also assumed that the subbands are broadened due to different scattering processes. The expression for gain coefficient is as follows:

$$g(\hbar\omega) = C_0 \sum_{n,m} \left|I_{h,m}^{c,n}\right|^2 \int_0^\infty dE_t \rho_r^{2D} \left|\hat{e}\cdot p_{cv}\right|^2 \times \frac{\gamma/(2\pi)}{\left(E_{h,m}^{c,n}(\mathbf{k}_t) - \hbar\omega\right)^2 + (\gamma/2)^2} \left[f_c^n(E_t) - f_v^m(E_t)\right]$$

$$\tag{15.19}$$

where:

$$C_0 = \frac{\pi q^2}{n_r c \varepsilon_0 m_0^2 \omega}$$

$$\rho_r^{2D} = \frac{m_r}{\pi \hbar^2 d}$$

$$I_{hm}^{cn} = \int dz \phi_n(z) g_m(z)$$

where:

ρ_r^{2D}	= the reduced density of states function for 2-D for quantized motion along the z-direction
d	= the width of the QW
I	= the overlap of envelope functions for electron (in the nth subband) and hole (in the mth subband)

$\hbar\omega$ = the photon energy
$|\hat{e}\cdot p_{cv}|^2$ = the momentum matrix element for QWs
γ = the finite linewidth of the spectrum

In Equation 15.19, c is the speed of light in free space, ε_0 is the permittivity of free space, n_r is the background refractive index, and m_0 is the free electron mass. The kinetic energy for 2-D free motion, E_t, is considered to be zero for the band edge transition. The Fermi occupation probabilities are expressed in terms of quasi-Fermi levels F_c and F_v as

$$f_c^n(E_t) = \frac{1}{1 + \exp\left\{\left[E_{cn} + (m_r/m_e)E_t - F_c / k_B T\right]\right\}} \qquad (15.20a)$$

$$f_v^m(E_t) = \frac{1}{1 + \exp\left\{\left[E_{hm} - (m_r/m_h)E_t - F_v\right]/ k_B T\right\}} \qquad (15.20b)$$

We first obtain the gain or absorption spectra for various injected carrier densities. The gain shows the usual peak around the energy corresponding to the separation between the first conduction subband and the first valence (heavy-hole) subband. The values of maximum gain are then plotted against injected carrier densities, and from the slope of the plot the value of a, the differential gain constant, is obtained. The gain or absorption spectra yield a value of N_{tr}, the transparency carrier density, which gives the value of current density at transparency via the relation $J_r = qdN_{tr}/\tau_s$. Then threshold current density is obtained using $J_{th} = J_{tr} + [qd/\Gamma\eta\tau_s]\cdot[\alpha + (1/2L)\ln(1/R_1 R_2)]$, where Γ is the optical confinement factor, η is the internal quantum efficiency, α is the loss coefficient, L is the length of the device, and R_1 and R_2 are the respective reflectivities of the front and back mirrors.

15.3.3 Charge Distribution and Threshold Base Current

The values of parameters used in the calculation are shown in Table 15.1. The values for mass, permittivity, etc. of the alloy are calculated by linear interpolation of the data given in Table 15.2.

The choice of time constants for escape and capture of electrons in the QW is somewhat difficult; however, we have chosen reasonable values after considering the simple analytical expressions. The subband energies are calculated by considering the finite barrier heights between InGaAs and GaAs. The band offsets (U in Figure 15.3) and subband energies are evaluated by following the approximate method used in [23].

We first show in Figure 15.4 a plot of the variation of injected carrier density in the base of width 100 nm in which the InGaAs QW is placed 70 nm away from the EB junction. The profile is obtained by joining two straight

TABLE 15.1

Values of Parameters Used in Calculation

Electron capture lifetime, τ_{cap}	10 ps	Escape lifetime, τ_{esc}	5 ns
Spontaneous emission time, τ_s	15 ns	Photon lifetime, τ_p	4 ps
Base recombination lifetime, τ_B	200 ps	Diffusion coefficient, D_n	$26 \times 10^{-4}\,\mathrm{m^2\,s^{-1}}$
Group velocity, v_g	$8.67 \times 10^7\,\mathrm{m\,s^{-1}}$	Loss coefficient/ unit length, α	$500\,\mathrm{m^{-1}}$
Confinement factor, Γ	0.03		

TABLE 15.2

Values of Parameters for Alloys Used in Linear Interpolation

Quantity	GaAs	InAs	Quantity	GaAs	InAs
Electron effective mass, m_e/m_0	0.063	0.023	Lattice constants, a_w and a_b	5.653	6.058
Split-off electron mass, m_{so}/m_0	0.15	0.049	Elastic constants, $C_{11} \times 10^{15}\,\mathrm{dyn\,m^{-2}}$	11.9	8.34
Heavy-hole mass, m_h/m_0	0.38	0.26	Elastic constants, $C_{12} \times 10^{15}\,\mathrm{dyn\,m^{-2}}$	5.34	4.526
Dielectric constant, ε	12.9	15.5	Hydrostatic deformation potential for CB a_c, eV	−7.17	−5.08
Bandgap, E_g, eV	1.424	0.354	Hydrostatic deformation potential for VB, a_v, eV	1.16	1
Split-off energy, Δ_{so}, eV	0.34	0.41	Shear deformation potential for VB, b_v, eV	−1.7	−1.8
Electron affinity, χ, eV	4.07	4.88			

lines of different slopes, the change of which occurs at the QW position. This plot shows the same nature as reported by Feng et al. in different publications [4,7]. We also obtained similar altered slopes for $z_Q = 20$ and 50 nm.

The calculated threshold carrier densities are compared with the corresponding experimental values for three different samples in Table 15.3. In addition, the dimensions of the QW and the base, the temperature of operation, and the calculated and experimental values of threshold base currents are shown in Table 15.3. A comparison is also made between the calculated and experimental values of the emission wavelengths.

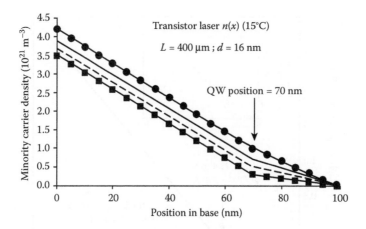

FIGURE 15.4
Variation of injected current in base with QW placed 70 nm from the EB junction for various virtual state carrier densities. ■: $3 \times 10^{20}\,\text{m}^{-3}$; dashed line: $5 \times 10^{20}\,\text{m}^{-3}$; solid line: $7 \times 10^{20}\,\text{m}^{-3}$; ●: $10 \times 10^{20}\,\text{m}^{-3}$.

It appears from the entries that good agreement is obtained for the calculated and experimental values of the emission wavelengths for all the three cases. This is also true for threshold base currents for the first two samples, but the discrepancy is quite large for the third sample. Note that, in spite of the smaller area of the third sample, the experimental threshold base current is higher (40 mA) in comparison with the other two. This large discrepancy may be due to the fact that the quality of the sample was not high at the earlier stage of research. For all the samples, the light output power shows the usual sharp rise as the base current exceeds its threshold value.

We have also estimated the values of threshold base current when the QW is placed at different positions in the base. Figure 15.5 shows the plot.

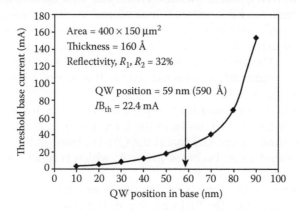

FIGURE 15.5
Variation of threshold base current for different positions of QW in base.

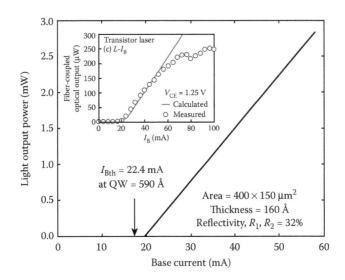

FIGURE 15.6
Variation of light output power for different base current. Inset shows the experimental results. (From Then, H. W., et al., *J. Appl. Phys.*, 107, 094509(1–7), 2010.)

It appears that the threshold base current is smaller the nearer the QW is to the EB junction. As the QW position approaches the BC junction, there is a sharp rise in the threshold base current. The reason for this increase may be understood as follows. The virtual state current density J_{VS} is proportional to the difference between the two slopes of electron distribution before and after the QW. As the QW moves closer to the BC junction, the difference between the two slopes is reduced. This requires higher injection, that is, higher base drive, to maintain the same value of J_{VS} needed to attain the threshold. This trend, that is, increased threshold base current for larger distance of QW from the EB junction, is in conformity with the results obtained by [15]. It was predicted that the bandwidth of the TL will increase as the QW is placed closer to the BC junction. The same prediction may be drawn from the expressions. The analysis provides the additional information that such an increase in bandwidth is obtained only at the cost of a larger threshold base current (Figure 15.5).

The threshold base current for Sample 2 in Table 15.3, for which the QW is placed at a distance of 59 nm from the EB junction, is 21.5 mA. It is interesting to note from Figure 15.5 that if the QW is placed at the middle of the base the calculated value becomes 15.6 mA, which is much lower than the experimental value.

We have also calculated the light output power when the QW is placed at a distance of 59 nm from the EB junction within a base of total width 88 nm. This corresponds to the experimental situation [4,7]. The calculated values of light output power for different base currents are shown in Figure 15.6. As

TABLE 15.3

Comparison of Calculated (This Work) and Experimental Values of Threshold Base Current and Emission Wavelengths

Sample No.	Temp (K)	L (μm), d (nm)	Area (μm²)	W_B (nm) and z_Q	I_{Bth} (calculated) (mA)	I_{Bth} (experiment) (mA)	Emission Wavelength (calculated) (nm)	Emission Wavelength (experiment) (nm)
1	213	450, 12	4500	85/59	7.3	7.4	975	980
2	288	400, 16	4000	88/59	21.5	22	990	1000
3	298	850, 12	1870	98/59	3.6	40	1000	1006

usual, the output power increases sharply once the threshold base current (~21.5 mA) is reached. Our calculated value of the threshold base current agrees completely with the experimentally obtained value. The calculated power output P increases linearly with base current according to the relation $P = (\hbar\omega/q)[\alpha_m/(\alpha_m + \alpha)][I_B - I_{Bth}]$, where α_m denotes mirror loss. Comparing our calculated values with the experimental values shown in the inset, it is found that both the theoretical and experimental values increase linearly up to a base current of ~55 mA. However, our calculated values of light power are ~12 times higher than the experimental values in the linear region. It is to be noted that the experimental values reported in [7] are obtained by coupling the light output from the TL to a fiber, the output end of which is connected to the power meter. The coupling losses between the TL and the fiber and between the fiber and the power meter in the experiment are responsible for the discrepancy between our calculated values and the experimental results, and a total coupling loss of about 10 dB may account for this discrepancy.

The experimental light power shows saturation behavior when the base current is above 55 mA. Our linear theory cannot explain this saturation effect, and gain compression and other phenomena should be invoked in the theory, even for qualitative agreement, above this current.

15.4 Gain Compression

The analytical model presented in Section 15.3 may explain the threshold base current, the light power output, and the base charge distribution. It is necessary in addition to develop a model that is capable of predicting the output characteristics of the HBT, that is, the family of I_C-V_{CE} curves and electrical gain. A thermionic emission diffusion model has been developed for this purpose, the details of which are given in [19]. The essential results are shown in Figures 15.7 and 15.8.

As may be seen from Figure 15.7, the output characteristics follow the usual pattern for a bipolar junction transistor (BJT) or HBT. The curves are

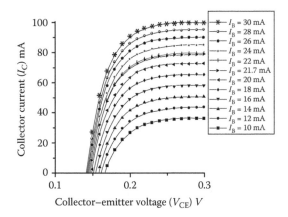

FIGURE 15.7
Variation of collector current as a function of collector–emitter voltage for different injected base currents.

parallel and almost equispaced. However, for higher base drive, the curves are closer with the same increment of the base current. In other words, the current gain decreases. This decrease occurs as the base current exceeds the threshold. Figure 15.8 illustrates the beta-compression. The current gain is high for low base drive; it then decreases near the base threshold, and, with a further increase in the base current, current gain is again more or less constant. The characteristics are in qualitative agreement with the experimental data. It is not possible to achieve an exact match due to the series resistance offered by the different layers.

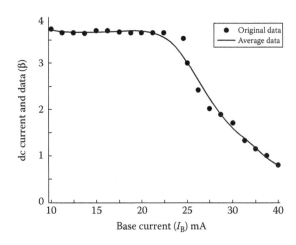

FIGURE 15.8
Variation of dc current gain as a function of base current.

15.5 Frequency Response

The frequency response and modulation bandwidth of TLs may be calculated by using the theory for QW lasers. The response shows the usual resonance peaks. However, an important observation is that the resonance peak disappears as the base current increases the threshold. The recombination is then dominated by the stimulated process, which causes the peak to disappear and the 3 dB bandwidth to increase [20].

Similar analytical results for dots in a well (DWELL)-TL [21] and for multiple quantum wells (MQWs) in the base [22] have also been reported.

Reading List

Then, H. W., M. Feng, and N. Holonyak, Jr., The transistor laser: Theory and experiment, *Proc. IEEE* 101, 10, 2271–2298, 2013.

Holonyak, Jr., N. and M. Feng, The transistor laser, *IEEE Spectrum* 43, 2, 50–55, 2006.

References

1. Feng, M., N. Holonyak, Jr., and W. Hafez, Light-emitting transistor: Light emission from InGaP/GaAs heterojunction bipolar transistors, *Appl. Phys. Lett.* 84, 151–153, 2004.
2. Walter, G., N. Holonyak, Jr., M. Feng, and R. Chan, Laser operation of a heterojunction bipolar light-emitting transistor, *Appl. Phys. Lett.* 85, 4768(1–3), 2004.
3. Feng, M., N. Holonyak, Jr., A. James, K. Cimino, G. Walter, and R. Chan, Carrier lifetime and modulation bandwidth of a quantum well AlGaAs/InGaP/GaAs/InGaAs transistor laser, *Appl. Phys. Lett.* 89, 131504(1–3), 2006.
4. Feng, M., N. Holonyak, Jr., H. W. Then, and G. Walter, Charge control analysis of transistor laser operation, *Appl. Phys. Lett.* 91, 053501(1–3), 2007.
5. Dixon, F., M. Feng, N. Holonyak, et al., Transistor laser with emission wavelength at 1544 nm, *Appl. Phys. Lett.* 93, 15, 021111-1–021111-3, 2008.
6. Feng, M., H. W. Then, N. Holonyak, Jr., G. Walter, and A. James, Resonance-free frequency response of a semiconductor laser, *Appl. Phys. Lett.* 95, 033509(1–3), 2009.
7. Then, H. W., M. Feng, and N. Holonyak, Jr., Microwave circuit model of the three-port transistor laser, *J. Appl. Phys.* 107, 094509(1–7), 2010.
8. Huang, Y., J. H. Ryou, R. D. Dupuis, F. Dixon, N. Feng, and N. Holonyak, Jr., InP/InAlGaAs light-emitting transistors and transistor lasers with a carbon-doped base layer, *J. Appl. Phys.* 109, 063106-1–063106-6, 2011.

9. Duan, Z., W. Shi, L. Chrostowsky, X. Huang, N. Zhou, and G. Chai, Design and epitaxy of 1.5 μm InGaAsP-InP MQW material for a transistor laser, *Optics Exp.* 18(2), 1501–1509, 2010.

10. Shirao, M., T. Sato, Y. Takino, N. Sato, N. Nishiyama, and S. Arai, Room-temperature continuous-wave operation of 1.3-μm transistor laser with AlGaInAs/InP quantum wells, *Appl. Phys. Express* 4(7), 072101-1–072101-3, 2011.

11. Shirao, M., T. Sato, N. Sato, N. Nishiyama, and S. Arai, Room temperature operation of npn-AlGaInAs/InP multiple quantum well transistor laser emitting at 1.3-μm wavelength, *Opt. Express* 20(4), 3983–3989, 2012.

12. Sato, N., M. Shirao, T. Sato, M. Yukinari, N. Nishiyama, T. Amemiya, and S. Arai, Room-temperature continuous-wave operation of npn-AlGaInAs transistor laser emitting at 1.3-μm wavelength, *IEEE Photon. Tech. Lett.* 25(8), 728–730, 2013.

13. Faraji, B., D. L. Pulfrey, and L. Chrostowski, Small-signal modeling of the transistor laser including the quantum capture and escape lifetimes, *Appl. Phys. Lett.* 93(10), 103509(1–3), 2008.

14. Faraji, B., W. Shi, D. L. Pulfrey, and L. Chrostowski, Analytical modeling of the transistor laser, *IEEE J. Sel. Top. Quantum Electron.* 13, 594–603, 2009.

15. Zhang, L. and J. P. Leburton, Modeling of the transient characteristics of heterojunction bipolar transistor lasers, *IEEE J. Quantum Electron.* 45(4), 359–366, 2009.

16. Taghavi, I., H. Kaatuzian, and J. P. Leburton, Bandwidth enhancement and optical performances of multiple quantum well transistor lasers, *Appl. Phys. Lett.* 100, 231114(1–5), 2012.

17. Shirao, M., S. Lee, N. Nishiyama, and S. Arai, Large signal analysis of a transistor laser, *IEEE J. Quantum Electron.* 47(3), 359–367, 2011.

18. Basu, R., B. Mukhopadhyay, and P. K. Basu, Estimated threshold base current and light power output of a transistor laser with InGaAs quantum well in GaAs base, *Semicond. Sci. Tech.* 26, 105014(1–6), 2011.

19. Asryan, L. V., N. A. Gun'Ko, A. S. Polkivnikov, G. G. Zegrya, R. A. Suris, P. K. Lau, and T. Makino, Threshold characteristics of InGaAsP/InP multiple quantum well lasers, *Semicond. Sci. Tech.* 15, 1131–1140, 2000.

20. Basu, R., B. Mukhopadhyay, and P. K. Basu, Modeling of current gain compression in common emitter mode of a transistor laser above threshold base current, *J. Appl. Phys.* 111, 083103(1–7), 2012.

21. Basu, R., B. Mukhopadhyay, and P. K. Basu, Modeling resonance free modulation response in transistor lasers with single and multiple quantum wells in the base, *IEEE Photonics J.* 4(5), 1572–1581, 2012.

22. Basu, R., B. Mukhopadhyay, and P. K. Basu, Analytical theory of a small signal modulation response of a transistor laser with dots-in-well in the base, *Semicond. Sci. Technol.* 27, 015022(1–7), 2012.

23. Basu, R., B. Mukhopadhyay, and P. K. Basu, Analytical model for threshold-base current of a transistor laser with multiple quantum wells in the base, *IET-Optoelectron.* 7, 71–76, 2013.

Appendix I: Heterojunction Lineup and Current–Voltage Characteristics

A1.1 Electron Affinity Rule

A model for an ideal heterojunction was first developed by Anderson (1962). When a heterojunction is formed, discontinuities occur in the energy bands as the Fermi levels line up at equilibrium. Anderson assumed that the vacuum energy level is continuous, and hence the conduction band discontinuity ΔE_c would be a difference in the electron affinities of the two semiconductors, and ΔE_v would be found from $\Delta E_g - \Delta E_c$, where ΔE_g is the bandgap difference of the two semiconductors. This means that the discontinuities in the conduction band and the valence band accommodate the difference in the bandgap between the two semiconductors. The built-in potential required to align the Fermi levels is divided between the two semiconductors. The resulting depletion region and built-in potential on each side of the junction can be determined by solving Poisson's equation with proper boundary condition at the junction and depletion approximation.

A1.2 Anisotype Heterojunction

Figure A1.1a shows the energy-band diagrams of two isolated semiconductors of opposite types, where the narrow-bandgap material is n-type. The two semiconductors are assumed to have different bandgaps E_g, different permittivities ε_s, different work functions ϕ_m, and different electron affinities χ. The different parameters for smaller-bandgap material are represented by the first symbol; for example, the work function of smaller n-type material is ϕ_{m1} for this case. The energy-band profile of the n-p heterojunction is shown in Figure A1.1b at equilibrium. Since the Fermi level must coincide on both sides in equilibrium, and the vacuum level is everywhere parallel to the band edges and is continuous, the discontinuity in the conduction band edges, ΔE_C, and valence band edges, ΔE_V, does not depend on doping for nondegenerate semiconductors where E_g and χ are not the functions of doping. The total built-in potential is given by

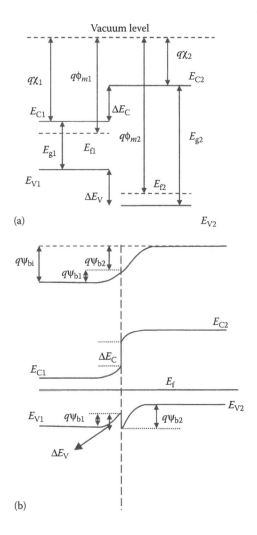

FIGURE A1.1
Energy-band diagram of n-p heterojunction with n-type smaller-bandgap material: (a) before contact, (b) after contact.

$$\Psi_{bi} = \Psi_{b1} + \Psi_{b2} \tag{A1.1}$$

where Ψ_{b1} and Ψ_{b2} are the electrostatic potentials supported by the smaller-bandgap and wide-bandgap semiconductors, respectively, at equilibrium. Also, at equilibrium, the Fermi levels in the two regions align, so that $E_{F1} = E_{F2}$, and this results in $\Psi_{bi} = |\phi_{m1} - \phi_{m2}|$.

The energy band profile for a p-n heterojunction with smaller-bandgap p-type semiconductor is illustrated in Figure A1.2.

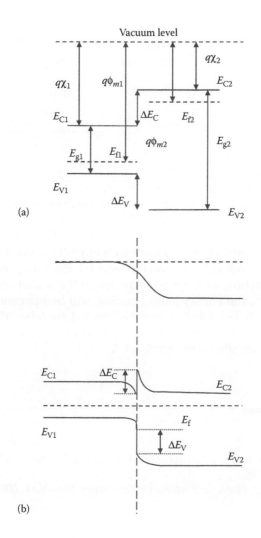

FIGURE A1.2
Energy-band diagram of p-n heterojunction with p-type smaller-bandgap material: (a) before contact, (b) after contact.

> **Example A.1:** The values of electron affinity are $\chi(\text{GaAs})=4.07$ eV and $\chi(\text{AlAs})=3.50$. Using the expression $\chi(x)=4.07-1.06x$ eV $(0<x<0.45)$, we obtain $\Delta E_c=1.06x$, and for $x=0.3$, $\Delta E_c=0.318$ eV.

A1.3 Isotype Heterojunction

When a heterojunction is formed with two different semiconductors, with both either n-type or p-type, the junction is called an isotype heterojunction.

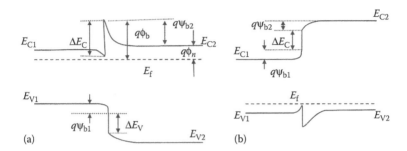

FIGURE A1.3
Energy-band diagram of isotype heterojunctions: (a) n-n junction, (b) p-p junction.

With the development of the high electron mobility transistor (HEMT), the theory of n-n heterojunction is very useful for modeling these devices. In an n-n heterojunction, as the work function of the wide-bandgap semiconductor is smaller, the energy-band bending will be opposite to that of the n-p heterojunction. For detailed analysis, readers are referred to the book by S. L. Chuang (2009). The energy-band diagrams for ideal n-n and p-p isotype heterojunctions are shown in Figure A1.3.

References

1. Anderson, R. L., Experiments on Ge-GaAs heterojunctions, *Solid State Electron.*, 5, 341–351, 1962.
2. Chuang, S. L., *Physics of Photonic Devices*, Wiley, New York, 2009.

Appendix II

Important Physical Parameters of a Few Semiconductors

Parameter	Si	Ge	Sn	GaAs	AlAs	InAs	InP	GaP
a (Å)	5.4307	5.6573	6.4892	5.6533	5.6600	6.0584	5.8688	5.4505
ε_r	11.9	16.2	24	13.1	10.9	15.15	12.5	11.11
C_{11} (10^{11} dyne cm^{-2})	16.577	12.853	6.9	11.879	12.5	8.329	10.11	14.05
C_{12} (10^{11} dyne cm^{-2})	6.393	4.826	2.9	5.376	5.34	4.526	5.61	6.203
m_c (m_0)	0.528	0.038	0.058	0.067	0.15	0.023	0.077	0.25
$m_{t,L}$ (m_0)	0.133	0.0807	0.0204	0.067	0.15	0.023	0.077	0.254
$m_{l,L}$ (m_0)	1.659	1.57	0.4072					4.8
γ_1	4.22	13.38	−15	6.85	3.45	20.4	4.95	4.05
γ_2	0.39	4.24	−11.45	2.1	0.68	8.3	1.65	0.49
γ_3	1.44	5.69	−8.55	2.9	1.29	9.1	2.35	1.25
$E_g\Gamma$ (eV)	4.185	0.7985	−0.413	1.424	2.16	0.42	1.35	2.350
$E_{g,L}$ (eV)	1.65	0.664	0.092					
Δ (eV)	0.044	0.29	0.8	0.34	0.28	0.38	0.11	0.08
a_c (eV)	1.98	−8.24	unknown	−7.17	−5.64	−5.08	−5.04	−7.14
a_l (eV)	−0.66	1.54	unknown					
a_v (eV)	2.46	1.24	−4.7	1.16	2.47	1.00	1.27	1.7
b_v (eV)	−2.1	−2.9	−2.7	−1.7	−1.5	−1.8	−1.7	−1.8

Index